D1186754

managing and shaping innovation

managing and shaping innovation

STEVE CONWAY with FRED STEWARD

OXFORD
UNIVERSITY PRESS

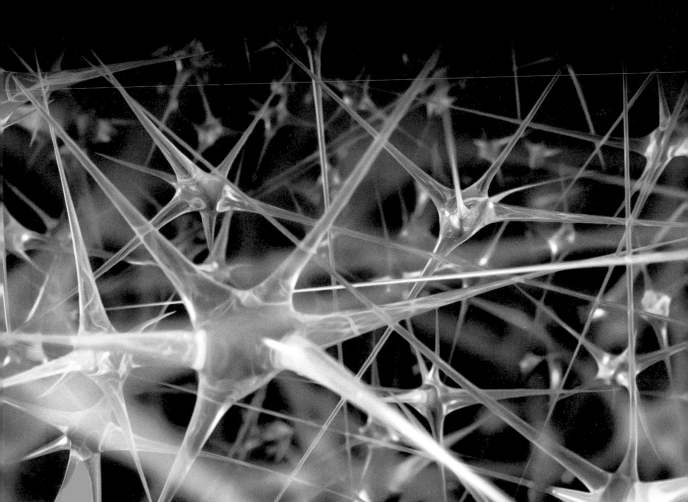

OXFORD
UNIVERSITY PRESS

Great Clarendon Street, Oxford OX2 6DP

Oxford University Press is a department of the University of Oxford.
It furthers the University's objective of excellence in research, scholarship,
and education by publishing worldwide in

Oxford New York

Auckland Cape Town Dar es Salaam Hong Kong Karachi
Kuala Lumpur Madrid Melbourne Mexico City Nairobi
New Delhi Shanghai Taipei Toronto

With offices in

Argentina Austria Brazil Chile Czech Republic France Greece
Guatemala Hungary Italy Japan Poland Portugal Singapore
South Korea Switzerland Thailand Turkey Ukraine Vietnam

Oxford is a registered trade mark of Oxford University Press
in the UK and in certain other countries

Published in the United States
by Oxford University Press Inc., New York

© Conway and Steward 2009

The moral rights of the authors have been asserted
Database right Oxford University Press (maker)

First published 2009

All rights reserved. No part of this publication may be reproduced,
stored in a retrieval system, or transmitted, in any form or by any means,
without the prior permission in writing of Oxford University Press,
or as expressly permitted by law, or under terms agreed with the appropriate
reprographics rights organization. Enquiries concerning reproduction
outside the scope of the above should be sent to the Rights Department,
Oxford University Press, at the address above

You must not circulate this book in any other binding or cover
and you must impose the same condition on any acquirer

British Library Cataloguing in Publication Data
Data available

Library of Congress Cataloging in Publication Data
Data available

Typeset in 8.8/13pt Stone Serif by Graphicraft Limited, Hong Kong
Printed in Italy on acid-free paper by Lego S.P.A.

ISBN 9780199262267

■ OUTLINE CONTENTS

◼ PREFACE

Over the last decade or so, innovation is a theme that has captured the imagination and entered the mainstream. It is a subject that has gained great currency among government ministers and policy-makers, managers and consultants, and the public at large. It has also received a great deal of attention among academics, across a wide range of disciplines. In response to this interest, an increasing number of books have emerged to serve both the classroom and the boardroom. This text is aimed at postgraduate and final-year undergraduate students undertaking core or elective modules in the management of innovation. Our aim has been to provide a distinctive text. We believe this has been achieved through the combination of:

- the breadth of the subject material. We explore innovation from multiple levels of analysis – the individual, the team, the organization, the sector, region, and nation-state. In doing so we ask 'What shapes the innovativeness of an individual, of a team, of an organization, of a sector, region, or nation?'

- the inter-disciplinary origin of the source material. We explicitly draw upon literature concerning innovation across a variety of disciplines, including economics, management, sociology, psychology, and geography, to provide multiple perspectives;

- within the text we have actively sought to connect ideas, concepts, and theories to individuals, disciplines, and time. Increasingly often, it seems, ideas have become disconnected from their time and place of origin, and somehow disembodied from their authors;

- we adopt a network perspective throughout the text, in order to highlight the importance of relationships between individuals and between organizations during the innovation process. This shifts the emphasis away from the creativity of individuals, teams, and organizations, and raises the profile of 'boundary-spanning' social networks, sectoral networks, regional clusters, and the 'distributed' and 'systemic' nature of innovation. In adopting this perspective, we highlight the importance of recognizing that innovation may be shaped by a wide range of stakeholders both within and beyond the boundaries of the organization;

- we also emphasis the iterative, emergent, and complex nature of the innovation process, rather than one that is characterized by linearity, rationality, and simplicity. Perhaps this is not surprising, given that the process is one that we view to be inherently social, political, and emotional, where the formal and planned is often less vital than the informal or serendipitous. From this orientation, the metaphor of a 'journey' is seen as more potent for understanding the innovation process, than the notion of a 'pipeline', where 'stages' can be methodically identified, planned, monitored, and signed-off;

- we challenge the reader, too, to reflect upon what might appear at first to be simple and well understood notions, such as 'novelty' and 'success', highlighting the importance of perspective;

- finally, we view the management of innovation as paradoxical and contradictory, rather than unproblematic, requiring managers and organizations to embrace 'and/both' rather than 'either/or' decisions.

Our approach then, is to provide a text that both reviews and critiques the literature on the management and shaping of innovation. Our intention is not to provide the reader with a collection of prescriptions, but to encourage reflection on the multifaceted and complex nature of innovative activity.

ACKNOWLEDGEMENTS

The origins of this book lie in the innovation modules taught on the MBA and under-graduate management programmes at Aston Business School during the late-1990s. Various iterations of these modules were developed and delivered by Ossie Jones (currently Professor of Entrepreneurship at the University of Liverpool Management School), Fred Steward, and myself. This work influenced both the breadth and orientation of the material and topics selected for this book.

I would like to thank the anonymous academic reviewers for their constructive and invaluable feedback on early drafts of the text. We have attempted to respond to this feedback as best we can, but clearly any remaining weaknesses or omissions are our own. I would also like to thank all those involved at Oxford University Press for their support, expert advice, and patience during the writing process, and in particular Kirsty Reade (Commissioning Editor) and Sacha Cook (Editor-in-Chief). And for transforming a text document into a smartly formatted book, I'd like to thank, among others, Helen Cook (permissions), Vanessa Plaister (copy-editing), Harriet Ayles (production editing), and Sarah Brett (on-line materials).

Finally, I would like to thank Annilee Game, and colleagues at the University of Leicester School of Management, in particular Gibson Burrell, Matthew Higgins, and Martin Parker, for their moral support during this writing project.

Steve Conway,
February 2009

■ DETAILED CONTENTS

5   Technological Regimes, Trajectories, Transitions, Discontinuity, and Long Waves   175

6   Innovation Strategies   208

◼ LIST OF FIGURES

■ LIST OF TABLES

■ LIST OF ABBREVIATIONS

AFRC	Agricultural and Food Research Council
AGC	Agricultural Genetics Company
ANT	actor network theory
APT	Advanced Passenger Train
ATM	automated teller machine
BBC	British Broadcasting Corporation
BCF	British Cycling Federation
BPR	business process re-engineering
CAD	computer-aided design
CAT	computed axial tomography
CD	compact disc
CEO	chief executive officer
CIM	computer-integrated manufacturing
CNC	computer numerical control
CoP	communities of practice
CPU	central processing unit
CSR	corporate social responsibility
CTA	constructive technology assessment
DAT	digital audio tape
DNA	deoxyribonucleic acid
DNDi	Drugs for Neglected Diseases Initiative
DRAM	dynamic random access memory
DSK	Dvorak simplified keyboard
DVD	digital versatile disc
ESRC	Economic and Social Research Council
FBI	Federal Bureau of Investigation
FMCGs	fast-moving consumer goods
GM	genetically modified
GUI	graphical user interface
HP	Hewlett-Packard
HRI	Horticultural Research International
IASC	International Accounting Standards Committee
ICT	information and communication technology
IKON	Innovation, Knowledge, and Organizational Networks (Warwick Business School, University of Warwick)
IT	information technology
KIBS	knowledge-intensive business service
KTN	knowledge transfer network
LED	light-emitting diode

LERU	League of European Research Universities
LMB	Laboratory of Molecular Biology
MAFF	Ministry of Agriculture, Fisheries, and Food
MBA	Master of Business Administration
MD	MiniDisc
MERIT	Maastricht Economic and Social Research and Training Centre on Innovation and Technology
MIPS	microprocessor speed
MIT	Massachusetts Institute of Technology
MLP	multiple-level perspective
MRC	Medical Research Council
MRS	Midlands Research Station
MSF	Médecins Sans Frontières
NAS	National Autistic Society
NFS	network file sharing
NHS	National Health Service
NPD	new product development
NSD	new service development
NSF	National Science Foundation
NSI	national systems of innovation
OST	objectives, strategy, and tactics
OTC	over-the-counter
P2P	peer-to-peer
PC	personal computers
PCC	product-customer centre
PLC	product life cycle
PTFE	polytetraflouroethylene
QFD	quality function deployment
R&D	research and development
RBV	resource-based view
RDA	regional development agency
RIAA	Recording Industry Association of America
RISC	reduced instruction set computing
ROI	return on investment
RSI	regional clusters or systems of innovation
RTTC	Road Time Trials Council
SECI	socialization–externalization–combination–internalization
SI	system of innovation
SME	small or medium-sized enterprise
SNA	social network analysis
SPARC	scalable processor architecture
SPRU	Science Policy Research Unit (Sussex University)
TDR	Special Programme for Research and Training in Tropical Diseases (UNDP/World Bank/WHO)

TENS	techno-economic system
TLC	technology life cycle
TQM	total quality management
TSB	Technology Strategy Board
UCI	Union Cycliste Internationale
UNDP	United Nations Development Programme
VR	virtual reality
WHO	World Health Organization

This book is dedicated to the memory of Jean Hutton
(1948–2006).

Part I

Building the Foundations

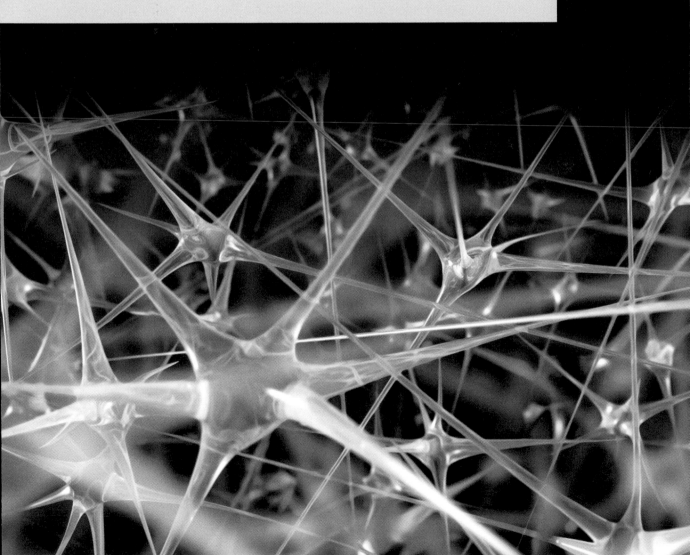

Introduction

The objective of Part I is to build the foundations for presenting a more sophisticated and nuanced interpretation of the management of innovation within organizations. We focus on three themes: in **Chapter 1**, we set about developing an understanding of the terms employed in the study of innovation; in **Chapter 2**, we focus our discussion on an exploration of the tensions, contradictions, and paradoxes that we consider to be inherent to the innovation process; finally, in **Chapter 3**, we provide an overview of the network perspective, and the nature and role of social and organizational networks in the innovation process.

Defining 'innovation'

Despite the dramatic rise in interest in innovation among academics, policymakers, and practitioners in recent years, and the accompanying increase in literature concerning the topic, Johannessen et al. (2001: 20) note that this has '*not yielded a widely-held consensus regarding how to define innovation*'. The objective of **Chapter 1** is to: define and discuss the distinctions between concepts such as 'discovery', 'invention', 'design', 'innovation', 'diffusion', and 'change'; identify different types of innovation, such as, product, process, service, and administrative; and explore different notions of novelty, success, and failure in innovation.

Paradox and contradiction in the management of innovation

A fascinating theme in the understanding of innovation in recent years has been an increased awareness that inherent to the innovation process is the management of a diversity of often-contradictory practices and goals. For example, some parts of an organization may be focused on novelty and exploration, whilst other parts may be more conservative and concerned with stability and exploitation. Indeed, the management of innovation can be seen as the capability to handle such tensions creatively.

The trigger for this shift arose from two prominent US studies that addressed the long-running theme of attempting to characterize successful innovative organizations. In their respective studies, Tom Peters and Bob Waterman, both partners at McKinsey, and Rosabeth Moss Kanter, of Harvard Business School, adopted an approach that focused on a systematic interpretation of successful managerial practice. In their 1982 text *In Search of Excellence*, Peters and Waterman concluded that '*the excellent companies have learned how to manage paradox*'. Kanter, in her 1985 book *The Change Masters*, similarly concluded that the ability to master 'innovation dilemmas' was at the heart of successful performance.

Such paradoxes and dilemmas are evident in our discussions throughout this text.

Innovation from a network perspective

A key feature of this text is its adoption of the network perspective as a 'lens' for discussing and exploring innovative activity. Networks are viewed as an important mechanism through which organizations source information, knowledge, technology, and resources during the innovation process. Social and organizational networks are also employed by individuals and organizations to 'shape' innovation.

Although the origins of the 'network' perspective in the study of science, technology, and innovation go back over forty years, the approach has received renewed interest in recent years. Since the mid-1990s, for example, managers, consultants, and policymakers, concerned with promoting the innovative capacity of organizations, regions, sectors, and nations have increasingly recognized the importance of regional and sectoral networks (Commission of the European Communities, 2003; Department for Innovation, Universities, and Skills, 2008). In part, this interest has been stimulated by the success of 'Silicon Valley'.

Whether or not networks are a new phenomenon or a panacea for innovation, however, is an issue that requires exploration. We address such issues in **Chapter 3**, along with an overview of the important theoretical, conceptual, and empirical contributions from the network literature that facilitate our understanding of innovation activity within and between organizations.

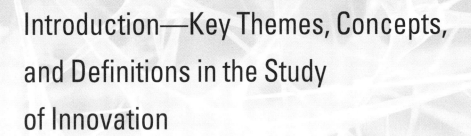

Introduction—Key Themes, Concepts, and Definitions in the Study of Innovation

Chapter overview

Learning objectives

This chapter will enable the reader to:

- recognize the different types of innovation, e.g. product, process, service, and administrative, and the range of innovative organizations, e.g. private sector, public sector, and voluntary sector organizations;

- explain the differences between discovery, invention, design, innovation, diffusion, and change;

- understand the differences between viewing innovation as an output, a process, and an organizational capability or competence;

- appreciate the nuances around the notions of novelty, radicalness, and success, with regard to innovation outputs;

- understand the different ways of classifying knowledge, and the distinction between knowledge, information, and data; and

- identify the dominant perspectives within the innovation literature.

1.1 Introduction

We will start with a brief and simple exercise. Think of five innovations and write them down on a piece of paper. Individual choices will no doubt vary, depending on personal background and experience, but the likelihood is that most of the innovations you have listed will have many of the following characteristics:

- they will be relatively novel or radical;
- they will be hi-tech;
- they will be tangible products;
- they will be discrete or stand-alone (i.e. not a component of another innovation);
- they will have been developed by private sector organizations; and
- they will be fairly recent developments.

Good examples of innovation that meet all of these criteria include the mobile phone, the digital camera, the compact disc (CD), the digital versatile disc (DVD), and the MP3 player.

But an innovation may have none, or only some, of these characteristics. For example, most innovations are, in fact, incremental developments, representing only minor changes, such as a new and improved washing powder or toothpaste. Innovation may be low-tech, employing very little, or at most, very established and mature technology, such as the Baygen radio (the battery-free radio). An innovation may be a service, such as telephone or Internet banking (see **Illustration 1.1**), or a new management practice, such as total quality management (TQM) or business process re-engineering (BPR). Furthermore, rather than being a discrete, stand-alone innovation, such as a 'Post-it note', for example, an innovation may be a component or subsystem, such as an 'airbag' in a modern automobile.

Illustration 1.1 Text banking—an innovative financial service

Launched in 1989 by the Midland Bank (now HSBC), 'First Direct' was the first telephone banking service in the UK. It rapidly attracted a loyal customer base—now over a million strong—typically young, technology literate, and 'money-rich', but 'time-poor'. Over the intervening years, the bank has continued to innovate, embracing emerging technologies—most notably, the Internet and mobile phones—to deliver innovative services such as Internet banking and text banking. First Direct launched text banking back in 1999 and has around a third of its customers registered for the service, which is 'the largest text messaging bank in the UK'. The novel service taps into the rapid rise in the use of text messaging and allows, for example, customers to receive text alerts when they are about to go overdrawn or over their authorized overdraft facility. The text alerts are set up by the customer via mobile phone or the Internet, and can be amended or removed at any time.[1]

Innovation may also originate within public sector or charitable organizations, such as NHS Direct (see **Illustration 1.2**). Whilst radical innovations of recent years may catch the imagination, such as virtual reality, the Internet, and genetically modified (GM) crops, it is useful to reflect on radical innovations of the past in order to understand how these may have impacted on individuals, organizations, societies, and nation states.

What can we learn from the development and diffusion of innovations such as steam power, electricity, and the microprocessor? The above discussion has hopefully opened your eyes to the breadth of what might constitute an innovation. The following sections will define more fully the nature of this diversity, in order that you will be equipped to distinguish between different innovations in a variety of ways.

Illustration 1.2 'NHS Direct'—an innovative public sector service

NHS Direct provides free and professional healthcare advice from trained nurses over the telephone, on a twenty-four-hour basis, seven days a week. Since it was launched in March 1998, NHS Direct has grown rapidly, from small-scale pilots, to a national service handling well over half a million calls a month by 2004—making it perhaps the world's largest 'e-health' service. It represents an innovative extension to the wide range of services already provided by the National Health Service (NHS) in England and Wales. (A similar service was introduced in Scotland in 2002.)

The service provides confidential advice for members of the public, who can call whenever they or family members are feeling ill, as well as information on specific health conditions, local healthcare services, and self-help and support organizations. The telephone service is now accompanied by NHS Direct Online—a website providing high-quality healthcare information,[2] which also had over half a million 'hits' a month by 2004—and a text-messaging version of NHS Direct for those who are deaf, hard of hearing, or who have speech difficulties.

More broadly, NHS Direct has helped to offset demand on other NHS services—in particular on 'out-of-hours' family doctors—and deal with the management of health scares. It has also empowered many patients by providing high-quality information to enable them to make decisions about their own health and that of their families.

Despite the dramatic rise in interest in innovation among academics, policymakers, and practitioners in recent years, and the accompanying increase in literature concerning the topic, Johannessen et al. note that this has '*not yielded a widely-held consensus regarding how to define innovation*'.[3] It is, however, important to disentangle and define the concept of 'innovation' as closely as possible, in order to develop a common understanding before we proceed. Thus, the following sections will distinguish between:

- 'discovery', 'invention', 'design', 'innovation', 'diffusion', and 'change';

- innovation as an output, a process, and a capability or competence;

- types of innovation, such as product, process, service, and administrative;

- notions of novelty and radicalness;

- notions of success and failure in innovation; and

- 're-innovation' and 'exnovation'.

1.2 Distinguishing between 'discovery', 'invention', 'design', 'innovation', 'diffusion', and 'change'

Many writers in the broad areas of management and strategy, as well as those more specifically in the niche areas of innovation and new product development (NPD), use the terms 'invention', 'design', 'innovation', and 'change' very loosely, and often interchangeably. Very broadly speaking, as one moves from 'discovery', to 'invention', to 'innovation', and finally to 'change', each may be viewed as an element, or subset, of the following term. Thus, innovation (the key focus of this book) may be seen as embracing discovery, invention, and design activities. Change, in its broadest sense, although encompassing innovation, is not a focus of this book.

1.2.1 Discovery

The term 'scientific discovery' usually refers to the process of recognizing or observing for the first time a particular natural phenomenon or object. Discovery may be viewed as part of the process of invention: for example, fire was 'discovered', but methods of creating fire had to be 'invented'. Although discovery may arise out of chance or serendipity, it is often the result of purposeful activity, even if the result was not one that had been anticipated: for example, the discovery of pasteurization by Louis Pasteur and the invention of the rubber vulcanization process by Charles Goodyear.

With such examples in mind, it has been argued that:

> chance or accidental observations come as a bonus to the perceptive researcher who has already done his 'homework'. As Louis Pasteur observed, 'Chance favours only the prepared minds'.[4]

Thus, perhaps one ought to view the discovery by Archimedes of a method for measuring volume not as an 'eureka' moment, as it is often portrayed in folklore, but as the culmination of a long and purposeful process of searching for a solution to a preconceived problem.

1.2.2 Invention

It is common for those writing about the management of innovation to distinguish between the concepts of 'invention' and 'innovation'. The origin of this important distinction is often attributed to the early work of Schumpeter.[5] Rothwell and Zegveld define invention as *'the creation of an idea and its reduction to practice'* and *'an act of technical creativeness involving the description of a novel new concept that would normally be suitable for patenting'*.[6] By 'reduction to practice', they mean the development of a laboratory test, or of a hand-built prototype to test the concept or principle. Similarly, Freeman and Soete define invention as *'an idea, a sketch or model for a new or improved device, product, process, or system. Such inventions may often (not always) be patented, but they do not necessarily lead to technical innovations'*.[7]

Whilst these are useful definitions, like many other definitions in the innovation literature, they are biased towards technological product and process innovation, as opposed to service or administrative innovation (the distinction between which will be addressed later in this chapter). What is key in these definitions, however, is the emphasis on 'novelty' or 'newness', and the implication of the purposeful pursuit of translating this concept into an artefact or process. It is also important to note here that the notion of 'invention' does not embrace the commercialization or 'bringing into common use' of a new idea. We will return later to the issues surrounding the relative degree of novelty of an innovation, and the different perceptions that different individuals and organizations might have as to the degree of novelty of a given invention or innovation.

1.2.3 Design

The term 'design' has a diversity of meanings. For Roy:

> Design . . . is the activity in which ideas or market requirements are given specific physical form, starting from the initial sketches or conceptual designs, through prototype development, to the detailed drawings and specifications needed to actually make the product.[8]

This definition is essentially referring to the process of product design, but by modifying or loosening this definition to allow for the incorporation of service and process innovations, design can be seen as a process of bringing form and order to both technical and non-technical solutions, as well as to the satisfying of user needs.

1.2.4 Innovation

In providing a distinction from invention, Rothwell and Zegveld define 'innovation' as involving *'the commercialisation of technological change'* and 'invention' as simply *'one element, albeit an important one, in the overall innovation process'*.[9] For Freeman and Soete, *'an innovation . . . is accomplished only with the first commercial transaction involving the new product, process, system, or device'*.[10] This second definition, in particular, is too narrow in focus for the purpose of this book, because it focuses on the point of entrance into the marketplace of a novel idea, rather than the broader processes of development and diffusion. Furthermore, as Trott points out, *'commercial failure . . . does not relegate an innovation to an invention'*.[11]

The following represent a selection of other useful definitions of innovation:

> Innovation is not a single action but a total process of interrelated subprocesses. It is not just the conception of a new idea, nor the invention of a new device, nor the development of a new market. The process is all these things acting in an integrated fashion.[12]

> Industrial innovation includes the technical, design, manufacturing, management and commercial activities involved in the [bringing to market] of a new (or improved) product or the first commercial use of a new (or improved) process or equipment.[13]

> Innovation refers to the process of bringing any new, problem-solving idea into use. Ideas for re-organizing, cutting costs, putting in new budgeting systems, improving communication, or assembling products in teams are also innovations. Innovation is the generation, acceptance, and implementation of new ideas, processes, products, or services. It can thus occur in any part of the corporation, and it can involve creative use as well as original invention. Application and implementation are central to this definition; it involves the capacity to change and adapt.[14]

> Innovation is the successful exploitation of ideas.[15]

> Innovation is the successful production, assimilation and exploitation of novelty in the economic and social spheres.[16]

All of these definitions share a number of key elements:

- innovation (like invention) concerns novelty;
- innovation (unlike invention) is concerned with the exploitation of new possibilities, through the bringing to market, or the bringing into practical use, of an idea or concept;
- innovation (like invention) is a process (as well as an output, such as a CD);
- innovation (unlike invention) is a broad concept that embraces the full range of activities from discovery and invention, through to development and commercialization.

As with the definitions of invention noted earlier, definitions of innovation also have a strong orientation towards technological products and processes. In this sense, the last two definitions, provided by the UK government's Department for Innovation, Universities, and Skills (DIUS), and the European Commission, respectively, are more embracing and thus useful working definitions for the purposes of this book.

From this synthesis of definitions in the innovation literature, it is easy to see why innovation is often expressed as:

innovation = invention + commercialization

But in order to embrace a wider range of innovations, such as those developed within the public or not-for-profit sectors, we will adopt a broader definition:

innovation = invention + bringing into common usage

1.2.5 Diffusion

The process of commercialization, or the bringing into common usage, of a new product, process, service, or practice is essentially one of 'diffusion': it concerns the spread over time of the consumption or adoption of an innovation among individuals or organizations.

One of the academics most associated with diffusion research is Everett Rogers, whose seminal text on the subject in the early 1960s, entitled *Diffusion of Innovations*, is now in its fifth edition. For Rogers, 'Diffusion *is the process by which an innovation is communicated through certain channels over time among the members of a social system*'.[17] The communication of an innovation around a marketplace or social system may lead to the adoption, adaptation, rejection, or discontinuance (i.e. rejection following initial adoption) of that innovation. Although the processes of launching and promoting a new innovation are important factors in the diffusion of an innovation, these are dealt with more fully in marketing texts and are not a particular focus in this text.

Diffusion *is* of interest to the concerns of this book in the following ways.

a) The diffusion of an innovation is influenced by the characteristics of the innovation itself, such as its complexity, i.e. the degree to which an innovation is perceived as difficult to understand and use.[18] (These characteristics are indicated in **4.7**.)

b) The diffusion of an innovation, through its implementation and usage by users, may bring about valuable feedback to the original innovator that can influence subsequent incremental development of that innovation. Indeed, it has been argued that user requirements and product characteristics can often only be discovered if the innovative product or process is actually used, sometimes for a long period of time.[19] (This process is discussed in **10.3.3**.)

c) The diffusion of an innovation, through its adaptation during implementation and usage by users, is itself a process of incremental innovation. This is sometimes referred to as 'user innovation'. Such adaptations may also be fed back to the original innovator and influence subsequent incremental development of the innovation.[20] (This process is covered within various subsections of **10.3**.)

d) Radical innovations, such as electricity and the Internet, can bring about major changes to the practices and organization of industry and society, whilst others can raise major ethical, moral, and environmental concerns, as we have seen with the genetic modification of arable crops. In such cases, diffusion can shape society, but at the same time, society—through a range of stakeholders—may attempt to shape the diffusion process, or even the innovation process itself. (These issues are addressed in **Chapters 5 and 11**.)

1.2.6 Change

Finally, it is important to differentiate the terms 'change' and 'innovation'. One way to achieve this is to focus on novelty. Indeed, for Zaltman et al., novelty is an important way of distinguishing between innovation and change, because whilst all innovation presupposes change, not all change presupposes innovation (i.e. novelty or newness).[21]

If we move away from such semantics, towards reviewing and contrasting the central issues addressed by those writers that deal with the 'management of innovation' and those that deal with the 'management of change', we see that the former are far more concerned with the earlier stages of the innovation process (i.e. the development and delivery of innovation to the marketplace), as opposed to the later stages of the innovation process (i.e. the adoption, adaptation, and, in particular, the implementation of innovation by end-users or customers), which are frequently the central concern of the latter.

1.2.7 Innovation as an output, a process, and a capability

It is worth briefly pointing out at this juncture that the term 'innovation' is used in a number of ways, although this variation in use is often implicit rather than explicit. Firstly, innovation is used to refer to a new product, process, or service, for example— that is, innovation is viewed as an 'end product' that is offered to the marketplace or end-user by an innovating organization. In this respect, innovation is characterized as an output.

Secondly, in defining innovation, many academics refer to the various activities that an innovating organization undertakes (e.g. idea conception, technical design, prototype testing, and commercialization) in translating an idea into an innovation that is then made available to the marketplace or end-user. In this respect, innovation is viewed as a process. If we are to develop insight into how innovative organizations bring innovative products, processes, or services to the marketplace or the end-user, then it is extremely important to 'prise open the black box' of the innovation process. Thus, it is worth noting briefly at this stage that the innovation process may be viewed from a number of perspectives:

- *as a 'management process'*—i.e. the focus is on the organization and management of the various activities and phases of the innovation process;

- *as a 'social process'*—i.e. highlighting the role, nature, and importance of social interaction during the innovation process;

- *as a 'political process'*—i.e. highlighting the contestation over finite resources and between alternative knowledge claims or solutions; and

- *as an 'emotional process'*—i.e. highlighting the importance of issues such as 'psychological safety' in allowing individuals to challenge long-held assumptions and present new alternatives.

We will expand upon each of these perspectives of the innovation process in **Chapters 8 and 9**.

Finally, as we will see in **Chapter 6** concerning strategy and core competence, innovation or the ability to innovate can be seen as an important organizational capability or 'soft' core competence. Innovative organizations, such as 3M, Hewlett-Packard, and Sony, are seen as not only possessing technological ('hard') competences, but also organizational and managerial ('soft') competences.

1.3 Distinguishing between different types of innovation output

1.3.1 Distinguishing between 'product', 'process', 'service', and 'administrative' innovation

The distinction is often made between 'product' innovation and 'process' innovation. For Tidd et al., product innovation refers to '*change . . . in the things (products/services) which an organization offers*' and process innovation to '*change in the ways in which they are created and delivered*'.[22] Similarly, Damanpour defines product innovation as '*new products or services introduced to meet an external and market need*' and process innovation as '*new elements introduced into an organization's production or service operation—input materials, task specifications, work and information flow mechanisms, and equipment used to produce a product or render a service*'.[23]

Other classifications of innovation distinguish between: 'administrative' innovation, i.e. new forms of organization, management, and administrative processes, which helps to highlight the distinction between organizational process innovations (e.g. TQM) and technological process innovations (e.g. a new chemical process or manufacturing process); 'delivery' innovation, i.e. new forms of delivery or distribution; and 'market' innovation, i.e. new forms of marketing or market behaviour. A clear distinction is also often made between 'service' innovation and 'product' innovation: unlike products, services are intangible, and typically produced and consumed simultaneously. **Table 1.1** provides a summary and examples of a broad range of innovation types.

But the distinction between product and process innovation, or between technological or organizational innovation, is problematic, because these are often intertwined or blended together during the implementation process.[24] For Clark and Staunton, for example:

> The separation may have possessed some merits when equipment was concentrated upon the shop-floor areas, but now that there are extensive usages of equipment in areas of coordination and of design the separation is confusing.

They provide the example of computer-aided production management innovation to illustrate the blending of technology and administrative facets.[25]

Furthermore, it is likely that the perception of the importance of the technological and administrative elements may vary: innovating organizations may place greater emphasis on the technological dimension, because this is likely to be most 'visible' during the innovation process, whilst adopting organizations might place greater emphasis on the administrative dimension, because this may be most 'visible' during usage, and in comparison with previous systems, procedures, or practices.

In some instances, the distinction between product and service innovation may also be problematic: an innovation may contain elements of both, or perceptions may vary between the innovator and the customer, as we will see in **Case Study 1.1** on the 'Laff Box'.

Research on innovation has traditionally focused on technological innovation and as such, has utilized a narrow working definition of product and process innovation.

Table 1.1 Types of innovation

Innovation type	Definition	Examples
Product	A novel tangible artefact, including materials and components, those based on high as well as low technology, and those aimed at individuals or organizations	From hi-tech (e.g. computers) to low-tech (e.g. ready-made meals), and from consumer products (e.g. mobile phones) to industrial products (e.g. new building equipment or materials)
Service	Intangible and involving the undertaking of a novel activity for another individual or organization	Online grocery shopping and home delivery offered by supermarkets
Process	Generally concerns novel technological processes, as distinct from organizational processes	DNA fingerprinting, frequently used in police work and paternity cases
Organizational/ administrative	Novelty in organizing or the undertaking of processes or tasks within an organization	TQM, BPR, 'hot-desking' and virtual team-working
Delivery	Novelty in the delivery of products or services, for example, from provider to consumer	Mobile breast cancer screening facilities, which shift provision out of hospitals and into local communities
Marketing	Novelty in the marketing of products or services, for example	'Viral' marketing or product placement in films
Business model	Novelty in the 'drivers' of an organization's activities or strategy	Low-cost airlines, as typified by EasyJet, and Internet firms, such as Google, which generate revenue through advertising rather than the services they provide
Institutions	The establishment of an organization with a novel role, whether within the private, public, or not-for-profit sectors	At their formation, institutions such as the United Nations, the World Trade Organization, and the British National Health Service were highly innovative

Product innovation in these studies commonly equates to tangible, technology-based, artefacts, such as medical equipment[26] and scientific equipment;[27] service innovation is generally neglected. Similarly, process innovation is often viewed as technology-based, such as that related to chemical processes.[28] Furthermore, much of the work on 'systems of innovation' focuses on technological product and process innovation.[29] Despite this bias, Sundbo argues that '*service firms often lag behind manufacturing ones in innovativeness*' and that, consequently, '*they can learn something from the manufacturing way of thinking about and organizing innovation*.[30] Nevertheless, there is an emerging body of research that is centred on service-sector innovation.[31]

1.3.2 Distinguishing between 'incremental' and 'radical' innovation

Studies of innovation often distinguish between 'incremental' and 'radical' innovation. Radical innovation is generally viewed as a '*major advance in the technological state-of-the-art*'.[32] As such, it is often seen to replace existing innovation based on prevailing technology, through the development and adoption of new technology to deliver new and improved functionality and performance. Examples of radical innovation include the development of steam power, which replaced water power, and the development of the semiconductor, which replaced the valve in computers.

In contrast, incremental innovation is seen as providing minor or major improvements in functionality and performance to an existing innovation. Clark and Staunton view the process of incremental innovation as similar to that embraced by the concepts of 'learning by doing' and the 'experience curve'.[33] Incremental innovation can be viewed as the evolutionary development of a radical innovation. For example, the technological progress of the semiconductor (as represented by the Intel 236, 386, 486, Pentium I, Pentium II, Pentium III, etc.) is characterized by a series of incremental innovations building upon the initial radical innovation of the semiconductor. Such incremental improvements may be seen as examples of 'entrenching innovation', because they have the result of reinforcing the usage of the original innovative solution through improving its utility, and through allowing for its diffusion into other sectors and applications.[34]

The sum total of such incremental improvements can dramatically improve the technical and commercial performance of the original radical innovation. Indeed, Rothwell and Zegveld argue that:

> radical and incremental innovation often will be closely linked, and a radical innovation will pave the way for an extended series of improvements, the sum of which can have as marked an influence on the innovation's commercial performance as did the original breakthrough . . . In competitive terms, while the introduction of a [radical innovation] . . . can afford the originator significant technical and market leads, continued [commercial] success depends on the ability of the firm to improve continuously the performance of a new device. The evolution of the innovation is part of a cumulative learning process within the firm and by customers.[35]

This point is well illustrated by Sony's innovation strategy following the launch of the Sony Walkman in 1979 (see **Illustration 2.3**). In order to sustain market leadership in the personal stereo market, Sony continued to develop the product through a series of incremental improvements (e.g. mini headphones in 1979, AM/FM stereo radio in 1980, a downsized unit in 1982, etc.).

Whilst it is useful to distinguish between radical and incremental innovation, Henderson and Clark argue that this categorization is '*incomplete and potentially misleading*'. Drawing upon case study evidence, they argue that it is important to distinguish between 'component' and 'architectural' knowledge and innovation, and the relationship between the two.[36]

1.3.3 Distinguishing between 'component' and 'architectural' innovation

As users and consumers of products such as personal stereo systems, washing machines, and automobiles, we often view these artefacts as stand-alone or discrete. Yet, each of these, in fact, incorporates many components and subsystems. Services too, such as Internet banking, can be thought of as a bundle of 'components', representing the different elements that make possible the delivery of the overall service, such as the 'behind-the-scenes' software and hardware systems.

In **Illustration 1.3**, concerning competitive ice speed skating, we see that the performance of the athlete is impacted by a series of different innovations, such as in the ice skate, which itself is comprised of other innovations, such as materials. Furthermore, these innovations are often interrelated, having 'knock-on' impacts, such as the interrelation between the ice skate, ice preparation, and skating techniques, for example—that is, an innovation in one may require innovation in another or a change in the way in which the 'components' are configured or linked to each other.

Illustration 1.3 Innovation and competitive ice speed skating

As in many sports, the improvement in performance within competitive ice speed skating has been marked over the last hundred years or so. Analysis of world records for both men (since the 1900s) and women (since the 1930s), across a range of distances, reveals performance improvements in speed of roughly 75 per cent. During this period there has been a great deal of innovation in the sport, such as in the ice skate, the specialist clothing, ice preparation, skating techniques, physical and psychological training methods, and nutrition, as well as in the organization of the sport itself. Each of these different areas of innovation may be viewed as individual 'components' in a system, impacting the overall performance of ice speed skaters.[37]

For Henderson and Clark, it is important to distinguish between the 'components' and 'architecture' of a product. To achieve this, they distinguish between 'the product as a whole—the system—and the product in its parts—the components', where they define a component as '*a physically distinct portion of the product that embodies a core design concept . . . and performs a well-defined function*'.[38] They refer to innovation that changes the way in which the different components are 'linked together', whilst leaving the core design concepts of the components intact, as 'architectural innovation'. They continue, by arguing that:

> successful product development requires two types of knowledge. First, it requires component knowledge, or knowledge about each of the core concepts and the way in which they are implemented in a particular component. Second, it requires architectural knowledge or knowledge about the ways in which the components are integrated and linked together into a coherent whole.

In **Figure 1.1**, the y-axis (vertical) concerns the impact of an innovation in a component upon the linkages between the different components of the overall product, whilst the x-axis (horizontal) concerns innovation in a component and its impact on the core concepts that 'lie behind' that component. The two boxes to the left of the figure relate to incremental innovation in a component (as defined in **1.3.2** above). Some incremental component innovations will, however, have an impact on the way in which different components are linked together and, where this is the case, the innovation is considered to be an 'architectural' innovation.

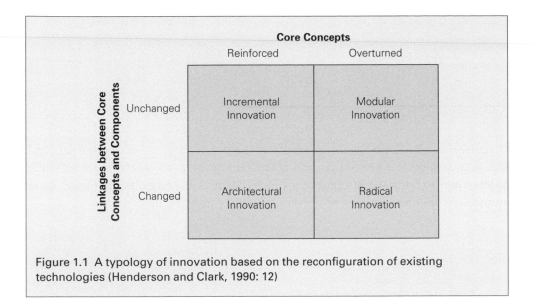

Figure 1.1 A typology of innovation based on the reconfiguration of existing technologies (Henderson and Clark, 1990: 12)

Similarly, the two boxes to the right of the figure relate to radical innovation (as defined in **1.3.2** above). But some radical component innovations will impact only the individual component and not the linkages between the components in the overall system, and, where this is the case, the innovation is considered to be a 'modular' innovation. Thus, what we see is a more nuanced interpretation of the radicalness of an innovation, which takes account of the broader impact that an innovation in a component might have on the product as a whole. For Henderson and Clark, these distinctions are important, because they help to '*account for the sometimes disastrous effects on industry incumbents of seemingly minor improvements in technological products*'.[39]

Figure 1.2 demonstrates the complexity that emerges when one considers innovations in relation to their component parts. We see here that a computer disk is made up of a number of components, whilst being at the same time one of many components that comprise a disk drive, which, in turn, is only one component of a mainframe computer. This 'nested' perspective also reveals the wide range of components and technologies that make up complex products, and the potential for interrelatedness between these components that might give rise to architectural innovation.

Although much of the discussion above and the research upon which it draws is based on product innovation, the concepts are also extremely useful for understanding other types of innovation, such as service or process innovation.

1.4 Assessing novelty and radicalness

Assessing the novelty or radicalness of an innovation output is not quite as simple as it might first appear. Our initial response might be to link radicalness solely and directly to the features of a new innovation, and to assess the innovation with regard to the degree to which it offers new and unique functionality or improved performance, or perhaps to

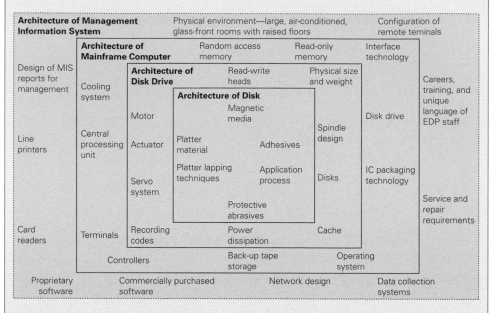

| Architecture of Management Information System | Physical environment—large, air-conditioned, glass-front rooms with raised floors | | | Configuration of remote teminals | |

Figure 1.2 A nested system of product architecture (Christensen and Rosenbloom, 1995: 238)

focus on the underlying technology embedded within the innovation. There are, however, other ways in which the radicalness of an innovation may be assessed. For example, radicalness may be viewed in relation to the degree to which an innovation impacts upon the activities or capabilities of the users of that innovation, or indeed upon the activities or capabilities of the innovating organizations.

Another way in which to view radicalness is to assess the impact that the innovation has had across business and society. An innovation that has been widely diffused, and which has thus had a wide impact across many sectors and nations, could be viewed as radical. Thus, the radicalness of an innovation may be considered from a range of perspectives.

1.4.1 Alternative perspectives on novelty and radicalness

The following discussion outlines five alternative perspectives for assessing the novelty or radicalness of an innovation. Because these perspectives are not mutually exclusive, we combine several of these perspectives to provide a more nuanced categorization of novelty and radicalness.

a) *The embedded characteristics of the innovation*—i.e. the degree to which the innovation offers new or improved functionality, form, and/or performance, as well as the underlying technology that allows the delivery of such features. Thus, those innovations perceived as highly novel are likely to be those offering new or substantial

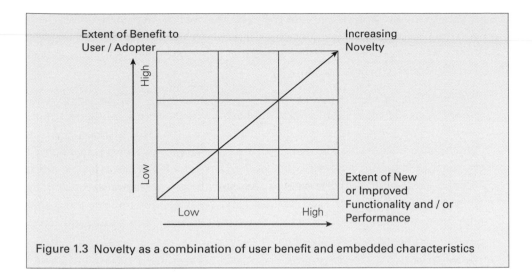

Figure 1.3 Novelty as a combination of user benefit and embedded characteristics

improvements in functionality, form, and/or performance, and delivering these through the application of a new technology (e.g. the CD was seen as highly novel in relation to the vinyl record). One might expect the greater the improvement in performance or functionality of the product, process, or service, the faster the erosion of the competitive position of alternatives.[40]

b) *Benefit to the user or adopter through usage or consumption of the innovation*—i.e. the novelty of an innovation is also likely to be influenced by the degree to which the embedded characteristics of the innovation provides the user or consumer with new benefits or possibilities, or changes their patterns and routines. For example, the Internet can be seen to provide radical new benefits to users, and has begun to change fundamental patterns and routines of life and work. As Clark and Staunton note, however, the impact of an innovation may vary *'depending upon the purposes for which it is used and the state of the contexts in which it is inserted'*.[41]

The above two perspectives might be usefully combined to provide an indicator of the novelty of an innovation (see **Figure 1.3**). From this new perspective, the highest novelty would be attributed to those innovations that provide both high user benefit (y-axis), and high functional and/or performance improvement (x-axis).

Space tourism (see **Illustration 1.4**) would be located in the top right-hand box of **Figure 1.3**, because it offers a highly novel service with highly novel opportunities for the user.

Illustration 1.4 Space tourism—a high-novelty service innovation

Space tourism is no longer a notion confined to science fiction. An alliance between the Russian Aviation and Space Agency and Space Adventures Ltd, a commercial US space exploration company founded in 1998, sent its first customer into space in May 2001. The tourist, Dennis Tito, a financier and former space scientist, paid $20m for an eight-day trip to the International Space Station. He was followed, in April 2002, by Mark Shuttleworth, the South African Internet businessman. Although still in its infancy, the stated mission of Space Adventures Ltd is *'to open spaceflight and the space frontier for private citizens'*.[42]

Refer back to the five innovations that you wrote down at the beginning of this chapter. Where might you locate these innovations in the above matrix? When you have done this, think about other examples of innovation that might be located in the segments of the matrix for which you have no examples.

c) *The breadth of diffusion of the innovation*—i.e. the more widely diffused an innovation, the greater the collective impact it may have on society and business. Thus, consequently, the more pervasive (i.e. widely diffused and used) an innovation, the more radical it is likely to be perceived. This notion of radicalness is highlighted by Johannessen et al., who argue that '*as the economic unit that adopts an innovation becomes more broadly defined or encompassing, the impact of the innovation is more likely to be radical*'.[43] Innovations such as electricity, the telephone, the computer, and the Internet can all be seen to have been widely diffused both within and across business sectors and geographical regions.

In this regard, Clark and Staunton distinguish between what they term 'generic' innovation and 'epochal' innovation. Generic innovation refers to clusters of innovations based on a new core technology (e.g. the internal combustion engine). As such, they collectively create a new technological paradigm through their wide diffusion across applications, industrial sectors, and nations. Such pervasive (i.e. widely diffused) innovations have the capacity to have major changes on society and industry (e.g. electricity). In contrast, epochal innovation refers to innovation that has the capacity to alter a particular sector dramatically (e.g. the float glass process for producing glass, or the development of automatic gear change in the automobile industry). Such innovation may be widely diffused within a particular sector, but because it does not extend to other sectors, it would not be regarded as pervasive.[44]

The above perspective might also be usefully combined with the notion of novelty, as developed in **Figure 1.3**, to provide an indicator of the radicalness of an innovation (see **Figure 1.4**). From this new perspective, the most radical innovations would be those that are both widely diffused (y-axis) and provide high novelty (x-axis).

Again, refer back to the five innovations that you wrote down earlier. Where might you locate these innovations in the above matrix? When you have done this, again think about other examples of innovation that might be located in the segments of the matrix for which you have no examples.

The notion of radicalness as developed in **Figure 1.4** is one that is largely market-orientated, with a focus on the extent of user benefit and user adoption. In addition to this framing of radicalness, two further perspectives of the radicalness of an innovation are useful to keep in mind: the impact on the innovator and the impact of time.

d) *The impact on an innovating organization's capabilities and competences*—from this perspective, radical innovation is that which requires new technological knowledge to exploit it, whilst rendering existing technological knowledge obsolete. In contrast, incremental innovation utilizes and builds upon existing technological knowledge. Thus, radical innovation can be viewed as competence destroying, whilst

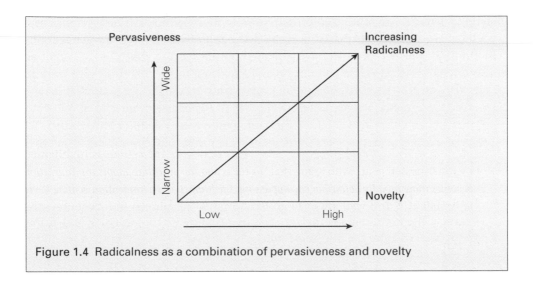

Figure 1.4 Radicalness as a combination of pervasiveness and novelty

incremental innovation can be seen as competence enhancing.[45] This perspective is developed further in **5.3.3**, concerning strategy and core competence.

e) *The time elapsed since the launch of the innovation*—i.e. the perceived novelty or radicalness of an innovation will decrease over time as it becomes diffused widely, and embedded in patterns and routines of life or work, or is superseded by more recent innovations. Consider, for example, the bicycle or steam power, both of which were once considered to be highly novel innovations.

1.4.2 Novel to whom?

Zaltman et al. define innovation as '*any idea, practice, or material artefact perceived to be new by the relevant unit of adoption*'.[46] Similarly, Rogers argues that '*it matters little . . . whether or not an idea is objectively new as measured by the lapse of time since its first use or discovery . . . If the idea seems new to the individual, it is an innovation*'.[47] What is of interest in these definitions of innovation is that newness or novelty is not considered to be a property of the innovation itself, but simply a perception of the individual or organization that adopts the innovation. Thus, an innovation may be: new to one consumer, but not to another; new to one organization, but not to another; or new to one industrial sector, but not to another. This touches upon a number of concepts that will be dealt with during this book: for example, the concept of 'lead users'[48] or 'early adopters'[49]—that is, consumers or user organizations that adopt innovations earlier than their peers.

A further example involves the concept of 'boundary spanning'[50]. In this instance, innovation may be cross-fertilized or transferred from one country, sector, organization, or discipline in which it is well established into another in which it is new and novel. A good example of cross-sector fertilization is provided by the case study of the Dyson cyclone vacuum cleaner. In the development of this novel vacuum cleaner in the mid-1980s, James Dyson, the inventor, employed cyclone technology that had long been used in industrial settings to extract dust and debris from the air.[51]

In summary, the perception of novelty of an innovation by an individual or organization is generally linked to the point in time of adoption, rather than the point in time of the discovery or commercialization of that innovation. This may lead different customers, for example, to have differing views as to the degree of novelty of the same innovation, even where it is used for the same purpose and in the same context. In developing a strategic perspective for the management of innovation by organizations, Sundbo develops a narrower definition:

> The delimitation of innovation [novelty] should exclude the introduction to a firm of an element that is widespread in an industry, but include the introduction of an element if the firm is among the very first to introduce it. The argument is that the introduction will give the firm an advantage in the market, even if one or a few other firms have already introduced it.[52]

Variations in the perception of novelty may also exist between different individuals and organizations along the supply chain at a single point in time[53]; what is considered to be a high-novelty innovation by the innovating organization might be perceived as a relatively low-novelty innovation by the end-user or customer—or vice versa. **Illustration 1.5** provides such an example.

Illustration 1.5 The DSK—a high-novelty innovation?

The Dvorak simplified keyboard (DSK) was developed by Dr August Dvorak in the 1930s as a more efficient and less fatiguing alternative to the QWERTY keyboard. To Dvorak, the innovator, and typewriter manufacturers, the DSK would have been considered an incremental innovation, because it simply required the rearranging of the keys of the keyboard. By typists (the end-users and customers of the innovation), however, it would have been considered a major innovation, because experiments indicated that the DSK was 20–40 per cent more productive than the existing QWERTY keyboard (which then, as today, was by far the dominant arrangement for the typewriter keyboard).

Despite such productivity improvements, the DSK was never adopted widely, not least because it required typists to relearn how to touch-type.[54]

We can see from the DSK illustration how the perception of novelty of an innovation by an individual or organization may be linked directly to the impact that it has on that individual or organization.

1.5 Assessing the success of an innovation output

Whilst it is very common for studies of innovation to focus on cases of success, it is far less common for these to outline explicitly what measure of success is being employed to distinguish the successful from the unsuccessful, or failed, innovations. Furthermore, in the majority of these studies, success is viewed from the perspective of the innovating organization.

As with assessing the radicalness or novelty of an innovation, however, determining the degree of success of an innovation is not unproblematic.[55] Firstly, success may be judged or measured by a range of criteria, such as by the degree of novelty of the innovation itself, or the extent to which it is diffused.[56]

Secondly, the success of an innovation may be judged quite differently, and in quite different ways, by different individuals, groups, organizations, sectors, and nations.

Thirdly, the distinction between success and failure can be blurred if one views innovation as an ongoing and iterative process: for example, the market failure of an innovation may lead to the creation of new market knowledge, re-innovation, and subsequent commercial success for the innovating organization.[57] For Leonard, unsuccessful projects:

> become invisible, and managers delude both themselves and others about the debt owed to failures. Only development team members know how much they individually gained from previous unsuccessful explorations.[58]

Indeed, organizations can learn much from failure and mistakes.[59]

Fourthly, evidence suggests that the measures employed by an organization to assess success or failure depend upon the time perspective employed: in the short term, measures such as 'speed to market' and timeliness of launch are often considered to be key, whilst in the long term, customer acceptance and financial performance are generally perceived as more important.[60]

Thus, in attempting to assess the 'success' of an innovation, it is important to ask three questions:

- what do we mean by success—i.e. what criteria are we using to judge or measure success?
- to whom is the innovation successful—the innovator or the user?
- of what are we measuring the success—a discrete innovation, or the longer-term output of a research project or team that could embrace subsequent improvements on an earlier innovation?

The first two questions are interrelated, because certain criteria are more likely to be employed by certain actors, e.g. sales and profit generated by the innovation are likely to be key indicators for measuring success by the innovating organization, whilst measures of user benefit are more likely to be of relevance to individual users. With regard to the third question, innovation studies have typically focused on discrete instances of innovation, but for Tidd et al.:

> the real test of innovation success is not a one-off success in the short-term but sustained growth through continuous invention and adaptation. It is relatively simple to succeed once with a lucky combination of new ideas and receptive market at the right time—but it is quite another thing to repeat that performance consistently.[61]

Perhaps not surprisingly, many texts on the management of innovation focus largely on the innovating organization and, as such, success is either implicitly or explicitly expressed in relation to the ability of these organizations to appropriate (i.e. capture) the commercial benefits of their innovations. Some authors take a broader perspective, focusing on the role of innovation at the sectoral or national level; here, success is expressed in relation to the emergence or renewal of industrial sectors or nations.[62]

From the perspective of the innovating organization, however, success might be evaluated in relation to a number of dimensions:

- *financial criteria*—e.g. the level of profit or turnover generated by the innovation, or the speed of the return on investment (ROI);

- *market criteria*—e.g. the rate of adoption of the innovation, or market penetration;

- *technical criteria*—e.g. the 'elegance' of the engineering design or improvement in performance and functionality;

- *strategic criteria*—e.g. the building or sustaining of competitive advantage through the development of superior product or service offerings, or the building of technical competences or capabilities; or

- *process criteria*—e.g. the compression of the time taken from idea conception to market launch.

For Johannessen et al., the success of an innovation '*is determined more by the extent of its adoption than by . . . how technologically advanced it is*'.[63] But these dimensions are not mutually exclusive and organizations might use a combination of measures across a range of the above dimensions. The choice of a particular dimension or combination of dimensions ought to reflect the strategic objectives of the innovating organization: for example, an organization aiming to achieve market penetration through a particular innovation may be more interested in the speed and breadth of its adoption, than in the speed of ROI.

The success of an innovation might also be judged by the impact or consequences that it brings through its usage to individual users, or through its wide diffusion, to society.[64] The recognition that innovation might have profound impacts on society, or other stakeholders such as the environment, is an important, although often neglected, aspect of the management of innovation: it highlights the link to other key emerging areas in the study of management and organizations—most notably, around notions of corporate social responsibility (CSR) and business ethics.

We will return to these issues in **Chapters 10 and 11**, which focus on the impact of context on the management of innovation and on the role of 'external' stakeholders in the shaping of innovation.

1.6 Innovation as an iterative process—from innovation to re-innovation

Innovation is rarely a one-off event. High-novelty innovation, as noted earlier, is generally followed by a series of lower-novelty innovations, which bring about a range of minor and major improvements to the original innovation—that is, innovation is often the result of an iterative process.

Such improvements or adaptations to the original innovation are sometimes referred to as 're-innovation'.[65] Re-innovation is generally viewed as a development activity undertaken by the original innovating organization, or by one of its competitors. But for those academics more interested in the diffusion of innovation, re-innovation also refers to the degree to which an innovation is changed or modified by a user in the process of its adoption or implementation.[66] Due to the intangible nature of service and administrative innovations, these might be seen as offering greater scope for user re-innovation than technological product or process innovations. Indeed, the examples of re-innovation provided by Rogers centre around innovations in educational programmes and computer-based administrative tools.[67] We discuss the important role of the user in the innovation process in **Chapter 10**.

The development and implementation of innovation—and, in particular, high-novelty innovation—not only requires the creation and adoption of new knowledge and practices by both developers and adopters, but often also the 'unlearning' of knowledge and practices associated with earlier innovations. This process of unlearning or disengagement with an earlier technology or knowledge base is sometimes termed 'exnovation' and can play an important part in the success of emerging innovation. Clark and Staunton argue that the mediation between the processes of exnovation and innovation is an important, but very neglected, area in innovation research.[68]

1.7 The origins and sources of innovation

The origins and sources of innovation have been of increasing interest to practitioners, academics, and policymakers since the 1950s. This interest has resulted in many studies of successful innovation over the last fifty years. In these studies, 'success' is generally taken as a combination of proven technological and commercial success. Seminal research projects include: 'Project SAPPHO I',[69] 'Wealth From Knowledge',[70] and 'Project SAPPHO II',[71] in the UK; and 'Project Hindsight'[72] and 'TRACES',[73] in the USA.

Although variations have been found to exist between sectors and different types of innovation (e.g. product versus process innovation, and high-novelty versus low-novelty innovation), these and subsequent studies have highlighted a number of common features of successful innovation:

- a substantial number of the key inputs into the idea-generation and problem-solving activities are derived from external sources (i.e. those that are external to the innovating team and innovating organization);

- informal intra- and inter-organizational networks spanning team, functional, and organizational boundaries provide an important conduit for the sourcing of such external inputs;

- a diverse range of external sources are employed in the innovation process, including users, suppliers, and universities, with users playing a particularly important role. This allows the innovator to 'couple' signals from both the technology base and marketplace;

- external sources complement rather than substitute indigenous innovative activities; and

- successful innovation projects are often 'sponsored' or 'championed' by one or more senior managers.

Most of the above characteristics of successful innovation highlight the importance of boundary-spanning interaction and communication. In his review of the innovation literature, Freeman noted that empirical studies of innovation since the 1950s had demonstrated *'the importance of both formal and informal networks, even if the expression* network *was less frequently used'*, and that *'multiple sources of information and pluralistic patterns of collaboration were the rule rather than the exception'*.[74] These are themes that permeate the whole of this text, and, in particular, **Parts III and IV** concerning the internal and external context, respectively.

1.8 Knowledge and the innovation process

The processes of knowledge creation and knowledge sharing are key elements of the innovation process. Thus, an understanding of the innovation process requires an engagement with the knowledge management literature. In this chapter, we will provide a brief overview of some of the important themes, such as the 'content', 'nature', and 'location' of knowledge. We will return to the subject of knowledge and knowledge management at various stages of this text—most notably, in relation to 'intellectual capital' in **Chapter 6** and the innovation process in **Chapter 8**.

We start by providing some context.

1.8.1 The information and knowledge explosion

Many authors have documented the explosion in information and knowledge since the second half of the twentieth century, fuelled by the dramatic rise in the number of scientists and engineers, and supported by rapidly expanding research and development (R&D) budgets. This has been accompanied by the exponential growth in scientific articles, journals, and books during the past two centuries.

This rapid expansion of data, information, and knowledge is not confined to science, but is also a prominent feature of technological endeavour. In many ways, the resulting problems are more serious in technology, because not only has it become extremely difficult for engineers to keep pace with the state-of-the-art in the many fast-growing technological fields, but they must also, at least partially, keep abreast of and absorb scientific information to maintain technological progress. Badaracco argues that *'these efforts and expenditures create not only more, but also increasingly specialized knowledge'*.[75] He notes that whilst, in the mid-1940s, there were only fifty-four scientific specialities, by the mid-1970s, this had risen to 900, and is considerably higher today. Badaracco notes that, of these specialities:

> Some were the result of increasingly fine distinctions in established fields, but others reflected the creation of new specialities. Knowledge proliferates further when scientists and engineers combine branches of knowledge . . . Just as new knowledge creates new technology, so new technology creates new knowledge.[76]

Indeed, the fusion of existing technologies often produces new technologies and innovations. The scientific genealogy of the video recorder, for example, reveals that it resulted from the bringing together of a number of technological and scientific advances, including magnetic and control theory, electronics, chemistry, and materials science. The CD player is another common example.

Dramatic advances in computer processing and telecommunications since the early 1980s, and the rapid rise in the use of the Internet since the mid-1990s, have enabled the rapid dissemination of the burgeoning information and knowledge base. With this in mind, Rogers and Kincaid argue that:

> Perhaps in recent decades *knowhow* was a major factor in the effectiveness of individuals in their daily lives. But at present the information explosion, facilitated by the widespread mass-media and by recent advances in communication technology (especially of the interactive sort), has created an information environment in which almost every individual possesses more knowhow than he can cope with. Such information overload is often handled by the structuring of interpersonal network links by individuals. *Knowwho* thus begins to replace *knowhow* as one of the main determinants of individual effectiveness.[77]

This observation highlights the importance of social networks and social interaction to the sourcing and sharing of information and knowledge. This is a theme with which we engage throughout the text, but in particular in **Chapters 3, 9, and 10.**

1.8.2 Distinguishing between 'data', 'information', and 'knowledge'

So far, we have employed the terms 'information' and 'knowledge' interchangeably. Whilst this is not uncommon, however, it is useful as a starting point in a discussion of knowledge to distinguish briefly between the concepts of 'data', 'information', and 'knowledge'.

'Data' may be viewed as raw, unconnected pieces of information. Each piece of data viewed separately might appear meaningless, but together, the whole represents 'information'. When this information is analyzed in a broader context and with the benefit of personal experience, this value-added information becomes 'knowledge'.[78]

1.8.3 The 'content' of knowledge

One useful way of classifying knowledge with respect to the management of innovation is to distinguish between 'technological' knowledge—that is, knowledge of the underlying technologies that enable the development, manufacture, and delivery of innovative products, processes, and services—and 'market' knowledge—that is, knowledge of customers and their needs, as well as of broader features of the marketplace.[79] Given our earlier reference to the importance of coupling technological and market signals in the innovation process, it is possible to envisage how incumbent (i.e. existing) firms are sometimes able to compete effectively with new entrants who enter a sector with superior technological knowledge, but little specific market knowledge.

Another way of classifying knowledge is to distinguish between 'component' know-ledge and 'architectural' knowledge. We have already discussed the difference between these two forms of innovation earlier in the chapter; in relation to knowledge, the former refers to knowledge of individual components, while the latter concerns knowledge of how these components are interconnected and configured together as a product, process, or service.[80]

1.8.4 **The 'nature' of knowledge**

One of the most common and useful ways of classifying knowledge is with regard to its 'tacitness'.[81] The term 'tacit' knowledge is used to embrace knowledge of methods, techniques, and designs, which are difficult to articulate, and thus to communicate and codify. Tacit knowledge may be defined as 'heuristic, subjective, and internalized know-ledge', which needs to be learned through practical examples, experience, and practice. Tacit knowledge and skills form a major component of the technological capability of an organization, and, through the process of R&D, is embodied in the materials, products, processes, and systems produced by organizations.

By contrast, 'codified', or 'articulated', knowledge may be transferred using formal and systematic language. Codified knowledge is used to encompass knowledge of general principles and laws of science, technology, and engineering, for example. These principles and laws are articulated in great detail within manuals and textbooks.

Many studies have identified tacit knowledge as an important component of the knowledge employed in the innovation process.[82] Freeman goes further, arguing that:

> it is now very generally recognized that in the technology accumulation process within firms and other organizations, tacit knowledge is often more important than codified formal specifications, blue-prints etc.[83]

The distinction between tacit knowledge and codified knowledge must, however, be treated with caution.

Orr distinguishes between 'uncodified' knowledge and 'non-codifiable' knowledge—that is, the tacitness of a piece of knowledge is related to the potentiality rather than the actuality of its codification.[84] In relation to this distinction, Tsoukas contends that the:

> interpretation of tacit knowledge as knowledge-not-yet-articulated—knowledge await-ing for its 'translation' or 'conversion' into explicit knowledge—an interpretation that has been widely adopted by management studies, is erroneous: it ignores the essential ineffability of tacit knowledge, thus reducing it to what can be articulated.[85]

Furthermore, Polanyi argues:

> These two (tacit and codified knowledge) are not sharply divided. While tacit knowledge can be possessed by itself, explicit [codified] knowledge must rely on being tacitly under-stood and applied. Hence all knowledge is either *tacit* or rooted in *tacit knowledge*.[86]

For some, the notion that codified knowledge requires tacit knowledge to be understood and applied relegates it to the status of information.

Knowledge is also highly 'context-bound' or 'context-specific', or what is sometimes termed 'sticky'[87]—that is, it does not 'travel' well from one project to another, from one organization to another, from one sector to another, or from one nation to another, for example. Brown and Duguid note that *markets work well with commodities*, although *'"sticky" knowledge isn't easily commodified'*.[88] Knowledge is also 'dynamic' and 'provisional',[89] as well as 'fragmented'.[90] In relation to this latter point, Brown and Duguid argue that the:

> picture of knowledge embedded in practice and communities does not dismiss the idea of personal, private knowledge. What people have by virtue of membership in a community of practice, however, is not so much personal, modular knowledge as shared, partial knowledge.[91]

All of these features of knowledge have major implications for the management and sharing of knowledge.

1.8.5 The 'location' of knowledge

The notion of the 'location' of knowledge may be interpreted in a number of ways. In the mainstream knowledge management literature, knowledge is generally assumed to be either located in physical or electronic form (e.g. in books or databases, or 'on the Internet') or within the heads of individuals. Whether located in the 'virtual' or 'physical' world, knowledge is generally seen as distributed, both within and beyond the boundaries of the organization.

On a more abstract level, knowledge may be seen as being held collectively by groups, rather than individuals, to be embedded in organizational systems, procedures, and processes, or embodied within artefacts themselves, such as a computer microprocessor.

For some, knowledge is located in practice and action, where it is both revealed and created.[92]

1.9 Perspectives on innovation in organizations

So far in this chapter, we have introduced and defined a number of core concepts and terms in the innovation literature, and highlighted a range of key themes in relation to innovative activity within organizations. We now turn to a brief discussion of the major perspectives that have been applied in the research of innovation within organizations. In doing so, we employ the typology of Slappendel—itself drawing upon earlier work[93]—that identifies three theoretical perspectives:

- 'individualist';
- 'structuralist';
- 'interactive process'.[94]

We will now deal with these each in turn.

1.9.1 The 'individualist' perspective

From the 'individualist' perspective, the actions of individuals are seen as a major source of innovation within organizations. Whilst context, such as organizational structure and culture, are largely downplayed, there is an assumption that there are certain personal characteristics that predispose individuals to be innovative. Slappendal notes that:

> the individualist perspective is most clearly expressed in those studies that identify individual-level antecedents of innovation . . . defined in terms of individual characteristics and individual-level concepts, such as age, sex, education level, values, personality, goals, creativity, and cognitive style.[95]

Much of the work concerning product champions, entrepreneurs and intraprenuers, and the creativity of individuals would fall within this category.

1.9.2 The 'structuralist' perspective

In contrast to the above approach, the 'structuralist' perspective is underpinned by the assumption that innovation is determined by organizational characteristics, such as size, formalization, and centralization. Attention is also brought to the interrelation between the organization and its environment. Thus, the unit of analysis of the structuralist perspective is at a 'higher' level—for example, at the organizational unit or sectoral level—as compared with the individual actor unit of analysis of the individualist perspective.

The 'contingency theory of innovation' would be embraced by this category: this theory predicts, for example, the varying impact of different organizational characteristics at different stages of the innovation process. (We will discuss this approach more fully in **Chapter 7**.)

1.9.3 The 'interactive process' perspective

It is increasingly argued from within the academic community that neither the individualist perspective, nor the structuralist perspective, can adequately reveal the complexity of the innovation process, because each illuminates only part of the story. The 'interactive process' perspective, however, '*attempts to account for both individual and structural factors through an analysis of their interconnection*'.[96] In doing so, it incorporates multiple levels of analysis. The perspective also more readily embraces the non-rational elements of organizational behaviour, such as the political and social dimensions of the innovation process, whilst rejecting the traditional 'rational economic' model of decision making. And there is an emphasis on the dynamic and 'unfolding' nature of both the innovation process and the innovation itself. Such a perspective requires different research methods; thus we are seeing a shift away from the over-reliance on quantitative cross-sectional surveys, for example, which dominate the individualist and structuralist perspectives, and a move toward qualitative longitudinal case studies and case histories.

Slappendel argues that whilst the individualist and structuralist perspectives '*have dominated the innovation field, they are being challenged by a growing interest in the interactive process perspective'.*[97] Indeed, although in writing this text we have drawn upon research from each of these three perspectives, we have most affinity with the third perspective—the interactive process perspective—with its focus on the interactive, iterative, emergent, complex, and contextual nature of innovative activity within organizations.

1.10 Concluding comments

Despite the dramatic rise in interest in innovation since the 1980s among academics, practitioners, and policymakers alike, there is still little consensus with regard to a 'working definition' of innovation. It would appear that many writers in the area of innovation assume that we have a common understanding of the term. This is not the case. Indeed, while some writers on innovation make a clear distinction between 'invention' and 'innovation', others have used these terms rather loosely and, to a certain extent, interchangeably. Furthermore, whilst the definitions of innovation discussed earlier in this chapter highlight a number of common threads, it is also evident that the breadth of the definitions varies enormously. This is true also of a number of related terms that are frequently to be found in the literature, such as, 'novelty', 'success', 'radical', and 'knowledge'; again, these terms are often employed very loosely.

This chapter has sought to review the different ways in which these terms are defined and utilized. In doing so, the discussion is intended to open the eyes of the reader to the nuances and the plurality of meanings of these core terms and concepts. Where appropriate, however, we have also highlighted what we believe to be the most useful 'working definitions' for the purposes of this book. This is not to say that these are 'right'; rather that they provide us—the authors, lecturers, and readers—with a common understanding and nomenclature. This is particularly important with regard to facilitating interaction and discussion around case studies of innovation, whether from this text or from elsewhere, such as through your own work experience. The discussion has also highlighted that when attempting to assess the 'novelty' or 'success' of an innovation—a practice that has equal resonance among academics, practitioners, and policymakers—it is important to first determine through whose 'eyes' we are making the assessment. Thus, for example, the question of 'how novel?', should always be accompanied by the question 'novel to whom?'.

This chapter has also sought to provide a brief introduction to the origins and sources of innovation, by drawing upon seminal studies of successful innovation, to knowledge and to the major perspectives employed in the innovation literature. All three of these areas will be developed during subsequent chapters.

CASE STUDY 1.1 THE DEVELOPMENT OF THE 'LAFF BOX' AND THE DIFFUSION OF 'CANNED LAUGHTER' ON TELEVISION

Introduction

'Canned laughter' is pre-recorded laughter; it has been employed by television producers and, to a much less extent, radio producers for over sixty years either to 'sweeten' (i.e. augment) the laughter and applause of a live studio audience, or as a substitute for a live audience. Canned laughter was particularly prevalent on television in US talk shows, sitcoms, and cartoons during the 1950s, 1960s, and 1970s (listen out next time you see past episodes of *The Brady Bunch*, *The Flintstones*, or *Scooby-Doo*).

Although not the originator of the concept of canned laughter—it had been used on radio and television prior to 1950—Charles Douglass is a central character in the story of canned laughter on television, as is his innovation, the 'Laff Box', which he developed in 1953. The original Laff Box contained a series of audiotape loops, which could be manipulated by a sound editor to mimic a range of audience responses, from murmurs and giggles, to explosions of laughter.

Charles Douglass and the development of the innovative 'Laff Box'

Charles Douglass was born in Guadalajara, Mexico, in 1910. Due to political unrest, his family moved to Nevada in the USA when he was 2 years old. His father was a pioneering electrical engineer and worked in the silver mines of Tonopah. During the Great Depression, Charles studied at the University of Nevada and obtained a degree in electrical engineering in 1933. He subsequently moved to Los Angeles, where he initially worked as a broadcast engineer for CBS radio, and later served in the Navy during the Second World War, where he worked on developing naval radar systems whilst based in Washington DC. After the war, Douglass returned to CBS, where he moved on to work as a technical director on a number of live television shows. It is during these years, up to the early 1950s, that he sought to find a more effective technical solution to producing canned laughter for live and recorded broadcasts.

As noted above, prior to the development of the Laff Box in 1953, canned laughter was already being employed to a limited extent on television and radio shows. The technology was, however, very crude: generally, a small sample of applause and laughter was recorded onto vinyl, and these records were played on cue at the appropriate times in a show, being faded in and out by the sound engineer. In contrast, the Laff Box contained a much larger range of laughter and applause recordings, but, far more importantly, it allowed these to be mixed with ease and in 'real time', to yield a far more varied range of 'audience responses'. Interestingly, some of the original laughter and applause recordings within the Laff Box were recorded at a Marcel Marceau mime performance, because the audience responses were consequently not 'tainted' by inadvertent dialogue from the performer.

Others in the industry were also trying to develop their own canned laughter machines at about the same time, but Douglass was the first to succeed. He quit his job at CBS and established his own company—Northridge Electronics—which went on to dominate the market for at least twenty years. He kept the inner workings of the Laff Box a closely guarded secret, even from colleagues whom he hired as the business grew.

But there was much more to the success of the Laff Box than simply the development of the artifact itself. Charles Douglass would himself be hired for the transmission and recording of a show. He would turn up with the Laff Box—which, due to its size (roughly a 60 cm cube) was clumsy to transport from studio to studio—and, drawing from his broadcasting experience, would himself manipulate the sound recordings within the Laff Box to produce the appropriate canned laughter and applause at the appropriate times for the show. For producers, the innovation allowed them to 'sweeten' laughter and applause when the audience did not respond in the desired manner, or had become tired from multiple takes in the recording of a show. It also allowed producers to 'insert' an audience when one may not have been present—important because research appears to indicate that home viewers and listeners often find a show more entertaining when it is accompanied by appropriate audience responses, suggesting that it makes them feel more 'connected' to the performance.

Over time, more and more recorded human sounds were added to the Laff Box, including audience responses from other cultures, whose sounds of laughter and applause can often be quite distinct from those of Americans. And although the innovation has been employed predominantly on television shows, it has also been used to sweeten live music recordings.

In 1992, Charles Douglass received an Emmy Award for his lifetime technical achievements and contributions to the industry. Bob Douglass, his son, says of the impact of the Laff Box: '*He* [Charles] *didn't realize how it would change the face of comedy*.' Douglass died in 2004, at the age of 93.

The future of canned laughter?

The use of canned laughter has long been criticized from both within and outside the industry for its artificiality, for its capacity to 'cover up' poor material, and even for its manipulation of television and radio audiences. Indeed, as television and radio audiences have became more sophisticated since the 1970s, shows increasingly bore the credit 'Filmed before a live studio audience' as a badge of honour. The 1970s also saw the emergence of a number of more serious sitcoms, such as *M.A.S.H.*, in which the use of canned laughter was far less appropriate.[98] Nevertheless, canned laughter is still used widely today, although arguably in a subtler manner. Northridge Electronics is still in business—although now run by Bob Douglass—and it is still a major player in the canned laughter marketplace. Today's version of the Laff Box is much smaller—now the size of a laptop computer—and the sound is digitized.

Questions

1. Is the Laff Box a product or service innovation, or both?

2. To what extent is the Laff Box a novel or radical innovation? Does your evaluation vary depending on whether you view the innovation through the eyes of the innovator, producer, or home audience?

3. Over time, the Laff Box has undergone re-innovation. Briefly outline these developments, and assess their degree of novelty and radicalness.

CASE STUDY 1.2 NAPSTER—THE RAPID RISE AND DECLINE OF A CONTROVERSIAL INNOVATION

Introduction

Napster is a piece of software that allows Internet users to browse and share MP3 files. Launched in mid-1999, it experienced rapid growth in its user base, with some estimates as high as 80 million users at its peak. It was not the first piece of software that allowed users to search for and transfer MP3 files on the Internet: music search engines, such as MP3.lycos.com and scour.com, were already available. Unlike its predecessors, however, Napster was not an Internet search engine, but more of a file transfer and messaging application.

For those not familiar with the original version of the software, it worked roughly as follows:

1. Users downloaded the Napster software.
2. The software scanned the hard drives of these users for MP3 files (music file format).
3. A list of MP3 files that each user was prepared to share was uploaded to the central Napster directory (the actual MP3 files were not uploaded or stored on the central server).
4. Users logged on to a central Napster directory, made a request by song or artist, and were provided with a list of relevant song titles, along with the link to computers on which each song was located.
5. By simply clicking on a selection from the list, the software initiated a 'peer-to-peer' (P2P) file transfer—i.e. a transfer directly from one user's computer to another.

Shawn Fanning and the development of Napster

Napster was not developed originally for its commercial potential; rather it had much humbler beginnings. During the autumn of 1998, whilst a freshman at Boston's North Eastern University in the USA, a friend of Shawn Fanning complained of the hassle of locating and downloading MP3 files using conventional Internet search engines. Shawn decided to try to write a bit of software that would solve the problem. He had soon dropped out of university to pursue the project, enlisted two friends whom he had met online (Sean Parker and Jordan Ritter), and set up a company along with his uncle, John Fanning. On 1 June 1999, Shawn handed over a test version of the software to thirty online friends, on condition that they would not tell anyone about the project. But they could not resist spreading the software and, within a few days, Napster had been downloaded by 10,000–15,000 people. And it was at this point that the company started to realize the huge potential of the application.

As noted earlier, Napster was not the first piece of software made available to search for and transfer MP3 files; nor was it, more generally, the first piece of software to allow P2P file transfer. What *was* new was the bringing together of these two elements—the application and the technical solution.

From commercialization . . .

Around January 1999, Shawn Fanning had approached his uncle, John Fanning, to seek his opinion on the commercial viability of Napster. John quickly realized the commercial potential of the innovation, helped to raise initial funding, incorporated the company, and moved its operations to Silicon Valley. By July 1999, the company was operational, but only had funding for a further six months. Nevertheless, the use of Napster grew at a phenomenal rate. Fuelled by further venture capital support—most notably, from Hummer Winbald, who invested US$15m in May 2000 and became chief executive officer (CEO) at Napster Inc.—things started to look promising.

Winbald soon negotiated an alliance with the German company Bertelsmann, a major player in the music industry.

. . . to controversy and bankruptcy

The record industry was quick to react to the rise of Napster. In December 1999, just a few months after Napster Inc. had been operational, the Recording Industry Association of America (RIAA) sued the company, claiming that the software violated copyright laws by facilitating the illegal sharing of music. In July 2000, a US court ruled in favour of the record industry. This decision was overturned the same month, but the final ruling in March 2001 called for Napster to stop operating. A few months later, following great acrimony between the various shareholders of Napster Inc., the company was declared bankrupt.

Postscript

Although Napster was able to attract successfully many millions of users and venture capital funding, as well as to build an alliance with a large record company, it never actually made a profit in its short life. Nevertheless, Napster made 'MP3' a household term, it popularized P2P networking, and it helped to change the rules of the music industry, taking power away from the record companies, and making music sampling unlimited in scope, free, and on demand, for those willing to break copyright rules.

Its legacy is also vast: it is estimated that several hundred million MP3 files are now shared every month around the world, the vast majority still being downloaded for free. Napster spawned a mass of new competitors—such as Aimster, AudioGalaxy, Bearshare, Gnutella, iMesh, KaZaA, Mojo Nation, and Napigator—which quickly boasted millions of users as a fickle user base moved on.

The story continues to unfold.

Questions

1. Is Napster a product, process, or service innovation, or all three?
2. To what extent was Napster a novel or radical innovation? Does your evaluation vary depending on whether you view the innovation through the eyes of the innovator, music fans, or the record industry?
3. Was Napster a successful innovation?

FURTHER READING

1. For a recent literature review of studies focused on identifying the factors that distinguish between success and failure in NPD projects, see Panne et al. (2003), and for new service development (NSD) projects, see Martin and Horne (1993; 1995).
2. For an introduction to some of the key problems faced in the management of innovation, see Van de Ven (1988).
3. For a discussion of newness and novelty, see Johannessen et al. (2001), and for a more challenging read, see Styhre (2006), which brings an interesting, alternative perspective on the subject.
4. For an overview of a wide range of tools employed by organizations to measure success in new products, see Griffin and Page (1993), and in new services, Johne and Storey (1998: 214–8).

NOTES

[1] See <http://www.firstdirect.com>.
[2] See <http://www.nhsdirect.nhs.uk>.
[3] Johannessen, Olsen, and Lumpkin (2001: 20).
[4] Kelly et al. (1986: 24).

[5] Schumpeter (1939).

[6] Rothwell and Zegveld (1985: 47).

[7] Freeman and Soete (1997: 6).

[8] Roy (1986b: 3).

[9] Rothwell and Zegveld (1985: 47).

[10] Freeman and Soete (1997: 6).

[11] Trott (2005: 16).

[12] Myers and Marquis (1969).

[13] Freeman (1974).

[14] Kanter (1985: 20–1).

[15] Department for Innovation, Universities, and Skills (2008: 12).

[16] Commission of the European Communities (2003: 7).

[17] Rogers (1995: 5).

[18] Rogers (1995: 16).

[19] Habermeier (1990).

[20] Hippel (1988), Rogers (2003).

[21] Zaltman, Duncan, and Holbeck (1973).

[22] Tidd, Bessant, and Pavitt (2001: 6).

[23] Damanpour (1991).

[24] Clark and Staunton (1989); Edquist (1997a); Tidd, Bessant, and Pavitt (2001).

[25] Clark and Staunton (1989: 54).

[26] Shaw (1985).

[27] Spital (1979).

[28] Freeman (1968).

[29] For example, Nelson and Rosenberg (1993); Carlsson and Stankiewicz (1995).

[30] Sundbo (2001: 3).

[31] Barras (1990); Gallouj (1998); Boden and Miles (2000); Aa and Elfring (2002); Alam (2003a); Thomke (2003).

[32] Rothwell and Zegveld (1985: 47).

[33] Clark and Staunton (1989: 81).

[34] Rothwell and Zegveld (1985); Clark and Staunton (1989: 79–82).

[35] Rothwell and Zegveld (1985: 47–8).

[36] Henderson and Clark (1990).

[37] Kuper and Sterken (2003); Versluis (2005).

[38] Henderson and Clark (1990: 11).

[39] Henderson and Clark (1990: 9).

[40] Afuah (1998).

[41] Clark and Staunton (1989: 80).

[42] Based on Standley (2003) and <http://www.spaceadventures.com>.

[43] Johannessen, Olsen, and Lumpkin (2001: 24).

[44] Clark and Staunton (1989: 10–11).

[45] Tushman and Anderson (1986).

[46] Zaltman, Duncan, and Holbeck (1973: 10).

[47] Rogers (1995: 11).

[48] Hippel (1986).

[49] Rogers (2003).

[50] Allen (1977); Tushman (1977); Conway (1995).

[51] Dyson (1997).

[52] Sundbo (2001: 17).

[53] Abernathy and Clark (1985); Afuah and Bahram (1995).

[54] Parkinson (1972); David (1985); Afuah (1998).

[55] Lovallo and Kahneman (2003).

[56] See Griffin and Page (1993) for a review of the measures employed to measure success or failure.

[57] Maidique and Zirger (1985).

[58] Leonard (1995: 121).

[59] Farson and Keyes (2002).

[60] Hultink and Robben (1995).

[61] Tidd, Bessant, and Pavitt (2001: 49).

[62] For example, Rothwell and Zegveld (1985); Freeman and Soete (1997).

[63] Johannessen, Olsen, and Lumpkin (2001: 28).

[64] Rogers (2003).

[65] Rothwell and Gardiner (1983).

[66] Rice and Rogers (1980); Rogers (2003).

[67] Rogers (2003).

[68] Clark and Staunton (1989: 12).

[69] Achilladis, Robertson, and Jervis (1971).

[70] Langrish et al. (1972).

[71] Rothwell et al. (1974).

[72] Sherwin and Isenson (1967).

[73] Illinois Institute of Technology (1968).

[74] Freeman (1991: 500).

[75] Badaracco (1991: 25).

[76] Badaracco (1991: 25–6).

[77] Rogers and Kincaid (1981: 343–4).

[78] Fuld (1991).

[79] Abernathy and Clark (1985).

[80] Henderson and Clark (1990).

[81] Polanyi (1969); Nonaka and Takeuchi (1995).

[82] For example, Conway (1995); Nonaka and Takeuchi (1995).

[83] Freeman (1991: 502–3).

[84] Orr (1990).

[85] Tsoukas (2003: 425).

[86] Polanyi (1969: 144).

[87] Hippel (1994).

[88] Brown and Duguid (2002: 28).

[89] Hayes and Walsham (2000); Orlikowski (2002).

[90] Brown and Duguid (2002).

[91] Brown and Duguid (2002: 25).

[92] Cook and Brown (1999); Brown and Duguid (2002); Orlikowski (2002).

[93] Pierce and Delbecq (1977).

[94] Slappendel (1996).

[95] Slappendal (1996: 110).

[96] Slappendel (1996: 123).

[97] Slappendel (1996: 109).

[98] *M.A.S.H.* was centred on the experiences of front-line medical staff during the Vietnam War.

Tensions, Paradox, and Contradictions in Managing Innovation

Chapter overview

Learning objectives

This chapter will enable the reader to:

- recognize the main ideas and authorities dealing with tension, contradiction, and paradox in the innovation process;

- understand the concepts of ambidexterity and metamorphosis as organizational practices to manage contradictions;

- apply the framework of paradox to specific cases of innovation;

- analyse the dynamics and tensions involved in innovation;

- synthesize the different meanings of tension, paradox, and contradiction in relation to both the theory and practice of innovation; and

- evaluate the explanatory power of different motors of change and alternative models of paradox management.

2.1 Introduction

A fascinating theme in our understanding of innovation in recent years has been an increased awareness that inherent to the innovation process is the management of a diversity of often-contradictory practices and goals. For example, some parts of an organization may be focused on novelty and exploration, whilst other parts may be more conservative, and concerned with stability and exploitation. Indeed, the management of innovation can be seen as the capability to handle these tensions creatively. Initially, this was hinted at through critiques of the 'linear' model of innovation and the promotion of the 'interactive' model. We discuss these models below in relation to 'coupling' and in more detail in **Chapter 3**.

More recently, it has led to theories of innovation that focus explicitly on these tensions and contradictions, and explore their implications both for organizational theory and managerial practice. The overall implications are profound and their imprint is found throughout this text, because not only do they lead to a shake-up in the concepts that we use to explore innovation, it also suggests that we need a radical rethink of what we regard as effective management.

In order to set the scene, in this chapter, we draw upon a range of seminal works since the 1960s to provide a historical perspective on the emergence of these themes.

2.2 Innovation as an interactive organizational process

In the 1960s and 1970s, a number of pioneering studies on innovation were undertaken that, for the first time, clearly recognized it as a process that was managed within organizations. This reconceptualized the essence of innovation. Prior to this, the implicit assumption was that innovation was an event that was either the creative achievement of an individual or was induced by impersonal economic forces.[1] The key studies in this period were undertaken within two distinct academic fields and pursued contrasting modes of inquiry.

One group of scholars was located in the emerging field of organization studies, drawing on the sociology of management. The unit of analysis that formed the focus of their investigations was the business organization—usually, an individual firm. Key figures were Tom Burns, a sociologist at Edinburgh University, who published *The Management of Innovation* (with Stalker) in 1961,[2] and Jay Lorsch of Harvard Business School, who published *Product Innovation and Organization* in 1965.[3]

The other group of academics belonged to the newly formed interdisciplinary research centres in science policy and technology management. Their unit of analysis was the innovation itself, such as individual cases of new products or manufacturing processes. Key figures were Don Marquis of the Massachusetts Institute of Technology (MIT), who published *The Anatomy of Successful Innovations* in the late 1960s, John Langrish, Freddie Jevons, and Mike Gibbons at Manchester University, who published *Wealth from Knowledge* in the early 1970s, and Chris Freeman and Roy Rothwell at Sussex University, whose SAPPHO study was reported in *The Economics of Industrial Innovation* during the same period.[4]

2.2.1 Cross-functional 'linking'

The overall thrust of the organization-focused studies was that firms with a dynamic business context showed a distinctive set of characteristics compared to those located in a more stable environment. The innovation-oriented organization, which was viewed as more suited for successful performance, was found to exhibit an internal structure that demonstrated effective horizontal links between different business functions. Burns and Stalker's analysis of these characteristics in electronics firms led them to define an interactive 'organic' structure, suitable for innovation, in sharp contrast to the specialized 'mechanistic' structure of the steady state firm (see **7.5.1** for a fuller discussion of these contrasting organizational structures).

The identification of cross-functional linking signified a recognition for the management of innovation of the importance of bringing together those different parts of the organization that pursued the contrasting roles of research, marketing, finance, and operations management, for example. It was demonstrated that innovation involved the integrative management of a diversity of sometimes-conflicting business functions.

These studies played a major role in the development of the 'contingency' approach in organization theory. This rejected the notion of 'one best way' in which to manage an organization, and recognized that the appropriateness of a mode of management was 'contingent' on the circumstances confronting the organization. This was very much associated with the recognition that a business oriented to innovation would need a distinctive mode of management in which effective horizontal cross-functional links were important.

The explanation of the process whereby a business achieved its 'appropriate' form of organization was the subject of debate. A strict environmental determinism would imply no scope for the influence of managerial action. John Child's concept of 'strategic choice' from the early 1970s, argued that 'environmental fit' did not preclude the significance of managerial choice in shaping the actual form that this takes.[5] In practice, this interpretation seems to have been widely shared: the contingency studies on innovation led to a range of proposals on 'organization design'—that is, how managers could consciously structure their organizations so that they could be more suited to the performance of innovation. These ranged from proposals for radical change to an organic matrix structure, to surveys of different types of linking mechanism that could be designed into a conventional functionally differentiated business structure.

2.2.2 Opportunity 'coupling'

The innovation-focused studies were particularly interested in the role that new scientific and technical knowledge played in the innovation process. Marquis examined the cases of 500 innovations in a study supported by the National Science Foundation (NSF).

The cases were selected as successful examples of 'normal', rather than radical, innovations that had been introduced during the previous ten years. He concluded that the characteristic of successful innovation was the simultaneous recognition of both technical feasibility and potential demand in order to initiate the innovation process.[6] Langrish and colleagues investigated fifty cases of award-winning innovations in the UK, also concluding that interaction between technology and market was crucial to success.[7]

It was Freeman and Rothwell, however, who articulated this insight in terms that had the clearest implications for conceptualizing innovation as a process in organizations that requires a particular type of management. Project SAPPHO undertaken at Sussex University's Science Policy Research Unit (SPRU), compared forty pairs of innovations—each pair, a 'success' matched with a 'failure'—in order to identify the organizational characteristics that distinguished the two. Freeman expressed the core finding as the ability to 'couple' a technological opportunity with a market opportunity. **Illustration 2.1** provides a good example of such coupling.

Illustration 2.1 DNA fingerprinting—the successful 'coupling' of technological advance with market opportunity

The pioneering work of Professor Sir Alec Jeffreys of the Department of Genetics at the University of Leicester provides a vivid and well-known example of the potential for 'coupling' technological advances with market possibilities. His research on inherited variation in human deoxyribonucleic acid (DNA) led to the discovery of regions of DNA, known as 'minisatellites', that revealed great variability in the genetic profile of individuals, but also similarities between individuals who were related. This basic research that demonstrated the uniqueness of an individual's DNA led to the development of DNA fingerprinting in 1984. This technology provided huge market opportunities when coupled with the wide range of potential applications, such as in legal cases concerning paternity and criminal cases.

As Professor Jeffreys recently recalled:

It was blindingly obvious that DNA fingerprinting not only had generated large numbers of highly informative genetic markers for medical genetic research, but also had accidentally solved another major problem in human genetics, namely the issue of biological identification and the establishment of family relationships in forensic and legal medicine. Within months of developing the first DNA fingerprint, we had applied this approach to resolving an immigration case and shortly afterwards a paternity dispute . . . [and] . . . our first forensic investigation in 1986 in the Enderby murder case. It was this case that triggered the application of molecular genetics to criminal investigations worldwide.[8]

When we consider the notion of opportunity 'coupling' more broadly, in relation to the management of innovation, a number of interesting pairings emerge. For example:

- basic scientific research with application and product development;
- technological possibilities with market opportunities;
- the 'inventor' and the 'entrepreneur'; and
- internal ideas and resources with external ideas and resources.

The verb 'to couple' is a very different one from 'to recognize'. Indeed, for Freeman:

The 'coupling' process is not merely one of matching or associating ideas in the original first flash; it is far more a continuous creative dialogue during the whole of

the experimental development work and introduction of the new product or process
... [it] ... involves linking and co-ordinating different sections, departments and
individuals.[9]

This was later expressed more fully by Roy Rothwell as the 'interactive model' of
innovation[10] (see **3.2.1** and **3.2.2** for a discussion of the linear and interactive models
of innovation).

The early innovation studies had two significant consequences for the management
of innovation: firstly, the linear 'science push' model of innovation was displaced;
secondly, the conceptualization of managing a coupling process in the organization was
introduced. The notion of 'management' in relation to this concept was expressed much
more strongly in terms of individual roles and communicative action, rather than structural form and managerial practice, which were associated with the organization studies
approach. In particular, the roles of entrepreneurs and 'product champions' were highlighted (see **9.5** for a discussion of these roles).

Overall, though, the major studies of this period from both research schools created
a new framework for the understanding of innovation that put at its heart the fundamental need for a managerial capability to 'link' and to 'couple' different, and sometimes
contradictory, business practices and goals.

2.2.3 Dualism

The new focus on linking and coupling was accompanied by different emphases on how
these might be most effectively achieved. Some interpretations of the organic innovative
organization stressed its integrative form, and argued for the need for fuzzy boundaries and non-linear channels of authority and communication through radical matrix
reorganization. Others, however, articulated the view that organizations should remain
dualistic in nature, because of the contradictory goals that they had to pursue. The
identification of core organizational attributes that could be regarded as generally
favourable to innovation also began to be challenged by scholars who felt that it over-
simplified the nature of innovation and its organizational requirements.

One approach suggested that different organizational characteristics could be pursued
simultaneously. Through their analysis of firms in the innovation-oriented plastics
sector, Lawrence and Lorsch concluded that although they often showed a high degree of
differentiation in their organizational structure (i.e. different segments of the organization were focused on different functions and tasks), they also showed a high presence of
effective integration devices that fostered horizontal links between the specialized functional units.[11] Duncan's work with Zaltman also argued the need for combining 'loosely
coupled' and 'tightly coupled' features of organization in the innovation process.[12]

This view represents a rather different approach to that of Burns and Stalker: rather
than a *choice* between *either* an organic type of structure *or* a mechanistic one, the sugges-
tion was of a distinctive *combination* of *both* integration *and* differentiation. This position
was supported by Nord and Tucker, whose study of routine and radical innovation in
the banking sector led them to question the prevalent interpretations of choice between
organic and mechanistic organizational forms. They found a 'mixture of organic and

mechanistic' in both routine and radical innovation processes. Although they drew on the organic/mechanistic terminology in their study, they concluded *'it may be productive either to abandon these terms or to become much more precise in using them'*. They call for more detailed analysis of their dimensions, combinations, and patterns of oscillation between the two, claiming that there is a risk of their usefulness being only as *'general, orienting concepts that lull people into a false sense of comprehension'*.[13]

Another approach proposed a sequential switching of organizational characteristics. Duncan argued that the process of innovation had two different phases—initiation and implementation—with contrasting characteristics and organizational needs. Initiation was seen as essentially a creative process assisted by organic structures, while implementation was seen as managerial and more amenable to mechanistic structures. As a result, Duncan argued that organizational design for innovation needed to be dualistic and to be able to switch between contrasting modes of organization. This notion has been expressed through the concept of the 'ambidextrous' organization—that is, one that has the ability to switch between alternative modes of organizing, or is capable of operating simultaneously in dual modes: in this instance, the mechanistic and organic modes. Some writers have made a clear distinction between the former, which refers to the asynchronous pursuit of dual modes and has been termed 'punctuated equilibrium' (i.e. it involves periodic shifts between the two modes of organizing),[14] and the latter, which is the synchronous pursuit of dual modes within different parts of the organization, termed 'ambidexterity'[15] (see **7.6** for a fuller discussion).

These discussions on organization were paralleled by research on inventiveness. The link with creativity was highlighted in a study of scientific discovery by Rothenburg in the late 1970s, where 'Janusian' thinking (named after the Greek god Janus who faced both ways), with contradictory elements, was found to be present.[16] This finding was in sharp contrast to earlier claims that so-called 'cognitive dissonance' was dysfunctional.[17]

2.3 Paradox—the new paradigm

Within a few years, from the early 1980s, there began to emerge a new and distinctive challenge to the earlier conceptions of the nature of innovation. It was increasingly acknowledged that the process of innovation presents inherent paradoxes for both its understanding and management. Different authors have referred to the inherent presence of contradiction, tension, paradox, dilemma, duality, dialectic, and dialogic in the innovation process.[18]

The trigger for this shift arose from two prominent US studies that addressed the long-running theme of attempting to characterize successful innovative organizations. In their respective studies, Tom Peters and Bob Waterman, both partners at McKinsey, and Rosabeth Moss Kanter, of Harvard Business School, adopted an approach that focused on a systematic interpretation of successful managerial practice as opposed to addressing explicitly existing academic theory of organization or innovation. In their text *In Search of Excellence*, Peters and Waterman concluded that *'the excellent companies have learned how to manage paradox'*,[19] whilst Kanter, in her book *The Change Masters*, similarly concluded that the ability to master 'innovation dilemmas' was at the heart of successful performance.[20] This is particularly true of innovation within the creative industries, as exemplified by **Illustration 2.2**.

Illustration 2.2 Balancing exploration and exploitation in movie production

Whilst innovation and creativity are the lifeblood of the film industry, project managers must manage the tensions and contradictions that emerge between the need for creativity *and* project effectiveness in the production of movies, on the one hand, and the trade-off between creativity *and* audience acceptance, on the other.

In the production of movies, for example, where budgets can often be tight, team composition becomes a prime consideration; 'newcomers' often enhance exploration and creativity, whilst experienced team members are important for exploitation and project effectiveness. For the commercial success of movies, it has been argued that '*consumers need familiarity to understand what they are offered, but they need novelty to enjoy it*'.[21] Thus, innovation in movie genre can be risky. To a certain extent, market failure can be tempered by employing '*well-established and bankable actors and actresses*', but where the audience is 'wrong-footed' by their typecasting outside of the 'normal rules', then this can backfire, and raises the risk of losing the approval and patronage of an audience.[22]

Responding to these insights, Van de Ven argued that the theme of paradox raised fundamental new challenges for explanation and theorizing. In particular, the simultaneity and co-location of apparent opposites in the management process meant that the traditional focus on consistency and contingency was inadequate, and new ideas were needed.[23] As a consequence, an important symposium on this theme was held at the 1985 Academy of Management conference in the USA. This resulted in the multi-authored *Paradox and Transformation: Toward a Theory of Change in Organisation and Management*. In their introduction to this text, editors Quinn and Cameron argue for '*the central place of contradiction in the effective performance of organizations*' around the key issue for innovation of stability versus change.

Quinn and Cameron use the concept of paradox as a 'mental construct' or a 'framework or metaphor', which expresses the presence of '*simultaneous opposites or contradictions in effective management*'. They distinguish it from other similar concepts—dilemma, irony, inconsistency, dialectic, ambivalence, conflict—by defining it as '*contradictory mutually exclusive elements that are present and operate equally at the same time*'.[24] At the heart of this new interest in paradox is the view that such contradictions may be seen as in a state of creative 'tension' rather than as a competitive split or 'schismogenesis',[25] and therefore 'need not be resolved'. In other words, it is suggested that we replace the traditional model of an 'either/or' choice with a management style that embraces a new 'both/and' approach. In particular, they argue for the power of paradox in explaining new ventures and innovations.

Good examples of such a management style, that endeavours to embrace the 'both/and' approach as opposed to the 'either/or' approach, include those that attempt to combine the benefits of and tensions between:

- formal organization (the organizational chart) *and* informal organization (the social networks between employees)—see **9.2** and **9.3**;

- planned activity *and* serendipity—see **9.3**;
- internal sources *and* external sources of ideas—a key theme of **Chapter 10**.

One of the aspects of the growing literature on the role of dilemmas and paradox in business strategy and management is the use of new metaphors. Charles Hampden-Turner, in *From Dilemma to Strategy*, argues that 'navigational' and 'steering' metaphors are much more appropriate than the 'heroic' and 'military' metaphors commonly found in traditional strategy texts.[26] Charles Handy, in *The Age of Paradox*, echoes this theme.[27] Pascale introduces the metaphor of the 'juggler'.[28] These metaphors are drawn upon in studies on strategies for innovation that embrace this new paradigm and address issues such as the '*capacity for reconciling dilemmas strategically*'[29] or the '*synchronisation of opponents*'[30].

Among the range of researchers sharing this broad approach, there is variation as to whether the successful management of contradictions and tension is seen as resolved through organizational design or whether it rests on the capability of the individual manager. For example, Pascale's study of Honda emphasizes that organizations must recognize the need to 'transcend' conflicts and look:

> to the tension (or dynamic synthesis) between contradictory opposites as the engine of self renewal . . . predicated on the notion that disequilibrium is a better strategy for adaptation and survival than order and equilibrium.

He continues, by arguing that the '*reconciliation of fit, split and contend factors contains a paradox and a new higher order (transcendent) approach to management is necessary to juggle them effectively*'.[31] This introduces the metaphor of the 'juggler' in contrast to that of the 'helmsman', which is frequently used by writers on paradox and tension.

For Schoonhoven and Jelinek, the 'dynamic tension' within innovative organizations is managed through the organizational structure.[32] In contrast, Clegg and colleagues argue for a 'relational' approach to paradox that stresses that the relationship between contradictions is established through a situated process of enactment and thus is not designed.[33] This fits closely with the approach of Cobbenhagen and his colleagues at MERIT on coping with dilemmas in managing innovation. They conclude that '*the art of managing innovations is to leave the dilemmas open, and depending on the situation, make temporary choices*'.[34] This attributes a greater role for the individual manager in the innovation process.

By the 1990s, this new paradigm of managing the tensions arising from contradictions inherent in the innovation process was being applied in a range of studies that addressed a variety of topics, from specific managerial practices in the organization and strategic aspects of innovation, to broad theoretical interpretations of change.

2.4 Managing tensions between contradictory practices

Much of the recent literature on innovation has explored the tensions between different, apparently contradictory, types of organizational practice that appear to be simultaneously present in the successful management of innovation. In the mid-1990s, Deborah

Dougherty used the concept of tension in a review of knowledge on organizing for innovation that focused on different activities necessary in innovation, and identified four tensions that occupied a lot of attention in the innovation literature:

1. the tension between the 'outside' and the 'inside' of the firm involved in the relationship between markets and technology;

2. that between the 'old' and the 'new' in the focus of creative problem solving;

3. that between 'emergence' and 'determination' in the innovation strategy process; and

4. that arising between 'freedom' and 'responsibility' in the involvement of staff in innovation.[35]

These can be summarized as in **Table 2.1**.

Table 2.1 '*The tensions, what perpetuates imbalances, and how to restore balance*' (Dougherty, 1996: 431)

Activity	Tension	Problem of normal functioning that disrupts tension	Particular practices which perpetuate disruption	Capacities to restore the balance in the tension
Market-technology linking	Outside vs. inside	Keeping operations efficient	Inward emphasis on dept thought worlds and units; fixed sense of business	Generate and maintain an identity based on the value provided to customers
Organizing for creative problem solving	New vs. old	Managing complexity	Segmentalist thinking and compartmentalization of work; power based on current work	See work of organization in terms of process, focusing on relationships among parts, and changes
Evaluating and monitoring innovation	Determined vs. emergent	Controlling multiple activities	Abstracting work into generic standards; no strategy making	Situated judgement, collective ability to be engaged in details of work but also appreciate unstructured problems
Developing commitment to innovation	Freedom vs. responsibility	Accounting for work, results	Illegitimacy of innovation; illegitimacy of inclusion	Collective accountability, accept and share responsibility, legitimise innovation and inclusion

Dougherty's review shows that the concept of tension enables a more effective focus on the practices that constitute the capacities of an organization for innovation.

In her review of a range of studies concerning the management of tensions within the innovation process, Marianne Lewis focuses on three paradoxes: 'learning', which deals with tensions between the old and the new; 'organizing', which addresses the tensions between control and flexibility; and 'belonging', which concerns the tensions between self and other. These can be summarized as in **Table 2.2**. Lewis concludes by outlining three, often interrelated, means of managing paradox:

1. *'acceptance'*, which is 'learning to live with paradox';

2. *'confrontation'* through *'socially constructing a more accommodating understanding or practice'*;

3. *'transcendance'*, which is the *'capacity to think paradoxically'*.[36]

Table 2.2 *'Exemplary organization studies of paradox'* (Lewis, 2000: 765)

Tensions	Reinforcing cycles	Management
Paradoxes of learning		
Learning requires using, critiquing, and often destroying past understandings and practices to construct new and more complicated frames of reference.		
Old/new	Defenses of repression, projection, and regression	Social reframing
• Sensemaking . . . • Innovation . . . • Transformation . . .	• Cognitive self-reference • Inertial actions and competencies • Simplification of systems, values, and structures	• Shocking crises and discrepancies • Open communications and experimentation • Paradoxical leadership
Paradoxes of organizing		
Organizing denotes an ongoing process of equilibrating opposing forces that encourage commitment, trust, and creativity while maintaining efficiency, discipline, and order.		
Contol/flexibility	Defenses of repression and reaction formation	Dynamic equilibration
• Performance . . . • Empowerment . . . • Formalization . . .	• Intense mistrust • Escalating resistance and alienation • Extreme chaos or rigidity	• Superordinate goals • Humour • Behavioral complexity
Paradoxes of belonging		
Groups become cohesive, influential, and distinctive by valuing the diversity of their members and their interconnections with other groups.		
Self/other	Defenses of ambivalence, projection, and splitting	Social acceptance
• Individuality . . . • Group boundaries . . . • Globalization . . .	• Destructive conflict • Group polarization • Tribalism	• Task focus • Valuing difference • Reduced power discrepancies

Thus, a wide review of innovation studies suggests that there are three broad thematic tensions evident in innovation practices, that have attracted the most attention from researchers and highlight the paradoxical nature of the innovation management process.

These tensions are those of practices associated with:

- *continuity and change*—maintaining stability and efficiency, while at the same time enabling dynamism and novelty;

- *constraint and choice*—giving direction and focus, while at the same time fostering initiative and diversity; and

- *complementarity and contrast*—integrating and harmonizing organizational activities, while at the same time facilitating differentiation and specialization.

2.4.1 Continuity and change

The most pervasive theme, addressed by many researchers, is that between stability and change—between the successful management of established practices and the creation of new ones. Whilst Schumpeter, in the 1930s, had identified the conflict between the established and the new as the key significance of innovation in broad economic terms,[37] its detailed implications for management were pursued much later.

One of the earliest conceptualizations of the management of a creative tension between these opposites was presented in the mid-1960s by Donald Schon, an exponent of a dialectical model of change.[38] Since the late 1970s, some of the key explorations of this tension have been undertaken by Harvard academics. For example, in his study of the state of the US automobile industry, James Abernathy highlighted the innovation-efficiency dilemma faced by US manufacturers. He argued that this dilemma had been resolved in favour of efficiency and that this had resulted in a 'roadblock to innovation' among US automobile manufacturers.[39]

In the 1990s, these tensions were explored by Dorothy Leonard-Barton in her contrast of 'core capabilities' and 'core rigidities' in the management of new product development (NPD),[40] and Clayton Christensen, in his discussion of the innovator's dilemma regarding 'disruptive' innovation.[41]

In her study, Leonard-Barton explores the role of core capability—i.e. *'the knowledge set that distinguishes and provides a competitive advantage'* for an organization—in relation to the management of innovation projects. She notes the paradox that core capabilities 'simultaneously enhance and inhibit' innovation projects. Her study investigated twenty new product and process development projects in five companies, including Ford and Hewlett-Packard. The research illustrates both the 'upside' of core capabilities in enhancing development and the 'downside' of 'core rigidities' in inhibiting development. The ability to manage this paradox varied considerably across the projects studied. In some cases, it led to the abandonment of the project or recidivism to familiar ground; in others, solutions were found that included the reorientation of the project through relocation, or the isolation of the project from the rest of the organization in order to enable new ground to be explored. More fundamentally, the ability of the organization to learn from its core rigidities requires a recognition of the multidimensional nature of core capability, and a willingness to address the conflict between the micro-level projects, where new requirements are identified, and the macro-level system of the organization, rooted in existing capabilities.[42]

James March of Stanford University has long been concerned with issues of ambiguity and choice in organizational decision making. But one of the key contributions made by March to innovation studies was the distinction that he made in the early 1990s between those organizational capabilities required for 'exploitation' and those for 'exploration'. These were articulated as contrasting modes of knowledge creation and acquisition, where 'exploitation' refers to the use of existing knowledge, and 'exploration' to learning and innovation through the pursuit and acquisition of 'new' knowledge.[43]

We have discussed the contrasting managerial solutions to the differing requirements of continuity and change earlier in this chapter. To summarize here, we draw upon the work of Michael Tushman and his colleagues, who have explored the different approaches to managing the tension between 'old streams' and 'new streams' of organizational activity through their separation by either time or space. The management of these distinct organizational activities by 'temporal differentiation' (i.e. time) may be regarded as a 'metamorphous' approach, because the organization changes form between periods of exploration and exploitation. In contrast, 'spatial differentiation' (i.e. space) involves managing synchronous streams of these contrasting activities and is characteristic of an 'ambidextrous' approach.[44] **Illustration 2.3** provides an example of the successful management of synchronous streams of incremental and radical innovation.

Illustration 2.3 Sony and the Sony 'Walkman'—managing synchronous streams of incremental and radical innovation

In 1979, Sony launched the Sony 'Walkman', the world's first compact personal stereo. Despite other major electronics companies entering the market, the 'Walkman' brand quickly became synonymous with the personal stereo and Sony remained a dominant player in the market right up until the launch of the Apple iPod in 2001. Between 1979 and 2004, Sony sold some 340 million Walkman stereos.

A key component in the longevity of Sony's success in this market has been its ability to develop and launch a series of incremental and radical innovations synchronously. For example, following the launch of the original Walkman in 1979, Sony introduced a series of incremental innovations to this product, such as, AM/FM radio in 1980, stereo recording in 1980, auto-reverse in 1981, Dolby in 1982, a water-resistant model in 1983, a graphic equalizer in 1985, a solar-powered model in 1986, and enhanced bass in 1988. In each of these incremental developments to the original product, Sony was 'first to market' with the innovation.[45] At the same time, Sony was also developing new formats for the Walkman, launching the world's first portable CD player in 1984, the first compact digital audio tape (DAT) player in 1990, and the first MiniDisc (MD) player in 1992.[46]

Since the launch of the iPod in 2001, Sony has fought back with the launch of its Walkman equivalent of the Apple iPod in 2004 and the Walkman mobile phone in 2005, developed and launched through Sony Ericsson (a joint venture established in 2001 with Ericsson of Sweden).

2.4.2 Constraint and choice

A second important theme is the tension between control over the direction of an organization and the creativity needed for innovation that is likely to arise from spontaneity and freedom. This has often been expressed in terms of the greater flexibility and autonomy found in the small organization compared with that found in the large, as encapsulated in the mid-1970s by Schumacher in his text *Small is Beautiful*.[47]

In the early 1980s, James Brian Quinn undertook a study of a number of large, innovative organizations, including Sony, Intel, Pilkington, and Honda, in order to identify their distinguishing characteristics. The results, published in *Harvard Business Review*, suggested that they were able to combine the flexibility and creativity of the small firm with the resources and capabilities of the large organization. This was achieved using a variety of management approaches—in particular, those that enabled the maintenance of a mixed portfolio of innovation projects under some general management direction. The label that he coined to embrace this observed behaviour was 'controlled chaos', to indicate a combination of flexibility through independent diverse project teams with control through periodic assessment and choice by management.[48] This solution is reminiscent of the concept of 'skunkworks'—that is, product development projects that work outside of the constraints of the formal organizational rules, structures, and decision-making processes (see **9.3.5** and **Illustration 9.2**).[49]

In their book *The Innovation Marathon*, published in 1990, Mariann Jelinek and Claudia Schoonhoven reported their study of five key US electronics companies, drawing from interviews with over a hundred managers. They aimed to explain how contemporary managers 'resolved the paradox' between bureaucracy and creativity, given that the sample firms were '*large complex organizations that do innovate successfully*'.[50] They found that through cultural, managerial, and systems approaches to the management of innovation, the companies were able to 'transcend the paradox' wherein the staff who were the creative 'source' of innovation were of 'low' authority.[51] The core dilemma of achieving '*permanence, precision and predictability*' in combination with a constant stream of repeated innovation requires '*a less dichotomous approach*' toward the apparent paradox of change and stability; it must acknowledge them as in 'dynamic tension', where '*freedom and control must be balanced*' to coordinate divergent specialists through organizational plans, and where '*the dilemmas of control and creativity must be ultimately reflected and resolved in management practices*'.[52]

Two of their case companies illustrate the different managerial approaches that were identified in the study. The solution at Texas Instruments was to run two parallel management systems: the product-customer centre (PCC) system focused on the 'efficiency' of current business operations, while the objectives, strategy, and tactics (OST) system addressed the 'effectiveness' of strategic activities. These were separate, formal, and explicit management systems.

Motorola's approach, in contrast, was much less structured, with operational and strategic activities mixed together. The tendency to chaos in this approach was counterbalanced by the introduction of highly systematic periodic reviews of activities with orderly follow-up meetings. This change was seen to be primarily about management 'style', rather than organizational structure. The Motorola approach was seen as more about 'simultaneous', rather than 'separate', management of the contradictory features of organization and innovation, and stability and change.[53]

The main point for Jellinek and Schoonhoven is that both companies illustrate the evolution of '*a middle-ground organizational response, characterised by both meticulous control and flexibility'*, in which the managers did not manage in dichotomous terms, but *'recognised, and explicitly managed'* a *'hybrid situation demanding both stability and change'*. They conclude that not only are rigid bureaucratic systems inappropriate in the rapidly changing environment of microelectronics, but so too are *'loosely structured, ad hoc "organic" approaches'* involving 'ambiguity' or 'uncertainty about responsibilities'. Indeed, for Jellinek and Schoonhoven, the companies instead illustrated that the *'crucial necessity of "both/ and" not "either/or" '* could be achieved through *'disciplined, systematic management of innovation, however apparently paradoxical'*.[54]

For Jellinek and Schoonhoven, it is this 'middle ground' of alternatives to *'the extremes of innovation by pure chance or innovation by meticulously planned, controlled effort alone'*, where the management of this paradox addresses the apparent contradiction between choice and the free flow of innovation in terms of idea creation and selection. The key is seen as a focus on 'process' rather than 'content', and *'ultimately rests upon individuals' exercise of choice as a front-end filter to the formal processes of innovation management'*. They go on to argue that the tension between choice to develop and choice not to develop innovative ideas is resolved by *'shared judgement, widely dispersed'*. The *'care and mainten- ance of this judgement is woven into the fabric'* of the firms and interweaves *'strategy making and innovation choice'*. In this sense, they argue that the companies were neither *'rigidly plan-driven automatons'* nor *'organic uncontrolled adhocracies'*.[55]

In conclusion, Jelinek and Schoonhoven criticize the commonly promoted inter- pretation of Burns and Stalker's organic structure as *'unstructured, loosely coupled, amorphous adhocracies'* that 'lack dynamic tension' and have a one-sided focus on *'constant free form flexibility through amorphous organizational forms'*. In contrast, they found clear evidence in their study of *'universally explicit reporting relationships and clear hierarchies'* and *'no evidence of amorphous reporting, unclear hierarchy, or fuzzy job respons- ibilities'*. They characterize this type of organization, with the dual ability to manage stability and change, as possessing *'dynamic tension between clear structures and frequent reorganization'*.[56]

2.4.3 Complementarity and contrast

The differentiation of an organization into different specialized functional departments and practices, and the problems of maintaining reasonable integration between them is a classic problem for innovation. We noted this earlier in the chapter through the dis- cussion of the work of Lawrence and Lorsch. Richard Pascale's study of the innovativeness of Honda, also discussed earlier, relabels this classic dilemma as that of 'fit and split'. He argues that the different functional areas 'contend' with each other in an ongoing process that recognizes that *'the presence and value of constructive conflict'* is an essential part of the successful business organization. He argues that *'some tensions in organizations should never be resolved once and for all'*: examples include cost control versus quality, and manufacturing efficiency versus customer service. Pascale's view is that the *'functional disciplines that advocate these points of view rub up against one another and generate debate'*. Contention across these functional boundaries is seen as 'inescapable' and, moreover, it can be 'productive'.[57]

Pascale illustrates his argument through the detailed analysis of Honda, which is structured as three interdependent companies and which:

> explicitly surfaces and manages contention in a constructive way . . . embraces contradictions that one might expect would blow the company apart . . . [and] not only live[s] with the paradox of these contradictions, but consciously and explicitly embrace[s] it as an operational tool of self renewal.[58]

He too sees a shift in managerial mode from the framework of 'either/or' to that of 'and/also'. But he emphasizes that success in embracing the paradox of reconciling opposites lies not through 'balance' (equilibrium), but by management 'orchestrating tension' through 'harnessing contending opposites'.[59]

The above three themes illustrate interesting directions in our understanding of the practices of management in the innovation process, as illustrated by a diverse range of empirical studies within this broad framework of contradiction and tension. Another development has been to suggest that we can apply the same framework to the different theories of innovation and embrace the tension between these in a creative fashion. We will now turn to this.

2.5 Innovation theory—using the tension between contradictory ideas

An interesting exploration of the implications of the paradox paradigm for conceptualizing the innovation process has been undertaken in a series of papers by Andrew Van de Ven and Marshall Scott Poole.[60] They arose from a major programme of research on innovation at the University of Minnesota in the 1980s and 1990s. As mentioned earlier, Van de Ven was a key instigator of the new paradox paradigm.

They argue that the weakness of traditional theories of innovation is that they fail to acknowledge the paradoxical nature of organizational change. Such paradoxes include the role of managerial action in contrast with that of contextual structure—i.e. is the course of innovation mainly influenced by managers' strategies and decisions, or is it primarily shaped by the technological and economic context? Another example is the contradiction between an internal dynamic of change in contrast with the influence of external factors—i.e. does the innovation process always need to go through a series of similar stages, or is the path shaped by a particular set of external circumstances? A third illustration is the paradox of explanations for stability in contrast with those of change—i.e. is it better to focus on the effective exploitation of the previous innovations on which an organization's success is based, or to embark on risky new exploratory paths for the organization's successful future? Each of these paradoxes has led to long-standing conflicts between theories that favour one or other side of these paradoxes.

In the mid-1990s, Van de Ven and Poole developed a metatheory of innovation that sought to accommodate in a positive way these traditionally competitive theories, through the '*juxtaposition of different theoretical perspectives*', with the position that '*relationships between such seemingly divergent views provides opportunity to develop new theory*'.[61] They stress, however, that although some integration is desirable:

it must preserve the distinctiveness of alternative theories . . . integration is possible if different perspectives are viewed as providing alternative pictures of the same organizational process without nullifying each other. This can be achieved by identifying the viewpoints from which each theory applies and the circumstances when these theories are interrelated. This approach preserves the authenticity of distinct theories and at the same time advances theory building, because it highlights circumstances when interplays among the theories may provide stronger and broader explanatory power.[62]

They attempt to do this by identifying ideal types of innovation theory and exploring how they figure in 'theoretical practice'. Specific theories of innovation are seen as more complicated than ideal types for two reasons: firstly, the organizational context (of innovation) extends over space and time; and secondly, the inherent incompleteness of any single 'motor'—each has one or more components the values of which are determined exogenously, e.g. variation, antithesis, dissatisfaction, or start-up.[63] The need therefore is for *composites . . . combinations . . . hybrids* of these contrasting theories.[64] The types of relationship in these hybrids may take different forms according to the mode of treating paradox—i.e. either explaining development and conditions under which different models hold through nesting (spatial) or sequencing (timing)—resolving by 'synthesis of contrasting models', or judging different types of complementarity.[65]

In developing their metatheory, Van de Ven and Poole reviewed a wide range of interdisciplinary literature on theories of change processes in social, biological, and physical systems. They explored the relevance of these to change, defined as '*a difference in form, quality or state over time in an organizational entity*'. This includes innovation in the sense discussed in this book. By 'process', they mean the progression (i.e. order and sequence) of events in an organizational entity's existence over time. They identified four basic types of process theory that explain how and why change unfolds in social or biological entities:

- 'life cycle' theory;
- 'teleological' theory;
- 'dialectical' theory; and
- 'evolutionary' theory.

These are seen to represent fundamentally different event sequences and generative mechanisms, which they call 'motors' of change.

An important feature of these motors is that they address the process for the generation of innovation and stand in a marked contrast to theories that use dimensions based on outcome, some of which were considered in **Chapter 1**: e.g. incremental/radical; continuous/discontinuous; first-order/second-order; competence enhancing/competence destroying.

We now turn to the four 'motors' of change of Van de Ven and Poole.

- *Prefigured path (life cycle)*—This approach emphasizes that the pattern of change is immanent within the innovation process, i.e. that there is a prefigured progression of events that follow a necessary sequence. In **Chapter 4**, we explore a number of such life-cycle models of innovation and technology, such as the 'technology

S-curve', which maps the technical performance trajectory of a technology in relation to R&D effort, and the 'diffusion curve', which maps the diffusion trajectory of an innovation over time. In **Chapter 8**, we discuss the typical activities of the innovation process, which are generally portrayed as a linear series of stages from idea generation to commercialization (see **Figure 8.1**, and **8.2.3** and **8.3**). Another example would be stage models of the new venture development.[66] These often imply that the main managerial issue is coping with the sequence of stages that will inevitably unfold in the innovation process. The explanation for prefiguration can range from concepts of inherent logic, through to determination by the socio-economic context.

- *Purposeful enactment (teleology)*—The cause of change in this model is viewed as the purpose or goal that is being pursued by actors within the innovation process. Rather than being predetermined, an actor has the *'freedom to enact whatever goal it likes'*. The approach suggests that the trajectory followed in the innovation process cannot be specified in general or in advance. A current trend within the innovation literature views innovation as arising from strategies pursued by managers: for example, Kanter defines goals to be pursued by innovative managers, whilst Pinch and Bijker explore the motives and goals of actors in shaping innovation.[67] They imply that the main managerial issue is the clear definition of goals and objectives. The explanation of purpose ranges across a wide spectrum from unbounded or bounded rationality, through to social constructivism. This notion of 'shaping' innovation is particularly evident in **Chapter 6**, with regard to innovation strategy, **Chapter 9**, in relation to the influence of informal networks within an organization, and **Chapters 3, 5, 10, and 11**, with respect to the role and influence of external actors and networks on the innovation process.

- *Conflict and synthesis (dialectics)*—This sees change as arising from oppositions of events, forces, and values expressed through entities that confront and engage one another in conflict. A novel construction (or synthesis) is created through this process. This approach is present in innovation studies such as those concerned with manufacturing process technologies and information technologies: for example, Williams and Edge explained the nature of computer numerical control (CNC) innovation as a result of conflict and negotiation between trade unions and managers,[68] whilst Lessig suggests that Internet innovations express conflicts between different social actors.[69] Such conflict shapes innovation. Again, we explore this theme in **Chapter 9**, in relation to the viewing of innovation as a social and political process, and in **Chapters 10 and 11**, concerning the role of external actors and networks in the innovation process. The implications for the management of innovation, are an emphasis on the political ability to manage conflict and negotiation. The explanation of conflict varies from concepts of fundamental social antagonisms to negotiation in a pluralist context.

- *Competitive selection (evolution)*—This approach sees change as arising through a process of variation, selection, and retention analogous to the evolution of species in the natural world. In this model, events unfold in a recurrent, cumulative, and probabilistic progression. Work on technological trajectories, by those such as Nelson and Winter, Dosi, and Levinthal, express this approach.[70] Examples of this orientation are particularly evident in **Chapter 5**. From a managerial perspective, the main implication is an emphasis on understanding the context or 'selection environment' within which innovation occurs. There is, however, a spectrum of views on the nature of variation and selection, ranging from the 'random', through 'general economic forces', to 'socially influenced'.

The authors suggest that these different motors have different emphases on:

- the unit of change or innovation (single or multiple);
- the mode of change or innovation (prescribed or constructed).

These theories can be summarized as in **Figure 2.1**.

Drawing on their own studies of innovation, such as their investigation into cochlear implants,[71] the authors argue strongly that effective explanation of innovation rarely draws on a singular motor, but instead creatively uses a combination of these motors of change, which bring together contrasting elements of single and multiple units of analysis or prescribed and constructed modes of innovation.

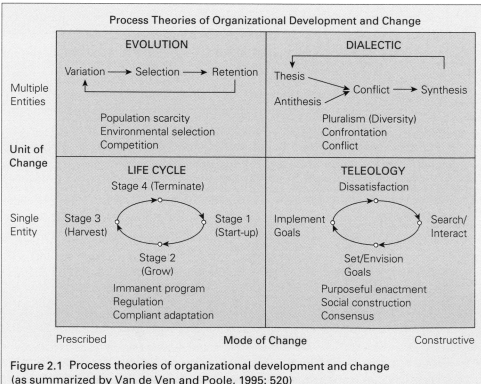

Figure 2.1 Process theories of organizational development and change (as summarized by Van de Ven and Poole, 1995: 520)

Thus, we see Van de Ven and his colleagues applying the notion of contradiction and tension to the theoretical explanation of innovation in its broadest sense. They suggest that a feature of some of the most creative and relevant areas of innovation theory is that they embrace contradictory fundamental concepts (motors) of change, but combine them in a creative fashion. This is clearly a very different approach to conventional efforts to construct theories of innovation.

2.6 Concluding comments

It is evident from the studies reviewed, that the original 'organization studies' model of the choice of an 'organic' structure as the route for successful innovation has been replaced by a model of 'tension' management, which combines contradictory practices necessary for successful innovation. But it is important to note that the concept of the 'organic' structure was a fundamental advance in the conceptualization of innovation as an organizational process; it enabled a range of subsequent investigations into innovation within organizations and provided the context for the emergence of 'tension' management as an important concept for the management of innovation.

We have seen from the studies discussed that there are three principal approaches to the meaning of the new paradigm of paradox and contradiction for the management of innovation. They emphasize the different dimensions of individual cognition, organizational configuration, and interpersonal communication.

The emphasis on cognition suggests that we need to break the hold of mental models of innovation that rely solely on one privileged motor of change. Instead, researchers and managers should learn to use frameworks for thinking about innovation that embrace contradictory motors of change. With this in mind, subsequent chapters will present and critique a range of models of the innovation process.

The research on organizational configuration indicates that managers need structures in their business that facilitate the creative management of contradiction and tension. While there is no simple prescription for this, many are moving to the view that this is accomplished by the clever juxtaposition of distinct and different bits of the organization, rather than through merging and mixing them up. In other words, the most effective approach is the design, recognition, and management of boundaries, rather than their blurring or dissolution. The 'metamorphous' organization and the 'ambidextrous' organization are examples of different approaches to this, in regard to combining the contradictory paths of exploration and exploitation. (We explore this theme in detail in **7.6**.)

There is widespread recognition that the culture and style of management is of critical importance for 'tension' management, and much of the literature on managing contradictions identifies the central importance of interpersonal communication in the process. This has been expressed by Howard Thomas as talking our way through ambiguity and change,[72] and by Ann Huff as a process of politics and argument.[73] Closer analysis of the nature of communication in the innovation process, with particular attention to conflict, is the subject of a number of studies. Indeed, communication and interaction within the organization is the central theme of **Part III**—i.e. **Chapters 7–9**—whilst communication and interaction between the organization and its environment is a key theme of **Chapters 3–6, 10, and 11.**

The framework of tension and contradiction is opening up new avenues for understanding the role of organizations and individual managers in the practices of the innovation process. In particular, it provides a valuable lens for appreciating the need to couple, for example, technological possibility with market opportunity, formality with informality, planned activity with serendipity, and internal resources with external resources. The tensions involved in the attempt to combine and reconcile such couplings are explored throughout this text.

CASE STUDY 2.1 THE GKN COMPOSITE SPRING—AN INNOVATION COMBINING TRADITION WITH NOVELTY

GKN was a long-established manufacturing firm focusing on steel components for the automotive industry. In the 1980s, it pioneered the innovation of a vehicle suspension leaf spring made not of metal, but of carbon fibre in epoxy resin. The management of the company had recognized, in the 1970s, that it needed to shift the basis of its manufacturing focus from steel, for two reasons: the performance offered by new composite materials, and the weight savings that they provided in the dawning era of oil and energy awareness. Such a shift was not easy to accomplish, given the fundamental importance to the success of the firm of its manufactured metal components.

A variety of managerial approaches were deployed to achieve this shift. One was to create a separate 'think tank' activity area called GKN Technologies, which had a clear mission of exploring new materials outside of the mainstream activities of the firm. In some ways, this was a protected space and projects on leisure products were pursued, but the team was also required to work closely with design engineers from the car industry to explore the real possibility for innovations in this sector. The other strand of the management approach was to pursue a partnership with a small, non-traditional firm, which was a pioneer in epoxy resin automotive applications.

The combination of these two approaches enabled the successful development of a new innovation stream, while retaining the successful commercial basis of the company in its traditional product area. In other words, it managed to be ambidextrous in its combination of the contradictory elements of old and new.

CASE STUDY 2.2 CLEARBLUE ONESTEP—AN INNOVATION COMBINING THE CONTRADICTORY STRENGTHS OF THE LARGE AND SMALL FIRM

Clearblue Onestep was the first successful over-the-counter (OTC) home pregnancy test kit in the UK. It was first marketed in 1989. It is a good illustration of the success of an innovation strategy that sought to combine the strengths of a large company, in terms of its financial and technical resources, with those of a small company, shown through its flexibility and responsiveness. The contrast between these material and behavioural properties has often been used to explain the relative advantages of large versus small firms.

Unilever, a large science-based multinational, took a decision in the early 1980s to seek to exploit its internal biotechnology capabilities in the emerging field of monoclonal antibodies in the market for home diagnostic kits. Unilever realized that health care in the 1980s was characterized by patients having to visit the doctor or hospital for pregnancy and other intimate tests. It recognized the potential for products that would allow patients to test themselves for certain conditions, but was aware that the newly emerging personal care business would require skilful innovation in order to succeed.

Rather than explore this opportunity as a big company R&D project, Unilever chose to establish a small company called Unipath as a more appropriate organizational form for innovation. Initially founded with only thirty staff, the small company was very effective in engaging with designers and users in order to develop a product that, as well as being technically effective, was easy to use and attractive to consumers in this personal product area.

The flexibility of the small firm was shown in its ability to draw upon technology from other sectors concerned with marker pens and plastic packaging to create a novel product, with reliability and ease of use as key requirements. It also gave much greater scope for new channels in design and distribution than the large company format would favour. On the other hand, it was also able to rely on the substantial technical expertise of Unilever with regard to biotechnology, as well as its financial backing.

This case is a good example of how the creation of a small venture within the framework of a large company enabled the direction of a clear strategy to be effectively combined with a diversity of technical and business solutions for the rapid development of a very innovative and very successful product.

CASE STUDY 2.3 QUANTEL PAINTBOX—AN INNOVATION COMBINING INTEGRATION AND DIFFERENTIATION OF KEY ACTIVITIES

The Quantel Paintbox was the first successful, totally digital, graphics hardware/software package with a user-friendly interface for the non-computer-literate artist. It was first marketed in the early 1980s and enabled the early exploitation of digital graphic techniques, initially through the broadcast media. The origins of digital technology lay in the military and industrial sectors, and although it was possible to foresee wider social application, the organizations with the technical capability were ill-suited to the development of such products. An early use of such a computer graphics system for the Superbowl broadcasts in the USA in 1978 hinted at the potential, but the system was both technically deficient and difficult to use.

The strength of Quantel's approach was that it successfully managed two very divergent streams of people and know-how: computer/software engineers, on the one hand, and graphic artists, on the other. These had been traditionally separated into distinct producer and user roles, and were further characterized by very different cultures and practices.

Quantel developed a management approach that enabled both of these streams to express their ideas fully, but also made them interact with each other. The consequence was a clear recognition of what the artists wanted—i.e. 'a device equivalent to a 6B pencil'—along with an awareness of what was technically required—a hardware/software combination. This not only enabled the successful development of a usable product, but the relationship between engineer and artist was used very effectively to market the product to organizations such as television companies, which themselves had a similar split between engineers and artists.

Within a few years, the Quantel Paintbox held 90 per cent of world market share in digital graphics and the artist David Hockney chose to explore its potential as a new medium in the public gaze. The case is a good illustration of management facilitating the expression of specialist areas while successfully integrating and harmonizing their activities.

Questions

1. Which of the three themes of contradiction management identified by Lewis (i.e. 'learning', 'organizing', and 'belonging') are expressed most strongly in each of these three case studies?

2. Give one example of the 'acceptance', 'confrontation', and 'transcendence' styles of contradiction management from the three case studies.

3. Explain a case of innovation with which you are familiar by using two of Van de Ven and Poole's 'motors of change'.

FURTHER READING

1. Lewis (2000) provides a useful review of studies concerning the management of paradox within organizations.

2. Gupta et al. (2006) provide an interesting review of the debate concerning the pursuit and interplay of the tasks of exploration and exploitation within organizations.

3. The paper by Van de Ven and Poole (1995) provides an excellent interdisciplinary review of the theories of change. From this review, they identify four basic types of process theory that explain how and why change unfolds in social or biological entities: life cycle; teleological; dialectical; and evolutionary. These are seen to represent fundamentally different event sequences and generative mechanisms, which they call 'motors of change'.

NOTES

[1] Schumpeter (1939); McClelland (1961); Kirzner (1973); Casson (1982).

[2] Burns and Stalker (1961).

[3] Lorsch (1965).

[4] Marquis (1988); Langrish et al. (1972); Freeman (1974).

[5] Child (1972).

[6] Marquis (1988).

[7] Langrish et al. (1972).

[8] Jeffreys (2003).

[9] Freeman (1974: 169).

[10] Rothwell (1983); Rothwell and Zegveld (1985).

[11] Lawrence and Lorsch (1967).

[12] Zaltman, Duncan, and Holbeck (1973).

[13] Nord and Tucker (1987: 344–7).

[14] Duncan (1976); Burgelman (2002).

[15] Kanter (1989a); Tushman, Anderson, and O'Reilly (1997); Benner and Tushman (2003).

[16] Rothenburg (1979).

[17] Festinger (1957).

[18] Eisenhardt and Westcott (1988); Quinn (1988c); Van de Ven and Poole (1988); Weick and Frances (1996).

[19] Peters and Waterman (1982).

[20] Kanter (1985).

[21] Lampel, Shamsie, and Lant (2006: 292).

[22] Perretti and Negro (2007); DeFillippi, Grabher, and Jones (2007).

[23] Van de Ven (1983).

[24] Quinn and Cameron (1988).

[25] Morgan (1981).

[26] Hampden-Turner (1990).

[27] Handy (1994).

[28] Pascale (1990).

[29] Hampden-Turner (1990).

[30] Grieco and Lilja (1996).

[31] Pascale (1990).

[32] Schoonhoven and Jelinek (1990).

[33] Clegg, Cunha, and Cunha (2002).

[34] Cobbenhagen, Philips, and Friso (1990).

[35] Dougherty (1996).

[36] Lewis (2000).

[37] Schumpeter (1939).

[38] Schon (1966; 1967).

[39] Abernathy (1978).

[40] Leonard-Barton (1992).

[41] Christensen (1997).

[42] Leonard-Barton (1992).

[43] March (1991; 1996).

[44] Tushman and Romanelli (1985); Tushman and O'Reilly (1996).

[45] Sanderson and Uzumeri (1995).

[46] <http://www.sony.net/SonyInfo/CorporateInfo/History/sonyhistory-e.html>, accessed 15 March 2007.

[47] Schumacher (1974).

[48] Quinn (1985).

[49] Peters (1988); Quinn (1985; 1988a).

[50] Jelinek and Schoonhoven (1990: 17).

[51] Jelinek and Schoonhoven (1990: 22).

[52] Jelinek and Schoonhoven (1990: 56–8).

[53] Jelinek and Schoonhoven (1990: 79–80).

[54] Jelinek and Schoonhoven (1990: 83–4).

[55] Jelinek and Schoonhoven (1990: 166–9).

[56] Jelinek and Schoonhoven (1990: 253–63).

[57] Pascale (1990: 24).

[58] Pascale (1990: 26).

[59] Pascale (1990: 33–4).

[60] Van de Ven and Poole (1988); Poole and Van de Ven (1989a; 1989b).

[61] Van de Ven and Poole (1995: 511).

[62] Van de Ven and Poole (1995: 511).

[63] Van de Ven and Poole (1995: 524–6).

[64] Van de Ven and Poole (1995: 527).

[65] Van de Ven and Poole (1995: 534).

[66] Burgelman and Sayles (1986).

[67] Kanter (1985), and Pinch and Bijker (1987), respectively.

[68] Williams and Edge (1996).

[69] Lessig (2001).

[70] Nelson and Winter (1977); Dosi (1982); Levinthal (1998).

[71] Van de Ven and Garud (1993).

[72] Thomas (1988).

[73] Huff (1988).

3

Innovation From a Network Perspective

Chapter overview

Learning objectives

This chapter will enable the reader to:

- distinguish between 'linear' and 'interactive' models of the innovation process;

- appreciate the nature and origins of the network perspective, and recognize the alternative approaches employed in researching and framing networks; and

- outline and explain the key dimensions and characteristics of relationships and networks, and their impact on the innovation process and innovative capacity of individuals, teams, and organizations;

- outline and explain important network concepts, such as the 'strength of weak ties', 'gatekeepers', and 'boundary-spanners', 'multiplexity', 'density', 'openness', and 'structural holes';

- recognize the role and importance of social and organizational networks to the innovation process within, and innovative capacity of, project teams, organizations, communities, sectors, and regions; and

- apply the network perspective to cases of innovation.

3.1 Introduction

Since the early 1990s, the network perspective has received a great deal of interest, not least among academics, managers, consultants, and policymakers concerned with promoting the innovative capacity of organizations, regions, sectors, and nations. Part of the success of the approach may be ascribed to the power of the network metaphor as a way of visualizing sets of links between individuals and organizations: it changes the imagery from a focus on pairs of relationships, to one of *'constellations, wheels, and systems of relationships'*[1] and of 'webs' of group affiliations.[2] For DeBresson and Amesse, *'the network approach has something original, useful, and durable to bring to innovation studies'*.[3] It has also been argued that:

> The utility of the network perspective is at least partially derived from the ease with which the concept can be expressed and applied. It is at once a concept and framework whose applicability is immediately recognisable by practitioners, whilst its academic pedigree has been firmly established . . . The flexibility of the network perspective, both in relation to the manner in which it may be applied . . . and the subject matter it may be applied to . . . has also been of great import to its success; it is a perspective that can throw light on phenomena at any unit of analysis, such as at the level of the individual . . . and at the level of the organisation.[4]

But the ubiquity of the approach often belies the variety of phenomena to which the term 'network' is used to refer, from the informal communities of scientists within a scientific discipline to the social connections of employees within an organization, and from global alliances of organizations to regional clusters. It also conceals the variety of configurations that networks may manifest themselves, from dense, closed, local networks, to loose, open networks, spanning great geographical and sociometric distance.

From a broad review of the innovation literature, it is possible to identify two distinct waves of academic interest in networks. The first wave of network studies, during the 1960s and 1970s, drew upon the network approach emerging from anthropology and sociology, and was focused on the communicative interactions between individuals. Three streams of research from this period continue to resonate in the network literature today:

- the first, centred on the communication networks of scientists, is typified by the work of Derek DeSolla Price and Diana Crane, and the notion of the 'invisible college';[5]

- the second, focusing on the interactions of researchers in research and development (R&D) departments, is best illustrated by the work of Thomas Allen, and the emergence of concepts such as 'gatekeepers' and 'boundary-spanners';[6]

- the third, concerning the diffusion, adoption, and adaptation of innovation by individuals, is captured by the research of Everett Rogers and his book *Diffusion of Innovations* (now in its fifth edition).[7]

These communication network studies have made an indelible impact on our understanding of the innovation process.

From the late 1980s, it is possible to observe the rise of a second wave of network studies. Whilst, in part, this second wave embraces a resurgence of the interests of the first wave, with its concern on the interactions and relationships between individuals,[8] it is perhaps characterized more by its focus on the exchanges and linkages between organizations. It may be argued that the avalanche of research concerning organizational networks that has emerged over the last decade has much to owe to the work of scholars such as Walter Powell and those connected to the 'Uppsala School', as represented by Håkan Håkansson, whose work from the mid-to-late 1980s has been particularly influential.[9] Although these two strands of research—that focused on networks of individuals and that centred on networks of organizations—have largely developed in parallel, but separately, there are some interesting areas of crossover: in particular, in relation to social capital, as exemplified by the work of Ronald Burt.[10]

During this chapter, we will draw upon and distil the key conceptual, theoretical, and empirical contributions from both of these 'waves' of network studies to our understanding of the innovation process and innovative capacity. Networks are an important mechanism through which individuals and organizations can 'shape' the innovation process, and thus the nature of the innovation outputs, such as the performance and functionality of new products, services, and processes.

3.2 Models of the innovation process

In this section, we will outline three models of the innovation process:

- the 'science-push' or 'technology-push' model;
- the 'need-pull' model;
- the 'coupling', or 'interactive', model.[11]

These models are relatively broad and abstract in nature, but they play an important role in aiding our conceptualization of the relationship between innovative organizations and their context, in particular with regard to their linkages with the marketplace, and the science and technology base. These models also provide a historical perspective of the way in which the practice of innovative organizations has evolved since the 1950s.[12]

We distinguish these models from what are sometimes referred to as 'stage models' of the innovation process. As we will see in **Chapter 8**, stage models are more explicit in nature, detailing the individual stages of the innovation process, and thus the ordering, activities, and linkages of each of these stages (e.g. idea generation, problem solving, prototype development, field testing). As such, stage models are more concerned with aiding the day-to-day management of the innovation process rather than its conceptualization.

Prior to the development of these models of innovation in the early 1980s, there had been a number of studies during the 1960s and early 1970s that had attempted to determine the nature and source of the stimuli leading to the initiation of successful innovation.[13] In his study, Utterback found that innovation within an organization could be classified as either 'means-stimulated' or 'need-stimulated',[14] terms that are synonymous with the concepts of 'technology-push' and 'need-pull', respectively. Myers and Marquis further subdivided need-stimulated events into those triggered by external market needs and those by internal production, or administrative, needs.[15] Indeed, the stimulus for a given innovation—whether means or need-driven—may originate from either within the innovative organization or externally, for example, from customers, suppliers, and universities.[16]

3.2.1 Linear models of the innovation process

Here, we will deal with the 'technology-push' and 'need-pull' models together: both are often referred to as 'linear' models of the innovation process, because they conceptualize the innovation process as a linear sequence of events.

The terms 'technology-push' and 'science-push' are often used interchangeably. This is perhaps rather confusing, given the gap that often exists between basic science, as undertaken within universities and government research laboratories, and technological advances within innovative organizations. Whilst the time gap between scientific discovery and technological innovation might be closing,[17] it is still of major concern to policymakers in many countries: in the UK and European Union, for example, this issue has been at the core of recent government reports on science and innovation.[18] Thus, it seems useful to make the distinction between 'science-push', where the underlying assumption is that innovation will originate from basic science activities within universities and government research laboratories, and 'technology-push', where the underlying assumption is that innovation will emerge from a focus on technological developments within innovative organizations. Nevertheless, what both of these models hold in common is the assumption that successful innovation will result from a focus on investing in the scientific and technological means rather than responding to the needs of the marketplace—that is, both models focus on the 'supply side' and emphasize the *causal role of scientific and technological advance*.[19] This perspective is, in part, supported by the contention of Kelly et al. that:

> 'Necessity is the mother of invention' is another half-truth . . . demand is a strong motive; but as a full explanation it fails, because many needs have not yet given rise to inventions, and many innovations arose from other causes . . . one could perhaps turn the adage around: 'Invention is the mother of necessity'.[20]

Rothwell refers to these 'push' models of innovation as 'first-generation' models, representing the practice, and embodying the underlying assumptions, of innovative organizations and governments between the early 1950s and mid-1960s. He describes this as a period of rapid economic growth, industrial expansion, and the emergence of new sectors, such as semiconductors, pharmaceuticals, and composite materials, driven by new technological advances—factors that would tend to support the successful operation of such 'push' models of innovation.[21]

Studies undertaken during the 1960s and 1970s indicated, however, that between two-thirds and three-quarters of innovations were need-stimulated.[22] Furthermore, Rosenberg argued that empirical evidence suggested that technological advance within many industrial sectors was occurring without a clear and full understanding of the underlying scientific principles and that scientific activity was often responding to fill this gap in scientific knowledge—that is, the innovative activity of organizations was found to be increasingly driving the scientific agenda of universities and government research laboratories. For Rosenberg, this was not surprising, given the economic and market opportunities afforded to technologically advanced firms.[23] What this research highlighted was a shift in the model of innovation, as practiced by innovative organizations, from 'science-push' or 'technology-push', to 'need-pull'[24] or 'demand-pull'.[25] **Figure 3.1** illustrates the difference between the science-push and need-pull models of the innovation process.

Rothwell refers to these 'pull' models of innovation as 'second-generation' models, representing the privileging of demand-side factors by innovative organizations from the mid-1960s to early 1970s. He characterizes this period as one of intensifying competition, the exploitation of existing technologies, and greater strategic emphasis on marketing to gain market share.[26] Studies during this period indicated that there was a positive correlation between need-stimulated innovation and commercial success.[27] But an over-emphasis of market needs as a driver of innovative activity within organizations can lead to both 'technological incrementalism' and a decline in novelty, and the neglect of long-term research and development.

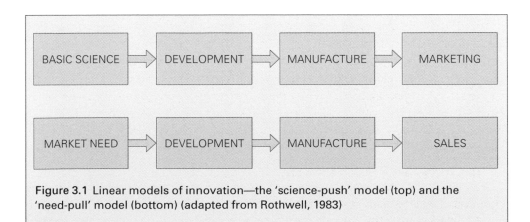

Figure 3.1 Linear models of innovation—the 'science-push' model (top) and the 'need-pull' model (bottom) (adapted from Rothwell, 1983)

Since the mid-1970s, these linear models of innovation have been viewed as increasingly untypical of the innovative behaviour of organizations and have been superseded by the 'coupling', or 'interactive', model of innovation.[28]

3.2.2 Coupling, or interactive, models of the innovation process

The 'coupling', or 'interactive', model of innovation represents a shift away from the linearity of the push and pull models of innovation; Rothwell refers to this as the 'third-generation' model of innovation. He argues that the key environmental factor encouraging the emergence of the interactive model was the severe resource constraints of the 1970s, and the consequent search for understanding the basis of successful innovation and reducing the incidence of failure.[29] Rothwell and Zegveld describe the basis of the model as follows:

> The overall pattern of the innovation process can be thought of as a complex net of communication paths, both intra-organizational and extra-organizational, linking together the various inhouse functions and linking the firm to the broader scientific and technological community and to the marketplace. In other words the process of innovation represents the confluence of technological capabilities and market needs within the framework of the innovating firm.[30]

Thus, at the heart of the interactive model of innovation lies a network perspective of the innovation process. As Rothwell points out, however, it is '*still essentially a sequential process . . . with feedback loops*'.[31] The interactive model is illustrated in **Figure 3.2**.

This conceptualization of the innovation process embodies two fundamental developments on previous models. Firstly, it stresses the need for the coupling of emerging technological possibilities and market needs, rather than privileging 'signals' from either the science/technology base or the marketplace; Schmookler likened this coupling activity to the blades of a pair of scissors.[32] For Rothwell and Zegveld:

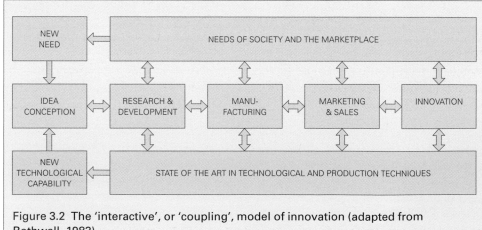

Figure 3.2 The 'interactive', or 'coupling', model of innovation (adapted from Rothwell, 1983)

> The linkages between science, technology and the market-place are complex, interactive and multidirectional, the dominant driving force varying over time and between one branch of industry and the next.[33]

Secondly, it stresses the importance of intra- (i.e. internal) and inter-organizational (i.e. external) links and interactions. In bringing together these two factors, Freeman argues that the:

> professionalization of industrial R&D represents an institutional response to the complex problem of organizing this 'matching' . . . The coupling process is not merely one of matching or associating ideas in the original first flash; it is far more a continuous creative dialogue during the whole of the experimental development work and introduction of the new product or process.[34]

The interactive model of innovation has implications for the way in which we conceptualize the relationship between scientific and innovative activity, traditionally seen as largely separate. It has been contended that the boundary between the 'scientific community' and 'technological activity' is fading. As a result of this trend, it has been proposed that the behaviour of academic researchers and industrial researchers should be studied in a more holistic way—that is, at the level of the 'technological community' or 'R&D community'. Debackere et al. view such a community as consisting of *'scientists and engineers working towards the solution of an inter-related problem set, who are dispersed across both private and public sector organisations'*, through which *'information and knowledge flow quite freely'*.[35]

The coupling, or interactive, model of innovation has clear implications for both managers and policymakers: managers must encourage the building and nurturing of formal and informal relationships and interaction across both internal and external organizational boundaries; policymakers should develop policies that promote the forging of relationships within and between the science base, innovative organizations, and the marketplace.

Rothwell introduces two subsequent generations of models of innovation. Whilst the fourth-and fifth-generation models embody increasingly more sophisticated responses by innovative organizations—to an environment increasingly characterized by the emergence of key new communication technologies, shortening product development times and product life cycles, the strategic use of generic technologies, integration along the supply chain, and extensive strategic alliance and networking activity—to some extent, they distract from the clarity of the core themes introduced within the third-generation model. Nevertheless, important features of the fourth- and fifth-generation models will be discussed during **Part III**, which concerns the managing and organizing of innovation activities.[36]

3.3 Introducing the network perspective

3.3.1 What is a 'network'?

The term 'network' was traditionally employed to refer to the set of relationships between a group of individuals, within a particular community or organization, for example.

Today, it is also frequently used in relation to the set of relationships between a group of organizations, such as those within a particular sector or geographical location. In this book, we will distinguish between these two types of network by referring to networks of individuals as 'social networks' and to networks of organizations as 'organizational networks'. Typically, however, connections between organizations are characterized by both individual-level relationships (i.e. those between people) and organizational-level relationships, such as through formal alliance agreements. Both social and organizational networks are important to our understanding of how innovation and knowledge emerge, are shaped, and are diffused. Indeed, for Law and Callon, a network, whether deliberating constructed or not, '*generates the space, a period of time, and a set of resources in which innovation may take place*'.[37]

Networks in the 'real world' lack convenient natural boundaries, stretching far beyond the confines of any organization, community, or sector.[38] Indeed, it has been hypothesized that any two individuals in the USA may be linked by a chain of up to six people; this is known as the 'small-world phenomenon'.[39] Whilst this hypothesis has received little empirical support, it nevertheless highlights one of the key problems associated with researching networks—that of boundary setting. It is therefore not surprising that network research frequently requires the setting of artificial boundaries in relation to which actors to include and which to exclude in a study. Similar decisions also need to be made about which relationships and which transactions are to be included. Thus, it is important to note that when we are looking at the results of a network study, what we are seeing is a 'partial' network—that is, one that has been abstracted from the 'total' network for the specific purposes of the study.[40]

There are a number of options available to the network researcher when setting such a boundary. Those focusing on the network of a specific actor, such as an entrepreneur, might only include those that are directly connected to that actor. When drawn, the resulting 'focal' or 'ego-centred' network looks like the hub and spokes of a wheel (**Figure 3.3** provides an example of a focal network). Those researchers focusing on a group of actors, such as a set of employees, a group of scientists, or organizations, might set the boundary around a particular organizational division, a particular scientific specialism, or a specific industrial sector, respectively. This focus on the relationships between a group of actors rather than those of a specific actor creates what is sometimes referred to as a 'socio-centred' network (**Figure 3.4** provides an example of a socio-centred network).[41]

Illustration 3.1 The R&D networks of large electronics firms

Figure 3.3 overleaf illustrates the set of R&D alliances and joint ventures of Philips and Siemens. It reveals the large number of relationships that they employ in their innovative activities, and the wide range of organizations from the USA, Europe, and Japan with which they are connected. Although focusing only on two firms within the industry, it is possible to see the degree to which the networks of organizations are often overlapping and interweaved—a good proportion of the organizations illustrated are linked to both Philips and Siemens. Mapping the alliances and joint ventures of an organization is important in any attempt to evaluate the competences, capabilities, and resources that they have at their disposal. We return to this theme in **Chapter 6**, in our discussion of the 'resource-based' view of the firm.

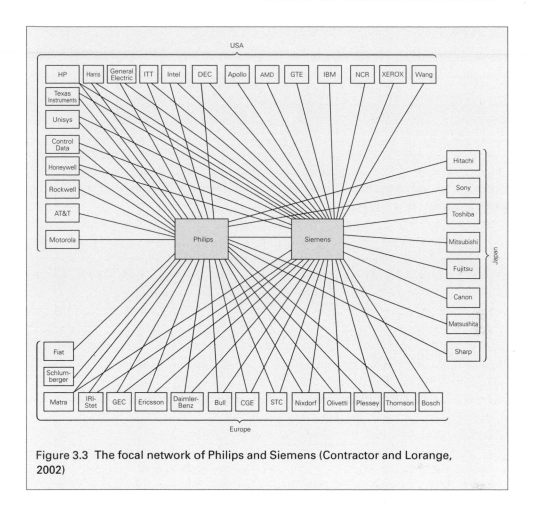

Figure 3.3 The focal network of Philips and Siemens (Contractor and Lorange, 2002)

Whether one is studying an ego-centred or socio-centred network, it is also common for network researchers to focus on specific types of exchange or transaction, such as information, friendship, or power; this would reveal the information network, the friendship network, and the influence network, respectively, where each network may be very different from the other even among the same group of actors. Researchers may wish to

Illustration 3.2 The highly connected world of Formula One and motor sport

Despite the highly competitive nature of Formula One and other motor sports, individuals in the sector are highly connected. In large part, this is a result of job mobility. The network map in **Figure 3.4** overleaf illustrates the connections in Formula One between key individuals who have worked in the same team, such as Williams, Benetton, or Ferrari, at some point in their career. Henry and Pinch, who undertook this fascinating study, also found that:

> one of the most important ways in which knowledge is spread within the motor sport industry is the rapid and continual transfer of staff . . . between the companies within the industry. Especially at the end of every racing season, there is an intense period of negotiation as designers, engineers, managers and, of course drivers move between teams.[42]

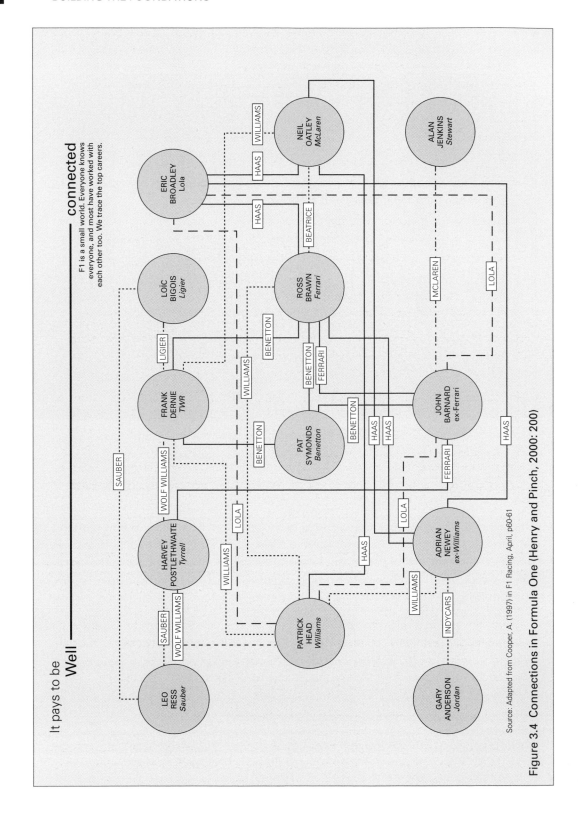

It pays to be

Well — connected

F1 is a small world. Everyone knows everyone, and most have worked with each other too. We trace the top careers.

Source: Adapted from Cooper, A. (1997) in F1 Racing, April, p60-61

Figure 3.4 Connections in Formula One (Henry and Pinch, 2000: 200)

focus still further, by considering only those actors, relationships, and exchanges that relate to a specific activity, such as the development of a new innovative service or product—this reveals what is sometimes referred to as an 'action set'.[43] All of the different boundary-setting approaches outlined above have been adopted by one researcher or another in the study of innovation.

Whilst a network is considered to be 'greater than the sum of its parts', it is nevertheless useful to highlight the component parts that comprise any given network:[44]

- *actors*—the individuals or organizations that are members of the network;

- *links*—the relationships between these individuals or organizations, including formal and informal links, as well as strong and weak ties;

- *flows*—the 'content' of exchanges or transactions that occur through the relationships between these individuals or organizations, such as, ideas, information, artefacts, money, and friendship. For Håkansson and Johanson, however:

  'content . . . is never equivocal. What one actor mainly considers as exchange of products may be viewed by another primarily as communication or by a third as a demonstration of power';[45]

- *mechanisms*—the modes of interaction employed by network members, from face-to-face contact, to that mediated through the Internet, for example, in enabling flows between actors and through the network.

There are many ways in which individuals can interact with one another within a network, including via telephone, email, documents, and face-to-face meetings and encounters. These mechanisms of interaction may be broadly divided into 'active' mechanisms—referring to those involving personal interaction, whether face-to-face or over the telephone, for example—and 'passive' mechanisms—essentially referring to documents and other textual material, where there is no direct interaction between the 'sender' and 'receiver' in the exchange.[46]

Networks do not emerge and grow without considerable effort, however, and thus we are not only interested in the mechanisms for the exchange of information and goods through the network, but also the mechanisms by which individuals and organizations build and nurture their networks. It is therefore useful to distinguish between 'networks' and 'networking': networks represent the set of relationships among network members, whilst networking can be seen as the interactions and flows between the individuals or organizations within a network, through which network relationships are built, nurtured, and mobilized. We thus see an interplay between the structure of a network and the networking that occurs within that structure: on the one hand, the network may constrain or liberate the patterns of interaction and exchanges between network members; on the other, networking behaviour may serve to either ossify (i.e. fix) the existing network membership and relationships, or create a dynamic in the membership and relationships within the network.

This interplay of the network and networking—that is, between structure and agency (or action)—is particularly well encapsulated by theories such as 'structuration'[47] and 'actor network theory' (ANT).[48] For example, Law and Callon, proponents of ANT, emphasize *'the process of mutual shaping'*, arguing that *'it is important to understand that actors are not simply shaped by the networks in which they are located (although this is certainly true), but they also influence the actors with which they interact'*.[49]

3.3.2 A new mode of organizing or a new analytical lens?

Although the network perspective has long been employed in developing our understanding of the emergence and diffusion of science[50] and innovation,[51] the period since the early 1990s has seen an avalanche of network research and the language of networks permeating the field of innovation and entrepreneurship. Whilst this research incorporates a variety of network approaches and theories, such as social network analysis (SNA),[52] ANT,[53] and structuration,[54] these all share a new focus on the patterns of interaction between a diversity of actors in the innovation process, rather than a focus on the characteristics and behaviour of an individual actor, whether an entrepreneur, a scientist or engineer, or an organization. Its usage, however—in particular, with regard to organizational networks—is laden with two quite different interpretations as to the fundamental significance of the concept of a 'network':

- that the network represents a new way of organizing; or
- that the network simply represents a new way of looking at practices that have long existed.

We will address these two positions in turn.

Networks as a new mode of governance

The first of these interpretations places emphasis on the network as a new and particular mode of governance. In his seminal and influential paper from 1990, 'Neither market nor hierarchy: network forms of organisation', Walter Powell presented the 'network' as an organizational form that was an alternative to the traditional 'hierarchy' (e.g. the vertically integrated organization) or 'market' modes of governance (see **3.4.2** for further discussion):

> firms appear to be changing in significant ways and forms of relational contracting appear to have assumed much greater importance. Firms are blurring their established boundaries and engaging in forms of collaboration that resemble neither the familiar alternative arms length market contracting nor the former ideal of vertical integration.[55]

Also around this time, Rod Rhodes and David Marsh employed the concept of network to identify a new 'policy space' as an alternative to the traditional categories of 'state' and 'society',[56] and in his influential trilogy *The Rise of the Network Society*, first published in 1996, the sociologist Manual Castells is clear in his view that the network represents a new organizational and economic form:

> A new economy emerged in the last quarter of the twentieth century on a world wide scale. I call it informational, global, and networked to identify its fundamental distinctive features and to emphasize their intertwining . . . It is *networked* because, under the new historical conditions, productivity is generated through and competition is played out in a global network of interaction between business networks.[57]

This 'new economy', Castells argues, has been enabled by the information technology revolution of the last thirty years, incorporating a converging set of technologies around microelectronics, computing (hardware and software), and telecommunications, as well as developments and applications in the emergent field of genetic engineering. Longitudinal research concerning links between organizations in a range of information and computing technology sectors, for example, highlights the emerging density of linkages between firms over time.[58] This is supported by research concerning the shift from 'linear' to 'coupling' modes of innovation, as discussed earlier in this chapter.

Such literature would appear to support the notion that organizational networks represent a new and increasingly important mode of organization.

Networks as a new analytical lens

The second interpretation regards the notion of network not as something new and specific, but as a general theoretical approach to the analysis and explanation of a long-existing phenomenon. Whilst there has long been recognition of the importance of social networks to science and technology,[59] the focus on the role of organizational networks is much more recent. Indeed, it is only since the mid-1990s that this new 'analytical lens' has been applied by business historians across a wide range of business settings to provide new insight.[60]

This research has clearly demonstrated the existence and importance of organizational networks through much of the last two hundred years of industrialization across business sectors as diverse as shipping, textiles, textile engineering, the railways, and banking, in Britain,[61] and textiles and pharmaceuticals[62] in the USA. A number of these studies provide particular evidence of the importance of these networks to innovation.[63] One of the most interesting of these studies is that of vaccine development at Merck, Sharp and Dohme, and Mulford, between 1895 and 1995, which demonstrates the importance of a series of complex and dynamic networks of scientific, governmental, and medical organizations in both the USA and abroad to the innovative activity of these companies.[64]

Illustration 3.3 highlights the extensive business networks among Quakers that played an important role in the industrialization of the UK.

Illustration 3.3 Quaker business networks in the eighteenth and nineteenth centuries

The Religious Society of Friends, commonly referred to as the 'Quakers', are a 'non-conformist' Christian religious denomination founded in the seventeenth century. Despite their small numbers, historians have highlighted the prominent role that they played in the industrialization of the UK during the eighteenth and early nineteenth centuries.

The Quaker movement was characterized by geographically dispersed, extended family-based networks that stretched throughout the UK and even overseas—in particular, to the USA. These networks were strengthened and reinforced through marriage, and were particularly prominent in the UK iron, textiles, and banking industries. The Meeting House system of worship was also an important ingredient, because it '*underwrote both commercial stability and confidence by facilitating capital flows . . .* [and acted] *as an internalised source of commercial support and advice*'.[65]

Prior and Kirby argue that:

> In the absence of institutional capital markets it was the Quakers' ability to exploit geographically dispersed pools of capital which helped create a chain of credit and finance . . . External networks with this degree of sophistication were arguably highly conducive to the formation and subsequent growth of the firm in the formative stages of Britain's industrialisation.[66]

Thus, whilst it may be the case that the new technological, competitive, and global context in which companies operate today is encouraging and facilitating the extension of the scale and scope of organizational networks, there is plenty of evidence to suggest that they represent a return to, or a continuation of, rather than a departure from, historical business practices. Perhaps this is not surprising: after all, as DeBresson and Amesse note, whilst '*Accelerating technological change can increase the importance of networking. Periods of technological turbulence in history are not new*'.[67]

3.3.3 The origins of the network perspective

John Scott argues that the origins of the network perspective may be traced to the structural concerns of the British anthropologist Radcliffe-Brown and the German 'gestalt' tradition of social psychology, principally associated with Kohler. Since this period, in the 1920s and 1930s, a diversity of strands have, at various times, intersected, fused, and once again diverged.

From this complex history, Scott identifies three important origins:

- the 'sociometric' analysts (typified by Moreno), who produced a number of technical advances by adopting the methods of 'graph theory' (a method of graphically representing networks);[68]

- the Harvard researchers of the 1930s who explored patterns of interpersonal relations and the formation of 'cliques'; and

- the University of Manchester social anthropologists of the 1950s and 1960s, such as Barnes and Mitchell.[69]

Over the last forty years, the network perspective has been adopted in a wide array of studies and a diverse range of disciplines have contributed to the conceptual development of the perspective, including sociology, anthropology, geography, political science, and organization theory.[70]

Rogers distinguishes between two main research traditions of network research: 'relational' and 'structural'.[71] Relational analysis evolved out of the Moreno-type network sociometry of the 1930s–50s, while scholarly interest in structural analysis was sparked by the development of 'block-modelling' techniques by Harrison White and others at Harvard University in the mid-1970s.[72]

A key difference between these two traditions is the contrasting emphasis that they place on the importance of the network structure in determining the actions of individual network actors. In the relational approach, the agency (or action) of actors in a network is considered to be shaped, but not determined, by the structure of the network of which they are a part. Furthermore, greater importance is given to the specific individual relationships and specific actors who lie within the network, such that different actors within the same network structure or same network positions might be expected to have different potentials to act.

In contrast, the structural approach focuses on patterns of similarity in networks. Thus, actors with the same set of relationships with the same set of actors are considered to be 'structurally equivalent', whilst those with a similar set of relationships, but with a different set of actors, are considered to be 'role equivalent'. Such 'equivalence' is considered to yield a similar potential to act by actors within such network positions[73]—that is, the agency (or action) of an actor is seen to be, in a large part, determined by the structure of the network in which they are located. But while Alba argues that *'neither approach excludes the other . . . Indeed, they ultimately complement each other'*,[74] Blau suggests that the relational approach is more useful in studying recently formed networks and the structural approach more useful in studying well-established networks.[75]

3.4 Alternative network approaches

3.4.1 Social network analysis (SNA)

In the 1950s, a small group of specialists began to concern themselves with devising more formal translations of the network metaphor, and, by the early 1970s, an avalanche of technical work and specialist analysis techniques had appeared that adopted a highly mathematical orientation.[76] This strand of the network perspective is termed 'social network analysis' (SNA). Network structure is the key focus and concern of this approach. Over the last thirty to forty years, SNA has been developed into a sophisticated set of concepts, tools, and techniques for the systematic collection, analysis, and visual display of networks. Today, the approach is employed across a wide range of disciplines and subject areas, from anthropology to geography, and from sociology to management studies.

Following the recent popularization of the approach by the likes of David Krackhardt, Morten Hansen, and Bob Cross, who have focused on the importance of revealing informal networks within organizations,[77] and Karen Stephenson, who has focused on human resource issues such as managing diversity,[78] SNA is being employed increasingly by organizations and consultants. It would appear that the popularity of the social network approach among practitioners has been bolstered by recent developments in the graphical 'front end' of SNA packages, which greatly aids the presentation of findings, as well as the very practical nature of the issues that it can help to address, such as those relating to knowledge flow, social support and inclusion, and the management of diversity.

3.4.2 Transaction cost economics

The 'transaction cost' approach in economics initially emerged as an attempt to explain why some transactions or exchanges were internalized within firms (often referred to as 'hierarchies'), whilst others occurred within markets. The notion was first conceived in the 1930s by Ronald Coase, but it was not until the seminal work of Oliver Williamson and others during the 1970s that the approach was developed and popularized.

Transaction costs include those concerned with the arranging, monitoring, and enforcing of contracts. Whilst markets are considered to be efficient in enabling simple discrete transactions, they are seen as a poor mechanism for learning or the exchange of technological know-how between organizations, for example. Thus, it is argued that as exchanges within a market become more complex and frequent, and the related transaction costs increase, there are increasing incentives to internalize the transactions, and thus the associated activities, within the organization.

Whilst such internalization enables greater control and accountability, however, it reduces the ability of an organization to be flexible and to adapt in rapidly changing environments. In this context, network forms of exchange and organizing that lie between those of the organization or market are considered to be more suitable.[79] It has been argued that the transaction cost approach is useful in developing an understanding of the presence of different governance structures—i.e. organization, market, and network—in different contexts. But a number of criticisms have also been raised, such as its focus on costs rather than the benefits of alternative forms of organizing, or on the location of capability, which are seen as reducing its explanatory power.[80]

3.4.3 Social capital

We have separated out 'social capital', because even though it draws heavily upon, and is inextricably linked to, the structural concerns of SNA, it is a concept that has drawn interest from a wide range of social science disciplines and extends well beyond SNA. Furthermore, since the mid-1990s, there has been a burgeoning literature within organization studies centred on the significance of social capital as a resource for social action[81]—that is, the manner in which relationships, as an asset, can be mobilized by an individual or an organization to obtain other resources or achieve particular ends.

Two key writers in this area, Janine Nahapiet and Sumantra Ghoshal, have defined social capital as:

> the sum of the actual and potential resources embedded within, available through, and derived from the network of relationships possessed by an individual or social unit . . . [it] comprises both the network and the assets that may be mobilized through that network.

In attempting to model the mechanisms and processes through which social capital is mobilized to create intellectual capital (i.e. knowledge and knowing), they distinguish between three dimensions of social capital:

- '*structural*', referring to the configuration of the network;
- '*cognitive*', referring to the shared meanings and understanding between network members; and
- '*relational*', regarding trust, obligations, and norms, between network members.[82]

It is worth highlighting here that, unlike other forms of asset such as money or goods, social capital is 'located' in the relationships that an individual has rather than within the individual him or herself. Indeed, Ronald Burt of the University of Chicago, a prominent network scholar, has argued: '*No one player has exclusive ownership rights to social capital. If you or your partner in a relationship withdraws, the connection dissolves with whatever social capital it contained.*'[83]

In organization studies, social capital has been used to refer to the network resources of individuals, such as employees and entrepreneurs, as well as collectives of individuals, such as cross-functional teams, organizational subunits, organizations, organizational networks, and even geographical regions. In relation to individuals, social capital has been shown to influence career success and job seeking,[84] for example, whilst at the level of the collective, research has indicated that social capital facilitates the knowledge transfer, and creation of value and intellectual capital, within organizations,[85] and knowledge transfer and the innovative capacity of regions, such as Silicon Valley.[86]

We will return to discuss social capital as an organizational resource in **Chapter 6**.

3.4.4 Communities of practice (CoPs)

A 'community of practice' (CoP) is concerned with the creation and sharing of knowledge and expertise within, or between, organizations. It is defined as a group of individuals who *'share cultural practices reflecting their collective learning'*[87] and who are *'informally bound together by shared expertise and passion for a joint enterprise'*.[88] Examples of such 'joint enterprise' around which CoPs might emerge and coalesce include geologists interested in techniques for locating mineral deposits below the sea, engineers involved in a particular manufacturing technique, or managers concerned with identifying and diffusing best practice. For Wenger, CoPs *'grow out of a convergent interplay of competence and experience that involves the mutual engagement* [of community members]', and are *'the basic building blocks of a social learning system because they are the social "containers" of the competences of that system'*.[89] A focus on identity and forms of belonging, for example, rather than structure, distinguish this approach from SNA.

Whilst such communities might be encouraged and nurtured by management, their *'organic, spontaneous, and informal nature'* makes them *'resistant to supervision and interference'*, and thus difficult to integrate with the formal organizational structures.[90] Thus, CoPs share some of the features of social networks: they too are emergent, dynamic, informal, and difficult to manage, bound, and control. But unlike social networks, CoPs are considered to be predominantly expertise or problem focused, and although they may span organizational boundaries, membership is often drawn from within an organization.

The concept is one that has gained great exposure in the area of knowledge management since its initial articulation in 1991 by Jean Lave and Etienne Wenger,[91] and its subsequent popularization by Wenger from the late 1990s.[92] Despite critiques relating, for example, to an insufficient recognition of the issues around power, trust, and individual dispositions,[93] the concept has gained considerable currency among practitioners, and has been adopted in many innovative and successful organizations, such as Hewlett-Packard, Shell, DaimlerChrysler, and the World Bank.

3.4.5 Actor network theory (ANT)

The late 1980s saw the emergence of two related perspectives within the sociology of science and technology field that offer useful insights into the way in which power and influence may shape the innovation process:

- 'actor network theory' (ANT),[94] which incorporates the concept of 'techno-economic systems' (TENS);[95] and

- the 'social construction of technology'.[96]

Although these two approaches, in common with other network approaches, seek to understand actors and their relationships, their emphasis is 'contextual' rather than 'structural'. Thus, unlike SNA, ANT, for example, is not primarily concerned with mapping interactions between individuals; rather it is concerned with mapping *'the way in which they* [individuals] *define and distribute roles, and mobilize or invent others to play these roles. Such roles may be social, political, technical, or bureaucratic'.*[97] That is, ANT highlights the dynamic and political nature of interactions between actors. Furthermore, Callon argues that *'the development of scientific knowledge and technical systems cannot be understood unless the simultaneous reconstruction of the social contexts of which they form a part is also studied'.*[98] A radical feature of ANT is the equivalent status afforded both human and 'non-human' actors, such as computers and other technological artefacts that act as 'intermediaries' in the interactions between human actors.

Two key concepts underlie the social construction of technology perspective: 'interpretative flexibility' and 'technological frame'. The notion of interpretative flexibility asserts that the social context in which a technology is embedded in its development and diffusion not only dictates the social meaning of that technology, but also its actual technical content—that is:

> technological artifacts are culturally constructed and interpreted . . . By this we mean not only that there is flexibility in how people think of or interpret artifacts but also that there is flexibility in how artifacts are *designed*. There is not just one possible way or one best way of designing an artifact.[99]

A technological frame is articulated as:

> the concepts and techniques employed by a community in its problem solving . . . encompassing within it the recognition of what counts as a problem as well as the strategies available for solving the problems and the requirements a solution has to meet.[100]

Although technological frames strongly influence how individuals within them define problems and solutions, this influence is not considered to be pervasive, because actors will have different degrees of inclusion within a frame and may well be members of more than one technological frame.[101] The 'notion of technological frame' is similar to that of 'technological paradigm' or 'technological regime', which we discuss in more detail in **Chapter 5**.

3.4.6 Alternative orientations within the network perspective

The network metaphor has long been employed as a powerful way of evoking the imagery of the complex web of relationships that characterize our societies and organizations.[102] For Gareth Morgan, who popularized the use of the metaphor as a tool within management studies in the mid-1980s:

metaphor is often just regarded as a device for embellishing discourse, but its significance is much greater than this. For the use of metaphor implies *a way of thinking* and *a way of seeing* that pervade how we understand our world generally.

Morgan also argues that by using a variety of metaphors to understand the complex and paradoxical character of organizational life, '*we are able to manage and design organisations in ways that we may not have thought possible*'.[103]

In stark contrast to this approach is the mathematical orientation of SNA. Developments in SNA over the last thirty or forty years have allowed the network perspective '*to move beyond the metaphorical stage*', providing sophisticated tools and techniques with which to collect, analyse, and display network data.[104] Influential network researchers such as Everett Rogers have, however, warned that '*far too much, I fear, we admire mathematical elegance in our network tools and tool-makers, while largely ignoring what useful objects we can dig up with these tools*'; Barry Wellman, meanwhile, has contended that network analysis should be viewed '*as a broad intellectual approach, and not as a narrow set of methods*'.[105] Indeed, our position is that the range of useful network concepts that have emerged from SNA over the years—such as network density and connectedness, the 'strength of weak ties', and 'structural holes', for example—are arguably the more important contribution to our understanding of networks from this orientation.

The middle ground that lies between these two extremes is occupied by what has been termed 'network mapping'[106]—that is, the use of graphics to represent and display the diversity of actors, links, and flows within a given network. It has been argued by the authors previously that such a graphical orientation '*has the potential to both amplify the imagery of the network metaphor whilst embracing a more systematic and explicit approach to collecting, analysing and presenting relational data*'.[107] Throughout this book you will find examples of network mapping.

A summary of alternative orientations within the network perspective can be found in **Figure 3.5** below.

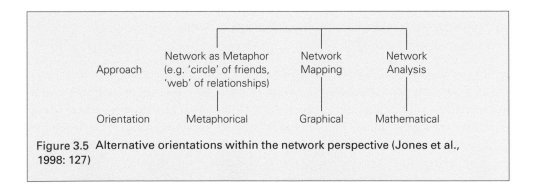

Figure 3.5 Alternative orientations within the network perspective (Jones et al., 1998: 127)

3.5 Key features of relationships and networks

In this section, we identify the key features or dimensions of both individual network relationships and of the overall network in relation to innovative activity. We also highlight the important network roles and network configurations. Examples of paradox and contradiction are evident throughout this discussion.

3.5.1 Key dimensions of network relationships

The relationship or link between two network actors, whether between two individuals or two organizations, may vary along a number of dimensions; the most relevant to our concerns include the following.

The nature of the relationship

A link between two individuals or organizations may, in part, be described and understood in terms of the nature of the 'tie' or 'bond' that maintains the relationship. In relation to individuals, we may distinguish between:

- 'instrumental ties', through which mutually rewarding 'economic' exchanges of information, goods, and money, for example, occur;
- 'affective ties', through which satisfying emotional sentiments, such as friendship, can be evoked; and
- 'moral ties', within which a 'code of fairness', 'social banking', and 'reciprocity' are the main binding forces.[108]

With respect to organizations, the following set of categories are useful:

- 'technical bonds', related to the technological links between two organizations;
- 'knowledge bonds', concerning the knowledge that two organizations have about each other's operations;
- 'social bonds', referring to the personal links and trust between individual members of two organizations;
- 'administrative bonds', related to the interconnectedness of administrative procedures and routines between two organizations; and
- 'legal bonds', in the form of contractual obligations between organizations.[109]

In both of these typologies, the categories are not mutually exclusive, such that a relationship between two individuals, for example, might include elements of instrumentality, friendship, and reciprocity. Furthermore, where more than one type of bond or tie is present, the relationship is likely to be stronger and more enduring. These typologies of tie or bond help us to move beyond what is exchanged between actors, to understanding why relationships are created, nurtured, and mobilized in order to facilitate such exchanges, as well as why they endure over long periods.

Formality

This factor refers to the degree to which a relationship between two individuals or organizations is formally sanctioned. In relation to links between individuals within an organization, for example, a formal link generally refers to those prescribed by management and illustrated through the organizational chart, whilst an informal link would emerge more organically through the development of friendships and acquaintances. With regard to organizational-level linkages, a formal relationship would be one that is enshrined within a contract or formal agreement, whilst informal links would represent those that have emerged between individuals and groups over time, but are not formally sanctioned by management. There has long been evidence that, despite a focus by managers and management academics on formal structures within organizations and formal contracts between organizations, informal relationships frequently operate in parallel, often serving to reinforce, supplement, and complement the formal.[110] In **Chapters 9 and 10**, we will discuss the nature and importance of informal links to the innovation process.

Intensity

The intensity of a relationship is indicated by the frequency of interaction and the number of transactions between two individuals or organizations over a given time period. Stronger links are often associated with higher levels of interaction and exchange.[111] But certain types of link, such as kinship ties, are intrinsically strong and therefore do not require the same intensity of interaction to be maintained over long periods of time.[112]

In their synthesis of a wide range of social network research, Rogers and Kincaid concluded that '*individuals tend to be linked to others who are close to them in physical distance and who are relatively homophilous* [i.e. similar] *in social characteristics*'; this is known as the 'homophily principle'.[113] Following on from the work of Zipf in the 1940s, who argued more generally that all human behaviour is guided by the principle of minimizing total action to achieve certain goals,[114] Rogers and Kincaid argue that '*both spatial and social proximity can be interpreted as indicators of* least-effort'.[115] Furthermore, repeated interactions over time are likely to lead to increasingly shared norms, beliefs, and knowledge bases.[116] As a result, linkages characterized by frequent interaction are less likely to yield novelty. A longitudinal study of social capital and knowledge creation among a large sample of scientists, for example, found that whilst relationship intensity had a positive impact on knowledge creation in the short term, this became negative in the longer term.[117]

Symmetry and reciprocity

This dimension refers to the balance of transactions or exchanges between two individuals or organizations over time. The relationship is considered to be 'asymmetric' where the flow is unbalanced (i.e. largely one-way), and 'symmetric' where the flow is balanced (i.e. two-way). The notion of 'reciprocity' is frequently used to refer to the degree of symmetry in a given relationship. Asymmetric linkages—that is, those with low reciprocity—tend to weaken over time, as the non-reciprocated actor becomes disinterested or disillusioned with the relationship. But asymmetric relationships may persist where they are characterized by an inequality in the power relations between the two actors.

In contrast, symmetric linkages are more likely to endure, in part, through the building of trust.[118]

Multiplexity

This signifies the degree to which two individuals or organizations are linked by multiple role relations. For example, two individuals might be connected as work colleagues, friends, football fans, and neighbours. It is argued that the greater the number of role relations linking two actors, the stronger the linkage, and that:

> there is a tendency for single-stranded relations to become many-stranded if they persist over time, and for many-stranded relations to be stronger than single-stranded ones, in the sense that one strand role reinforces others.[119]

In relation to organizational relationships, such multiplexity might manifest itself through the emergence of social links between individuals in two different firms that arise from them working together under a formal alliance or joint project. Formal interactions between organizations can lead to the formation of informal ties, such as those identified by Håkansson in his study of the technology-related interactions between 130 Swedish companies. In this research, he found that *'people on both sides* [of a link between two firms] *can gradually build up confidence and trust in one another, and in this way an important social element enters the interactions'*.[120]

From their study of informal collaboration in the Dutch biotechnology industry, Kreiner and Schultz, for example, note the importance of serendipity in the emergence of personal relationships. Such informal relationships between individuals may subsequently lead to formal linkages between organizations, through the establishment of joint R&D projects.[121] Thus, over time, formal organizational-level links may arise from informal individual-level relationships, and vice versa.

Trust

An important element of relationships is trust. Trust is central to the maintenance and development of relationships, and to the sharing of information, knowledge, and other resources between actors. Trust may be viewed as 'a state of mind' regarding the expectation that 'the other' will act reliably and fairly, and exhibit goodwill when unforeseen circumstances arise.[122]

Within social networks, trust is higher where the ties between individuals are strong rather than weak and is reinforced in dense networks where an individual has more to lose from a reputation of low trustworthiness.[123] Higher trust is associated with a greater willingness to share and disclose information and knowledge.[124] In relation to trust within CoPs, Wenger argues that:

> People must know each other well enough to know how to interact productively and who to call for help or advice. They must trust each other, not just personally, but also in their ability to contribute to the enterprize of the community, so they feel comfortable addressing real problems together and speaking truthfully.[125]

Research has also highlighted the importance of trust for inter-organizational relationships.[126] Trust between organizations has been defined and categorized in various ways.[127] But the well-cited typology of Sako provides a useful distinction between:

- 'contractual' trust—that is, the expectation that a trading partner will adhere to agreements and promises;
- 'competence' trust, concerning the expectation that a partner will perform its role competently; and

- 'goodwill' trust, referring to mutual expectations of open commitment between trading partners.[128]

For Dodgson, high trust between two organizations facilitates the communication of tacit, commercially sensitive, and proprietary information and knowledge, which he argues is necessary for the generation of learning and innovation.[129]

Recent research has found, however, that in uncertain environments, high trust between partners can be a disadvantage, because it generates overconfidence in the information received from others, whilst lowering vigilance in environmental scanning.[130] As such, whilst high trust and repeated partnerships between two organizations

Table 3.1 A summary of the key features of relationships

Feature	Definition	Relevance to innovation
Nature of relationship	The types of 'tie' or 'bond' that help to explain the creation and persistence of a relationship between two individuals or organizations	Understanding the underlying logic of a relationship helps individuals and organizations to appreciate the risks associated with relying on 'the other', and of exchanging sensitive or valuable information and knowledge
Formality	The degree of formalization of the relationship between two individuals or organizations	Whilst formal relationships help to provide a framework for interaction, informal relationships 'breath life' into the formal linkage and are important for transferring tacit knowledge
Intensity	Concerns the frequency of interaction and exchange between two individuals or organizations	Although frequent interaction promotes strong relationships, they can also lead to the emergence of a common belief and value system that may undermine the novelty in the exchanges between two actors
Symmetry/ reciprocity	The degree to which the exchanges within a relationship are reciprocated	Reciprocity is important for building linkages, and for facilitating the exchange of information and knowledge
Multiplexity	Multiplex relationships are those that are based on more than one type of 'tie' and/or role-relation between two individuals or organizations	Multiplex relationships promote the persistence of linkages and help in the building of trust—both of which are important for stability and communication
Trust	Trust is the expectation in a relationship that 'the other' will act reliably, fairly, and exhibit goodwill	Higher trust is associated with a greater willingness to share and disclose information and knowledge, especially if it is sensitive or of value

can yield efficiency benefits at the level of the alliance, it can negatively impact perform-ance at the level of the organization.[131]

3.5.2 Key dimensions of networks

The overall network may also vary along a number of dimensions; the most relevant to our concerns include the following.

Size

This dimension simply refers to the number of actors participating in a network, although in network studies, this is often influenced by some arbitrary boundary set by the researcher.[132] Intuitively, it may seem that membership of a large network rather than a small network is better for exposure to novel information. But network size cannot be considered in isolation from dimensions such as network density and network diversity, or network configuration, which we will discuss subsequently. For example, large, high-density, low-diversity networks are likely to yield less diverse or novel information and high opportunity costs associated with the maintenance of high numbers of relationships with low information benefits, than a similar-sized network with low density and high diversity.[133]

Density

The density of a network refers to the number of actual linkages in a network as a pro-portion of the total number of possible linkages (i.e. if every actor in the network were to be connected to each other).[134] Boissevain warns against misinterpretations of density measurements, however, stressing that *'networks with the same density can have very differ-ent configurations'* and that *'network density is simply an index of the potential not of the actual flow of information'*—that is, it is a measure of network structure, not of networking activity nor of the exchange of information.[135] High-density networks include many 'redundant' links—that is, links that *'lead to the same people, and so provide the same information benefits'*;[136] such networks are useful for the rapid diffusion of information and knowledge around a network, but less useful for the sourcing of novel information and knowledge.

Network graphics are a powerful way of identifying high-density and low-density areas in a network, as well as alternative configurations. Turn briefly to **Figure 3.9**, later in this chapter. In this figure, the shapes represent organizations and the lines, alliances between them. Viewing the top drawing, we immediately see a dense set of relationships between a subset of organizations to the right, with the remaining organizations only very loosely connected—that is, exhibiting low density. In contrast, in the lower drawing, we observe two dense networks—one to the left and one to the right—with few organizations not connected to one of these two densely connected groupings.

Reachability

A dimension closely related to density is that of reachability, which is defined as the num-ber of links separating two actors in a network,[137] or more generally, the average number of links between any two actors in a given network.[138] The denser the network, the lower the average number of links between two actors will be. Recent research concerning the relationship between knowledge complexity, ease of transfer, and the number of links between knowledge source and recipient (i.e. distance) has found that whilst simple knowledge travels equally well through either long or short distances, very complex knowledge 'resists diffusion' over even short distances in the network. Dense networks, with many redundant links and hence shorter average distances between individuals than low-density networks, are, however, advantageous for the transfer of knowledge of moderate complexity.[139]

Diversity

This network dimension refers to the diversity of the actors within a network. With regard to networks of individuals, this might relate to one or more characteristics, such as profession, age, gender, or education, for example. With respect to networks of organizations, this might relate to characteristics such as business sector, size, or type of organization (e.g. private sector company, public sector body, or pressure group). High diversity in a network increases the likelihood of the presence of different information, knowledge, and perspectives, all of which raise the possibility of the emergence of novelty and innovation. Furthermore, research has indicated the importance of team diversity for accessing a diverse array of external networks and perspectives,[140] although this can be constrained by the diversity of the organizations or environment in which they are embedded.[141]

Openness

This dimension refers to the degree to which the actors within a network are connected to actors in other groups and networks. Closed networks are those with a group of actors who interact predominantly with each other and, as such, are characterized by the presence of many strong ties and few weak ties. They are sometimes referred to as 'interlocking' networks, because of the overlapping and interlocking nature of the links of the network members.

In contrast, open networks embrace both strong ties between a core group of actors, as well as a set of weak ties with a subset of individuals from that core group and other groups. Such networks are sometimes referred to as 'radial' networks, because their links radiate out to other groups.

Importantly, openness is often associated with creativity and innovativeness,[142] whilst closed networks 'may simply facilitate the pooling of ignorance among the individual members'.[143]

Stability

Many network studies represent a snapshot of a network at a particular point in time—but networks are not static. Indeed, many networks are highly dynamic in relation to their size, membership, and density, for example. Good examples of such network instability or dynamism are provided by Hagedoorn and Schakenraad, in their longitudinal study of strategic alliances and joint ventures in a number of information technology (IT) sectors (again see **Figure 3.9**), and by Steward and Conway, in their longitudinal study of three technology diffusion projects funded by the European Union.[144] In contrast, recent research has also revealed factors that might create inertia in organizational networks, such as the desire of network members to limit the disruption to established structures and routines that arise from changing partnerships and ties.[145]

In a study by Steward and Conway, three clear network stages were identified—network formation, network development, and network extension—at which changes in network size, diversity, and configuration were observed as the projects matured. Other research has suggested that networks respond in relation to changes in environmental uncertainty and resource availability; in periods of uncertainty, but improved resource availability, network size and diversity are likely to increase as organizations seek new alliances to source critical resources to compete in the new environment, as well as to combat new entrants attracted by the availability of resource. In contrast, periods of greater certainty, but reduced resource availability, are likely to be accompanied by a decrease in network size and diversity as organizations respond to greater clarity and reduced opportunities in the environment.[146]

Table 3.2 A summary of the key features of networks

Feature	Definition	Relevance to innovation
Size	The number of individuals or organizations in a network	Other things being equal, a large network is better than a small network for exposing members to novelty
Density	The number of linkages between the various members of a network as a proportion of the total number of possible linkages	Dense networks facilitate the rapid diffusion of knowledge and aid exploitation, but are much less useful for sourcing novel knowledge
Reachability	The average number of linkages between any two members in the network	There is an inverse relationship between knowledge 'complexity' and the links through which it can be diffused
Diversity	The diversity in the membership of a network. This might be in relation to age or gender for individuals, or size or sector for organizations.	High diversity increases the likelihood of the presence of novel information, knowledge, and perspectives, which all raise the potential for novelty and innovation
Openness	The degree to which individuals or organizations in the network are connected to those in other networks	Open networks aid exploration and creativity, whilst closed networks can lead to 'the pooling of ignorance'
Stability	The degree to which the size, membership, diversity, or density of a network is stable over time	Stable networks promote exploitation, but, in the long term, can lead to a decline in novelty and innovation

3.5.3 Key types of network configuration

As we noted above, general measures of the overall network, such as size, density, and diversity, may conceal important variations in the actual configuration within two seemingly similar networks. Thus, in addition to observing such network features, it is necessary to assess a network in relation to the number, size, and location of 'clusters' and 'structural holes', the number and positioning of 'weak ties' and 'strong ties', and 'embeddedness'. We now turn to these network configurations.

Clusters and cliques

The densely interconnected regions in networks are often termed 'clusters'. In addition to prescribed clusters, such as work groups and committees, two types of emergent cluster are often distinguished: 'coalitions', viewed as temporary, narrowly based alliances, and 'cliques', generally longer lived and pursuing a broader range of purposes.[147]

Warner and Lunt define a clique as *'an informal association of people among whom there is a degree of group feeling and intimacy and in which certain group norms of behaviour have been established'.*[148]

'Strength of weak ties' versus 'strength of strong ties'

The 'strength of weak ties' concept, advanced by Mark Granovetter in 1973, has been one of the most cited in network research over the last thirty years. In this concept, the 'strength' refers to the novelty of the information that we receive from those we have 'weak ties'—that is, those with whom we are acquainted, but with whom we interact relatively infrequently and who are not otherwise connected to our network. In contrast, those with whom we have 'strong ties' are likely to be highly connected to each other and thus information shared in this dense network is less likely to be novel.[149]

Although originally concerning job seeking, the concept has been widely adopted in innovation studies regarding, for example, developing innovation,[150] knowledge sharing,[151] and the diffusion of innovation.[152] In much of this work, 'weak ties' are viewed as a valuable source of novelty and innovation. But research has also indicated that those with whom we have 'strong ties' are more likely to be motivated to assist and more likely to be easily available,[153] as well as being much more likely to provide socio-emotional support and to be trusted.[154] Furthermore, research indicates that low-conflict organizations are characterized by higher numbers of inter-group strong ties, because such ties help to mitigate against disruptive conflict.[155] This dilemma is captured in Hansen's conclusion to his study of intra-organizational knowledge search and transfer: '. . . *a strong tie will constrain search, whereas a weak tie will hamper the transfer of complex knowledge. Thus there appears to be no obvious design solution to the search-transfer problem.'*[156]

Structural holes

One of the most influential scholars in network theory over the last decade or so has been Ronald Burt. He is perhaps best known for his book *Structural Holes: The Social Structure of Competition*, first published in 1992. Burt shifts the emphasis away from the strength of ties, as discussed above, to focus on the 'holes' within networks and the bridges that span these. Here, a 'structural hole' refers to the absence of links between two clusters, or cliques. Potential information and control benefits, as well as entrepreneurial opportunities, are considered to accrue to those individuals or organizations that are able to build the connection or bridge between two such clusters or cliques, because they are able to act as brokers between them.[157]

Embeddedness

The notion of 'embeddedness', as presented by Mark Granovetter, refers to the overlap between work-related transactions and social relations. Organizational activity is said to be embedded in social networks to the extent that transactions within and between organizations are shaped by friendship or kinship ties, rather than the '*economic logic of the market*'.[158] Highly embedded organizations can accrue benefits in relation to organizational survival, learning, risk-sharing, and the speed of bringing new products and services to market. It is also possible, however, for organizations to be over-embedded, such that they face decreasing opportunities and inertia in the face of changing markets and technologies. Research indicates that differences in national culture regarding, for example, the willingness to cooperate and to trust 'outsiders' can greatly impact the degree of embeddedness, leading some economies, such as those in Japan and South Korea, to be deeply embedded.[159]

Table 3.3 A summary of the key types of network configuration

Feature	Definition	Relevance to innovation
Clusters and cliques	The densely connected regions in a network	Clusters are useful for facilitating the rapid diffusion of knowledge and exploitation. Their innovativeness depends on whether they are 'open'
Strength of weak ties/strength of strong ties	'Weak' ties' refer to linkages between individuals or organizations where there is low overlap in their networks	Weak ties are associated with novelty, whilst strong ties are associated with trust and aiding knowledge transfer
Structural holes	A 'structural hole' is present where there is an absence of linkages between two networks	Structural holes present opportunities for entrepreneurial and innovative activity
Embeddedness	Relates to the degree to which organizational relationships are reinforced by social relationships	Highly embedded organizations can accrue benefits in relation to survival, learning, risk sharing, and the speed of bringing an innovation to market

3.5.4 Key network roles and positions

'Boundary-spanners', 'gatekeepers', and 'brokers'

The terms 'boundary-spanner' and 'gatekeeper' are often used interchangeably, and refer to individuals whose relationships span boundaries, often both those within and beyond the organization. Through their social networks, such individuals have the potential to facilitate communication and integration between projects, functions, and divisions, and to provide access to novel information, solutions, and ideas from outside the organizational unit. In effect, they provide 'bridges' between different networks or groups of individuals. To span boundaries successfully, however, individuals must be capable of understanding and translating the 'languages' and perspectives present within different domains, and of acting as intermediaries between, for example, the R&D, marketing, and accounting functions, or between product developers, marketers, and product users. Gatekeepers are also associated with other roles, such as 'information filter' to prevent information overload within the communication channels of the network, and 'uncertainty absorber'—that is, drawing and communicating inferences where there exists ambiguity and uncertainty in information received.[160] Given the importance of boundary-spanners and gatekeepers to innovation and knowledge sharing, we will return to discuss these roles in more detail in **3.6.1** and in **9.5.2**.

The 'broker' role is associated with individuals who have relationships that span structural holes and are thus able to act as intermediaries between two or more networks or groups.[161] In a study of ideas and brokerage, Burt found that the 'best' ideas generated by a group of managers within an organization arose from those with the least 'constrained' networks—i.e. those with larger, open networks—whilst the 'worst' ideas arose from those with the most 'constrained' networks—i.e. those with small, closed, and dense networks. In **Figure 3.6**, the x-axis (horizontal) represents network constraint, with

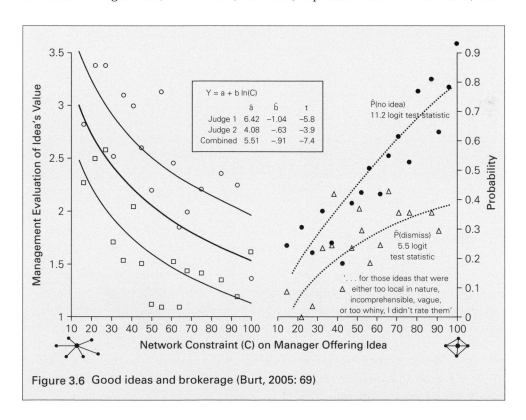

Figure 3.6 Good ideas and brokerage (Burt, 2005: 69)

low constraint to the left and high constraint to the right. The y-axis (vertical) represents the 'value' of the idea, from 'high' or 'good' at the top, down to the 'low' or 'poor' at the bottom.[162]

Stars

In the network literature, a 'star' is defined as an individual with a high number of links to others in a given network. The term 'star' is employed because, in graphical depictions of networks, such actors are easily identifiable by the many links radiating from them.[163] Such individuals can play an important role in diffusing information rapidly throughout a community or organization. Gatekeepers are often stars.

Isolates

In sharp contrast to stars, an 'isolate' is an actor who has no relationships with others in the network—that is, an actor who is 'uncoupled from the network'. Individuals may become isolates for many reasons and it is not necessarily a measure of their potential or value.[164]

Table 3.4 A summary of the key network roles and positions

Feature	Definition	Relevance to innovation
Boundary-spanners/ gatekeepers/brokers	Individuals who, due to their linkages, provide bridges between networks	The boundary-spanning and brokering activity of these individuals is vital to the innovation process
Stars	Individuals who are highly connected within a network	These individuals can play an important role in diffusing information
Isolates	Individuals who are disconnected within a network	Such disconnected individuals are unable to contribute their ideas

3.6 Innovation networks

In the preceding sections of this chapter, our discussion has focused on outlining the important dimensions of relationships and networks, and identifying key network roles and configurations. During this discussion, we highlighted the impact of these individual dimensions and features on novelty and knowledge transfer. In the remaining sections of this chapter, we shift our focus to providing an overview of the range of empirical studies of social and organizational networks in the innovation process, and highlight some of the key findings and implications for the management of innovation. We start by summarizing this work in **Table 3.5**.

Table 3.5 A sample of innovation studies adopting a network perspective

	Focus on specific innovative activity (e.g. speed and innovativeness of projects)	Focus on general innovative activity (e.g. communication and knowledge transfer)
Organizational networks	• Håkansson (1987)—product development in drill machinery in a Swedish industrial network • Steward and Conway (1998)—network dynamics in innovation diffusion	• Hagedoorn and Schakenraad (1992)—network configuration and dynamics in a range of ICT sectors • Roijakkers and Hagedoorn (2006)—network configuration and dynamics in pharmaceutical biotechnology
Combination of organizational and social networks	• Law and Callon (1992)—an ANT study of a British aerospace project • Galambos and Sewell (1995)—a longitudinal study of social and organizational networks in vaccine development at three US firms	• Saxenian (1994)—a comparison of the social and organizational networks in Silicon Valley and Route 128 • Castilla et al. (2000) and Assimakopolous et al. (2003)—emergence and role of social and organizational networks in Silicon Valley
Social networks within and between organizations	• Conway (1995)—informal sources of ideas and solutions in a sample of successful UK innovations • Blundel (2006)—a longitudinal study of radical innovation in the design and manufacture of dinghies in the UK • Jones and Conway (2004)—social networks and the case of James Dyson in the UK	• Crane (1972)—social organization in scientific communities • Kreiner and Schultz (1993)—informal collaboration in Danish biotechnology sector • Perry-Smith (2006)—social networks and their impact on individual creativity • Aldrich and Zimmer (1986), and Johannisson (2000)—social networks and entrepreneurship
Social networks within organizations	• Hansen (1999)—the role of weak ties in knowledge search and transfer between subunits of a large electronics firm and impact on speed of project completion • Hansen et al. (2001)—the impact of network structure on speed of completion of exploratory and exploitation-tasked projects	• Allen (1977)—communication patterns among engineers for the transfer of technological information and technology in a R&D organization • Krackhardt and Hanson (1993)—the nature and role of social networks • Tsai (2002)—knowledge sharing within a multiunit organization

New ideas seldom appear fully formed and articulated from a single source, and thus it is perhaps not surprising that innovation generally arises from a portfolio or network of actors and relationships. Studies of successful technological innovation, such as those in **Table 3.5**, have long highlighted the importance of a number of characteristics of such 'innovation networks':

- the importance of an open, rather than a closed, network configuration;
- the presence of bridges and boundary-spanning activity;
- a diversity of internal and external actors involved in the development process; and
- the prevalence of informal or personal relationships in supplementing and 'breathing life' into formal relationships (i.e. the organizational chart) and links at the level of the organization (e.g. joint development projects).

3.6.1 Social networks and 'invisible colleges'

Drawing principally from the studies listed in **Table 3.5** and, in particular, the seminal studies noted in the introduction of this chapter, in this section, we seek to outline some of the key insights and management implications from research on social networks in relation to innovation and knowledge sharing. We will build on this discussion in subsequent chapters and, in particular, in **Chapter 9**, concerning social networks and informality in the innovation process, **Chapter 10**, with regard to the external sources of innovation, and **Chapter 11**, in relation to systems of innovation.

Social networks, informal communication, and boundary-spanning

During the 1970s, work began to emerge concerning the social networks of engineers. A key academic in this area was Thomas Allen, whose research programme at the MIT Sloan School of Management and his book *Managing the Flow of Technology: Technology Transfer and the Dissemination of Technological Information within the R&D Organization* (published in 1977) were particularly influential.

An important management implication that arose from this research was the role of informal networks. Individuals working in a research laboratory were asked to identify social interactions with colleagues, as well as technical communication interactions with them. It was then possible to map out the two types of network and explore their relationship with the formal organizational structure.

The findings were interesting in a number of ways. Firstly, there seemed to be a close relationship between the networks of social interaction and technical communication, suggesting that personal links were associated with work-related activities. Secondly, the overlap between these networks and the formal organization varied markedly in different settings. In some cases, the networks reflected the formal organizational boundaries, whilst in other cases, they transcended these formal boundaries.

An example of a communication network from this research is shown in **Figure 3.7**. In this network map, the circles represent engineers, the solid lines represent information flow, the dotted lines represent social interaction, and, finally, the straight lines separate the engineers from different sections within the laboratory. A closer look at the figure reveals that the majority of the communication occurs within, rather than between, the different sections of the laboratory, with only a small number of boundary-spanning interactions, which interestingly, are largely social interactions.

From this research, Allen concluded that:

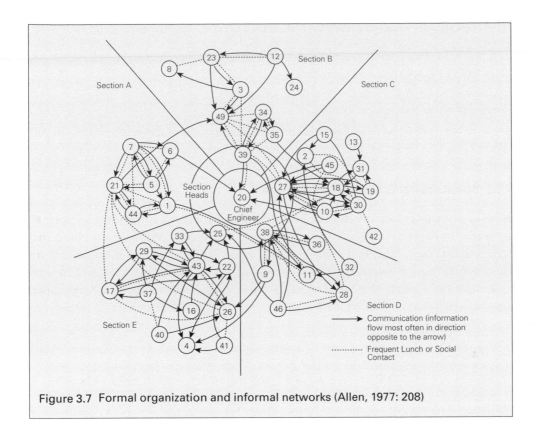

Figure 3.7 Formal organization and informal networks (Allen, 1977: 208)

although formal organization may be the more important of the two determinants of communication, informal organization makes its own independent contribution of nearly equal magnitude.[165]

Indeed, subsequent studies relating to successful innovation,[166] informal learning,[167] and knowledge sharing and creation,[168] have underlined the importance of this conclusion. This has important implications for the management of innovation. While accepting that managers 'cannot dictate friendships', however, Allen argued that they can create the conditions that facilitate the development of informal contacts through organizational and spatial arrangements. We will discuss the relationship between informal networks and organizational structure, culture, and the physical work environment in **Chapter 7**.

In our earlier discussion of the interactive model of the innovation process (see **3.2.2**), we highlighted the importance of interaction across boundaries both within and beyond the organization. Studies have long indicated the importance to successful innovation of building and mobilizing such 'boundary-spanning' relationships between project groups, functional departments, and divisions. In particular, research has highlighted the importance of interactions that span the internal marketing and R&D interface,[169] and informal relationships that span the organizational boundary, connecting the firm to a diverse range of external sources during the innovation process, such as users and consumers, competitors, research organizations, suppliers, and distributors.[170] Informal boundary-spanning interaction is also seen as particularly valuable to the transfer of tacit and complex knowledge.[171]

Gatekeepers and gatekeeper networks

Another key finding revealed by Allen through his mapping of the communication patterns of engineers was that the degree of interaction varied substantially among different staff, ranging from highly connected 'stars' to unconnected 'isolates'. Furthermore, it was found that many of the engineers obtained information from outside of the organization indirectly through such network 'stars' acting as network intermediaries. Allen labelled these individuals as technological 'gatekeepers'. These 'gatekeepers' combined a range of network links outside the organization with a set of extensive internal connections and were found to be an important route for bringing new information into the organization. Deeper investigation showed that these individuals were often strongly connected with each other through a highly efficient and effective internal communication 'gatekeeper' network, such that:

> new technology generally enters an organization through one of the members of the [gatekeeper network], is readily disseminated among them, and then spreads to the remainder of the organization through contacts that they have with colleagues.[172]

A 'gatekeeper' network is illustrated in **Figure 3.8**. Here, the circles represent the gatekeepers, the squares indicate others in the organization, and the solid lines represent the flow of externally sourced information around the organization.

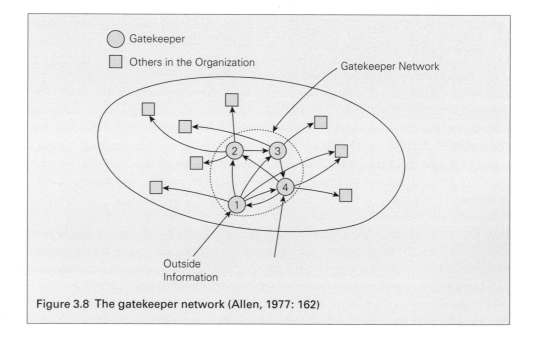

Figure 3.8 The gatekeeper network (Allen, 1977: 162)

The 'gatekeeper' was therefore identified as a key role in the activities of R&D and, by implication, in the innovation process. Allen was struck by the fact that this aspect of the organizational communication network had *developed spontaneously with no managerial intervention*. Nevertheless, given the 'vital role' of the gatekeeper, he felt that there was both a need for management to be aware of their value and a case for the identification of gatekeepers in order *'to reward them, organize information dissemination around them, and if they are to be lost through promotion, replace them'*, whilst warning against the 'formalization' of the role as 'unnecessary' or even 'undesirable'.[173]

Social networks and knowledge sharing across large organizations

A particular problem for large, divisionalized organizations is the sharing of knowledge between its various subunits. Recent research has found, however, that informal communication via social networks can provide an important conduit for knowledge search and knowledge transfer between divisions.[174] To this should be added a number of mitigating factors, such as the matching of ties strength and knowledge complexity, the degree of centralization, and the perceived competition between organizational subunits.

Morten Hansen, in an interesting study of 120 new product development (NPD) projects within 41 divisions of a large electronics company, found that whilst weak ties can help project teams to search for useful knowledge from other parts of an organization, where knowledge is complex (i.e. uncodified, tacit, and interrelated to other knowledge), they can impede knowledge transfer and negatively impact the completion time of projects. In his study of knowledge sharing within a large multiunit organization, Tsai found that centralization reduced the willingness to share knowledge between subunits.[175] Findings concerning the impact of perceived competition between subunits (both in relation to market competition and for internal resources) are mixed, with evidence of both knowledge sharing[176] and knowledge hoarding.[177]

Extended international assignments by researchers have traditionally been employed by research-intensive multinational firms as a way of encouraging knowledge transfer and social networks between subunits. But the cost of such assignments, and the disruption to family life of relocation, have meant that such assignments are being increasingly replaced by short-term visits. Whether such visits are as conducive to knowledge transfer and the building of strong relationships is still to be seen.[178]

Social networks and knowledge sharing between organizations

Seminal research concerning the social networks of scientists can be traced back to the 1960s. Of particular interest is the work of Derek DeSolla Price and his book *Little Science, Big Science*, published in 1963, and that of Diana Crane and her classic text *Invisible Colleges: Diffusion of Knowledge in Scientific Communities*, published in 1972. This research revealed the crucial role of organizational boundary-spanning social networks and informal exchange to the science information system. Indeed, for university scientists in particular, it was contended that such social networks were a far more important social system than that located within the organization that employed them.

Subsequent studies have reinforced these findings. Two types of social grouping were identified: 'solidarity groups' and 'invisible colleges'. A solidarity group consists of a group of researchers under the leadership of one or two highly productive scientists. Such groups are found within a typical research laboratory. The 'invisible college', meanwhile, is the apex or elite of a scientific community who communicate with each other and transmit information informally across the whole field. In this way, the invisible college forms a communication network that links the various solidarity groups within a given speciality in much the same way as the gatekeeper network operates within organizations.[179]

In contrast, research during the 1970s by Allen and others, concerning the interactions of engineers and technologists, found quite a different pattern of communication. Typically, interactions were contained within the boundary of the organization. Allen has termed this behaviour 'enforced localism'.[180] This is perhaps not surprising given that organizations from the private sector might be expected to be more protective of their intellectual capital, because it is viewed as a key ingredient of competitive advantage. More recent research across a number of industrial sectors has, however, found increasing evidence of social networks and informal exchange among engineers and technologists across the organizational boundary.[181] Such social networks are built through the career histories of individuals whose paths may have crossed during their education, past research projects, conferences, or previous employment, for example.

Personal networks appear to be especially important for the transfer of tacit knowledge[182] and '*superior conduits for knowledge flow*' between geographically distant locations.[183] Freeman argues that '*behind every formal network, giving it the breath of life, are usually various informal networks*'.[184] This is perhaps best exemplified by the case of Silicon Valley. Here, a number of in-depth studies have highlighted the importance of extensive friendship networks and high job mobility to the '*free-wheeling information exchange*' between engineers in competitor firms within Silicon Valley. This is largely tolerated because the benefits are seen to outweigh the disbenefits of constraining exchange.[185]

Although far more limited in scope, there is also emerging evidence that such informal boundary-spanning social networks are important for knowledge transfer, learning, and new service development (NSD) in a range of knowledge-intensive service industries, such as banking and professional services.[186]

Social networks and knowledge sharing—overlaying national culture

Much of the research on social networks and knowledge sharing is based on Anglo-American experience; there is, however, an emerging body of empirical work focused on knowledge sharing in a range of other countries, such as China, Taiwan, Japan, and Russia. In contrast, to the UK, USA, and Australia, for example, these countries are characterized by 'collectivism'—that is, the norm is for group goals to have priority over individual goals and there is an emphasis on social relations over rules. These cultural dimensions are considered to lead to intensive social relationships (known as '*guanxi*' in China) and knowledge sharing within 'in-groups' present in organizations.[187] Comparative research between US and Chinese managers, for example, has shown that the willingness to share is significantly higher among Chinese in circumstances under which such sharing would benefit the company, but potentially damage the self-interests of the sharer.[188] Similar results were found in a comparative study between Australian and

Taiwanese managers, respectively.[189] Research has indicated, however, that knowledge sharing beyond 'in-groups' within Russian organizations is constrained by a number of factors, such as the fear of making mistakes, a lack of trust and exclusion of 'out-group' members, and a tendency for individuals to work with those with whom they are familiar, leading both to a strong reticence to share knowledge widely and a suspicion of knowledge from 'outsiders'.[190]

3.6.2 Industrial and sectoral networks

Since the early 1990s, there has been a great deal of interest and research focused on the incidence of R&D partnering and alliances between firms. Mapped out collectively, often between large numbers of organizations, these represent increasingly complex networks of innovation within a particular sector or region—and, as we have noted above, these organizational networks are often overlaid by extensive social networks. It is, however, important to note that the pattern and nature of partnering can vary considerably between sectors, regions, and nations, as well as over time. For example, national differences in historical socio-economic preferences, regulation, economic policies, and existing inter-firm relationships can impact the propensity or willingness of organizations to establish new partnerships.[191]

In his analysis of R&D partnerships since 1960, Hagedoorn identified a number of important trends, as follows.[192]

- There has been a clear pattern of growth of R&D partnering since the 1960s;
- This growth has been accompanied by a notable shift since the mid-1970s from equity-based partnerships (e.g. joint ventures) to non-equity contractual forms of partnership (e.g. joint development agreements), which are seen as more suited to periods of technological instability and uncertainty;
- Since the 1980s, an increasingly larger proportion of R&D partnerships have occurred in high-technology sectors, and, in particular, within the biotechnology and information and communication technology (ICT) sectors, now accounting for over 80 per cent of the total; and
- The patterns of international R&D partnering are a bit more mixed, although, rather unsurprisingly, they have been dominated by links between the highly developed economies of Europe, North America (including Canada), Japan, and South Korea, although this trend is likely to change dramatically with the emergence of the Chinese and Indian economies, for example. Perhaps more surprising is the lower prominence of high-technology sectors and the higher prominence of medium-technology sectors, such as chemicals, in international partnering.

Although the above analysis by Hagedoorn focuses on technology sectors, such as biotechnology and ICT, there is also emerging evidence from recent research that there has been a similar trend toward partnerships and networking in a variety of knowledge-intensive service sectors, such as banking, and capital-intensive service sectors, such as hotels.[193]

How might this rise in organizational networks be explained? Such networks appear to thrive best when expertise is widely dispersed, and the knowledge base is complex and expanding.[194] This would certainly be typical of the early stages of the life cycle of a new technology, for example. Thus, their utility may, to some extent, be contingent upon context. Networks are viewed as being 'light on their feet' and able to adapt relatively rapidly to changes in technology and market, enabling network members to reposition themselves more speedily. They are particularly adept at bringing together distributed resources, knowledge, and competences in novel ways. Networks also enable network members to share risk, spread costs, shorten development cycles, and enter new markets more rapidly.[195]

But there is also some evidence that not all networks provide equal benefits to all of their members and that some networks are more successful than others. Baum et al., for example, found that the composition of a network explained differences in the performance of a sample of Canadian start-up firms in the biotechnology sector.[196] Other research has indicated that a minority of network members might dominate the network and appropriate the greatest benefits.[197]

In subsequent sections, we distinguish between 'industrial networks', 'sectoral networks', and 'regional clusters'. Broadly speaking, research concerning sectoral networks and regional clusters adopt a socio-centric approach (i.e. they focus on the set of links between a group of organizations), whilst much of the research concerning industrial networks has adopted an action-set approach (i.e. their focus is limited to the set of links and actors associated with the development of a particular innovation). In addition, for regional clusters, in contrast to sectoral and industrial networks, geographical proximity is a key feature of network members. Although there are clearly overlaps in these three types of organizational network, this classification allows us to reveal similarities and differences between each.

Industrial networks

The notion of 'industrial networks' is largely associated with the Uppsala School of Sweden, as represented by researchers such as Håkan Håkansson, which emerged from a series of studies during the 1970s and 1980s concerning technological development and collaboration among Swedish companies. The research programme at the University of Uppsala played a key role in shifting the emphasis away from the prevailing focus on internal resources and activities in technological development, to one that recognized the importance of collaboration with a pluralistic set of external actors, such as customers, suppliers, competitors, and universities. The maxim that captured the essence of the approach and guided the research programme was '*No business is an island*'—that is, all companies are closely interconnected with other companies and organizations.

In support of this new research orientation, Håkansson argued at the time that:

> This is easy to accept and understand when looking at a real company and its problems. But it is not reflected at all in the accepted theoretical view of the company . . . [which] is usually seen as a unit with clear boundaries . . . The question is not how the company manages its technological development *per se* but how it manages to relate its technological development to what is happening inside and between other organizations.[198]

The term 'industrial network' was originally coined largely because the initial research at Uppsala was centred on industrial companies. It quickly became apparent, however, that such networks *'may be the rule rather than the exception for a wider population of business organizations in general'*, but one assumes that, by then, the term had stuck. The implications for the management of innovation of the Uppsala studies on industrial networks were and are important: principally, they laid emphasis on the need to link activities and resources together within a network of organizations when developing new products.[199]

Sectoral networks

Since the late 1990s, there has been a great deal of academic interest in sectoral networks. Much of the resulting research has tended to focus on high-technology sectors—in particular, on ICTs[200] and biotechnology[201]—but nevertheless has highlighted a number of interesting features of sectoral networks. Taking the lead from work such as that by John Hagedoorn and Jos Schakenraad (see **Figure 3.9**), many studies have attempted to map the networks graphically, as a way of capturing and representing the configuration of the networks, and where more than one snapshot in time is illustrated, revealing the dynamics within the network, such as changes in size, morphology (i.e. shape or configuration), and the network position of organizations.

In studies employing time-series data, such as the seminal study by Hagedoorn and Schakenraad of strategic alliances in a range of ICT sectors (i.e. information technologies, computers, industrial automation, microelectonics, software, and telecommunications) between 1980 and 1989, and the detailed study by Roijakkers and Hagedoorn of the pharmaceutical biotechnology sector between 1975 and 1999, it is clear that the density of networks in many high-technology sectors has grown significantly over the last twenty or so years. Perhaps not surprisingly, clear cliques have emerged in these sectors, often around the leading companies. Whilst the leading companies are well connected, however, they do not appear to dominate the overall network. An interesting finding from the study by Roijakkers and Hagedoorn is the shift in the role of small entrepreneurial firms: during the 1980s, these firms played a crucial role in developing the network and often formed important bridges between the various cliques within the network; by the 1990s, with the emergence of a more prominent role for large companies in the network, such entrepreneurial firms were found to play a declining role in both R&D partnering and bridging.

Although research has highlighted the importance of networks across a range of high-technology sectors, in an interesting comparative study of the network in two computer-component technology sectors—that of dynamic random access memory (DRAM), a general-purpose memory chip, and reduced instruction set computing (RISC), a newer technology initially employed within powerful workstations used for applications such as computer-aided design (CAD)—Duysters and Vanhaverbeke found significant differences in the configuration of the sector networks. In the DRAM network, the structure is open with no obvious cliques (see **Figure 3.10**), whilst, in contrast, the RISC network is characterized by three clear cliques around 'focal' actors, with few relationships between these groups (see **Figure 3.11**).

Duysters and Vanhaverbeke argue that these differences relate to the different contexts of the two technologies, which yield different roles for strategic alliances: for DRAM,

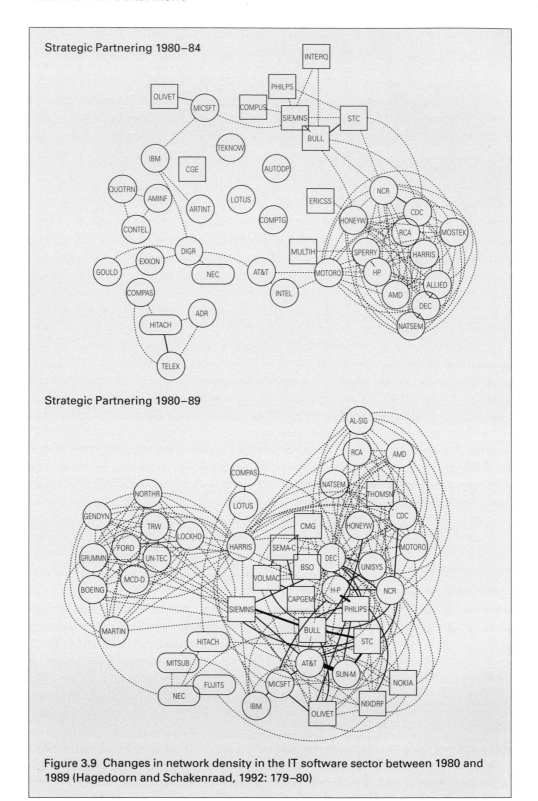

Figure 3.9 Changes in network density in the IT software sector between 1980 and 1989 (Hagedoorn and Schakenraad, 1992: 179–80)

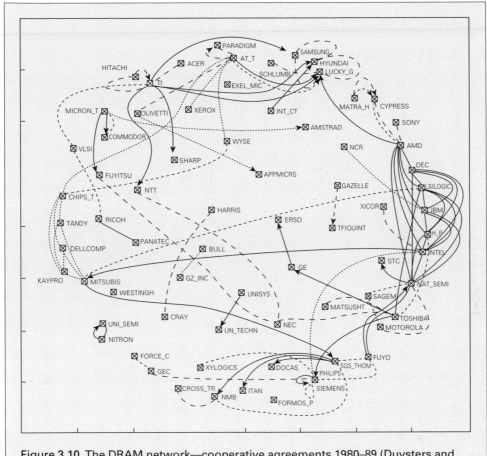

Figure 3.10 The DRAM network—cooperative agreements 1980–89 (Duysters and Vanhaverbeke, 1996: 446)

despite high competition, the rapid pace of technological change and huge R&D budgets pushes even the largest players together; in contrast, for RISC, which, at the time of the study, was a relatively new technology, the need to establish an industry standard from those available led to the emergence of different cliques around the possible contenders.

We now turn our attention to a discussion of regional clusters and industrial districts. As we noted above, unlike sectoral networks, regional clusters and industrial districts are characterized by linkages between geographically close organizations.

3.6.3 Regional clusters and industrial districts

During the 1980s, there was growing interest in the innovativeness and economic success of particular geographical regions; populated by innovative small and medium-sized enterprises (SMEs), these regions did not fit the classic 1960s focus on the large R&D

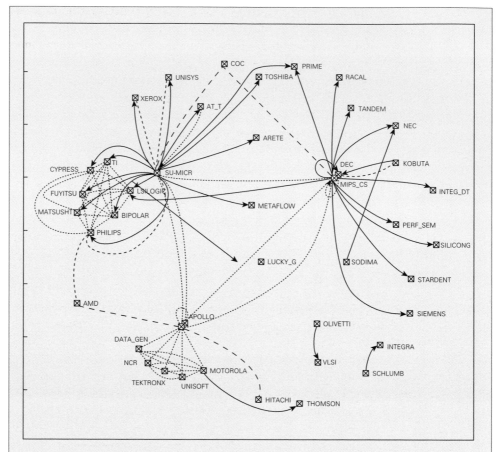

Figure 3.11 The RISC network—cooperative agreements 1980–89 (Duysters and Vanhaverbeke, 1996: 445)

intensive corporation as the main site of innovation. Seminal work from this period includes that of Michael Piore and Charles Sabel, their text *The Second Industrial Divide* highlights the emerging importance of industrial districts characterized by small, flexible, and specialized companies, and *The Competitive Advantage of Nations* by Michael Porter.[202] From the early 1990s, this phenomenon began to be explained more explicitly in terms of network concepts through the work of scholars such as Roberto Camagni, in relation to innovation networks within Italy, and Anna Lee Saxenian, with her focus on Silicon Valley in California. These studies placed emphasis on spatial proximity as the basis for dense and embedded networks of interaction and trust.[203]

Popularized by the phenomenal success of Silicon Valley and scholarly work such as that mentioned above, the concept of regional clusters has gained a great deal of interest and currency among both scholars and policymakers since the early 1990s. Indeed, para-doxically, in a world where production and markets are becoming increasingly global, it appears that regions can offer an important source of competitive advantage.

A 'regional cluster', then, may be defined as a geographical concentration of inter-connected companies and associated institutions, including end producers, universities, research laboratories, and service providers, all focused on a specialized area of economic activity. They are frequently characterized by extensive inter-firm social networks and job mobility.[204] Such co-located companies and institutions are often highly embedded, benefiting from infrastructure, knowledge 'spillovers' and a highly specialized pool of labour. Prime examples of regional clusters include Silicon Valley (for IT) in the USA, Motor Sport Valley (for racing car production) in the UK (see **Illustration 3.4**), the Baden-Wurttemburg region (for advanced production technology and electronics) in south-western Germany, Hsinchu Science-Based Industry Park (for IT) located southwest of Taipei, Taiwan, and the Emilia Romagna region (for knitwear, ceramic tiles, motorcycles, and food-processing equipment) in north-central Italy.[205]

The regional cluster is not, however, a new phenomenon. Indeed, the term 'regional cluster' is often used synonymously with that of 'industrial district', first articulated by the economist Alfred Marshall in the late nineteenth century.[207] The concept thus represents a rediscovery or reinvigoration of the industrial districts typical of northern Europe during the nineteenth century: for example, the clustering of textile mills in the towns and cities of Yorkshire, England, such as Leeds and Halifax; the concentration of cutlery production in Sheffield, England; and the concentration of silk production in Lyon, France. Whilst the fortunes of such regional clusters have ebbed and flowed, many successful examples persist or have emerged.

Perhaps the best known and certainly one of the most researched regional clusters is 'Silicon Valley' in northern California.[208] Indeed, one of the more interesting and influential studies of regional clusters was undertaken by AnnaLee Saxenian, who, in the mid-1990s, published her comparative study of the emergence and organization of 'Silicon Valley' and 'Route 128' (near Boston, Massachusetts). During the 1970s, these two US regions represented world-leading centres of innovation in electronics. By the early 1980s, however, both were experiencing major downturns: the former due to Japanese competition; the latter due to shifts in technology. Saxenian attributed the subsequent divergence in the performance of these two regional economies—Silicon Valley adapted and subsequently grew rapidly whilst Route 128 stalled—to their contrasting industrial systems: Silicon Valley being characterized by dense social and organizational networks, as well as an open 'free-wheeling' labour market, whilst in contrast, Route 128, was dominated by a relatively small number of large independent corporations. She concluded:

Illustration 3.4 'Motor Sport Valley'—a world-leading cluster in the UK

The British motor sport industry dominates the global production of racing cars, with approximately 75 per cent of all single-seater racing cars being designed and assembled in the UK, including the majority of the cars for Formula One and Indy racing. Perhaps more astonishing is that this industry is primarily located within a 50-mile radius in Oxfordshire (west of London), with the famous Silverstone circuit located at its epicentre. In this small geographical area, some 30,000 individuals are employed in many specialist small and medium-sized engineering firms. **Figure 3.12** reveals the geographical concentration of racing and rally car teams. A similar geographical concentration is evident for over 150 suppliers.[206]

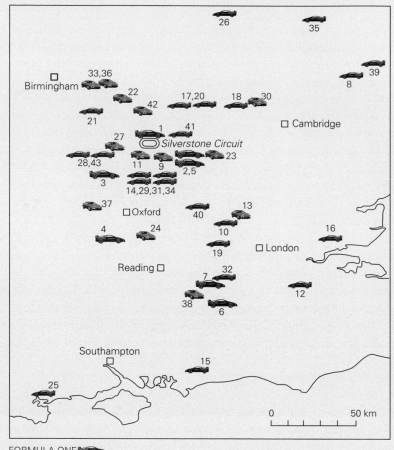

26

35

39
8

33,36
Birmingham

22
42
17,20   18   30

21

□ Cambridge

27   1   41

Silverstone Circuit

28,43   11   9   23
3   2,5

14,29,31,34

37
□ Oxford   40   13

10   16
4   24   19   □ London

Reading □

7   32
38   12
6

Southampton

15

25

0   50 km

FORMULA ONE

1   Benson and Hedges Total Jordan Peugeot
2   Danka Arrows Yamaha
3   Mild Seven Benetton Renault
4   Rothmans Williams Renault
5   Stewart Ford
6   Tyrrell Racing Organization
7   West McClaren Mercedes

OTHER FORMULAE   and TOURING/RALLY CARS

8   Argo Cars
9   Audi Sport
10   Bowman Cars
11   HMV Team Schnitzer
12   Elden Racing Cars
13   Ford Motorsport
14   Galmer Engineering
15   G Force Precision Engineering
16   Hawke Racing Cars
17   Jedi
18   Lola Cars Ltd.
19   Lyncar
20   Magnum
21   Marrow-Jon Morris Designs
22   Mitsubishi Ralliart
23   Motor Sport Developments
24   Nissan Motorsport Europe
25   Penske Cars Ltd.

26   Pilbeam Racing Design Ltd.
27   Prodrive
28   Pro Sport Engineering Ltd.
29   Ralt Engineering
30   Ray Mallock
31   Reynard Racing Cars Ltd.
32   Ronta
33   Rouse Sport
34   Spice Racing Cars
35   Spider
36   Total Team Peugeot
37   TWR Racing
38   Valvoline Team Mondeo
39   Van Dieman International
40   Vector Racing Car Constructors
41   Vision
42   Volkswagen - SBG Sport
43   Zeus Motorsport Engineering

Figure 3.12 Motor Sport Valley in the UK (Henry and Pinch, 2000: 192)

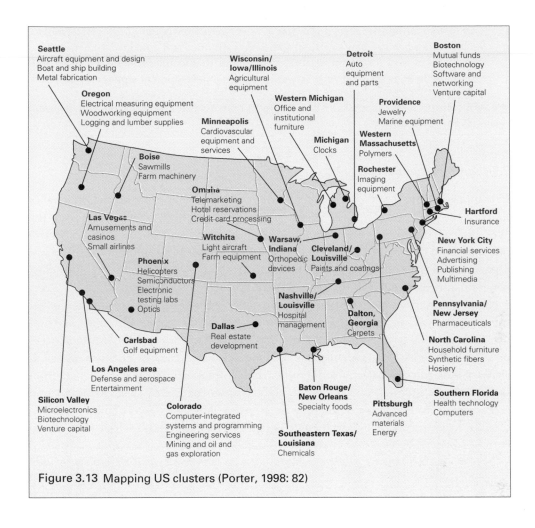

Figure 3.13 Mapping US clusters (Porter, 1998: 82)

The contrasting experiences of Silicon Valley and Route 128 suggest that industrial systems built on regional networks are more flexible and technologically dynamic than those in which experimentation and learning are confined to individual firms. Silicon Valley continues to reinvent itself as its specialized producers learn collectively and adjust to one another's needs through shifting patterns of competition and collaboration. The separate and self-sufficient organizational structures of Route 128, in contrast, hinder adaptation by isolating the process of technological change within corporate boundaries.[209]

A key question regarding regional clusters relates to the degree to which they can be created and fashioned by government intervention. The story of Silicon Valley would seem to suggest not, having emerged over a prolonged period of time through spontaneous private sector interactions, albeit seeded and nurtured by the activities of Stanford University, for example.

The case of Hsinchu Science-Based Industry Park, located south west of Taipei, however, provides conflicting evidence. For Mathews, Hsinchu was '*achieved* as a deliberate matter of public policy. *It was not a development so much as a* creation', because:

The Taiwan authorities recognized that as latecomers they could not hope to emulate Silicon Valley achievements without some extra stimulus and guidance; they sought to create a framework for private-sector developments that would facilitate, promote, and discipline them, in keeping with Taiwan's industrial development strategy overall.[210]

We will return to this discussion in **Chapter 11**, in relation to national and regional systems of innovation.

3.6.4 Entrepreneurial networks

In contrast to earlier work on entrepreneurship, which often focused on the personality traits of the entrepreneur,[211] the adoption of a social network perspective during the 1980s helped to shift the view of entrepreneurship from an individualistic to a collective phenomenon. The seminal work in the area of social networks and entrepreneurship was undertaken in the early to mid-1980s by academics such as Howard Aldrich, Sue Birley, Bengt Johannisson, and Dorothy Leonard-Barton.[212]

For Jones and Conway, the network perspective provides a '*conceptualization of the entrepreneurial process as a complex and pluralistic pattern of interactions, exchanges, and relationships between actors*'.[213] This is important because, paradoxically, whilst entrepreneurs may be characterized by their autonomy and independence, they are also '*very dependent on ties of trust and cooperation*'.[214] On the one hand, entrepreneurial networks benefit from spanning structural holes, in order that the entrepreneur is able to act as broker and access opportunities; on the other, they are served well by an extensive and diverse web of social relationships that allow them to tap into a range of resources. Indeed, Leonard-Barton argues that '*entrepreneurs who, for geographic, cultural or social reasons, lack access to* free *information through personal networks, operate with less* capital *than do their well-connected peers*'.[215]

3.7 Concluding comments—networks: a panacea for innovation?

During this chapter, we have sought to provide an overview of the network perspective as a distinctive lens and approach for studying and understanding innovative activity within and between organizations. In so doing, we have explored the extent to which different network features, such as density, openness, diversity, and configuration, as well as network roles, such as gatekeepers and brokers, favour or constrain the identification, circulation, choice, and utilization of new ideas and information in the innovation process. From a review of the literature, it is clear that there is plenty of empirical evidence to suggest that social networks and organizational networks are a prevalent and important feature of innovative activity within and between many modern organizations, sectors, and regions. Much of this literature emphasizes the positive and functional, rather than the negative and dysfunctional, aspect of networks. But it is important to raise the question regarding the extent to which networks represent a panacea for managers and organizations in pursuit of innovation.

Social networks can play an important role in the sourcing and disseminating of information, ideas, and knowledge, particularly where they span boundaries between functions, divisions, CoPs, and organizations. They can also play an important role in complementing the formal organizational structures, and in promoting integration and coordination between activities within and beyond the boundaries of the organization. By their very nature, however, social networks are dynamic, personal, and unrecorded, and, as a result, they are difficult to manage and direct. We will return to the important theme of social networks and informality in **Chapter 9**.

Organizational networks also play an important role in the innovation process. We noted earlier, for example, that networks are flexible, enabling network members to reposition themselves more speedily in the face of changes in technology and market. They also bring together distributed resources, knowledge, and competences, often in novel ways, as well as help to share risk, spread costs, shorten development cycles, and speed up market entry. But we also noted that such benefits may not be evenly distributed among network members. Furthermore, networks, in some instances, may create dis-benefits and 'lock-in' effects. For example, well-established networks may create barriers to entry for new members, whether intentionally or unintentionally, leading to the formation of closed networks and the locking out of cutting-edge knowledge and techno-logies.[216] To remain vibrant, then, networks must be continually renewed and reoriented to face changing economic and technological challenges.

For Powell, it is inaccurate *'to characterize networks solely in terms of collaboration and concord. Each point of contact in the network can be a source of conflict as well as harmony'*, or to assume that reciprocity and cooperation preclude the exertion of power by network members.[217] More fundamentally, Hobday argues that whilst networks such as those that characterize Silicon Valley are effective at generating new and innovative products and services during the early stages of the product life cycle, they lack the 'complementary assets' (i.e. the large-scale production, marketing, distribution, and financial resources) that are necessary to exploit the growth and maturity stages of mass-market innovations, and hence are poor at capturing the rewards of innovation.[218] We will revisit these issues in subsequent chapters.

CASE STUDY 3.1 HORTICULTURAL RESEARCH INTERNATIONAL (HRI) AND THE AGRICULTURAL GENETICS COMPANY (AGC)—THE DEVELOPMENT OF 'NEMASYS' AND 'NEMASYS H'[219]

Introduction

'Nemasys' and 'Nemasys H' are agricultural products that act as biological alternatives to synthetic pesticides: they constitute tiny insects called 'nematodes', who seek, and digest, other insects. The research at Horticultural Research International (HRI) brought about essential devel-opments in the mass rearing, formulation, and storing of these nematodes.

Organizational background

HRI was formed by the amalgamation of a number of well-established public-sector horticultural research centres in the UK. It represents a national and international centre of excellence in a number of fields, employing around four hundred scientists. The Agricultural and Food Research Council (AFRC) funds basic research at HRI, while near-market technology development is supported by a combination of funding from the Ministry of Agriculture, Fisheries, and Food (MAFF) and the private sector. The Agricultural Genetics Company (AGC) was set up by the UK government in the mid-1980s, to aid the commercialization of biotechnology in the horticultural sector. It is privately funded. At the time, a number of formal and personal links existed between HRI and AGC: for example, Professor Payne (a head of department within HRI) had been a scientific consultant to AGC since its conception.

The innovation process

Research on nematodes as a pest control agent had been undertaken at the HRI Littlehampton site and at a number of other horticultural institutes throughout the world (particularly in Australia and the USA) for a number of years prior to the initiation of the research project. While a number of parasitic nematodes had been identified, however, there existed a fundamental obstacle to scaling up the worms for commercial usage—that is, those that had been identified could only be reared within other insects. Then, in 1983, Paul Richardson, who had studied nematodes at HRI since 1971, came across a paper from Australia whilst browsing in the library. The paper detailed how the scientist had scaled up production of a totally different worm to 'flask level' with an artificial diet. The HRI scientist, with a family background in the horticultural industry, immediately recognized the commercial application to the high-value North West European 'glasshouse' industry.

Having obtained the nematode strain from the Australian scientist, Paul Richardson started initially by attempting to find and isolate the bacteria on which this particular nematode fed. At this stage, it was pure research. Having got as far as rearing the nematodes, he undertook the first tests on mushrooms, having received MAFF funding for mushroom pest control. At this time, the AFRC and MAFF were particularly interested in funding environmentally friendly horticultural research and technology projects. The tests proved successful. This work was then presented at an open day at HRI for professional horticulturalists in 1985 and received particular interest from Fisons. At that time, AGC had first right of refusal for all HRI's developments. AGC decided to fund the project, allowing staff to be taken onto the team at HRI. AGC also funded a soil survey around 1987, which was undertaken by a highly regarded academic at Imperial College. This survey was necessary to identify a British nematode of the same strain as the Australian worm that had been used in the tests to date, and was required under the UK's Wildlife and Countryside Act of 1981, with regard to releasing non-indigenous creatures into the UK. Following the basic research at HRI, with inputs from the worldwide nematode community either personally or via journal articles, the project moved to the development stage and shifted from HRI to AGC.

The scientific community interface

The initial inspiration for the project originated from within the scientific community, of which Richardson was a part, in the form of a scientific journal article by Robin Bedding. Indeed, scientific journals proved an important mechanism for sourcing knowledge and information during the project. Having read the paper, Richardson approached the Australian scientist who had undertaken the research for samples of the nematode and bacterial culture; these proved fundamental to the early work at HRI. Richardson and other project members tapped into two important research communities during the life of this project: the nematological community, which studies nematodes; and the broader entomological community, which studies insects (these are represented by the links to the top-left quadrant in **Figure 3.14**, which maps out the innovation network for the development of Nemasys). Interaction with these networks was aided by Richardson's profile within the respective communities. Richardson noted:

> If you're in a place like this [HRI] that's well known . . . for a period of over twenty years, you do know all the entomologists, all the nematologists, not only in this country but abroad . . . I think I'm quite widely known and know everybody else that's widely known.

Similar research was being undertaken in Australia and the USA using different nematodes under different climatic conditions, and knowledge of such work allowed the development of general research principles. Richardson recalled: '*If there was a clue there, we would use it.*'

As commercial alternatives to chemical pesticides have been increasingly sought, so these research communities have increasingly attracted scientists and funding from commercial organizations such as AGC. As with others in the research sphere, Richardson's personal research network was built largely through old university contacts (in particular, Imperial College where he did his degree and masters), conferences, editorial refereeing, and journal papers. Most of this interaction with the research community was informal and ad hoc, although, in some instances, work was formally contracted, such as in the case of the British soil and nematode survey, which sought viable British strains of the nematode. This survey was undertaken by Dr Bill Hominik (an eminent entomologist and Richardson's PhD supervisor) and his group at

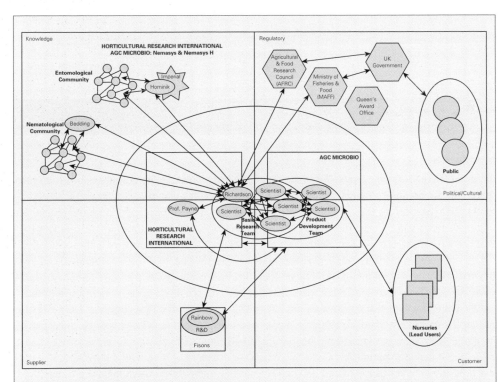

Figure 3.14 Actors and links mobilized in the development of nemasys and nemasys H (Steward et al., 1996)

Imperial College. Richardson noted that '*they* [Imperial College] *have been our closest allies in this area because they have experienced entomological and nematological staff . . . I've always tended to go back to the people I know*'.

The market interface

AGC subsequently acted as the intermediary between the research project at HRI and the marketplace. But AGC was far more than simply a commercializer, with much of the later fermentation technology required for scaling up undertaken by AGC scientists, such as John Godleman, Mark Sheppard, and Jeremy Pearce. Indeed, once the team managed by Richardson was up to speed, most were relocated to AGC, where the research increasingly focused on the scaling up and commercialization of the technology. This shift in location was largely to emphasize the shift from research to product development. Although Richardson continued to undertake research for AGC, his role in the project shifted from manager to consultant, within which he attended monthly meetings. With respect to the working relationship between Richardson and the AGC development team, Richardson recalled that: '*There was never ever a problem. Very informal and a tremendously relaxed relationship. I think it contributed a lot to the success of the whole thing.*'

Questions

1. How important were informal relationships and contacts to the development of Nemasys and what roles did they play during the innovation process?

2. What were the distinctive roles of HRI and AGC in the development and commercialization of Nemasys? What do you think are the benefits and disbenefits of such a separation of roles between such organizations?

CASE STUDY 3.2 THE BREAKING OF THE CYCLING HOUR RECORD—THE TRIUMPH OF ISOLATION AND NOVELTY[220]

Introduction

On the 17 July 1993, an obscure Scottish amateur cyclist, Graeme Obree, broke the world hour record for the greatest distance achieved by a cyclist in one hour on a velodrome (banked cycling track). This was a feat that, although largely ignored in Britain, gave him immediate celebrity status elsewhere in Europe. The French sporting newspaper *L'Équipe*, for example, relegated the first French stage victory in that year's Tour de France to p. 4 to make way for Obree's story.[221]

The hour record holds a unique, mythical status within cycling. While renowned races such as the Tour de France afford their winners great status, they produce winners every year who have competed only against their contemporaries. The hour record, in contrast, is rarely attempted— and only then by the greatest names in the sport while at the peak of their powers. Obree's background contrasted markedly with that of other hour-record holders. Obree was an unemployed amateur, who had never raced internationally, and had constructed his own, very unconventional, bicycle that included parts from a washing machine. In stark contrast, with no exceptions, all of the post-war holders of the record had come from the mainstream Continental European cycle racing scene and were established professionals. Of particular note are Eddie Merckx, who held the hour record between 1972 and 1984, and Francesco Moser, who held the record from 1984 until Obree's attempt in 1993.

Graeme Obree's unorthodox background and isolation as a cyclist

Obree's isolation from the cycling mainstream operated on several levels. Firstly, British cycling has been isolated historically from that of the rest of Europe. Despite the popularity of cycling in the UK, racing had been marginalized for the first half of the twentieth century by a law banning cycle racing on the highways. The result of this was that while cycle racing was becoming established as a spectator sport in Continental Europe, British racing cyclists adapted to their environment by competing in clandestine time trials. Time trials involve an individual rider completing a precisely measured course (10, 25, 50, or 100 miles), with the rider posting the shortest time being declared the winner. When the law relating to cycle racing on the highway was subsequently relaxed, the impact of this period of 'prohibition' was to continue: time trialling, for example, remained overwhelmingly popular in the UK. Furthermore, because the distance of events was standardized, riders increasingly competed not against each other, but against themselves in an attempt to produce the fastest time possible for each of the standard distances, which, in turn, encouraged the use of flat courses rather than the undulating minor roads favoured elsewhere. Thus, British racing cyclists tended to be time trial specialists, ill suited to making the jump to becoming professionals in Continental Europe, with the result that the UK cycle racing scene remained isolated.

Obree's isolation had a second dimension: he was not a part of the British cycling mainstream either. The British Cycling Federation (BCF) is the internationally recognized body for cycle racing in Britain and, as such, selects riders to represent Britain at international meetings, and nurtures new and existing talent through training programmes and other forms of support. Domestic time trialling is administered by the Road Time Trials Council (RTTC), which does not need international recognition, because time trialling in the British sense does not exist elsewhere. Obree had little contact with the BCF, because his main interest was time trialling, and his idiosyncratic and solitary training methods did not fit with the BCF's model of how high-calibre racing cyclists should be developed. These training methods, coupled with his highly unusual self-designed bicycles, set him apart from other riders.

Finally, Obree was geographically isolated. Such cycle racing that exists in Britain takes place primarily in England, so Obree's residency in Scotland made racing and interaction with others in the British cycle racing scene more problematic.

Graeme Obree's unorthodox bicycle

Perhaps more controversial than Obree's background was the unorthodox bicycle that he used to break the hour record. The most significant innovation was the handlebar arrangement, which allowed him to assume the position of a downhill skier; which is far more aerodynamic than a conventional cycling position. Obree's saddle position was further forward than is usual, which allowed him to pedal a higher gear than would normally be the case. Finally, he developed a technique of constructing the bicycle such that the pedals could be much closer together than on a conventional machine (it was this innovation that required the use of washing machine parts); this also improved aerodynamics and, Obree claimed, allowed a more physiologically efficient pedalling action. To accommodate this, the frame of the bicycle also had to be of a non-standard design.

Postscript

The date of Obree's first hour-record attempt had not been selected randomly. Chris Boardman, a well-known British cyclist and time trial specialist, had made known his intention to make an attempt on the hour record, and on the 23 July 1993, successfully claimed the hour record from

Obree. Shortly after his hour record, Obree rode the world pursuit championship in August 1993, using the same machine that he had used for the hour record. Not only did he win the championship, but he also did so in a world record time. Then, in April 1994, Obree reclaimed his hour record from Boardman. But by the time that it had come for Obree to defend his world pursuit title in August 1994, the Union Cycliste Internationale (UCI) had become convinced that Obree's riding position was giving him an unfair advantage.[222] Nevertheless, his bicycle conformed to UCI regulations and so it was obliged to allow him to compete. Despite this, during the championship, the UCI hurriedly altered its definition of a bicycle, which rendered Obree's riding position illegal and prevented him from riding in the final.

It appears that Obree's isolation from both the UCI and the BCF resulted in Obree gaining little support from either organization. The story became the subject of a successful movie in 2006, *The Flying Scotsman*, directed by Douglas MacKinnon and starring Jonny Lee Miller.

Question

1. With reference to the case study, and with regards to novelty and diffusion, what do you think are the benefits and disbenefits of being an isolate or at the periphery of a network?

FURTHER READING

1. For a complementary review of the literature concerning networking and innovation, see Pittaway et al. (2004).

2. For an overview of social capital, see Alder and Kwon (2002), and for an overview to communities of practice, see Wenger and Snyder (2000).

3. For a useful overview of the major trends in sectoral and international inter-firm R&D partnerships since the 1960s, see Hagedoorn (2002).

4. See Roijakkers and Hagedoorn (2006) for an interesting study of the R&D network within the pharmaceutical biotechnology sector captured at five points in time between 1975 and 1999: through a set of network graphics, the article clearly illustrates the dynamism and growth of the network over time. Also see Hagedoorn and Schakenraad (1992) on networks in a range of IT sectors, and a comparative study by Duysters and Vanhaverbeke (1996) of the networks in two computer-component technologies (DRAM and RISC).

5. See Porter (1998) for a good overview of the role and nature of regional clusters, and a good range of examples from a number of countries and sectors.

6. See Saxenian (1994) for an excellent comparative study of the origins, organization, and success of Silicon Valley and Route 128. Also see Casper (2007) for an interesting study of social networks and job mobility in the San Diego biotechnology cluster in California. For an interesting critique of the 'Silicon Valley' network model of innovation, see Hobday (1994).

NOTES

[1] Auster (1990: 65).
[2] Simmel (1955).
[3] DeBresson and Amesse (1991: 363).
[4] Conway et al. (2001: 351–2).

[5] DeSolla Price (1963); Crane (1972).

[6] Allen (1977).

[7] Rogers (1962; 2003).

[8] Kreiner and Shultz (1993); Conway (1997); Hansen (2002).

[9] Håkansson (1987; 1989); Powell (1990).

[10] Burt (1992).

[11] Rothwell (1983); Rothwell and Zegveld (1985).

[12] Tidd et al. (2001).

[13] For example, Myers and Marquis (1969); Utterback (1971); Rothwell et al. (1974).

[14] Utterback (1971).

[15] Myers and Marquis (1969).

[16] Hippel (1988); Conway (1995).

[17] Rothwell and Zegveld (1985).

[18] Commission of the European Communities (2003); Department for Innovation, Universities, and Skills (2008).

[19] Rothwell and Zegveld (1985: 49).

[20] Kelly et al. (1986: 18).

[21] Rothwell (2002).

[22] Utterback (1971); Rothwell et al. (1974).

[23] Rosenberg (1982).

[24] Rothwell (1983); Rothwell and Zegveld (1985).

[25] Langrish et al. (1972).

[26] Rothwell (2002).

[27] Utterback (1971); Rothwell et al. (1974).

[28] Rothwell (1983); Rothwell and Zegveld (1985).

[29] Rothwell (2002).

[30] Rothwell and Zegveld (1985: 50).

[31] Rothwell (2002: 117).

[32] Schmookler (1966).

[33] Rothwell and Zegveld (1985: 65).

[34] Freeman (1986: 31–3).

[35] Rappa and Debackere (1992); Debackere et al. (1994: 22).

[36] Rothwell (2002).

[37] Law and Callon (1992: 21).

[38] Barnes (1979); Mitchell (1969).

[39] Milgram (1967).

[40] Mitchell (1969).

[41] Scott (2000).

[42] Henry and Pinch (2000: 195).

[43] Mitchell (1969); Aldrich and Whetten (1981); Fombrun (1982).

[44] Conway et al. (2001: 355).

[45] Håkansson and Johanson (1990: 462).

[46] Kelley and Brooks (1991).

[47] Giddens (1984).

[48] Callon and Latour (1981); Callon (1986); Law and Callon (1992).

[49] Law and Callon (1992: 25).

[50] For example, Crane (1972).

[51] For example, Allen (1977).

[52] For example, Hagedoorn and Schakenraad (1992); Assimakopoulos et al. (2003).

[53] For example, Law and Callon (1992).

[54] For example, Jones et al. (2000).

[55] Powell (1990: 297).

[56] Rhodes and Marsh (1992).

[57] Castells (2000: 77).

[58] Hagedoorn and Schakenraad (1992); Assimakopoulos et al. (2003).

[59] Burns and Stalker (1961: 24–5); DeSolla Price (1963); Crane (1972).

[60] Munro and Slaven (2001).

[61] Munro and Slaven (2001); Rose (2000); Cookson (1997); Prior and Kirby (1993); Divall (2006).

[62] Rose (2000); Galambos and Sewell (1995).

[63] Galambos and Sewell (1995); Cookson (1997); Divall (2006).

[64] Galambos and Sewell (1995).

[65] Prior and Kirby (1993: 68).

[66] Prior and Kirby (1993: 67).

[67] DeBresson and Amesse (1991: 370).

[68] Moreno (1934; 1953).

[69] Mitchell (1969); Barnes (1979).

[70] Scott (2000: 7–37) provides a detailed chronology of the development of SNA.

[71] Rogers (1987).

[72] Boorman and White (1976); White et al. (1976).

[73] Kilduff and Tsai (2003: 59, 141).

[74] Alba (1982: 63).

[75] Blau (1982).

[76] Boorman and White (1976); White et al. (1976).

[77] Krackhardt and Hansen (1993); Hansen (2002); Cross et al. (2002b).

[78] Stephenson and Krebs (1993); Stephenson and Lewin (1996).

[79] Williamson (1975; 1985); Powell (1990).

[80] Walker and Weber (1984); Dietrich (1994).

[81] See Adler and Kwon (2002) for a review and synthesis.

[82] Nahapiet and Ghoshal (1998: 243, 251).

[83] Burt (1992: 58).

[84] Granovetter (1973); Burt (1992).

[85] Nahapiet and Ghoshal (1998); Tsai and Ghoshal (1998); Inkpen and Tsang (2005).

[86] Cohen and Fields (1999); Inkpen and Tsang (2005).

[87] Wenger (2000: 229).

[88] Wenger and Synder (2000: 139).

[89] Wenger (2000: 229).

[90] Wenger and Synder (2000: 140).

[91] Lave and Wenger (1991).

[92] Wenger (1998; 2000); Wenger and Synder (2000); Wenger et al. (2001).

[93] Contu and Willmott (2003); Mutch (2003); Roberts (2006); Handley et al. (2006).

[94] Callon and Latour (1981); Callon (1986); Law and Callon (1988).

[95] Callon (1992).

[96] Bijker (1987); Pinch and Bijker (1987).

[97] Law and Callon (1988: 225).

[98] Callon (1986: 20).

[99] Pinch and Bijker (1987: 40).

[100] Bijker (1987: 168).

[101] Bijker (1987).

[102] For example, Simmel (1955).

[103] Morgan (1986: 12–13).

[104] Tichy et al. (1979: 507).

[105] Wellman (1983: 156).

[106] Conway and Steward (1998); Jones et al. (1998).

[107] Conway et al. (2001: 354).

[108] Kanter (1972).

[109] Håkansson and Johanson (1990).

[110] Allen (1977); Cunningham and Homse (1984); Kreiner and Schultz (1993); Conway (1995).

[111] Aldrich (1979); Tichy et al. (1979).

[112] Aldrich (1979); Aldrich and Whetten (1981); Tichy (1981).

[113] Rogers and Kincaid (1981: 298).

[114] Zipf (1949).

[115] Rogers and Kincaid (1981: 298).

[116] Rogers and Bhowmik (1971).

[117] McFadyen and Cannella (2004: 744).

[118] Boissevain (1974); Aldrich (1979); Tichy et al. (1979).

[119] Boissevain (1974: 30).

[120] Kreiner and Schultz (1993).

[121] Håkansson (1989: 115).

[122] Sako (1991); Dyer and Chu (2003).

[123] Krackhardt (1992); Castilla et al. (2000).

[124] Krackhardt (1992).

[125] Wenger (2000: 230).

[126] Jarillo (1988); Sako (1991); Dyer and Chu (2003).

[127] See Newell and Swan (2000: 1292–7) for overview.

[128] Sako (1991).

[129] Dodgson (1993).

[130] Krishnan and Martin (2006).

[131] Goerzen (2007).

[132] Tichy et al. (1979); Auster (1990).

[133] Burt (1992: 16–25).

[134] Aldrich and Whetten (1981).

[135] Boissevain (1974: 40, 37).

[136] Burt (1992: 17).

[137] Aldrich and Whetten (1981); Tichy (1981).

[138] Tichy et al. (1979).

[139] Sorenson et al. (2006).

[140] Ancona and Caldwell (1998); Joshi and Jackson (2003).

[141] Joshi (2006).

[142] Conway (1997).

[143] Rogers and Kincaid (1981: 136).

[144] Hagedoorn and Schakenraad (1992); Steward and Conway (1998).

[145] Kim et al. (2006).

[146] Koka and Madhavan (2006).

[147] Tichy (1981).

[148] Warner and Lunt (1941: 32).

[149] Granovetter (1973).

[150] For example, Conway (1997).

[151] For example, Hansen (1999; 2002).

[152] For example, Rogers and Kincaid (1981).

[153] Granovetter (1983: 209).

[154] Krackhardt (1992).

[155] Nelson (1989).

[156] Hansen (1999: 109).

[157] Burt (1992).

[158] Granovetter (1985).

[159] Uzzi (1996).

[160] Allen (1977); Tushman (1977); Aldrich (1979); Tushman and Katz (1980).

[161] Burt (1992; 2005).

[162] Burt (2005: 66–71).

[163] Tichy et al. (1979).

[164] Tichy et al. (1979: 508).

[165] Allen (1977: 223).

[166] Conway (1995); Hansen (1999); Hansen et al. (2001).

[167] Marsick and Watkins (1997).

[168] Nonaka and Takeuchi (1995).

[169] Rothwell et al. (1974); Moenaert et al. (1994a; 1994b).

[170] Langrish et al. (1972); Hippel (1988); Kreiner and Schultz (1993); Conway (1995).

[171] MacDonald and Williams (1993a; 1993b); Hansen (1999).

[172] Allen (1977: 157).

[173] Allen (1977: 161, 180).

[174] Hansen (1999); Tsai (2002); Hansen et al. (2005).

[175] Tsai (2002).

[176] Ibid.

[177] Hansen et al. (2005).

[178] Criscuolo (2005).

[179] DeSolla Price (1963); Crane (1972).

[180] Allen (1977); Frost and Whitley (1971).

[181] Hippel (1988); Schrader (1991); Kreiner and Schultz (1993); Conway (1995); Dahl and Pedersen (2004).

[182] Conway (1995); Fleming et al. (2007).

[183] Bell and Zaheer (2007: 955).

[184] Freeman (1991: 503).

[185] Rogers and Larson (1984); Saxenian (1985; 1991; 1994); Castilla et al. (2000); Assimakopolous et al. (2003).

[186] Lievens et al. (1999); Lindsay et al. (2003); Freeman et al. (2007).

[187] Michailova and Hutchings (2006).

[188] Chow et al. (2000).

[189] Chow et al. (1999).

[190] Michailova and Husted (2003); Michailova and Hutchings (2006).

[191] Sakakibara and Dodgson (2003).

[192] Hagedoorn (2002).

[193] Grangsjo and Gummesson (2006); Ul-Haq and Howcroft (2007).

[194] Powell et al. (1996).

[195] Powell (1990); Hagedoorn (1993); Saxenian (1994); Grandori (1997); Lavie (2006); Lee (2007).

[196] Baum et al. (2000).

[197] Gulati (1998); Dhanaraj and Parkhe (2006).

[198] Håkansson (1990: 371).

[199] Håkansson (1987; 1989; 1990); Håkansson and Snehota (1989).

[200] Hagedoorn and Schakenraad (1992); Duysters and Vanhaverbeke (1996); Okamura and Vonortas (2006).

[201] Powell et al. (1996); Gay and Dousset (2005); Roijakkers and Hagedoorn (2006).

[202] Porter (1990); Piore and Sabel (1984).

[203] Saxenian (1991); Camagni (1991).

[204] Rogers and Larson (1984); Saxenian (1985; 1991; 1994); Castilla et al. (2000); Casper (2007).

[205] Saxenian (1985); Henry et al. (1996); Sabel et al. (1987); Mathews (1997); Brusco (1982).

[206] Henry and Pinch (2000: 192–3).

[207] Marshall (1961).

[208] Rogers and Larson (1984); Saxenian (1985; 1991; 1994); Castilla et al. (2000); Assimakopolous et al. (2003).

[209] Saxenian (1994: 161).

[210] Mathews (1997: 27).

[211] For example, McClelland (1961).

[212] Johannisson and Peterson (1984); Leornard-Barton (1984); Birley (1985); Aldrich and Zimmer (1986).

[213] Jones and Conway (2004: 91).

[214] Johannisson and Peterson (1984: 1).

[215] Leonard-Barton (1984: 113).

[216] Powell (1990); DeBresson and Amesse (1991); Glasmeier (1991); Gulati (1999); Goerzen (2007).

[217] Powell (1990: 305).

[218] Hobday (1994).

[219] Based on interviews at HRI by Steve Conway.

[220] Based on an interview with Graeme Obree by Steve Conway and Peter Quaife, and a seminar presentation.

[221] *L'Equipe* (19 July 1993).

[222] *Cycling Weekly* (7 May 1994).

Part II

Strategy and the Mapping of Innovation and Technological Progress

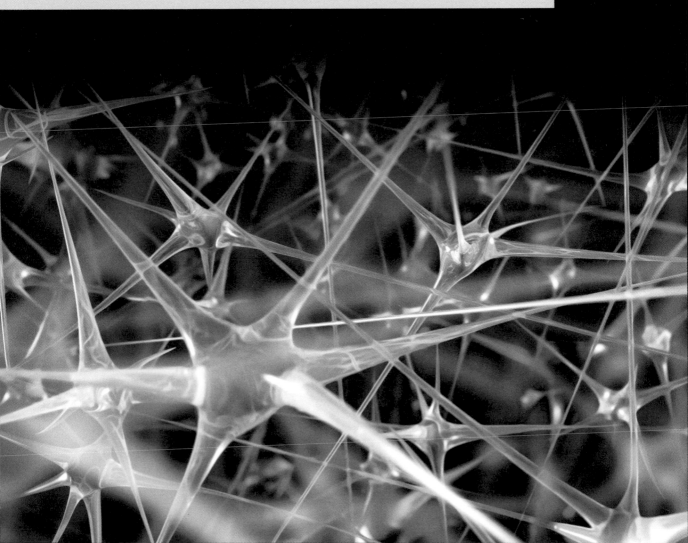

Introduction

Much of Part II is dedicated to a discussion and evaluation of a series of models that have been developed to map the progress of a technology over its life cycle. Such models are seen as important for both managers and policymakers alike, because they aid the comprehension and structuring of the past, and attempts to predict the future. This is crucial for the development of appropriate capabilities and strategies at the organizational level, as well as at the regional and national levels. At a minimum, such models provide us with '*a language and a facility for talking about and directing technology*' (Foster, 1986). Christensen et al. (2004: xx–xxi), go much further, however, arguing that:

> the only way to look into the future is to use these sorts of theories, because conclusive data is only available about the past . . . the best way to make accurate sense of the present, and the best way to look into the future, is through the lens of theory. Good theory provides a robust way to understand important developments, even when data is limited.

Part II embraces three themes: the first concerns the 'mapping' of technological progress and the associated patterns of innovation (**Chapter 4**); the second concerns the 'shaping' of technology and draws upon a range of concepts that help us explain why technologies tend to follow 'natural' trajectories, whilst also experiencing periodic discontinuities (**Chapter 5**); and the third and final theme is that of innovation strategy (**Chapter 6**).

Mapping innovation and technological progress

Technology can be mapped in relation to its market performance (i.e. the demand side) or its technical performance (i.e. the supply side). For example, models such as the 'technology life cycle' and the 'diffusion curve' map the sales and diffusion rates of a technology. In contrast, the 'technology S-curve', the 'product-process cycle', and the 'dominant design' model plot improvements in technical performance, the shift from radical to incremental and product to process innovation, and the diversity of product design, respectively. Although each of these models focuses on a different dimension of 'progress', they are interrelated and complementary.

Much of the literature and theorizing in this area has been undertaken in relation to technological product and process innovation. There has, however, been a growing interest in the degree to which such concepts are of relevance to the service sector. Increasing mechanization, standardization, and modularization within the service sector may all be considered to be core elements of a 'paradigm' that is common across many service sectors, and which guides the 'progress' of service innovation along a 'natural' or 'common' trajectory in much the same way as innovation in technological products and processes.

Shaping innovation and technological progress

Technologies exhibit periods of continuity, during which they appear to progress along fairly predictable trajectories, as well as periods of discontinuity, during which the prevailing technology becomes obsolete and is replaced. We introduce the notion of 'technological regimes' or 'technological paradigms', and discuss the role that they play in shaping the trajectory of a technology and, ultimately, in its transformation or obsolescence. In developing our argument, we draw upon the philosophy of science (through the work of Kuhn), evolutionary economics (through the work of economists such as Nelson and Winter), and the sociology of technology (through the work of sociologists such as Rip, Kemp, and Geels). Each of these disciplines have made valuable contributions to our understanding of how technology develops and progresses over time.

The strategic management of innovation

In our discussion of strategy, we focus on a number of key themes that are especially salient to the strategic management of innovation. Building on our exploration of the role of social and organizational networks in **Chapter 3**, we adopt an extended version of the 'resource-based view' (RBV) to accommodate these important 'relational' assets. We discuss the various 'timing-to-market' strategies (e.g. 'first to market' or 'late to market') that an organization might adopt and the associated resources that each entails. In light of our discussion of technological trajectories and the ways in which they may be shaped or disrupted, we look at the options open to organizations for entering a market, or defending a market position in the face of discontinuity or disruption.

Following on from our discussion of paradox in **Chapter 2**, we will see how strategic decision making in relation to innovation must often embrace a 'both/and', rather than an 'either/or', approach, for example, with regard to whether an organization should outsource or develop in-house capability, or whether it should focus on radical or incremental innovation.

4

The Patterns of Innovation Within the Life Cycle of a Technology

Chapter overview

Learning objectives

This chapter will enable the reader to:

- identify and outline the different dimensions along which 'progress' can be expressed and projected during the life cycle of a technology, such as through the improvement in its technical performance or its rate of diffusion;

- discuss the key features of the range of models that have been developed to conceptualize and map the trajectory of technological progress along these different dimensions; and

- identify and outline the implications for the management of innovation of each of these conceptual models;

- provide a critique of the utility of each of these conceptual models, through an appreciation of the complications that arise in their application, particularly with regard to complex or 'nested' technologies, and a recognition of their limitations;

- evaluate the relevance of these conceptual models to other sectors and, in particular, the service sector; and

- apply the various models of technological progress.

4.1 Introduction

The potential impact of innovation and technological progress on the nature and pattern of competition within a sector highlights the importance of incorporating such analyses into the strategy decision-making process. This and the subsequent chapter outline a number of 'dynamic' models that have been developed to aid our understanding of the patterns of innovation and technological progress, and, as such, they are potentially useful aids to strategic analysis, formulation, and choice. These models help us to visualize and map the potential trajectory of a technology or innovation over time. Each model has empirical support—but each model also has its weaknesses and limitations, and these will be highlighted as each of the models are discussed. Perhaps the most fundamental critique of these models as a whole is that they appear overly deterministic and thus under-play the potential for innovating organizations themselves, as well as users, governments, and pressure groups, to influence the 'shape' and 'direction' of innovation or technology. The starting point for this chapter then rests with the following two questions:

- can we identify patterns of innovation in the evolution of a technology?

- if so, what messages do these patterns highlight for the management of innovation, and, in particular, in relation to strategic decision making around innovation and technology?

In addressing these two key questions, we discuss the following conceptual models:

1. The 'technology life cycle' (TLC) model—this maps the sales volume trajectory of a technology over time.

2. The 'technology S-curve' model—this maps the technical performance trajectory of a technology in relation to research and development (R&D) effort.

3. The 'product-process cycle' model—this maps the interrelated trajectories over time of product and process innovations within a sector.

4. The 'dominant design' model—this maps the emergence of a dominant design of an innovation or technology over time.

5. The 'diffusion curve' model—this maps the diffusion trajectory of an innovation or technology over time.

Each of the models listed above are interrelated. Each focuses on a different dimension of the 'progress' of a technology, however, such as technical performance or the rate of adoption and diffusion.

We will start with a brief discussion of the TLC, which outlines the various stages or phases of the life cycle of a technology. This lays down the foundation for the more substantive discussion of the technology S-curve, the product-process cycle, the dominant design model, and the diffusion curve.

4.2 Understanding the broad features of models for mapping innovation and technological progress

Before we move on to discuss each model individually, it is useful to highlight some of their broad features.

All of the models discussed in this chapter and the next are dynamic in nature—that is, they adopt a longitudinal perspective and attempt to map the path or evolution of a technology. In doing so, a number of the models adopt explicitly the biological metaphor of the 'life cycle', incorporating life stages akin to birth, growth, maturity, and death.

The following models can be grouped according to whether they attempt to map the 'market' performance trajectory (i.e. demand side) or the technical performance trajectory (i.e. supply side) of a technology. Thus, the TLC, the diffusion curve, and technology 'long waves' (see **5.5**) map the diffusion, market performance, or economic impact of a technology. In contrast, the technology S-curve, the product-process cycle, the dominant design model, and 'technological discontinuity' (see **5.3**) all plot the trajectory of some technical aspect of the innovation or technology.

Each of the models typically adopts either a cumulative or a non-cumulative approach to expressing and mapping progress. To help to explain this feature, consider **Figure 4.1**, which maps the number of new adopters of an innovation over a number of years. This figure plots the same data in two ways: the first—represented by the curve that looks a bit like a letter 'S'—adds together or cumulates the number of new adopters over time, month on month; the second—the lower bell-shaped curve—plots only the number of new adopters for each particular month, i.e. it is non-cumulative. A number of the models that we discuss in this chapter may be mapped either cumulatively or non-cumulatively, i.e. the TLC, the technology S-curve, and the diffusion curve. The TLC is, however, typically presented as a bell-shaped (i.e. non-cumulative) curve and the technology S-curve, perhaps not surprisingly, is typically plotted as a 'S'-shaped (i.e. cumulative) curve. In contrast, the product-process cycle is universally presented as a non-cumulative curve.

The final feature of interest here is the classification of the models in relation to the four 'motors of change' that were introduced in **Chapter 2** (see **2.5**). At first sight, all of the models that we discuss in this chapter might be expected to be classified as predicated on the 'prefigured path' motor of change—that is, that the patterns of progress mapped out by the models are an inherent, in-built, or 'prefigured', feature of the technology. But as we look at the models more closely and critically, we see that other motors of change are

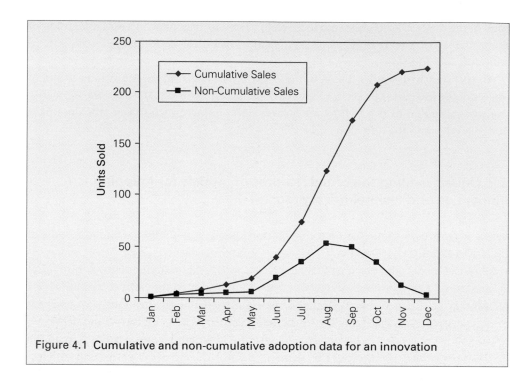

Figure 4.1 Cumulative and non-cumulative adoption data for an innovation

also highly relevant. For example, the 'competitive selection' motor of change is important for explaining the emergence of a dominant design (see **Illustration 4.3**), whilst the 'purposeful enactment' motor of change is central for our understanding of the variations in the speed and pattern of diffusion of a technology (see **Illustration 4.4**).

4.3 The technology life cycle model

The technology life cycle (TLC) model maps the life cycle of a technology, in much the same way as the 'product life cycle' (PLC) model maps the life of a product (see **Figure 4.2**), highlighting the trajectory of sales over the lifetime of a technology. Both the TLC and the PLC generally incorporate four stages or phases: 'introduction'; 'growth'; 'maturity'; and 'decline'. In most instances, a TLC will embrace many PLCs, derived from the various versions and 'generations' of a product, and thus represents the aggregated sales of products that employ the technology.[1]

During the 'introduction' phase, the slope of the curve is shallow, representing the gradual adoption of the new technology. The 'growth' stage, at which sales begin to accelerate, is indicated by the steep incline of the curve. 'Maturity' is the stage at which sales reach their peak, and the curve begins to tail off, flatten, and, ultimately, to fall. Finally, the 'decline' phase in the life of a technology occurs when it is made obsolete by the emergence of a new substituting technology (e.g. the emergence of digital recording

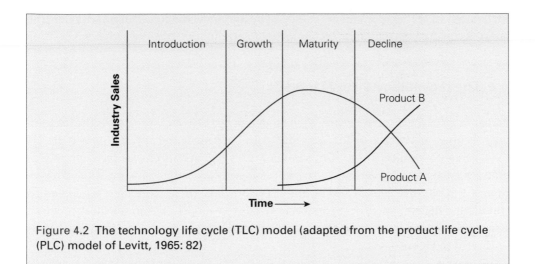

Figure 4.2 The technology life cycle (TLC) model (adapted from the product life cycle (PLC) model of Levitt, 1965: 82)

technology to replace analogue recording technology); the faster the process of substitution, the more rapid the decline and thus the steeper the downward slope of the TLC. The rationale that lies behind the PLC concept,[2] and thus the TLC concept, emanates from empirical work on the diffusion and adoption of innovations.[3] (We will discuss the diffusion curve later in this chapter.)

Nieto argues that such life cycle models embody a number of conceptual inconsistencies, most notably:

> (1) their quasi-tautologic nature: sales define the phases of the . . . life cycle that, at the same time, explain the sales, and (2) their large determinist component: it tries to describe a pattern of evolution that will invariably occur.[4]

Underlying and influencing the trajectory of the TLC of a given technology are a number of other patterns—principally:

- the rapid increase and subsequent levelling off of the improvement in technical performance of the technology through a series of incremental developments (incorporating both major and minor improvements);

- the switch from product innovation (i.e. a focus on product novelty and technical performance) in the earlier stages of the TLC, to process innovation (i.e. a focus on quality and efficient mass production);

- the switch from a variety of available designs in the earlier stages of the TLC, to the emergence of one or a small number of dominant designs (e.g. the IBM-compatible in home computers; the VHS format in home video recorders; and Boeing and Airbus designs in large-bodied passenger aircraft);

- the increase and subsequent levelling off of the diffusion and adoption of the technology; and

- the shift to a new technological paradigm (e.g. analogue to digital recording of music), leading to the (often rapid) decline phase of the existing technology.

All but the last of these patterns are embraced by the models discussed in this chapter. The last, referring to the concept of 'technological discontinuity', is addressed in **5.3**.

4.4 **The technology S-curve model**

Perhaps one of the most frequently cited models of technological progress is the 'technology S-curve', developed in the early 1980s by Richard Foster, of McKinsey—the US management consulting firm.[5] Christensen argues that it *'has become the centrepiece in thinking about technology strategy'*.[6] The technology S-curve represents the typical trajectory of the improvement in the technical performance of a technology in relation to cumulative R&D effort. It is an inductively derived model of technological progress. The term 'S-curve' was coined from the resulting shape of the curve when plotted, as illustrated in **Figure 4.3**.

In **Figure 4.3**, the vertical axis represents the technical performance of a technology. It is likely, however, that the technical performance of a given technology can be measured in a number of ways. Taking the automobile engine, for example, technical performance might be measured against acceleration, top speed, fuel efficiency, or emissions. In the S-curve model, technical performance may represent a single performance measure, or a combination of performance measures weighted equally or differently. For Foster, what is important is that the technical performance measure adopted should represent something that is both valued by the customer, and something that can be expressed in terms that make sense to the scientists and engineers who are developing the technology—i.e. it is measurable. The horizontal axis in **Figure 4.3** represents the cumulative R&D effort, and not time, as in many other models, because, as Foster argues, *'It is not the passage of time that leads to progress, but the application of effort'*.[7] There are two common proxies that are employed for representing R&D effort: R&D expenditure and R&D man years.

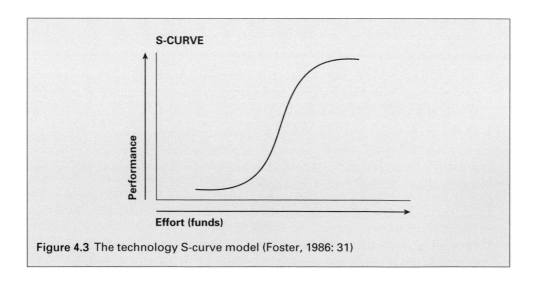

Figure 4.3 The technology S-curve model (Foster, 1986: 31)

The technology S-curve is comprised of three phases, which are sometimes labelled the 'emergent', 'growth', and 'maturity' stages. During the emergent stage, the rate of improvement in the technical performance of a technology is slow, because much of the R&D effort is required to develop the basic knowledge underpinning the technology. Thus, the technical performance returns from R&D effort are low; this stage is represented by the shallow slope at the bottom of the S-curve.

During the growth stage, as knowledge about the technology accumulates, and is diffused and applied, the rate of progress in the improvement of the technology begins to accelerate. Here the technical performance returns from R&D effort are high; this stage is represented by the steep central segment of the S-curve. The 'point of inflection' is the point on the S-curve (roughly the middle) at which the yield is at its highest. Prior to this point, the innovating organization benefits from increasing returns per unit of R&D effort; after this point, however, the firm begins to suffer from decreasing returns. Sahal explains this decline in the rate of technical progress in relation to either complexity or scale phenomena (i.e. things become too small, or too large) that arise as a technology matures.[8]

During the maturity stage, the yield on R&D effort begins to decline at an increasing rate as it approaches the 'natural' or 'technical' limits of the technology; this is represented by the shallow slope at the top of the S-curve.

An important element of the S-curve model is the notion of a 'technical limit', beyond which the performance of a technology cannot be improved irrespective of the R&D effort employed. A 'technical limit' is a physical constraint that is set by the laws of nature, such as the number of transistors that can be placed on a square centimetre of silicon, which is limited by the crystal structure of silicon. Foster argues that '*most industries are far from these ultimate natural barriers, and much more likely to come up against practical technological limits*'[9]—that is, many of the technical barriers that innovative organizations hit are located more in the prevailing design and assumptions held by the firm, rather than in any naturally occurring technical limit. This last point is well illustrated by the case of reduced instruction set computing (RISC). RISC was developed from the mid-1970s, at a time when many believed that the technical performance of microprocessors had reached their limit; by using simpler instructions, RISC was able to improve the speed of microprocessors considerably.

Another key aspect of the S-curve model is the concept of the 'technical potential' of a technology. This is the gap between the current state-of-the-art of a technology and its technical limit. Represented visually, the technical potential is the gap between the current position on the S-curve for a given technology and the top of the S-curve.

The technology S-curve approach can be used at the organizational level, or aggregated and employed at the industry level. Examples of S-curve analyses exist at both levels: at the organizational level, in automobile tyre cord[10] and computer disk-drive components;[11] at the industry level, in artificial hearts and pocket watches,[12] computer disk drives,[13] aircraft engines,[14] foam rubber,[15] agricultural insecticides,[16] computer-integrated manufacturing (CIM),[17] and permanent magnets.[18]

At first sight, then, the technology S-curve model appears to be intuitive and seductively simple. Furthermore, the wide variety of empirical studies, including those noted above, appear to support its efficacy. As a result, it is argued that the technology S-curve combines descriptive, explanatory, and predictive potential.[19] It thus represents a potentially useful tool for forecasting technological progress, and for informing and supporting strategic decision making in relation to innovation and technology strategy. This view is supported by Foster, who argues that:

> observations, coupled with the underlying theory of why it is happening, seem to me to be convincing evidence that these [S-] curves describe reality . . . If that is true and if the limit for an s-curve can be predicted then the s-curve can yield valuable insights. If we can define important parameters, trace the early days of progress of these parameters versus the effort to make the progress, and develop a point of view on what the limits of these performance parameters are, then we have a basis for foreseeing how much further current products can be improved and how much effort it will take to get them to higher levels of performance.[20]

Illustration 4.1 Intel and the performance trajectory of the microprocessor

Over the last forty years, dramatic increases in the technical performance and density of microprocessors have been achieved through a combination of major product and process innovations. Perhaps one of the best-known innovators and manufacturers of microprocessors is Intel, the processors of which are installed within many millions of personal computers (PCs) throughout the world. The exponential progress in the speed and density of this series of microprocessors illustrates that the technology is still in its growth phase (see **Figures 4.4** and **4.5**): such a trajectory of technical progress was predicted by Gordon Moore forty years ago.[21] Whilst such progress is unsustainable in the long term, there appears to be little consensus as to when the technology might reach its maturity phase, or indeed, its technical limit (see **Illustration 4.2**).

Having described the features of the technology S-curve, the following two subsections assess the utility of the technology S-curve framework: firstly, by looking at the implications of the model for the management of innovation, and secondly, by reviewing some of the limitations and complications of the model that arise through its application.

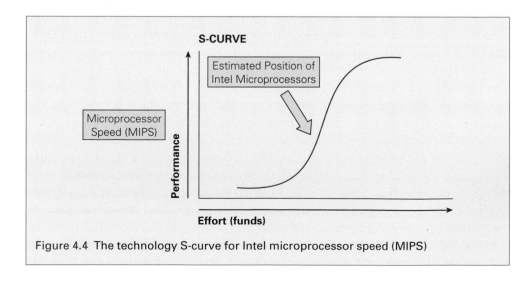

Figure 4.4 The technology S-curve for Intel microprocessor speed (MIPS)

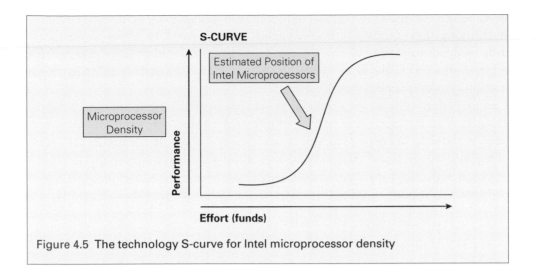

Figure 4.5 The technology S-curve for Intel microprocessor density

4.4.1 The innovation management implications of the technology S-curve model

Before using the S-curve, it is important that the measures employed to monitor and map improvements in the technical performance of a technology are 'meaningful', and thus should not be established in isolation: they need to incorporate customer needs and emerging legislation, for example. This is one way in which the trajectory of a technology is shaped by exogenous factors. Even apparently sensible measures of technical perform- ance can sometimes prove difficult to get agreement on: a good example of this is pro- vided by Afuah: *'The MIP (million instructions per second) became a joke with some computer designers who renamed it "meaningless indicator of performance"'*.[22]

The model indicates that the early stages of the development of a technology require a great deal of R&D effort simply to establish the basic knowledge and expertise to make subsequent progress—that is, in the emergent phase of a technology, R&D effort may yield very little tangible progress in improving the performance of a technology. Thus, patience is required, because this 'ground work' is vital if the organization is able to move on to the growth stage of the technology, at which R&D effort will begin to yield rapidly increasing returns. Consider the extraordinary coordinated R&D effort of laboratories around the world in mapping the human genetic structure through the 'Genome' project—a necessary precursor to the growth stage of the S-curve for the development of human gene therapies.

The model also highlights the importance of determining the technical limits and thus the technical potential of a technology. Innovative organizations therefore need to invest in understanding the scientific basis of phenomena and materials behind the technology. Foster provides the example of the development of rayon, the first synthetic automobile tyre cord, to illustrate the importance of determining technical limits in advance: of the US$100m invested in developing rayon, the first US$60m brought about an 800 per cent improvement in technical performance of the material; as the technology moved towards maturity and, ultimately, towards its technical limits, unbeknown to the rayon developers, the subsequent US$15m investment in R&D resulted in only a 25 per cent improvement and the final US$25m, only 5 per cent. He argues convincingly that had the organizations developing rayon known the technical limits, then their R&D investment strategies might have been very different.

As the technical limits for a technology are approached and the technical potential diminishes, each increment of R&D effort will yield a lower increment in the improvement in the technical performance of the technology—as was demonstrated in the case of rayon. Indeed, rapidly diminishing returns on R&D effort signal the imminent approach of the technical limit. Foster suggests that there are a number of subtle signals that indicate the onset of 'technological decay' (i.e. diminishing returns), including:

- a shift towards missing R&D deadlines;
- a shift in the company or sector towards process-orientated R&D; and
- a rise in dissatisfaction among R&D staff.[23]

Many managers confuse the 'economic health' with the 'technological health' of their organization. Foster notes that:

> Since most companies don't know how to measure their technological health, they measure their economic health. The trouble is that economic health is a result of many things that essentially are independent of the underlying technological health of the firm.[24]

That is, an innovative organization may still be generating healthy sales and profit from a given technology even as that technology begins to approach its technical limit. Thus, a focus on the technological health of the organization, using tools such as the technology S-curve, can also encourage organizations to take a longer-term perspective of their economic health.

Perhaps one of the most important messages of the technology S-curve model is that even before the technical limits of a technology are approached, innovative organizations should begin searching for, and investing in, alternative technologies that will allow for further performance improvement in relation to a particular application or market need. The shift from one technology to another, and thus from one technology S-curve to another, is known as a 'technological discontinuity'. An example of such a technological discontinuity is provided by the shift from vinyl records to digital compact discs (CDs). The shift from one technology to another is a difficult and painful journey for many incumbent organizations (i.e. those currently located within a sector), not least because it entails making obsolete existing core technological competences that may still be the basis of the economic success of the organization. Consider the position of traditional camera manufacturers, such as Kodak and Olympus: how do they react to the technological discontinuity created in their sector by the emergence of digital cameras, and the rapid entrance of computer companies, such as Hewlett-Packard?

As noted earlier, we will return to discuss the important concept of 'technological discontinuity' in the next chapter (see **5.3**).

4.4.2 A critique of the technology S-curve model

Thus far, we have presented a fairly 'non-problematized' account of the technology S-curve. As we will see below, however, despite the potential utility of the approach, a number of issues exist, some of which are inherent to the model itself, whilst others arise through the attempt to apply the model to a dynamic and complex technological environment. We separate these issues very roughly into:

- the inherent limitations of the technology S-curve model; and
- the complications in the application of the technology S-curve model.

The limitations of the technology S-curve model

The S-curve essentially represents a crude input–output model: rough proxies for R&D effort are generally employed, and there is the assumption that the relationship between input (R&D effort) and output (technical performance) is 'undisturbed' by the process and context of transformation of these inputs into outputs. As we will see from **Part III**, the nature of the transformation (i.e. innovation) process and the internal context (e.g. the configuration of organizational structure and the allocation of resources) are key influencers on the 'efficiency' and 'effectiveness' of R&D effort.

It is also very difficult to determine or measure with any accuracy many of the fundamental elements of the model, such as:

- the nature of the relationship between R&D effort and technical performance improvement of a technology (as noted above), such that it could be known that a specific level of R&D effort (input) led to a specific improvement in the performance in a technology (output);
- the current position of an individual organization or the state-of-the-art of a technology on the S-curve; and
- the technical limit, and as a consequence, the technical potential of a technology.

Concerning such measurement issues, Becker and Speltz argue that '*prior to the midpoint, it is impossible to forecast definitely when the slope will begin to decline*',[25] whilst Afuah notes: '*it is difficult to measure effort. Is it a firm's own R&D spending, or that of the whole industry since it benefits from the industry's spillovers?*'[26]

More fundamentally, Christensen appears deeply sceptical about the notion of technical limits:

> *nobody knows* what the natural, physical performance limit is in complex engineered products, such as disc drives and their components. Since engineers do not know what they will discover or develop in the future, since the physical laws (and the relationships between laws) governing performance are imperfectly understood, and since possibilities for circumventing known physical limits cannot be well foreseen, the natural or physical limits cited by scholars of technological maturity. . . . may in practice be moving targets rather than immovable barriers.[27]

> **Illustration 4.2 IBM achieves new breakthrough in microprocessor speed—shifting the technical limit**
>
> Recent research undertaken at IBM and the Georgia Institute of Technology has achieved a new record-breaking performance speed for commercial silicon-based technology. Although the speeds of half a trillion cycles per second were obtained at very low temperatures (−268°C), the scientists argue that: '*This work redefines the upper bounds of what is possible using silicon-germanium nanotechnology techniques.*'[28] This case reinforces the point made above by Christensen that, in practice, technical limits may '*be moving targets rather than immovable barriers*'.

Problems also arise in attempting to map an S-curve for a technology where those measures of technical performance considered to be the most important by the customer are not able to be expressed in technical terms (i.e. are not easily measurable). Foster notes that this is a particular issue with regard to consumer products, rather than industrial products, in relation to which sophisticated industrial customers are used to focusing on measurable criteria.[29] Consider a consumer product such as a detergent for cleaning clothes: how do scientists measure 'brightness', 'freshness', 'cleanness', or 'softness'?

Even where the desired criteria of technical performance for a technology can be identified and measured, a subsequent focus on moving up the S-curve towards the technical limit for these criteria can lead to a rather 'blind' technology-push orientation to R&D (see **3.2.1** regarding the limitations of the technology-push model of innovation). Returning to the example of detergent for cleaning clothes, the important goal for criteria such as 'brightness' should not be the technical limit achievable through 'optical brighteners' (i.e. the chemicals that reflect light and give clothes a brighter, and, in the eyes of the consumer, cleaner appearance), but the perception of the consumer that the desired performance has been achieved.[30]

Finally, Pogany questions the simplicity of the strategy that is suggested by the S-curve model—that of the abandonment of research on 'obsolescent' technologies to focus on emerging technologies in order to maximize the return on R&D investment—arguing that this approach '*ignores the fact that as an industry matures it also grows in total volume* [and revenue] . . . *A large turn-over can easily pay for* some *continued research*'.[31]

Complications in applying the technology S-curve model

A major complication in applying the technology S-curve arises where the important technical performance measures for a given technology are dynamic. Indeed, it is not uncommon for key performance criteria for a technology to change again and again in response to evolving exogenous factors. In the case of the automobile, for example, factors such as developing consumer tastes and lifestyles, the price of fuel, or environmental legislation have raised the priority of technical performance measures relating to safety, fuel efficiency, and emissions, alongside traditional performance measures of speed and acceleration.[32] Problems may also arise when customers demand improvements in multiple, but conflicting, performance criteria, such as acceleration and fuel efficiency in relation to automobiles, or hard-disk capacity and weight in relation to laptop computers. Trade-offs between such technical performance criteria are commonplace in developing technologies and products, and are exacerbated where customers will not accept a price increase.[33]

The shape of the S-curve may well vary between different technologies. Pogany argues that S-curves:

> are by no means identical in shape. Not only does the inflexion point of different technologies fall at different places . . . the two halves of the curve [above and below the midpoint] need not be symmetrical either.[34]

Thus, it is important to remember that the model is stylized and generalized.

As was noted in **1.3.3**, many innovations or technologies are comprised of a set of 'components' that are configured together in a particular architecture. Because innovation can occur at both the component level and the architectural level, the overall technical performance of an innovation or technology is influenced by performance improvement brought about at either level of innovation. Christensen observed, in his study of computer disk-drive components, that:

> Although one component's performance may be on a plateau—an actual or perceived physical limit—engineers can continue to improve system performance by applying effort to less mature elements of the system design.[35]

Thus, in applying the technology S-curve model in such cases, S-curves would need to be drawn for each of the components (which may be large in number), as well as for the overall architecture or system.

Whilst Foster recognizes that an innovation may result from the confluence of scores of technologies, however, he argues that:

> usually, there are one, two or several technologies that are crucial to a product or its production (the semiconductor chip in a computer or the pole in a vaulter's hands), and these are the technologies with which manager's, inventors and all of us ought to be concerned.[36]

A further complexity in applying the technology S-curve model arises from the fact that many innovations, such as computers, automobiles, and aircraft, are comprised of a whole series of nested levels of technologies. This is best illustrated by an example: Christensen describes the set of nested innovations of a computer disk-drive as follows (see also **Figure 1.2**):

> a read-write head can be viewed at one level as a complex system architecture, comprising component parts and materials that interact with each other within an architected system. At the next level, the head is a component in a disk drive, which itself is a complex architected system, composed of a variety of components. At a higher level, the disk drive is a component in a computer, in which a central processing unit, semiconductor memory, rigid and floppy drives, and input-output peripherals interact within a designed architecture.[37]

Thus, the technical performance of a technology that is comprised of such a set of nested levels of technologies is determined not only by improvements in the technical performance of the components and architecture at the 'upper' level of that nested series, but also by improvements in the technical performance of the components and architecture at each of the 'lower' levels. This would imply that a deep understanding of the technical limits and potential of such a 'nested' technology requires the mapping of an equivalent set of nested S-curves, each in turn with their set of component and architecture S-curves. In light of this, Christensen argues that:

> Limits to performance improvement, while often clear in retrospect, are changing, dynamic concepts in the world of the operating [R&D] manager. Since there are many different component and system [architecture] technology levers to pull in the pursuit of performance improvement.[38]

A similar and overlapping concept is that of 'technological interdependence'.[39] Unlike the previous two points, however, an interdependent or complementary technology may not actually be embodied within the innovation itself: for example, 'enabling' technologies such as computer-aided design (CAD) have had a crucial impact on the rapid development of other technologies, such as large-scale integrated circuits, whilst itself relying heavily on earlier developments in semiconductors.[40]

For Rosenberg:

> The growing productivity of industrial economies is the complex outcome of large numbers of interlocking, mutually reinforcing technologies, the individual components of which are of very limited economic consequence by themselves. The smallest relevant unit of observation, therefore, is seldom a single innovation but, more typically, an interrelated clustering of innovations.[41]

Large variations may also exist in the practical technical limits between different competitors in developing the same technology, due to differences either in proprietary (i.e. in-house) design or technological competence. This can lead to substantial differences between organizations of the appearance of the maturing of a technology, i.e. differences in the point at which diminishing returns on R&D effort in relation to improvements in technical improvement are experienced. With this issue in mind, Christensen suggests that benchmarking against the technical performance of competitors might provide a clearer view of the technical potential of a technology.[42]

Despite the underlying limitations of the model, and the potential complications that exist in its application to complex products and technologies, and whether or not an accurate picture of a technology's S-curve can be achieved, the very process of attempting to apply the S-curve can be valuable: the process of attempting to identify relevant technical performance measures, R&D productivity, technical limits, and technical potential can itself lead to important knowledge creation within an organization about its key technologies, internal technological competences, and the needs of its customers, for example. In this vein, Becker and Speltz have argued that the S-curve *'provides a vehicle for discussion . . . forcing us to ask the right questions'*.[43] Furthermore, it is a process that is likely to bring together strategy makers and technical experts, such as scientists, engineers, and technicians, from other parts of the organization, to develop a more informed technology and innovation strategy. In summary, for Foster: *'Precision isn't as important as point of view. It's enough to know the rough shape of a technology's approach in order to make good judgements.'*[44]

Table 4.1 An overview of the technology S-curve model

Key features	The model highlights the changing relationship between R&D effort (inputs) and improvement in the technical performance of a technology (outputs) over its life cycle. The 'technical limit' and the 'technical potential' (i.e. the gap between the current position on the curve of an organization or the state-of-the-art and the technical limit) are key elements of the model.
Implications for managing innovation	Early on in the life of a technology, patience is required, because R&D effort yields are low as the organization and sector build the foundational knowledge and skills. The diminishing returns on R&D are important for signalling the narrowing of technical potential and the approach of the technical limit, and thus are key for highlighting the need to look for alternative technologies.
Critique—limitations	It represents a fairly crude input–output model, which 'black boxes' important aspects of process and context. It is also very difficult to determine or measure with any accuracy the key elements of the model, such as the technical limit or the current position on the curve of the organization or state-of-the-art.
Critique—complications in application	The important performance criteria for a technology are subject to change and are often in tension with one another. A technology may be comprised of many components, the performance of which will be influenced by their configuration (i.e. architecture). Furthermore, a complex technology will have 'nested' levels of technologies and many interdependencies.

4.5 The product-process cycle model

Seminal work concerning the theorizing and empirical investigation of the relationship between the life cycle of a technology and the life cycle of a sector or 'productive unit' (e.g. a firm or a division of a firm) was undertaken in the mid-1970s by academics at Harvard Business School. Principal among this group of academics were William Abernathy, James Utterback, and Kim Clark. Despite subsequent extensions and revisions, perhaps one of the most enduring models that resulted from this work was the 'product-process cycle' model.[45] In this chapter, we will focus predominantly on the original version of the model.[46]

The product-process cycle incorporates three phases, each with particular patterns of innovation, competition, industry structure, and organization:

- the 'fluid' phase;
- the 'transitional' phase; and
- the 'specific' phase.

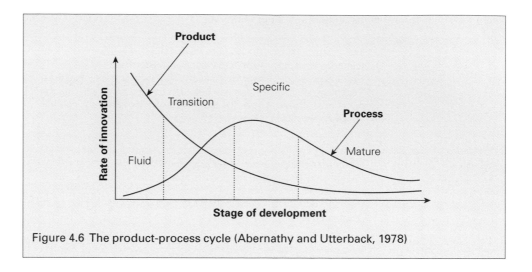

Figure 4.6 The product-process cycle (Abernathy and Utterback, 1978)

These three phases might also be usefully labelled the 'era of ferment', the 'era of emergence of a dominant design', and the 'era of incremental change', respectively.[47] This model is illustrated in **Figure 4.6**.

The fluid phase is associated with the emergence of a new market need or a new way of meeting an existing market need. Initially, in meeting this need, there is often vagueness concerning the appropriate performance criteria of the new product offering. Consequently, this phase is characterized by frequent major changes to product offerings, as well as the presence of a diversity of alternatives in the marketplace. To accommodate such product dynamism and diversity, production facilities need to be general purpose and staffed by highly skilled staff, although they are often inefficient as a result. Organizations able to operate effectively in this environment tend to be small, informal, flexible, and entrepreneurial.

In order to meet rising demand for the new product, the transitional phase sees a shift away from frequent major product innovation to major process innovation. This phase is associated with the emergence of one or a small number of stable product designs, which allows for significant production volumes and the development of 'islands of automation' as production facilities necessarily become more specialized and rigid. Organizations effective in this environment rely increasingly on formalized project or task groups, and on vertical and horizontal organizational communication systems.

In sharp contrast to the fluid phase in particular, the specific phase is characterized by a focus on cost reduction and product quality, and a shift towards incremental product and process innovation to bring about such improvements. It is worth noting that the gradual and cumulative impact of 'countless' minor incremental product and process innovations can have a dramatic impact on productivity.[48] With competition based on cost rather than differentiation, products become increasingly standardized and production facilities are required to be large-scale, capital-intensive, highly specific, and rigid. Organizations able to operate effectively in this phase tend to be large, inflexible, and bureaucratic.

Thus, we see a number of important changes between the 'fluid' phase and the 'specific' phase—most notably:

- the shift from major product innovation to major process innovation, and eventually to incremental product and process innovation (as highlighted in **Figure 4.6**);

- the shift from product diversity to a standardized dominant product design as both market and technical uncertainty are reduced (we will discuss the process of the emergence of a dominant design later in this chapter);

- the shift from competition based on product differentiation to competition based on cost;

- the increasing focus on the efficiency, capital-intensity, and specificity of large-scale production facilities;

- the growth in size of firms within the sector and their subsequent rationalization to a small number of large manufacturers through market competition based on cost; and

- the shift from entrepreneurial ('organic') to bureaucratic ('mechanistic') modes of organizing (a central theme in **Chapter 7**).

At the sectoral level, the model would appear to be a useful tool for both describing and explaining the shift in the patterns of innovation, competition, industry structure, and effective organizational design during the evolution of the underlying technology within that sector.

For Abernathy and Utterback, the model has predictive capacity, because they view these patterns as occurring in a 'consistent' and 'predictable' manner. Where the sector experiences a technological discontinuity, a further iteration of this product-process cycle would be expected to unfold. Furthermore, the model also indicates the evolution of incumbent organizations within a sector, encapsulated by the notion of increasing 'specificity' of resources, processes, and organization.[49] In this regard, the product-process cycle model helps to highlight, in instances of technological discontinuity, why new entrants to a particular sector might be better positioned to exploit a new phase of 'fluidity' than incumbent organizations, which have grown large, bureaucratic, and inflexible. We will return to this issue in **Chapter 5**.

Having provided a broad overview of the main features of the product-process cycle model, we now move on to discuss some of its key implications for the management of innovation.

4.5.1 **The innovation management implications of the product-process cycle model**

At the heart of the product-process cycle model is the shift in the pattern of innovation within a sector during the life of a technology, from major product innovation during the 'fluid' phase, to major process innovation during the 'transition' phase, and eventually to a combination of incremental product and process innovation in the 'specific' phase.

A major implication of this dimension of the model is that for an organization to survive these shifts within a sector, it must evolve the nature of its innovative activity appropriately (i.e. from product to process innovation, and from major to incremental innovation), such that there is a continuing match with the 'requirements' of its environment.

At the organizational level, each phase of the model requires a different strategic orientation, centred on the balance between innovation and efficiency, and a different mode of organizing, concerning the balance between the 'organic' and 'mechanistic' modes (see **7.5.1**). Each phase also requires quite different organizational competences and endowments. For example, in the 'fluid' phase, competences related to obtaining patents and deciphering ill-defined customer needs are important in order to succeed; in the 'transitional' phase, meanwhile, competences in developing distribution channels and networks of complementary are key; and finally, in contrast, in the 'specific' phase, competences around designing products for manufacturability and reducing development cycle times are vital.[50] The overall pattern at the level of the 'productive unit' is one of increasing specificity of its processes, systems, competences, and resources, as the sector evolves.

Combining the above two points, a third emerges: the evolution of a productive unit to match the shifting requirements during the evolution of the sector increasingly 'locks' it into the prevailing technological paradigm and market requirements. The nature and negative consequences of this 'lock-in' are highlighted by Abernathy and Utterback, who note that:

> The productive unit loses its flexibility, becoming increasingly dependent on high-volume to cover its fixed costs and increasingly vulnerable to changed demand and technical obsolescence.[51]

At the sectoral level, each phase of the model is characterized by different market conditions and competitive pressures. The 'fluid' phase, for example, is characterized by highly differentiated products and services. Rivalry between competitors is likely to be less than in later phases, although barriers to new entrants are difficult to erect due to low technical and market certainty. The threat of substitution may also be high, especially where there still exist products based on an earlier technology that remains viable. Lead users and early adopters are also likely to exhibit high bargaining power.

In the 'transitional' phase, rivalry is often increased through the attempt by competitors to establish a dominant design and, where this is achieved, there tends to be a 'shake-out' in the industry. The emergence of a dominant design is also likely to be accompanied by a reduction in the threat of new entrants, but as materials and product offerings become more standardized, the power of suppliers and buyers often increases.

With the shift to the 'specific' phase, rivalry among competitors competing largely on cost is likely to be high, although the threat of new entrants is low. This phase is often accompanied by high bargaining power of suppliers and buyers, and with the maturity of the underlying technology, the threat of substitutes fuelled by a technological discontinuity begins to rise.[52]

4.5.2 A critique of the product-process cycle model

We now turn to a discussion of the main limitations of the model and the key complications that arise in its application.

The limitations of the product-process cycle model

The original work by Abernathy and Utterback presented the trajectory of the product-process cycle as 'unilinear' and 'irreversible'.[53] But whilst research based on detailed archival records of Ford established the utility of the framework, it also highlighted instances of 'reversals' in the cycle.[54] So what brings about such reversals? Abernathy argues that *'the direction of normal development can be reversed through severe changes in the market environment'*;[55] for the automobile industry this might include rising fuel costs, emission regulations, and changing consumer lifestyles. Subsequent research comparing the innovative approach of US and Japanese automobile firms led to an extension of the model through the addition of a 'de-maturity' phase.[56] 'De-maturity' refers to the re-emergence of a period of 'ferment' similar to that which characterizes the early stage of the development of a sector (i.e. the 'fluid' phase). In summarizing the changes to the original model, Clark and Starkey note that:

> The most obvious feature of the revision is the replacement of the original, deterministic, unilinear model anchored in the biological metaphor of the life cycle with a framework of the life course which permits reversals and which requires considerable judgement in its application.[57]

Adner and Levinthal argue that models such as the product-process cycle are essentially focused on the 'supply-side' of technical change.[58] Adopting a 'demand-side' perspective, they offer an alternative interpretation of the dynamics of product and process innovation forwarded by Abernathy and Utterback:

> Viewing the evolution of a technology through a demand-based lens suggests that the early evolution of technologies is guided by responding to the unsatisfied needs of the market . . . product maturity may be as much a function of satisfied needs as it is of exhausted technologies.

Thus, they contend the shift from major product to major process innovation occurs when competitive firms respond to 'technologically satisfied' consumers.[59]

Whilst the product-process cycle would appear to offer a plausible model for the evolution of a manufacturing sector, such as for digital video discs (DVDs), computers, and automobiles, it has little efficacy when applied to a service sector. Following empirical investigations of innovation in the insurance, accountancy, local government, and retail banking service sectors, Barras offers an alternative model: the 'reverse product-process' cycle (see **Figure 4.7**).[60]

In summary, the model has three phases:

1. the application of new technology in a service sector allows for the increased efficiency of existing services;

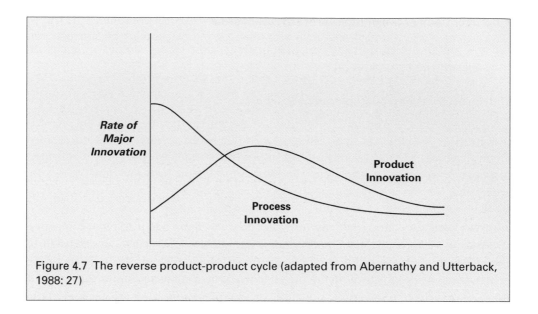

Figure 4.7 The reverse product-product cycle (adapted from Abernathy and Utterback, 1988: 27)

2. the application of new technology facilitates the improvement in the quality of existing services in that sector; and

3. the new technology allows for the development of new services.

Thus, in contrast to the original model in which major product innovation proceeds major process innovation, in the reverse product-process cycle, Barras argues that major process innovation proceeds and facilitates the development of major service (product) innovation.

Good examples of this 'reverse' cycle include the development of automated teller machines (ATMs—i.e. 'cash machines'), telephone and Internet banking and insurance, and Internet retailing (e.g. Amazon.com). In all of these instances, major service (product) innovation was proceeded by major process innovation facilitated by the application of new computing and network technologies—sometimes referred to as 'enabling techno-logies'. Whilst these examples demonstrate the utility of the reverse product-product cycle in conceptualizing the evolution of innovation in the service sector, Gallouj warns that the model *applies mainly to those vanguard services . . . which are most affected by techno-logical evolution* and thus, in orientation, *remains fundamentally technologist*. As a con-sequence, he notes that, from this perspective *innovation is not really considered to occur outside of "technological possibilities"*;[61] yet many innovative services are not facilitated by technology—consider the emergence of the range of innovative new 'lifestyle' services designed for busy executives, for example.

It is also important to note that the process of transition between the different phases of the model is not 'frictionless' for the incumbent organizations. Neither the original nor revised models emphasize the problems, however, or the process by which incumbent firms successfully achieve a transition between the distinct phases of the product-process cycle. This is important, because as Clark and Starkey note:

> The temporal unfolding of sectors is . . . characterized by discontinuities which . . . require existing firms to engage in a transition of their corporate policies and pre-existing structural repertoires [organizational routines].[62]

Furthermore, dominant designs do not always emerge in a sector.[63] Whilst Abernathy and Utterback argue that the *'more interesting cases are those where the transition . . . though predicted, has not come about'*, this interest lies *'in identifying barriers and pinpointing appropriate responses'*;[64] they do not explicitly note that dominant designs may sometimes not emerge.

For DeBresson and Townsend, the assumption that lies behind the clear ordering between major product and major process innovation—i.e. that firms will largely ignore the production process and the control of costs until the product design is stable—is problematic.[65] This view is supported by Adner and Levinthal, who argue that:

> While design improvements are a primary driver of market acceptance, in many instances cost savings provide a critical motivation for technological substitution and are therefore an early focus of innovative effort.[66]

Clark and Starkey also note that the model *'neglects the survival and prosperity of specialist manufacturers'*, arguing that complex sectors, such as the automobile industry, *'tend to contain a diversity of types of enterprise'*.[67]

Complications in applying the product-process cycle model

One of the key problems in applying the model was highlighted early on by Abernathy and Utterback, who noted that: *'Identifying the evolutionary transition from product to process innovation is sometimes troublesome. In some cases the transition may have occurred so rapidly as to be unrecognized.'*[68] The problem of identifying transitions between phases of the model is compounded by a number of other issues: Afuah argues that the *'boundaries [between the different phases] are at best, fuzzy'*, that *'the duration of each phase also varies from product to product'*, and that *'it is not always easy to tell which innovation is process and which is product'*.[69]

As we noted in **4.4.2**, in relation to the complications in applying the technology S-curve, complex products such as computers, automobiles, and aircraft are comprised of a whole series of nested levels of technologies—each of which might be attached to different productive units. Application of the model across a portfolio of productive units within a sector might highlight, as did the research of Ford by Abernathy,[70] that different productive units (e.g. producing engines, brakes, suspension systems, and body) may, at any point in time, be located within different phases and perhaps moving in different directions (e.g. towards maturity or de-maturity).

Despite the various limitations and complications in applying the product-process cycle framework, the model is useful in its attempt to link the trajectories of innovation (e.g. product and process), sectoral characteristics (e.g. the nature of competition), and features of the productive unit (e.g. mode of organizing). But as Abernathy and Clark argue, *'industry development may be richer and more varied than simple life cycle notions might suggest'*; their revised framework—the 'transilience map'—consequently adopts a 'life course' rather than 'life cycle' metaphor.[71]

Table 4.2 An overview of the product-process cycle model

Key features	The product-process cycle incorporates three phases, each with particular patterns of innovation, competition, industry structure, and organization: 1. the 'fluid' phase; 2. the 'transitional' phase; and 3. the 'specific' phase. During the life cycle, there is a shift from product to process innovation, major improvement to incremental innovation, entrepreneurial to large 'mechanistic' organizations, and competition based on differentiation to that based on cost.
Implications for managing innovation	At the organizational level, each phase of the model requires a different strategic orientation, centred on the balance between innovation and efficiency; and a different mode of organizing, concerning the balance between the 'organic' and 'mechanistic' modes, as well as different competences and resources. At the sectoral level, different market conditions and competitive pressures characterize each phase of the model.
Critique—limitations	The original model is 'unidirectional' and 'irreversible', however, subsequent versions allow for de-maturity during the 'life course' rather than 'life cycle' of a technology. The model has little efficacy when applied to a service sector, hence the development of the reverse product-process cycle. It underemphasizes problems of transition for incumbent organizations.
Critique—complications in application	Identifying the evolutionary transition between phases of the model is problematic, in part due to the 'fuzziness' of their boundaries. It is difficult to apply to complex products, such as computers, automobiles, and aircraft, which are comprised of nested levels of technology.

4.6 The dominant design model

The concept of 'dominant design' is generally attributed to William Abernathy and James Utterback. As we saw above, it is a key element in their model of the evolution of a technology in relation to the unfolding pattern of innovation and industry structure. Utterback defines a dominant design in a product class as '*one that wins the allegiance of the marketplace, the one that competitors and innovators must adhere to if they hope to command significant market following*'.[72] For Tushman et al., the emergence of a dominant design or industry standard in a marketplace signals the ending of a period of technological turbulence and competitive ferment.[73] Examples of dominant designs can be seen in diverse product classes such as refrigerators, bicycles, PCs (software and hardware), electric light bulbs, and passenger aircraft. Dominant designs are also prevalent within a diverse range of service sectors, including insurance, mortgages, personal banking, and fast food, for example.

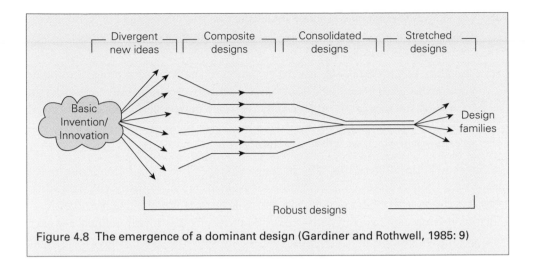

Figure 4.8 The emergence of a dominant design (Gardiner and Rothwell, 1985: 9)

In operationalizing the concept in their historical analyses of the US minicomputer, cement, and glass industries, Anderson and Tushman considered a dominant design to have emerged when one design had accounted for 50 per cent or more of new sales for three years in a row: this provides a more explicit measure of design dominance.[74] Not all dominant designs will be owned and controlled by individual organizations, but those that are, such as Microsoft Windows and Microsoft Office, will provide these organizations with the potential to wield tremendous power in their respective marketplaces.

Gardiner and Rothwell developed the original ideas of Abernathy and Utterback to provide a more detailed model of the process that leads up to, and follows, the emergence of a dominant design (see **Figure 4.8**). They identify six phases, as follows.[75]

1. *The invention phase*—that is, the invention of an artefact or process that brings about the emergence of a new technology: for example, the telephone by Graham Bell in 1876, the 'flying machine' by the Wright brothers in 1903, and the simultaneous development of the semiconductor silicon chip by Jack Kilby and Robert Noyce in the late 1950s. Often, such inventions bring about a technological discontinuity.

2. *The divergent design phase*—the emergence of a new technology leads to a great variety of competing innovations, with 'divergent' designs in delivering that technology to the marketplace. This is described as a period of 'competitive ferment' and an 'era of design competition'.

3. *The shake-out phase*—there is a 'shake-out' among the divergent set of designs that are initially available in the marketplace, followed by the evolutionary improvement of a narrow range of 'composite' designs.

4. *The dominant design phase*—as the market matures, an even narrower set of 'consolidated' designs emerge, with the possibility of the emergence of a single 'dominant' design.

5. *The design stretching phase*—once established in the marketplace, the dominant design can then be 'stretched', through the development of new materials, components, and accessories.

6. *The design family phase*—subsequent evolution of the dominant design can occur through the development of 'design families'. The development of a range of design families off the back of the original Sony Walkman design platform has been one of the key reasons for Sony's continued dominance of the personal stereo market[76]—an example to which we will return shortly.

Gardiner and Rothwell argue that 'robust' designs emerge when a technology has evolved through the various stages of composition, consolidation, and subsequent stretching. They also argue that robust designs are more likely to be commercially successful—and thus become dominant designs—than 'lean' designs (i.e. highly refined designs), because they are more flexible and adaptable. To illustrate the point, compare the huge commercial success of the Boeing 747 'Jumbo' Jet (i.e. a 'robust' design) with that of 'Concorde' (i.e. a 'lean' design).[77]

The 'groundwork' for the emergence of a dominant design often occurs during the early stages of a technology, during which 'standardization' activity focuses on the creation of a common language—'nomenclature and symbols'—and begins to address issues of technical performance expectations and certification.[78] Such activity aids the processes of knowledge creation, accumulation, and diffusion. Thus, whilst tangible signs of the emergence of a dominant design might appear during the latter phases of the third stage of the Gardiner and Rothwell model, more intangible signs are likely to be present during the second stage. Furthermore, basic design choices made early on—particularly around the overall 'architecture' of a technology—have numerous impacts on subsequent design choices.

Illustration 4.3 The bicycle and the emergence of a dominant design

The above six phases of the emergence of a dominant design can be applied to the development of the bicycle as follows.[79]

1. *The invention phase*—the first bicycle, the 'Hobby Horse', was developed in 1817.

2. *The divergent design phase*—there emerged literally hundreds of competing designs between the 1830s and 1880s, among which were many rather extraordinary-looking designs including two-, three-, and four-wheel configurations.

3. *The shake-out phase*—by 1880, through trial and error, three bicycle designs had emerged as front-runners: the 'Penny Farthing', the tricycle, and the rear chain-driven bicycle.

4. *The dominant design phase*—the Rover 'Safety' bicycle (1885), equipped with pneumatic tyres (1888), displaced all others to become the dominant design. This design is much like the bicycle that we recognize today.

5. *The design stretching phase*—new and improved materials (e.g. alloy steels), components (e.g. hub gears), and accessories (e.g. dynamo lights) 'stretched' the dominant design, and allowed innovative organizations to modify the core design for a variety of uses, such as track racing, off-road cycling, and touring.

6. *The design family phase*—the introduction of the 'Moulton small-wheeler', in 1962, spawned a whole family of new designs, such as the 'Chopper', 'Shopper', and 'BMX'.

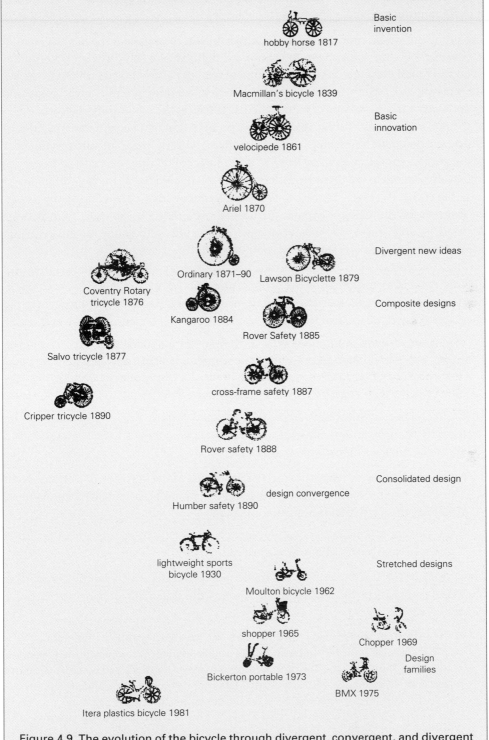

Figure 4.9 The evolution of the bicycle through divergent, convergent, and divergent phases between 1860–1980 (adapted from Roy and Cross, 1983)

4.6.1 The innovation management implications of the emergence of a dominant design

The dominant design model provides a number of clear messages for the management of innovation over the life cycle of a technology, highlighting the varying opportunities open to incumbent firms as compared to new entrants at each of the different stages. For example, following the emergence of a new technology, there is great scope and opportunity for organizations to enter the marketplace, because during the early stages of the life cycle of a technology, there exists a proliferation of divergent designs, none of which yet dominate the market. If one considers the early stages of the development of the home computer during the late 1970s and early 1980s, there existed a wide array of design alternatives to the IBM-compatible, which did not emerge as the dominant design until the end of the 1980s. In the UK, for example, design alternatives during the divergence phase for the home computer included those offered by Sinclair, Commodore, Amstrad, Acorn, and Apple—many of these companies were young, small, entrepreneurial firms, and new entrants to the computer industry.

Following the early phase of the life of a new technology, the dominant design model highlights the subsequent narrowing of opportunity and design alternatives. As competition between designs grows, organizations need to seek ways to survive the shake-out in the marketplace. But to what extent is survival the result of superior design? Tushman et al. argue:

> Dominant designs and technological discontinuities are not technically determined; rather they are windows of competitive opportunity where managerial action can shape market rules and subsequent innovation patterns . . . During eras of ferment, management can move to shape the closing on a dominant design.[80]

Thus, after an initial period of diverging technological designs, organizations need to act strategically to shape both the marketplace and the trajectory of the technology. Such a strategy must incorporate two key strands:

- the establishment of diffusion mechanisms, such as the licensing of in-house technology designs to other manufacturers, the opening up of distribution channels, and advertising to build strong brands; and

- the attempt to influence industry standards, especially those that are set by either governmental bodies or industry consortia.

Drawing from the case history of the home video recorder, Cusumano et al. outline the manner in which JVC shaped the closing of the dominant design in this major consumer market:

> A few important moves made the difference. JVC created a winning alignment of VCR [video recorders] producers in Japan by the way its managers conducted the formation of alliances, showing versatility and humility, whereas Sony pressed commitment and reputation. The alliance with the giant Matsushita brought huge added benefits . . . JVC's early success in aligning itself with Matsushita and other Japanese producers allowed the company to gain a decisive edge in the race for distribution rights. Sony's reluctance to be an OEM supplier, and its underestimation of the threat from VHS, left Beta in a minority position for potential market power in North America and Western Europe as well as Japan.[81]

As the market evolves and is shaped by various stakeholders, a small number of consolidated designs are selected. In such a market—where the focus is shifting from product differentiation to product cost, as indicated earlier by the product-process cycle model—late entrants will find it increasingly difficult to enter with new designs that compete directly with the well-established, dominant designs of incumbent firms.

Innovation opportunities, however, do exist for new entrants and incumbent firms alike, via the stretching of the dominant design (e.g. through the improvement of components, accessories, and materials) or the development of market niches. For example, despite the dominance of the IBM-compatible, Apple went on to create a number of niche markets for its computers in areas such as publishing, whilst Acorn Computers carved out a niche in the UK primary education sector. Where there is no patent protection for a dominant design—such as in the case of the Sony Walkman, which relies largely on existing technologies—opportunities for stretching the dominant design allow companies to maintain their dominant position in the marketplace. Indeed, a key factor in the continuing dominance of Sony in the personal stereo market has been its 'variety-intensive product strategy'.

Following their in-depth study of the history of the Sony Walkman, Sanderson and Uzumeri argue:

> In the case of the Walkman, moreover, Sony held no determining patents and was unable to defend any technological barriers to entry. While Sony's brand name undoubtedly carried weight, its competitors (Panasonic, Toshiba, and Sanyo) were also well known and highly regarded . . . In examining Sony's decade-long dominance, we believe there is reason to suspect the existence of an important new form of competitive advantage, namely a firm's skill at managing the evolution of its product families . . .[82]

Thus, through the stretching of the dominant design and the development of a number of product families, Sony has both maintained dominance in its original market segment and opened up a number of new market segments, including the market for outdoor lifestyles (i.e. 'Sports Walkman'), for young children (i.e. 'My First Sony'), and for audiophiles (e.g. the 'WMD6C' and other direct-drive models).

Fleck brings an interesting perspective to bear on the issue of design dynamism, by focusing on what he terms the 'artefact–activity couple'—that is, the relationship between the innovation and its use. This allows for changes in either the supply-side or the demand-side, through changes in either the innovation (i.e. the 'artefact') or its use(s) (i.e. the 'activity'). Thus, from this perspective, a change in the use of an innovation, for example, may induce a variation in the design of the innovation, perhaps following a period of design stability; alternatively, a change in the innovation may induce new uses or perhaps undermine existing uses. In this way, Fleck argues that *'if contextual changes are relatively minor, the artefact-activity couple may be able to track them to a new stable point'*.[83]

4.6.2 A critique of the dominant design model

Having described the dominant design model and highlighted the key implications that it raises for the management of innovation, we now turn to a discussion of the limitations of the model and the complications that arise through its application.

The limitations of the dominant design model

As noted earlier, dominant designs do not always emerge. Furthermore, many of the dominant designs that do emerge in a new technology are not necessarily better than competing designs (e.g. the IBM-compatible PC, which triumphed over the Apple PC, and the VHS home video recorder format that succeeded at the expense of the Beta format), and they frequently lag behind the state-of-the-art at the time of their introduction. Nevertheless, a dominant design comes to embody the expectations of the marketplace in relation to technical performance and functionality, and sets the benchmark for all subsequent designs.[84] Whilst Gardiner and Rothwell's model provides insight into the stages in which a dominant design emerges, it does not tell us *why* a dominant design emerges.

So why do dominant designs emerge? Reference to other work in strategy and innovation studies can help to throw some light on this question. Porter suggests that the development of brand loyalty and switching costs from 'first-to-market' innovations can enable organizations to establish dominant designs in a marketplace and sustain competitive advantage.[85] But research has indicated that such 'first-mover' advantages are far from guaranteed.[86]

Alternatively, Afuah posits that manufacturers are forced to interact with customers, suppliers, and competitors to resolve the technological and market uncertainties that pervade in the early stages of the evolution of a new technology; this, in turn, leads to a process of isomorphism in design.[87]

For Abernathy and Utterback, the emergence of a dominant design is an inherent and logical transition in the evolution of a technology, which allows for the shift to efficient mass production.[88]

Perhaps one of the more convincing explanations is that of 'network externalities'.[89] The concept of 'network externalities' refers to whether the usage patterns of a technology are linked to:

- the availability of complementary products (e.g. such as applications software for PCs, or recorded videos for home video recorders); or
- the need for compatibility between different users of the technology (e.g. such as between mobile phone or email users).

A dominant design is more likely to emerge where one or both of these network externalities is important to the user in using the product.

How then do dominant designs emerge? Gardiner and Rothwell's model also tells us little about the actual market, regulatory, or social processes by which a dominant design emerges. Grant usefully distinguishes between dominant designs that are imposed by governments or regulatory bodies through the setting of technical standards (e.g. standards for television broadcasting and wireless telecommunications) and de facto standards that emerge out of competitive rivalry (e.g. the 35 mm camera, the VHS format for home video recorders, and the MP3 format for audio computer files).[90]

Shapiro and Varian argue that one mechanism by which dominant designs emerge is through 'bandwagon' effects in the take-up of one particular product over and above another. 'Bandwagon' effects occur where early sales in an innovation lead to rising interest in that product, and consequently to a momentum that builds among distributors and customers to support the product, regardless of whether or not it is technically superior to alternative designs.[91]

Tushman et al. suggest a more complex and combative context—an 'era of ferment' and 'design competition' enacted among a range of stakeholders:

> Except for the most simple, nonassembled products (e.g. cement), the closing on a dominant design is not technologically driven. Rather, dominant designs emerge out of competition between alternative technological trajectories initiated and pushed by competitors, alliance groups, and governmental regulators, each with their own political, social, and economic agendas.[92]

But whilst research has highlighted cases of standard setting achieved through *either* competition[93] *or* cooperation,[94] recent research indicates a trend toward a hybrid of *both* competition *and* cooperation.[95]

Overall, the dominant design model represents technological progress as a single, unidirectional cycle—a key limitation that we noted earlier in relation to the product-process cycle model. Thus, a further limitation of the dominant design model is that although it allows for the stretching of the dominant design during maturity, it does not explicitly accommodate the possibility of a reversal of the overall process (i.e. 'de-maturity') where, for example, the existing dominant design is usurped by another design, or where there is a further period of design divergence before a new dominant design emerges.

Complications in applying the dominant design model

We noted in our discussion of the complications in applying the technology S-curve that many innovations, such as computers, automobiles, and aircraft, are comprised of a whole series of technologies. This also raises complications for the dominant design model, because within such innovations, a variety of dominant designs may exist at either the component or architectural levels. If we take the case of the IBM-compatible PC, we see that there are many dominant designs: for example, the central processing unit (i.e. Intel microprocessor); the user-interface (i.e. Microsoft Windows); and various applications software (e.g. Microsoft Word and Microsoft PowerPoint). Thus, with the development of such complex innovations, there is likely to be fierce competitive manoeuvring in a number of different technological arenas to establish various dominant designs.

The notion that a complex technology may incorporate a number of dominant designs in each of its constituent technologies is further complicated by the fact that their emergence might run in parallel or may be staggered over time. For example, in their discussion of the evolution of the watch industry between 1970 and 1985, Tushman et al. reveal the dynamic and unfolding nature of the competitive manoeuvring for each of the key technological subsystems (e.g. movement, energy source, watch face), noting:

> where oscillation was the key strategic battlefield through the early 1970s, once the quartz movement became the dominant design, the locus of strategic innovation shifted to the face, energy, and transmission subsystems.[96]

In summary, whilst the model is useful in highlighting the various phases of the emergence of a dominant design, it says little of the complex interactional processes that are played out in its emergence. In this knowledge, Tushman et al. contend that:

> While one can know dominant designs and successful product substitutes ex post facto . . . management teams cannot know the 'right' decisions on either dominant designs or substitution events ex ante.[97]

Table 4.3 An overview of the dominant design model

Key features	In the early phase of a new technology, a range of divergent designs are present. After a time, however, there is a 'shakeout' as the market converges around one or a small number of designs that eventually dominate. Opportunities remain in relation to 'stretching' the design through innovation in materials and components, or in the development of new design families.
Implications for managing innovation	The emergence of a dominant design or industry standard in a market signals the ending of a period of technological turbulence and competitive ferment. With the emergence of a dominant design comes the narrowing of opportunities for new entrants, and the need for incumbent organizations to influence the choice of design in the market and/or the industry standard.
Critique—limitations	The model is 'unidirectional', despite the recognition of niche opportunities. The model itself says little about why or how dominant designs emerge. But subsequent research has highlighted non-technological factors, such as developing loyalty among the user base, building wide distribution networks, and influencing standard-setting processes.
Critique—complications in application	It is time-consuming to apply to complex products, such as computers, automobiles, and aircraft, which are comprised of nested levels of technology; dominant designs may exist within each of these levels, and emerge at different times and rates.

4.7 The diffusion curve model

The origins of diffusion research are often sourced to Gabriel Tarde—one of the forefathers of sociology and social psychology—who made a number of observations on the 'imitation' of innovations at the turn of the last century.[98] Although there were a number of anthropological and sociological studies from the 1920s, the roots of the diffusion curve and associated methodology are often attributed to the empirical work conducted by Bryce Ryan and Neal Gross in the 1940s, concerning the adoption of hybrid seed corn.[99] One of the academics most associated with diffusion research is, however, Everett Rogers, who published a seminal text on the subject in the early 1960s entitled *Diffusion of Innovations*,[100] now in its fifth edition.[101] It is to this research that we will primarily be referring.

It is, however, important to highlight here that there has also been a strong tradition of research on technology diffusion undertaken in the area of industrial economics,[102] initially through Kuznets' work in the 1920s,[103] and later through economists such as Griliches and Mansfield in the 1950s and 1960s.[104] There has also been a long tradition of diffusion research from a marketing perspective.[105]

The diffusion curve represents the trajectory of the diffusion of an innovation or technology over time. It is worth noting at this point that 'diffusion' and 'adoption' are 'two sides of the same coin'—that is, the diffusion of an innovation occurs through its adoption by users. As noted towards the start of this chapter, diffusion data can be plotted as either a bell-shaped (non-cumulative or frequency) curve, or an 'S'-shaped (cumulative) curve. To avoid confusion with the technology S-curve, which, as you will recall, measures the improvement in technical performance of a technology in relation to R&D effort, we will plot the diffusion curve as a bell-shaped curve. Such a visual representation also aids the presentation of the different 'adopter' categories (see **Figure 4.10**).

The five 'ideal-type' categories shown below partition individuals in a social system according to the degree to which they are relatively early or late in their adoption of a new innovation.[106]

- *Innovators*—these are individuals with a keen interest in new ideas and they are the first to adopt an innovation in a given target market or social system. They constitute only a small minority of the overall group (i.e. generally only about 2.5 per cent). Sometimes, 'innovators' are proactive in importing an innovation into a social system. In such cases, they act as 'bridges' between social systems that may be geographically or socially distant. Innovators need to be able to deal with both the uncertainty and complexity that often accompany the introduction phase of a novel new innovation. Innovators are not, however, necessarily representative of the broader market or social system, because, as Rogers notes: '*This interest in*

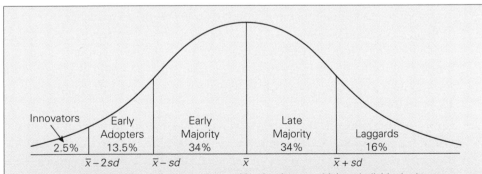

The innovativeness dimension, as measured by the time at which an individual adopts an innovation or innovations, is continuous. The innovativeness variable is partitioned into five adopter categories by laying off standard deviations from the average time of adoption (*x*).

Figure 4.10 Adopter categorization on the basis of time of adoption (Rogers, 2003: 281)

new ideas leads them [innovators] *out of a local circle of peer networks and into more cosmopolite social relationships.'*[107]

- *Early adopters*—unlike 'innovators', this group of individuals are embedded in their local social system. They are 'localites', rather than 'cosmopolites', and represent a larger proportion of the overall group than the innovator group—although still a minority (i.e. around 13.5 per cent). As a result of their central position in the communication networks of a social system, 'early adopters' can have a major impact on the diffusion of an innovation within that social system. They are often considered to be opinion leaders by their peers, but to be continued to be held in esteem, they must adopt innovations prior to their peers and convey their subjective evaluations through their personal networks. This process helps to reduce uncertainty about an innovation in the social system.

- *The early majority*—this group of individuals adopts an innovation just before the average person in a social system. Such individuals make up about one third of the overall group. Although they take longer to deliberate before adopting or rejecting an innovation than innovators and early adopters, the 'early majority' have an important role to play in the diffusion process, through:
  - providing interconnectedness between individuals in the social system;
  - creating momentum in the diffusion of the innovation;
  - creating social pressure on non-adopters to adopt by altering norms in the social system; and
  - building a critical mass of adopters.

This last aspect is particularly important with regard to innovations the utility of which requires others also to adopt (e.g. the telephone or email)—known as 'reciprocal' interdependence, to which we will return shortly, in our critique of the model. The concept of reciprocal interdependence has strong resonance with the concept of network externalities mentioned earlier in this chapter.

- *The late majority*—this group of individuals adopts innovations after the average person in a social system. Such individuals also make up about one third of the overall group. Late adoption in this group arises as a result of a tendency to be more cautious or sceptical of new ideas, or through a scarcity of resources to adopt. Either way, adoption is more likely to occur in this group when the uncertainty about the innovation has been largely removed through the earlier adoption by others in the social system. Often peer pressure is an important stimulus for the 'late majority' to adopt an innovation.

- *Laggards*—these individuals are the last in a social system to adopt. They constitute about one sixth of the overall group (i.e. 16 per cent). Individuals in this group are generally suspicious of new ideas and change, and their point of reference tends to be the past, rather than the future. Many 'laggards' are poorly connected in the social system and their awareness of new innovations often lags far behind that of others. This group may however, also incorporate individuals who are well informed, in whom non-adoption may be the result of a rational and principled position: for example, an environmentalist might decide not to own an automobile.

There has been a great deal of diffusion research aimed at identifying the different characteristics of individuals that fall into the above adoption categories. This research has highlighted important differences between early and late adopters in relation to their socio-economic status, personality traits, and communication behaviour.

In summary, this research has indicated that earlier adopters tend to:

- have a higher socio-economic status;
- be better educated;
- be less dogmatic and open to change;
- be more able or willing to cope with uncertainty and risk;
- be more interconnected in their social system; and
- be actively engaged in information-seeking activities (e.g. surfing the Internet).

Perhaps surprisingly, however, there is inconsistent evidence concerning the relationship between the age of an individual and his or her adoption behaviour, with around 50 per cent of studies showing no relationship.[108] These findings provide useful pointers to innovators, with regard to the selection of potential users to aid the processes of need recognition, idea generation, and prototype testing, for example. We will return to the role of the user in the innovation process in **Chapter 10**. The findings also provide important pointers to marketers or policymakers (see **Illustration 4.4** concerning the switchover to digital broadcasting in Australia) for the targeting of advertising at the different stages of the diffusion process. This particular aspect of the diffusion process is, however, beyond the scope of this text.

Both the early work of Kuznets, and that of Ryan and Gross, highlighted the normal distribution curve for the diffusion and adoption of new technology and innovations.[109] Many studies since the 1950s—of agricultural, consumer, industrial, and educational innovations—have confirmed this finding. But the question is: why does the diffusion of an innovation follow such a pattern? Rogers offers the following explanation:

> We expect a normal adopter distribution for an innovation because of the cumulatively increasing influences upon an individual to adopt or reject an innovation, resulting from the activation of peer networks about the innovation in a system . . . We know that the adoption of a new idea results from information exchange through interpersonal networks. If the first adopter of an innovation discusses it with two other members of the system, and each of these two adopters passes the new idea along to two peers, and so forth, the resulting distribution follows a binomial expansion, a mathematical function that follows a normal shape when plotted over a series of successive generations. The process is similar to that of an unchecked infectious epidemic . . . [diffusion] begins to level off after half the individuals in a social system have adopted, because each new adopter finds it increasingly difficult to tell the new idea to a peer who has not yet adopted, for such nonknowers become increasingly scarce.[110]

Whilst Rogers provides this as a theoretical explanation to demonstrate the nature of the process of diffusion, he does recognize that there are a number of underlying assumptions that can interfere with such a pattern of diffusion. We will return to these assumptions shortly, when we discuss the limitations of the diffusion curve.

Whilst diffusion studies have revealed a 'normal' adopter distribution pattern for a range of innovations, such research has also indicated that the speed of adoption can vary between innovations, as well as between social systems for the same innovation. Based on earlier diffusion studies, Rogers identifies five attributes of innovations that he argues helps explain variations in the rates of adoption, as follows.[111]

- *Relative advantage*—that is, the degree to which an innovation is perceived by potential adopters to be 'better' than existing alternatives. This perception may be in relation to one or more different measures, including 'objective' factors (e.g. price, technical performance, or functionality), as well as more 'subjective' aspects (e.g. convenience, or prestige). The rapid diffusion of the mobile phone can be attributed largely to its perceived relative advantage in relation to convenience and social status, over traditional 'land line' telephones and public telephone boxes. Thus, the perceived relative advantage of an innovation by a target group is positively related to its rate of adoption within that group.

- *Compatibility*—that is, the degree to which an innovation is perceived by potential adopters to be aligned or consistent with their prevailing needs and values, or with their experience with earlier innovations. An often-quoted example of an incompatible innovation is that of birth control in Catholic countries. Thus, the perceived compatibility between an innovation and the values, needs, and experiences of a target group is positively related to its rate of adoption within that group.

- *Complexity*—that is, the degree to which an innovation is perceived by potential adopters to be difficult to understand or use. Novel financial services and products are often perceived by consumers to be complex, and this can dramatically slow down their diffusion. Thus, the perceived complexity of an innovation by a target group is negatively related to its rate of adoption within that group.

- *Trialability*—that is, the degree to which an innovation can be experimented with on a limited basis by potential adopters prior to 'full' adoption. Even where potential adopters may be able to trial the innovation, different types of innovation require different lengths of time to trial. Consider the difference in time frame for the test-driving of a new automobile by a consumer (i.e. maybe around 30 minutes), as compared to the trialling of a new type of seed or crop by an arable farmer (i.e. several months, in order to incorporate at least one full growing cycle or season). Thus, the ability to trial an innovation, combined with the shorter the time frame required to trial that innovation by a target group, is positively related to its rate of adoption within that group.

- *Observability*—that is, the degree to which the benefits of an innovation can be observed by potential adopters. Hoover exploited this attribute of the vacuum cleaner to great effect in the first half of the twentieth century, when travelling salesmen would go from house to house to demonstrate the innovation. Thus, the observability of the benefits of an innovation to a target group is positively related to its rate of adoption within that group.

There are also many potential dimensions of variation between social systems that might help to explain differences in the rate of diffusion and adoption of the same innovation between two social systems: for example, variations in social structure, social norms, geographical distribution, and technological infrastructure.

4.7.1 The innovation management implications of the diffusion curve model

The diffusion curve model and diffusion research, more generally, have a number of major implications for the management of the innovation and commercialization process. These can be grouped under three broad categories:

- the role of key adopter groups;
- the attributes of the innovation; and
- the support for the commercialization process.

Organizations can influence the rate of diffusion and adoption of their innovations by addressing the following issues.

Firstly, innovative organizations need to identify and interact with three key adopter groups:

- *innovators*—because they recognize new needs earlier than the rest of the market and can be an important source of ideas and feedback during the idea-generation and alpha-testing stages of the innovation process, and because they are proactive in adopting new innovations, they often act as 'bridges' to other marketplaces and 'importers' of innovations;

- *early adopters*—because this group tend to be much better embedded into the local social system than innovators (i.e. they are 'localites', rather than 'cosmopolites'), they can thus play a very important role in the diffusion of an innovation to the early majority. Two features of early adopters—their central position in the communication network and their role as 'trendsetters' and 'opinion leaders' within a social system—are valuable resources that can potentially be tapped into by the innovative organization; and

- *the early majority*—because they create momentum in the diffusion of an innovation and build a critical mass of adopters, which is important for organizations because it allows them to increase sales volumes and reduce unit costs. It is also vital for innovations that require others to adopt (e.g. the telephone)—that is, in relation to which there exists 'reciprocal' interdependence.

Secondly, organizations also need to be acutely aware of the various attributes of an innovation that may impact its rate of diffusion, particularly in relation to:

- user perceptions of its relative advantage over existing alternatives;
- user perceptions of its compatibility with prevailing user needs and values;
- user perceptions of the complexity in understanding or using the innovation;
- the degree to which the innovation can be trialled prior to 'full' adoption; and
- the degree to which the benefits of the innovation can be observed by users.

Given that the design of an innovation, as much as any inherent feature of the underlying technology, can influence any one of these attributes, it is important that such dimensions are evaluated throughout the innovation process—that is, an innovation can be designed to influence the diffusion process.

Finally, we noted earlier in this chapter (see **4.6.2**) that the rate of diffusion of an innovation is greatly impacted by the degree to which it becomes, or is compatible with, an industry standard or dominant design. The rate of diffusion of an innovation is also greatly influenced by factors that support the commercialization process, such as the nature and extent of the marketing and advertising effort, and the breadth and speed with which an effective distribution network can be established. These latter issues are beyond the scope of this text; there are, however, a multitude of introductory marketing texts that can provide you with an overview of these activities and direct you towards further reading.[112]

Illustration 4.4 The diffusion of digital television and the 'switchover' to digital broadcasting—the case of Australia

Digital television was first introduced in Australia on 1 January 2001. The relatively low levels of adoption have, however, meant that the Australian government, which has overall responsibility for managing the 'broadcasting airwaves', has already had to delay the final switchover to digital transmission three times. The original date set for the switching off of analogue television transmissions was 1 January 2005. This deadline was first postponed until 1 January 2009, and was later delayed further to a gradual phase-out during 2010 to 2012.[113] More recently, this has been extended until the end of 2013.[114]

Despite the tangible performance benefits of digital over analogue television, price and low awareness, in particular, have had a negative impact on its diffusion. Furthermore, as in many other countries, this has been predominantly a supply-side rather than a demand-side driven affair. In a survey in October 2006, 14.7 per cent of respondents said that they were 'currently satisfied' or saw 'no need' to adopt. This would indicate that a sizeable group sees no comparative advantage of the new technology.[115]

In April 2008, Senator Conroy, the Australian Minister for Broadband, Communications, and the Digital Economy, said '*we still need to convert the nearly 60 per cent of Australians who currently do not receive digital free to air broadcasts*'.[116] This would indicate that digital television was still to penetrate the 'late adopter' and 'laggard' groups, which typically account for the last 50 per cent of adopters.[117] The switching off of analogue television transmissions may well require diffusion levels much closer to 95–100 per cent. To achieve this will require a concerted and integrated approach from the government, the broadcasters, and the technology providers.

4.7.2 A critique of the diffusion curve model

We now turn to a critique of the diffusion curve, once again by assessing the limitations of the model and some of the more notable complications that arise from applying it to 'real world' examples of innovation and technology.

The limitations of the diffusion curve model

Rogers observes that:

> The s-curve of diffusion is so ubiquitous that students of diffusion may expect every innovation to be adopted over time in an S-shaped pattern. However, some innovations do not display an S-shaped rate of adoption.[118]

Why, then, might an innovation not display the expected pattern of diffusion?

The answer to this question lies in challenging one of the core assumptions underpinning the diffusion model—that is, the assumption that interaction between individuals in a market or societal system is unfettered. In reality, in many markets and social systems, members do not necessarily have the freedom to interact with each other; interaction might be limited by geographical barriers or differences in status, and there may be taboos that exist to restrain members from discussing the new innovation openly (e.g. new methods of contraception). All of these factors might impact the rate and extent of the adoption of an innovation.

Furthermore, when analysing the diffusion pattern for an innovation, the adopter categories, and the proportion of the members of the targeted market or system to be allocated to these adopter categories, are defined by the researcher; those illustrated earlier (e.g. 'innovators' accounting for the first 2.5 per cent of adopters; 'early adopters' the next 13.5 per cent, etc.) are those most widely employed in diffusion studies, but they are nonetheless subjective.

The model also does not explicitly take account of subsequent improvements in either the performance or functionality of the innovation that might alter the pattern of its diffusion. Nor does the model take into account the impact of separate, but interrelated, innovations—referring back to the concept of 'technological interdependence' that was mentioned earlier in this chapter. Thus, for example, improvements in manufacturing processes might considerably reduce the cost of a product, making it accessible to a much wider group of potential customers and resulting in a widening of the diffusion of an innovation. Hence Rothwell and Zegveld argue that *'There is no reason, of course, for companies to accept the inevitability of this pattern* [of decline] *and positive action can be taken to stave off the decline'.*[119] Of particular relevance here is the empirical work concerning the PLC (which measures sales of a product over time), which has long demonstrated the multiplicity of different shapes for the PLC curve (see **Figure 4.11**) and highlighted a range of influencing factors.[120]

Rogers also points out that: *'The S-shaped* [diffusion] *curve only describes cases of successful innovation, in which an innovation spreads to almost all of the potential adopters in a social system.'*[121] This highlights two key problems in employing the diffusion model *ex ante* (i.e. before or during the process of diffusion): it is only really possible to determine the potential number of adopters in a system, and thus the extent of success of an innovation, *ex post facto* (i.e. after the event).

This point can be illustrated by taking the example of the diffusion of home video recorders. From its launch in the mid-1970s through to the early 1980s, the home video recorder market was perceived to be a niche product, appealing to specific demographic groups. Forecasts typically projected a levelling off of market penetration at around 15–30 per cent in most advanced countries by the late 1980s. These forecasts proved to be way off the mark, with 29 per cent of UK households owning a home video recorder by 1983, and 30 per cent of US households by 1985;[122] the levels of market penetration for home videos was substantially higher during the 1990s. It is obvious from this example that, if the projected number of potential adopters of an innovation in a market or social system is inaccurate (i.e. either too high or too low), then this makes *ex ante* analysis highly problematic. Thus, the model relies on reliable market analysis of the potential

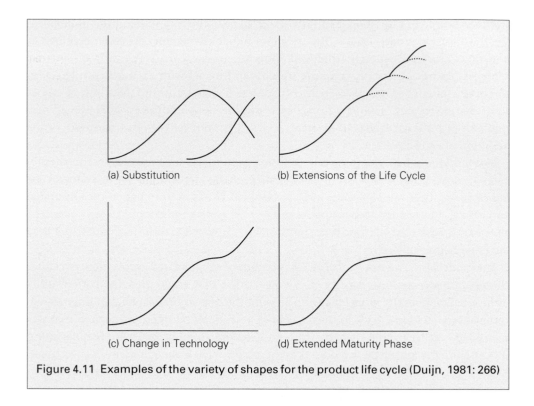

Figure 4.11 Examples of the variety of shapes for the product life cycle (Duijn, 1981: 266)

groups that are likely to adopt the innovation and thus the potential number of adopters; without this, the diffusion curve model essentially becomes relegated to a reporting tool rather than a forecasting tool.

Complications in applying the diffusion curve model

One of the central criticisms of diffusion models based on the diffusion of information through personal contact is that access to information about an innovation is only the first step in a complex process of its adoption, adaptation, or rejection by an individual.[123] Rogers clearly recognizes this: his book includes a chapter on the 'innovation-decision' process. Nevertheless, this is an important message to keep in mind when applying such diffusion models.

It is also worth noting that the diffusion pattern of 'interactive' innovations, such as the telephone, electronic messaging systems, fax, and teleconferencing, differs from the diffusion pattern of discrete or non-interactive innovations, because the utility of such an innovation to an individual is influenced by the adoption of the same innovation by those with whom that individual wishes to interact. Thus, unlike non-interactive innovations, which exhibit 'sequential' interdependence (i.e. earlier adopters influence later adopters by affecting their perception of the benefits of adoption), interactive innovation experiences 'reciprocal' interdependence (i.e. earlier adopters influence later adopters, and visa versa), because each additional adoption increases the utility of the innovation for both current and future users.[124]

Finally, there is a common assumption in innovation diffusion studies that potential adopters choose either to reject an innovation, or to adopt it intact. There are, however, two other important and distinctive possibilities that need to be accounted for: the adaptation of the innovation, or its adoption and subsequent discontinuance at a later stage, either through dissatisfaction with the innovation or the emergence of a better alternative. In its current form, the diffusion curve model appears to conflate adoption and adaptation, into adoption, and to ignore discontinuance.

Interestingly, Rogers provides an even more fundamental critique, levelled at the very nature of prevailing diffusion research:

> One of the most serious shortcomings of diffusion research is the innovation bias . . . [This] is the implication in diffusion research that an innovation should be diffused and adopted by all members of a social system . . . the bias is assumed and implied . . . The bias leads diffusion researchers to ignore the study of ignorance about innovations, to underemphasize the rejection and discontinuance of innovations, to overlook re-invention, and to fail to study antidiffusion programs designed to prevent the diffusion of 'bad' innovations (like crack cocaine or cigarettes, for example).[125]

Table 4.4 An overview of the diffusion curve model

Key features	The diffusion curve represents the typical diffusion pattern for an innovation. The model identifies five adopter categories, as well as five innovation attributes, which help to identify features of both users and innovations that impact the speed of diffusion/adoption of an innovation.
Implications for managing innovation	Different adopter categories have different roles to play in the diffusion/ adoption process, as well as in the innovation process. Shaping the innovation with regard to the five innovation attributes identified will help promote the rate of diffusion/adoption of an innovation.
Critique—limitations	Without reliable market data for the potential groups that are likely to adopt the innovation, and thus the overall potential number of adopters, the diffusion curve model essentially becomes relegated to a reporting tool rather than a forecasting tool. The model does not explicitly take account of subsequent major improvements in the performance, cost, or functionality of the innovation that might alter the pattern of its diffusion.
Critique—complications in application	The diffusion pattern of 'interactive' innovations, such as the telephone, fax, email, and teleconferencing, differs from the diffusion pattern of discrete or non-interactive innovations. An innovation may be adopted, rejected, adopted and adapted, or adopted and then later abandoned.

In summary, the diffusion curve model is a rather simplistic representation of the diffusion of an innovation or technology and suffers from much the same criticisms as those levelled against the PLC. Nevertheless, it does alert us to the different types of adopter at the different stages of the diffusion of an innovation or technology: 'innovators', 'early adopters', and the 'early majority', each have a potentially important, although differing, role to play in the innovation and commercialization processes. As noted earlier, we will return to the role of the user in the innovation process in **Chapter 10**.

The diffusion literature also highlights the various attributes of an innovation and the features of the commercialization process that can impact the rate of diffusion of an innovation; these are important contributions to the literature on innovation.

4.8 Concluding comments

This chapter opened by noting the importance of incorporating technological analyses into the strategy formulation process within organizations. Two key questions were then raised:

- can we identify patterns of innovation in the evolution of a technology?
- if so, what messages do these patterns highlight for the management of innovation, and, in particular, in relation to strategic decision making around innovation and technology?

In addressing these two key questions, we discussed the following conceptual models:

- the 'technology life cycle' (TLC) model, which maps the sales volume trajectory of a technology over time;
- the 'technology S-curve' model, which maps the technical performance trajectory of a technology in relation to R&D effort;
- the 'product-process cycle' model, which maps the interrelated trajectories over time of product and process innovations within a sector;
- the 'dominant design' model, which maps the emergence of a dominant design of an innovation or technology over time; and
- the 'diffusion curve' model, which maps the diffusion trajectory of an innovation or technology over time.

Each model was described, its implications for the management of innovation were noted, and its limitations, regarding its underlying assumptions and difficulties of application, were discussed. **Table 4.5** summarizes and compares the key phases of each. The purpose of such a review is to aid the reader to develop a more informed view as to the utility of each of these models when considering their application.

For Anderson and Tushman, a technology cycle model, such as those discussed in this chapter, allows managers to view their industry from an historical and temporal perspective, rather than through a snapshot of the 'here and now'. As such, it allows managers to compare the impacts of various patterns of innovation on the structure of their

Table 4.5 A summary of the key features of the models of technological progress at the various stages of the technology life cycle

Model	Introduction stage	Growth stage	Maturity stage
Technology S-curve	The rate of improvement in the technical performance of a technology is slow, because much of the R&D effort is required to develop the basic knowledge underpinning the technology. As a result, the technical performance returns from R&D effort are low.	As knowledge about a technology accumulates and is diffused and applied, the rate of progress in the improvement of a technology begins to accelerate. Here, the technical performance returns from R&D effort are high. But after the 'point of inflection' (the middle of the S-curve) at which the yield is at its highest, the innovator begins to suffer from decreasing returns.	The yield on R&D effort begins to decline at an increasing rate as it approaches the 'natural' or 'technical' limits of the technology.
Product-process cycle	Associated with the emergence of a new market need or a new way of meeting an existing market need. Initially, in meeting this need, there is often vagueness concerning the appropriate performance criteria of the new product offering. As a result, this phase is characterized by frequent major changes to products as well diversity of products in the marketplace.	In order to meet rising demand for the new product, there is a shift away from frequent major product innovation to major process innovation. This phase is associated with the emergence of one or a small number of stable product designs to allow significant production volumes, and the development of 'islands of automation'.	There is a focus on cost reduction and product quality, and a shift towards incremental product and process innovation to bring about such improvements. The gradual and cumulative impact of 'countless' minor incremental product and process innovations can have a dramatic impact on productivity.

Table 4.5 *(cont'd)*

Model	Introduction stage	Growth stage	Maturity stage
Dominant design	The emergence of a new technology leads to a great variety of competing innovations with divergent designs in delivering the technology to the marketplace.	There is a shakeout among the divergent set of designs initially available in the marketplace, followed by the evolutionary improvement of a narrow range of 'composite' designs.	As the market matures, an even narrower set of consolidated designs emerges, with the possibility of the emergence of a single dominant design. Once established in the marketplace, the dominant design can then be 'stretched', through the development of new materials, components, and accessories, or through the development of design families.
Diffusion curve	'Innovators', with a keen interest in new ideas, are the first to adopt an innovation. They are a small minority of the overall group (i.e. only about 2.5 per cent). 'Early adopters', unlike innovators, are embedded in their local social system; they are 'localites' rather than 'cosmopolites'. This group represents a larger proportion of the overall group, although still a minority (i.e. around 13.5 per cent).	The 'early majority' are individuals who adopt an innovation just before the average person. They make up about one third of the overall group. Although they take longer to deliberate before adopting or rejecting an innovation than innovators and early adopters, they have an important role to play in the diffusion process, through creating momentum in the diffusion of the innovation and social pressure on non-adopters to adopt, and help to build a critical mass.	The 'late majority' are individuals who adopt innovations after the average person. They also make up about one third of the overall group. This group tends to be cautious or sceptical of new ideas. Often, peer pressure is an important stimulus for this group to adopt. 'Laggards' are the last to adopt. They constitute about one sixth of the overall group. They are generally suspicious of new ideas and change, and their point of reference tends to be the past.

industry.[126] But whilst such models are useful in highlighting some of the common patterns that occur during the life of a technology, perhaps the most fundamental critique of these models is that they are overly deterministic and that, as such, they underplay the potential for innovating organizations themselves, as well as users, governments, and pressure groups, for example, to influence the shape of, and speed of movement along, the trajectory of a given technology.

Whilst the predictive capacity of such models may be much weaker than is often portrayed, they do provide the language and tools for discussing the past, current, and future technological environment; this is an important prerequisite to the formulation of a considered innovation and technology strategy. Nevertheless, commenting on the evolution of such models, Clark and Staunton observe that:

> The revelation of long-term empirical patterns seems to pass through two phases. First there is the claim to have discovered an empirical regularity whose basis in statistical data is portrayed as conclusive and whose wave-like profiles are evident . . . Second, subsequent thinkers—and even the original contributors—qualify the earlier position by suggesting 'the patterns' are principally of heuristic value as guiding principles in the development of a processual, dynamic perspective.[127]

A further criticism is raised by Adner and Levinthal, who argue that:

> by far the larger portion of the work on technological change is concentrated on the 'supply-side' dynamics . . . Relatively underexplored in these discussions of innovation is the effect of the demand environment on the development and evolution of technology.[128]

This criticism could be levelled at a number of the models discussed in this chapter—in particular, the technology S-curve and the product-process cycle. This (implicit) supply-side orientation has a major impact on the interpretation of an observed pattern of technical change, as well as on the range of variables that are embraced by the analysis. It is important to keep this in mind when using such models.

Building on the above foundations, in **Chapter 5**, we turn our attention to the process of transition from one technology to another—that is, the notion of 'technological discontinuity, and to the relationship between economic long waves and innovation.

CASE STUDY 4.1 VIRTUAL REALITY—THE SLOW DIFFUSION OF A PROMISING TECHNOLOGY

Introduction

Virtual reality (VR) emerged as a mainstream business tool in the early 1990s, amid huge expectation and promise from VR pioneers and industry commentators alike. For example, Hamel and Prahalad, in their popular strategy text *Competing for the Future* of the mid-1990s, described VR as '*a technology with profound implications for almost every industry*'.[129] Indeed, at around this time, nine of the fifteen 'Technology Foresight' sector panels established by the British government recognized the strong potential of VR in sectors as diverse as construction, defence, aerospace, financial services, leisure, and manufacturing. Swann also notes from his analysis of the business press during the 1990s that, whilst there were no articles concerning VR before 1990, by 1996, VR was being discussed as much as biotechnology.[130]

We will adopt the definition of VR as a '*collective term for those computing technologies which enable the user to interact in real-time with 3D computer-generated environments*'.[131] As we will see later, this rather broad definition incorporates a range of different categories of VR, some of which VR 'purists' would not classify as such.

By the late 1990s, VR had been employed successfully across a range of sectors and in a variety of novel applications—for example:[132]

- by the British Broadcasting Corporation (BBC), to develop studio sets and explore the feasibility of new programme concepts;
- by Boeing and McDonnell Douglas, to support engineers in evaluating the maintainability of pre-manufacture aircraft designs;
- by Rank Xerox, as a sales support tool;
- by British Airways, as a tool for modelling passenger flow through airport terminal buildings; and
- by John Deere, to create virtual prototypes of earth-moving equipment.

Nevertheless, despite such successes, there has been a notable shift in tone, among users in particular, from enthusiasm to scepticism. Many commentators have blamed this shift on the excessive hype of the technology during the 1990s and the associated unrealized promises attributed to VR. Does VR simply represent a classic case study of an over-hyped technology, or are there other explanations as to the slow diffusion of such a versatile and promising technology?

The ambiguity surrounding an emergent technology

Many of the features of the prevailing state of VR are typical of an emergent technology, such as the ambiguity concerning design trajectories and user needs. Indeed, there is still disagreement among the VR community as to what constitutes VR and what does not. Some of this ambiguity is attributable to the immaturity of the technology, but the diverse origins of VR—from robotics and computer-aided-design (CAD), media and the arts, to various scientific disciplines—also bring together quite different traditions and languages.[133]

The following features can be identified.

- *Contestation concerning what constitutes VR*—Swann distinguishes between three types of VR system: (1) 'immersive' VR, where the user is immersed in a 3D environment and interacts with it through the use of hardware interfaces such as 'head-sets' and 'gloves'; (2) 'widescreen' VR, where the virtual world is projected onto a large screen in order that a number of individuals can view the interaction between an individual and that world; and (3) 'desktop' VR, where the user interacts with the virtual environment via a 'mouse' and views the virtual world on a standard PC monitor. Whilst the last of these three is the fastest growing market, only the first of these categories is considered to be VR by 'purists' within the VR community.[134]
- *The 'paradox of ubiquity'*—there exists a range of technological possibilities open to VR providers that are also acceptable to potential VR users. Whilst this may initially appear favourable to the future of the VR market, this may actually prove to be counterproductive if it delays the emergence of a smaller number of designs around

which a critical mass of activity can accumulate. This has been termed the 'paradox of ubiquity'.[135]

- *The lack of a common language*—the language surrounding VR is still underdeveloped and fragmented, not least due to the multiple origins of the technology noted above. There is also some ambivalence towards the term 'virtual reality' itself, due to its association with the hype of the 1990s.[136]

- *The lack of common design and interface standards*—whilst these standardization issues are being recognized and addressed by technology providers, Watts et al. have argued that:

> the lack of well-defined technical standards for VR has been a major obstacle to developing useful VR models. In particular, some have spoken of considerable difficulties with interfaces between existing . . . simulation tools and new VR tools.[137]

Factors such as those described above indicate that VR is still in what may be termed a 'pre-paradigmatic stage'. Such a stage of development of a technology is generally associated with ambiguity and risk for both innovator and user, and this is likely to stall the 'take-off' of the diffusion of the technology.[138]

The characteristics of VR technology

There are also a number of factors that have negatively impacted the diffusion of VR, which might be considered to be inherent to the technology itself. For example:

- *the difficulty and cost in constructing virtual environments*—whilst it is relatively easy for non-experts to learn how to navigate around virtual environments, Watts et al. argue that *'exploiting VR applications is not simply an issue of making the technology available for people to use . . . Learning how to construct a useful virtual environment is much harder'*.[139] Furthermore, there are high costs involved in the construction of virtual worlds, particularly when building 'immersive' VR applications;

- *the difficulty in constructing user-specific virtual environments*—building on the more general point above, the construction of virtual environments for specific user applications requires a great deal of interaction and understanding between technology experts, human factor experts, and end-users if 'useful' content is to be embedded within the virtual world.

But it is highly possible that future developments in both technology and process will help reduce both the cost and difficulties of constructing virtual environments. Such developments have the potential to encourage the wider diffusion of VR.

The marketing and selling challenge for VR

Despite the high profile afforded VR in the business press, Watts et al. observed that there still exists:

> very limited awareness of the capabilities of VR systems and their relevance to business issues. Many people still associate VR purely with games and entertainment and do not see it as a serious business tool.[140]

This lack of awareness can, in part, be explained by the poor quality of selling materials; prospective users have found it difficult to appreciate the value of VR from existing VR applications that are not specific to their own needs.[141]

As we alluded to above, however, user-specific materials require in-depth knowledge of user needs and applications, and can be very costly to prepare. Furthermore, whilst academics and experts might make clear distinctions between technologies such as CAD, multimedia, animation, and VR, the boundaries are far more blurred than is often portrayed, which can only serve to confuse the non-expert as to the added value of VR over other technologies.

Scepticism has been compounded by the fallout from the hype of VR in the early 1990s—labelled by some commentators as 'poor sales practice'. On this point, Swann argues that:

> In the early days of the VR market, around 1990, many VR pioneers and VR vendors were making extravagant claims about the potential of VR. But it soon became clear that the business benefits from adoption of VR were not so easily accessible as those pioneers and vendors had suggested. Some users became disenchanted and abandoned VR, and that spread a very sceptical attitude towards the technology which has deterred subsequent adopters.[142]

Such lack of awareness, on the one hand, and scepticism, on the other, have created major barriers to the wider diffusion of VR, and pose a major challenge to the future marketing and selling of VR into the marketplace.

Questions

1. Employing the dominant design model, analyse the current position of the VR market. What are the implications of your analysis for the diffusion of VR?

2. Employing the diffusion curve model, analyse the current position of the VR market. What are the implications of your analysis for the diffusion of VR?

3. Evaluate VR along the five attributes of innovations identified by Rogers (i.e. relative advantage, compatibility, complexity, trialability, and observability). Using your analysis, comment on the impact of each attribute to the rate of diffusion of VR.

EXERCISE 4.1

Select one novel product innovation and one novel service innovation. These should be innovations with which you are reasonably familiar, and ones that have moved beyond the initial introduction phase such that they have had time to develop and diffuse.

Questions

1. Attempt to apply the five models discussed in this chapter to both of your selected innovations.

2. From your application of the various models, comment on the current position and trajectory of your two selected innovations.

3. What were the main problems that you encountered in attempting to apply these models to your selected innovations? Were these problems the same for both your selected product innovation and service innovation?

4. Despite the problems you may have encountered in the application of the various models, in what ways did their application help you develop insight into the trajectory of your selected innovations?

FURTHER READING

1. For an insightful application and critique of the technology S-curve model, see Christensen (2001a; 2001b).

2. For an overview of recent trends in standard setting, see Oshri and Weeber (2006). The article draws upon the case of the recent emergence of a dominant design in wireless device technology to illustrate the movement toward hybrid modes of standard setting involving both cooperation and competition. Also, see Blundel (2006) for an interesting case study illustrating the co-evolution and competition between two designs of dinghy.

3. For a detailed illustration (based on the Sony Walkman) of the way in which an innovative organization can stretch a dominant design, see Sanderson and Uzumeri (1995).

4. See Barras (1986; 1990) for a more detailed discussion of the reverse product-process cycle model and its application to the service sector.

NOTES

[1] Betz (1987: 74).

[2] Dean (1950); Levitt (1965).

[3] For example, Ryan and Gross (1943); Rogers (1962).

[4] Nieto (1997: 126).

[5] Foster (1986).

[6] Christensen (2001a: 124).

[7] Foster (1986: 100).

[8] Sahal (1981).

[9] Foster (1986: 216).

[10] Foster (1986).

[11] Christensen (2001a).

[12] Foster (1986).

[13] Christensen (2001a).

[14] Constant (1980).

[15] Roussel (1984).

[16] Becker and Speltz (1983).

[17] Tchijov and Norov (1989).

[18] Van Wyk et al. (1991).

[19] Foster (1986); Betz (1987); Nieto (1997); Christensen (2001a).

[20] Foster (1986: 98–100).

[21] Moore (1965).

[22] Afuah (1998: 121).

[23] Foster (1986).

[24] Foster (1986: 153).

[25] Becker and Speltz (1986: 21).

[26] Afuah (1998: 121).

[27] Christensen (2001a: 131).

[28] *R&D Magazine* (2006: 10).

[29] Foster (1986).

[30] Foster (1986).

[31] Pogany (1986: 24).

[32] Foster (1986).

[33] Foster (1986).

[34] Pogany (1986: 24).

[35] Christensen (2001a: 131).

[36] Foster (1986: 33).

[37] Christensen (2001a: 129).

[38] Christensen (2001a: 137).

[39] Rosenberg (1979).

[40] Rothwell and Zegveld (1985).

[41] Rosenberg (1979: xx).

[42] Christensen (2001b).

[43] Becker and Speltz (1986: 23).

[44] Foster (1986: 43).

[45] Utterback and Abernathy (1975); Abernathy and Utterback (1978).

[46] For a fuller discussion and critique, see Clark and Starkey (1988: 22–36); Clark and Staunton (1989: 118–20).

[47] Afuah (1998: 137–8), adapted from Tushman and Rosenkopf (1992).

[48] Abernathy and Utterback (1978).

[49] Abernathy and Utterback (1978).

[50] See Afuah (1998: 145–9) for a fuller discussion.

[51] Abernathy and Utterback (1978: 41).

[52] See Porter (1980), in particular, chs 8, 10, 11, and 12; Afuah (1998: 139–45), for a fuller discussion.

[53] Utterback and Abernathy (1975); Abernathy and Utterback (1978).

[54] Abernathy (1978); Abernathy et al. (1981).

[55] Abernathy (1978: 113).

[56] Abernathy et al. (1983).

[57] Clark and Starkey (1988: 35).

[58] Adner and Levinthal (2001).

[59] Adner and Levinthal (2001: 627).

[60] Barras (1986; 1990).

[61] Gallouj (1998: 127, 136).

[62] Clark and Starkey (1988: 25).

[63] Afuah (1998); Grant (2002).

[64] Abernathy and Utterback (1978: 46).

[65] DeBresson and Townsend (1981).

[66] Adner and Levinthal (2001: 614).

[67] Clark and Starkey (1988: 36, 35).

[68] Abernathy and Utterback (1978: 46).

[69] Afuah (1998: 149).

[70] Abernathy (1978).

[71] Abernathy and Clark (1985: 14).

[72] Utterback (1994: xx).

[73] Tushman et al. (1997).

[74] Anderson and Tushman (1991).

[75] Gardiner and Rothwell (1985).

[76] Sanderson and Uzumeri (1995).

[77] Gardiner and Rothwell (1985).

[78] Reddy et al. (1989).

[79] Roy (1986a).

[80] Tushman et al. (1997: 5–17).

[81] Cusumano et al. (1997: 91–2).

[82] Sanderson and Uzumeri (1995: 778).

[83] Fleck (2000: 263).

[84] Anderson and Tushman (1991).

[85] Porter (1985).

[86] Lieberman and Montgomery (1988).

[87] Afuah (1998).

[88] Abernathy and Utterback (1978).

[89] Katz and Shapiro (1986).

[90] Grant (2002).

[91] Shapiro and Varian (1999).

[92] Tushman et al. (1997: 9–10).

[93] Shapiro and Varian (1999); Blundel (2006).

[94] Axelrod et al. (1995).

[95] Oshri and Weeber (2006).

[96] Tushman et al. (1997: 8).

[97] Tushman et al. (1997: 17).

[98] Tarde (1903).

[99] Ryan and Gross (1943).

[100] Rogers (1962).

[101] Rogers (2003).

[102] For a review of the diffusion literature from an economics perspective, see Baptista (1999).

[103] Kuznets (1930).

[104] Griliches (1957); Mansfield (1961).

[105] For a review of the diffusion literature from a marketing perspective, see Mahajan et al. (1990).

[106] Rogers (1995: 263–6).

[107] Rogers (1995: 263).

[108] Rogers (1995: 268–74).

[109] Kuznets (1930); Ryan and Gross (1943).

[110] Rogers (1995: 259).

[111] Rogers (1995: 204–51).

[112] For example, Slack et al. (2006); Kotler (2008).

[113] Weekakkoody (2007).

[114] <http://www.minister.dbcde.gov.au/media/media_releases/2008/022>, accessed on 2 October 2008.

[115] Weekakkoody (2007).

[116] <http://www.minister.dbcde.gov.au/media/media_releases/2008/022>, accessed on 2 October 2008.

[117] Weekakkoody (2007).

[118] Rogers (1995: 260).

[119] Rothwell and Zegveld (1985: 17–18).

[120] Rink and Swan (1979); Duijn (1983).

[121] Rogers (1995: 260).

[122] Cusumano et al. (1997).

[123] Nieto (1997).

[124] Markus (1990).

[125] Rogers (1995: 100).

[126] Anderson and Tushman (1991).

[127] Clark and Staunton (1989: 105).

[128] Adner and Levinthal (2001: 611).

[129] Hamel and Prahalad (1994: 103).

[130] Swann (2001).

[131] Watts et al. (1998: 46).

[132] Watts et al. (1998).

[133] Swann and Watts (2000).

[134] Swann (2001).

[135] Swann and Watts (2000).

[136] Swann and Watts (2000).

[137] Watts et al. (1998: 53).

[138] Swann and Watts (2000).

[139] Watts et al. (1998: 52).

[140] Watts et al. (1998: 52).

[141] Swann (2001).

[142] Swann (2001: 1120).

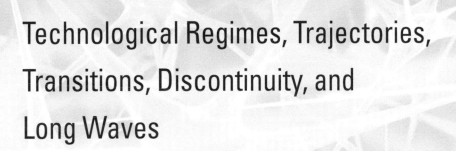

Technological Regimes, Trajectories, Transitions, Discontinuity, and Long Waves

5

Chapter overview

Learning objectives

This chapter will enable the reader to:

- outline the Kuhnian concepts of 'scientific paradigm', 'normal science', 'extraordinary science', and 'scientific revolutions', and identify and explain their impact on scientific endeavour and progress;

- outline the concept of 'technological paradigm' or 'technological regime', and identify and explain its impact on innovative activity within organizations and sectors; and

- explain the relationship between technological regimes and technological trajectories;

- outline the concept of 'technological discontinuity', and identify and explain its impact on organizational competence and industrial structure within a sector;

- appreciate the notion of transitions and transformations in technologies and technological regimes, and identify the various pathways of such transitions and transformations;

- outline the concept of long waves in relation to economic activity and patterns of innovation, and identify and explain the relation between the two;

- apply the concepts from this chapter to cases of innovation, technology, and science; and

- assess the applicability of the concepts to the service sector.

5.1 Introduction

In the previous chapter, our focus was on discussing the patterns of innovation during the life of a technology: the shift over time from radical to incremental innovation, from product to process innovation, and from divergent to convergent designs, for example. Building on this foundation, this chapter turns its attention to the process of transition from one technology to another. In doing so, we introduce the notion of 'technological regimes' or 'technological paradigms', and their role in shaping the trajectory of a technology and, ultimately, in its transformation or obsolescence. In developing our argument, we draw heavily upon the seminal work of Thomas Kuhn and his text *The Structure of Scientific Revolutions*, first published in 1962. His concepts of 'scientific paradigms', 'normal science', 'scientific revolutions', and 'extraordinary science' are all important for developing an understanding of trajectories and discontinuities in technological progress.

The substitution of one technology by another brings about a discontinuity between the 'old' and the 'new'. In relation to the models discussed in the previous chapter, such as the technology S-curve, this involves the shift to a new S-curve and the start of a new cycle of innovation. In **Table 5.1**, we compare and contrast the features of technological continuity (i.e. the movement along a trajectory) with those of technological discontinuity (i.e. the shift from one trajectory to another).

In the latter part of this chapter, we draw upon the seminal work of Kondratiev and Schumpeter in our exploration of the relationship between patterns of innovation and economic cycles. This work inspired a great deal of subsequent research and debate during the 1970s and 1980s among 'evolutionary' economists, such as Chris Freeman, Jacob Duijn, Alfred Kleinknecht, Gerhard Mensch, and Carlota Perez. Important questions that they sought to address included, for example, whether or not innovations occur in waves or clusters and, if so, whether these are synchronous with long waves in economic activity. Debate also centred heavily on what such findings might mean with regard to the nature of the relationship between innovative activity and economic

Table 5.1 A comparison of the features of continuity and discontinuity

Continuity	Discontinuity
Focus on innovation tied to an existing technology and related to its movement along an existing technological trajectory	Focus on innovation that gives rise to a new technology and a shift to a new technological trajectory
1. Focuses on a single cycle of technology	1. Focuses on multiple cycles of technology
2. Is characterized by stability	2. Is characterized by instability
3. Involves knowledge accumulation	3. Involves 'creative destruction'
4. Is competence enhancing/strengthening	4. Is often competence destroying/disrupting
5. Involves incremental innovation	5. Involves radical innovation

activity, and the mediating role of the broader socio-economic and institutional system in which this activity is embedded.

5.2 Technological regimes and trajectories

In their seminal paper, 'Towards a useful theory of innovation' published in 1977, Richard Nelson and Sidney Winter of Yale University sought to bridge the gap between the 'micro' (e.g. the level of the research project or organization) and 'macro' (e.g. the level of industrial sector or technology). In doing so, they developed a number of key concepts, principal among them 'technological regimes', 'natural trajectories', and 'selection environments'. Each of these concepts is important to our understanding of the progress of technology and the patterns of innovation, and each has created great interest and debate in academic circles.

The following sections will explore these concepts, with a particular focus on the first two. We will start with a discussion of the influential work of Kuhn.

5.2.1 Scientific paradigms and 'normal' science

Any discussion of technological paradigms or regimes first requires an introduction to the seminal work of Thomas Kuhn, who was emeritus professor of linguistics and philosophy at MIT. His book, *The Structure of Scientific Revolutions*, first published in 1962 and now in its third edition, has been highly influential in shaping our understanding of scientific progress.[1]

In this text, Kuhn argues that the growth of scientific knowledge is characterized by periods of continuous cumulative growth interspersed with periods of discontinuity. He proposes that the growth and development of scientific knowledge in a particular field occurs as a result of the development of a paradigm—a set of accepted scientific laws, theories, and practices—around which scientists coalesce. Kuhn provides a number of examples of paradigms, among them Darwinian evolution, Newtonian mechanics, and

Freudian psychology. Such paradigms provide the guidelines for research in their respective fields, influencing the 'puzzles'[2] that are considered to be worthy of pursuit and those that are not. This process leads to the limitation of novelty and a focus on *the scope and precision with which the paradigm can be applied*. As a result, the attention of scientists in a given field of scientific endeavour is focused, and scientific knowledge grows in a systematic and cumulative fashion. This pattern of activity is associated with what Kuhn terms 'normal science'.

In relation to the emergence of a paradigm, Kuhn argues that:

> Paradigms gain their status because they are more successful than their competitors in solving a few problems that the group of practitioners has come to recognize as acute. To be more successful is not, however, to be either completely successful with a single problem or notably successful with any large number. The success of a paradigm . . . is at the start largely a promise of success . . . Normal science consists in the actualization of that promise, an actualization achieved by extending the knowledge of those facts that the paradigm displays as particularly revealing, by increasing the extent of match between those facts and the paradigm's predictions, and by further articulation of the paradigm itself.[3]

In fields that lack paradigms, or which experience lengthy periods of confrontation between paradigms, members fail to agree about the important problems and the methods for solving them, and the resulting knowledge is not cumulative. In addition, Crane contends that *'the less precise the paradigm, the more room for disagreement, and the less likely that large numbers of researchers will accept it and choose to work on it'*.[4] Nevertheless, it is worth noting that whilst the paradigm of a mature scientific community is likely to be identifiable with relative ease, that of an emergent scientific community might initially be very limited in both its scope and precision.

Similarly to Kuhn, Lakatos highlights the importance of the coalescence of scientists around a set of core hypotheses, theories, and models, which drive and shape progressive research programs. Commitment to these core hypotheses, theories, and models is thus perpetuated and reinforced by this research, which seeks to extend the scope and precision of the 'core', and hence its robustness. The core hypotheses are also buffered by a set of auxiliary hypotheses that are modifiable or expendable, and thus act as a 'protective belt' to the core. The end result of this process is the narrowing and focusing of scientific endeavour within a particular field, which promotes stability along the lines of Kuhn's 'normal science'.[5]

Although it is well beyond the scope of this text, it is worth noting that Kuhn's conception of the workings of science and, in particular, his notion of 'normal science' generated intense, and sometimes heated, philosophical debate. Well-known contemporaries, such as Karl Popper and Paul Feyerabend, for example, raised objections concerning conservatism and authoritarianism in science.[6]

5.2.2 Technological paradigms, regimes, and frames

Similar concepts to those discussed above have also been developed to try to understand why technology appears to advance in a somewhat predictable manner. Formative work, from the late 1960s to the early 1980s, was undertaken by economists such as Nathan Rosenberg, Richard Nelson, Sidney Winter, and Giovanni Dosi. A common theme to emerge here was the importance of a shared outlook or set of cognitive routines within an engineering community, not dissimilar to Kuhn's notion of a scientific paradigm, in shaping technological choices and progress. Such cognitive routines, it was argued, not only focused the attentions of engineers in certain directions, but they also had 'a powerful exclusion effect' on other technological possibilities.

Illustration 5.1 Paradigms in classical music—the case of the sonata form

Many non-science and non-technological fields of endeavour exhibit dominant paradigms. The 'sonata form', as practised by Haydn and Mozart, was introduced in the middle of the eighteenth century and provided composers with '*a solid framework on which to construct and arrange their musical ideas . . . a mold . . .* [to] *pour their ideas*'. Slowik outlines the components of this framework as follows:

> Sonata form incorporates a definite pattern: a short introduction, the exposition (which presents the main 'subjects', melodies, and themes of the piece), the development (which explores these ideas, often in dramatic fashion), the recapitulation (a repeat of the exposition), and a short concluding coda.

This sonata form was perpetuated through the tutelage of composers during long apprenticeships. In this way, Haydn, for example, passed this technique on to his gifted student, Beethoven.[7]

Rosenberg expressed this shared outlook as a set of 'technological imperatives', such as bottlenecks in connected processes or weaknesses in a product, which provide clear signals for the prioritization of technical problems in need of resolution. In contrast, Nelson and Winter developed their concept of 'technological regime', which focused on the beliefs of engineers, in relation to '*what is* [considered] *feasible or at least worth attempting*'. They contended that the research strategies of an organization were often intimately linked to the prevailing technological regime within its sector.

Dosi, drawing explicitly upon the work of Kuhn, defined his notion of 'technological paradigm' as an 'outlook' and set of procedures that embody '*strong prescriptions on the* directions *of technical change to pursue and those to neglect*'. Although less well developed than later work, Dosi recognized the broader influences on, and embedded nature of, a paradigm.[8] Whilst there are similarities and overlaps in these concepts there are also differences (see **Table 5.2** for a definition of each).

Dosi illustrates the focusing influence of such cognitive routines in relation to the development of semiconductors:

> The identification of a technological paradigm relates to generic tasks to which it [in this case semiconductors] is applied . . . to the material technology it selects (e.g. . . . silicon), to the physical / chemical properties it exploits (e.g. the 'transistor effect' and the 'field effect' of semiconductor materials), to the technological and economic dimensions and trade-offs it focuses upon (e.g. density of the circuits, speed, noise-immunity, dispersion, frequency range, unit costs etc.). Once given these technological and economic dimensions, it is possible to obtain, broadly speaking, an idea of 'progress' as the improvement of the trade-offs related to those dimensions.[9]

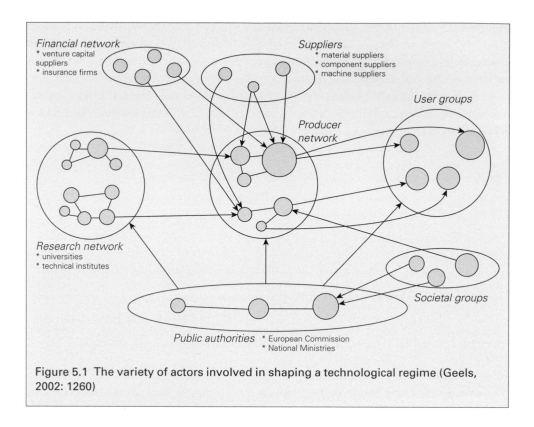

Figure 5.1 The variety of actors involved in shaping a technological regime (Geels, 2002: 1260)

This early work has been overlaid and nuanced since the 1990s by sociologists such as Arie Rip, Wiebe Bijker, Frank Geels, and René Kemp, who have highlighted, for example, the key role played by a much broader constituency of actors, including users, suppliers, scientists, banks, policymakers, and special-interest groups, in the shaping of the outlook of engineers and organizations, and hence technological choices and trajectories. This wider constituency is illustrated in **Figure 5.1**.

The rules and practices that comprise and shape a technological paradigm or regime, therefore, may originate or be influenced by actors outside the engineering or technological community. Ende and Kemp, for example, note that:

> These rules . . . consist of formal rules in the form of compatibility standards, product standards, regulatory demands (such as emission limit values) . . . and less formal rules that are implied in engineering (research) practices . . . management styles and systems, and in market conditions . . . The less formal rules are often more important than the formal . . . the nature and strength of these rules varies between regimes.[10]

Similar to the core and peripheral hypotheses of Lakatos, noted above, technological paradigms or regimes embrace both core and peripheral rules. Guiding principles, such as the focus on the efficiency or flexibility of a technology, which influence the types of problem that are tackled and the solutions considered are examples of core rules. These are typically stable and robust. Rules indicating the types of design tool and practice that should be employed, for example, are more peripheral and are open to change, although such would involve training and the implementation of new processes.[11] Whilst technological regimes vary between sectors, rules might be shared between technological regimes.

A number of overlapping concepts and definitions have emerged from this work on the sociology of technology: Rip, Ende, and Kemp are associated with a socially embedded version of 'technological paradigm'; Geels with 'sociotechnical regimes'; Orlikowski and Gash, and more recently, Kaplan and Tripas, with 'technological frames'. (See also **Table 5.2**.) This broader and more nuanced notion of a technological paradigm or regime, and how it is shaped, resonates well with the network perspective adopted in this text. We will thus adopt this position from here onwards and use the term 'technological regime' to cover the range of terms.

A technological regime is neither fixed nor entirely coherent; this provides space for tension to emerge between actors, especially between innovators (who have reached their technical limits) and customers (with unsatisfied demands), or between competitors with differing, but competing, innovations.[12] They are also 'less well articulated' than scientific paradigms, because they contain *'a high proportion of tacit, non-verbal know-ledges'*, more loosely defined and more readily shaped by the micro-politics within organizations. Furthermore, although a technological regime will shape the research activities of organizations within the sector, the regime itself is likely to vary between different organizations, because each will internalize and shape the regime through its own distinctive in-house expertise *'coupled with patents, technological know-how, and a network of interfirm linkages'*. This gives rise to the notion that a technological regime may be viewed at the level of the organization and, when aggregated, at the level of a sector.[13]

Kaplan and Tripas support this possibility, by making a clear distinction between individual (i.e. users, producers, institutions, etc.) and collective technological frames, as do Ende and Kemp, who argue that:

> the research activities of companies . . . are shaped importantly by the problems of existing regimes . . . This is not to say that everything is fixed. There is still a great deal of variety, but this variety is bounded.[14]

Nevertheless, Kemp et al. argue that the 'engineering imagination' is restricted since the embedding of existing technologies *'creates economic, technological, cognitive, and social barriers to new technologies'*.[15]

5.2.3 Technological trajectories and pathways

As we have noted earlier in the book, the majority of innovation is incremental rather than radical: minor modifications and improvements in the functionality or performance of a product, process, or service—perhaps slightly faster, more efficient, cheaper, lighter, smaller, or perhaps yielding more choice. Also, as Nelson and Winter observe, *'advances seem to follow in a way that appears somewhat "inevitable"'*.[16] Indeed, in **Chapter 4**, we illustrated how rather than being a random and ad hoc occurrence, innovation tended to push a technology incrementally in certain directions and thus along certain trajectories.

Table 5.2 Concepts concerning the shaping of choices of innovators

Concept	Proponent(s)	Definition	Orientation
Technological imperatives	Rosenberg (1969)	R&D is directed toward the imperfections of an existing technology, such as bottlenecks in processes or weaknesses in products, which provide signals for prioritizing R&D projects. Pursuing such 'technological imperatives' pushes a technology in a particular direction.	Inducements and signals
Technological regime	Malerba and Orsenigo (1993)	'We defined technological regimes as combinations of opportunity and appropriability conditions and degrees of cummulativeness of technological advances . . . Opportunity conditions refer to the ease of innovation by would-be innovators, and are related to the potential for innovation of each technology. Appropriability conditions refer to the ability of innovators to protect their innovations from imitation, and therefore to reap profits from their innovations. Cumulativeness conditions refer to the degree to which new technology builds on existing technology.' (Malerba and Orsenigo, 1996: 453)	Inducements and signals
Technological paradigm	Dosi (1982)	'We shall define "technological paradigm" . . . as an "outlook", a set of procedures, a definition of the "relevant" problems and of the specific knowledge related to their solution . . . each "technological paradigm" defines its own concept of "progress" based on its specific technological and economic trade-offs . . . [and as such] embodies strong prescriptions on the directions of technical change to pursue and those to neglect.' (Dosi, 1982: 148, 152–3)	Cognitive
Technological regime	Nelson and Winter (1977)	'Our concept is more cognitive, relating to technicians' beliefs about what is feasible or at least worth attempting . . . The sense of potential, of constraints, and of not yet exploited opportunities, implicit in a regime focuses the attention of engineers on certain directions in which progress is possible, and provides strong guidance as to the tactics likely to be fruitful.' (Nelson and Winter, 1977: 57)	Cognitive
Technological regime	Rip and Kemp (1998) Kemp et al. (1998) Ende and Kemp (1999)	'Technological regimes, in the way we use the term, are a broader, socially embedded version of technological paradigms.' (Kemp et al., 1998: 182) 'The notion of regime helps to focus the attention on the structure of which the actors and technologies are a part of . . . on rules and practices, embedded in a web of interrelations,	Socially embedded

Table 5.2 (cont'd)

Concept	Proponent(s	Definition	Orientation
		and ongoing trends as a backdrop . . . We are not saying . . . either individuals and companies are unimportant . . . [but they] . . . are bounded in a historic sense: they are equipped with a certain outlook, capabilities and role, which influences both what they will do and can do at any given time . . . the research activities of companies . . . are shaped importantly by the problems of existing regimes and the accumulated knowledge, capital stock, established consumption patterns, and the norms at the macro level.' (Ende and Kemp, 1999: 848–9)	Socially embedded cognition
Socio-technical regime	Geels (2002)	'While the cognitive routines of Nelson and Winter are embedded in the practices and minds of engineers, these rules are [also] embedded more widely in the knowledge base, engineering practices, corporate governance structures, manufacturing processes and product characteristics. This widening also means that more social groups are taken on board . . . users, policy makers, societal groups, suppliers, scientists, capital banks etc. Because the activities of these groups are also guided by rules, I will use the term 'sociotechnical [ST] regimes' to refer to the semi-coherent set of rules carried by different social groups . . . ST regimes thus function as [a] selection and retention mechanism.' (Geels, 2002: 1260)	Socially embedded cognition
Technological frame	Orlikowski and Gash (1994) Kaplan and Tripas (2008)	'A "technological frame" . . . captures how actors make sense of technology . . . Specifically, technological frames shape how actors categorize a technology relative to other technologies and which performance criteria they use to evaluate the technology . . . [it] guides the actor's interpretation of what a technology is and whether it does anything useful . . . [they] do not spring up randomly, but rather are the encoding of . . . [an actor's] prior history, including both idiosyncratic organizational experiences and industry affiliations . . . with industry associations, customer sets, competitive groups, user groups, etc. . . . Technological frames do not influence technologies directly but rather through the interpretive processes of these actors . . . Thus, the interpretative process is the mechanism that connects technological frames to technological outcomes.' (Kaplan and Tripas, 2008: 791–2)	Socially embedded cognition

For example, in the gradual improvement of performance, a technology moves up the technology S-curve; in the convergence of designs, a technology moves toward a dominant design; in the shift of emphasis from novelty and variety to efficiency and quality, innovation shifts from product to process innovation and pushes the technology along the product-process cycle. The presence of a technological regime in an industry or sector helps to explain why such patterns of innovation unfold.

Firstly, technological regimes focus the attention of innovators on certain problems and certain solutions with a particular notion of what constitutes progress. This yields cumulative, rather than ad hoc, advances in technology and brings about a period of continuity or 'normal' problem-solving activity akin to Kuhn's notion of 'normal science'.[17] This is well demonstrated in **Illustration 5.2**.

Illustration 5.2 Advancing the performance of a technology through incremental innovation—the case of Intel and the microprocessor

One of the key preoccupations in the development of the microprocessor has been the attainment of ever-greater densities of transistor on a microprocessor. **Figure 5.2** illustrates this trajectory for the Intel microprocessor—note the phenomenal (exponential) improvement in performance.[18]

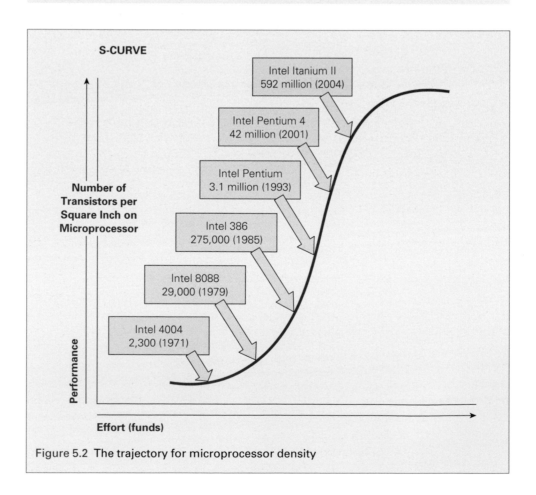

Figure 5.2 The trajectory for microprocessor density

Geels expresses this more directly:

> Technological regimes result in technological trajectories, because the community of engineers searches in the same direction. Technological regimes create stability because they guide the innovative activity towards incremental improvements along trajectories.[19]

The cycle is stabilizing because it is also self-reinforcing; incremental innovation not only seeks improvements in the technology, but it is also *'aimed at regime optimization rather than regime transformation'*.[20] Stability is also enhanced by the emergence of regulations and standards, in the adaptation of lifestyles to the technology, and in the sunk investment in equipment and infrastructure, for example.[21] You might consider all of these factors in relation to the mobile phone to appreciate their stabilizing affect. Technological regimes and the trajectories that they yield are also sustained by vested interests.[22]

Secondly, studies have revealed that although a technology typically has aspects that influence its own direction of development, which are encoded within the prevailing technological regime, there also exist certain common trajectories along which many technologies travel—most notably, *'those of the progressive exploitation of scale economies and increasing mechanisation'*, which Nelson and Winter have termed 'natural' trajectories.[23] Innovation that seeks to exploit scale economies and increase mechanization is most likely to shift a technology towards convergence, and one or a small number of dominant designs, as well as to progress it along the product-process cycle, where the emphasis shifts from product to process innovation.

Research by Pavitt, however, reveals different 'natural' trajectories for different types of sector: he distinguishes between sectors that are 'supplier-dominated' (e.g. agriculture and professional services), 'scale-intensive' (e.g. bulk materials, such as steel, and assembly, such as automobiles and consumer products), 'specialized suppliers' (e.g. machinery and instrumentation), and 'science-based' (e.g. chemicals). In 'supplier-dominated' sectors, Pavitt argues that innovation originates largely from suppliers and is focused on process innovation to reduce costs. This is in sharp contrast to sectors characterized by 'specialist suppliers', in which innovation is largely in-house and focused on product development.[24]

In our view, the term 'natural' trajectory is likely to conjure up overly deterministic notions of technological progress, so perhaps 'common' is a more suitable term than 'natural'. Clark and Staunton have also critiqued the term 'trajectory' itself, which they perceive as invoking 'an over-precise image' of the progress of a technology; they employ the term 'pathway' in its place. The imagery of the pathway allows for the possibility that, in the development of a technology, progress may include many branches and offshoots, and that these may lead into mazes and cul-de-sacs. In this sense, the pathway, unlike the trajectory, may not always be smooth or travel in the same direction. Nevertheless, pathways, as with trajectories, represent the journey to an intended destination, and as such *'reflect cumulative efforts . . .* [that] *will possess a highish degree of momentum in certain directions'*, such that *'switching from the existing paradigm . . . to another paradigm can be extremely difficult and also expensive'*.[25]

We noted above that a technological regime is likely to vary between different organizations within a sector, nuanced by their distinctive in-house experiences, resources, competencies, and histories. Accordingly, individual organizations will pursue different innovation agendas and strategies, and, as a result, often travel along different and sometimes divergent pathways. Thus, the overall pathway of a technology should be viewed as the aggregated pathways of individual organizations. Indeed, for Clark and Staunton, '*pathways are the outcome of multiple firm-specific approaches to innovation*'.[26] As we also noted earlier, however, whilst there is some variation in the technological regimes of organizations, with the resultant divergence in their technological trajectories, this degree of variation is constrained by the broader social, cultural, economic, political, environmental, and material (e.g. the physical arrangements of cities, and the communication, transport, and energy infrastructures) context. Geels refers to this context as the 'socio-technical landscape', intentionally conjuring up imagery of durability and hardness; indeed, such landscapes are even slower to change than technological regimes.

5.2.4 Paradigms and trajectories in the service sector

Much of the literature and theorizing in this area has been undertaken in relation to technological product and process innovation. Over the last ten years or so, however, there has been a growing interest in the degree to which concepts such as technological paradigms and trajectories are of relevance to the service sector. Work by Jon Sundbo, and Camal Gallouj and Fïaz Gallouj, has been particularly influential in this regard.

Traditionally, innovation within the service sector has typically been unsystematic and ad hoc. Over the last two decades, however, there has been increasing pressure to 'industrialize' service production across a range of service sectors, such as retailing, banking, transportation, and leisure. This pressure has arisen largely as a result of increasing price competition and a focus on quality, and has lead to increasing mechanization, especially through the application of information and communication technologies (ICT), standardization (to aid quality control), and modularization (to allow mass customization of services).[27] Similar patterns of standardization and modularization are also emerging in the delivery of professional services, such as in consultancy, accountancy, and legal support, despite the need to provide solutions that meet client-specific problems and needs; this is particularly the case for larger and more 'company-like' professional service firms.[28]

Thus, mechanization, standardization, and modularization may all be considered to be core elements of a 'paradigm' or 'regime' that is common across many service sectors, and which guides service development along a 'natural' or 'common' trajectory in much the same way as innovation in technological products and processes. It is important to point out, however, that much of the innovation within the service sector is non-technological in nature. Nevertheless, Gallouj and Gallouj argue that the technological paradigm concept is '*sufficiently broad to leave room for services*', especially if innovation is viewed loosely in terms of 'problem-solving activity' rather than narrowly in terms of technological innovation.[29] In this regard, instances of service innovation might be seen to follow an 'institutional trajectory', based on institutional innovations,[30] or a 'professional trajectory', bounded and guided by well-established professional codes of practice, methods, and values.[31]

5.3 **Technological discontinuity and creative destruction**

In the discussion above, we have focused on the features of stability and continuity in the progress of technology. We now turn our attention to disruption and discontinuity, or what Schumpeter termed 'creative destruction'. For Schumpeter:

> Creative Destruction is the essential fact about capitalism . . . it is not [price] competition which counts but the competition from . . . *the new technology* . . . competition which strikes not at the margins of the profits . . . of existing firms but at their foundations and their very lives.[32]

In the late 1960s, Drucker declared that we live in *The Age of Discontinuity*.[33] This is a view supported by Foster in his work on technology S-curves; drawing upon examples across a breadth of sectors, such as chemicals, sugar, tyres, and electronics, he argues that:

> Discontinuities do occur more frequently than most of us realize. And if anything, their frequency is on the increase. It's hard to find an industry where they're not happening or are looming on the horizon. And their ramifications can be enormous . . . fortunes can change dramatically. The leaders in the current technology rarely survive to become the leader in the new technology . . . This leaders-to-losers story has been played out not just by companies but by whole industries.[34]

Others are more circumspect. Abernathy and Clark, for example, note that historical evidence suggests that technology advances through long periods of incremental innovation, which is punctuated by periods of ferment and discontinuity.[35] Tushman and Anderson argue that '*discontinuities are generally uncommon, and their frequency varies greatly by industry*';[36] they draw upon their detailed historical studies of innovation within airline transportation, cement production, glass production, and minicomputers to support their argument.[37] There is also increasing evidence to suggest that many firms survive discontinuities in their industries, depending upon whether the discontinuity is 'competence destroying' or 'competence enhancing'—an important distinction to which we return later.[38]

We begin this section, as we did the previous, by turning to the work of Kuhn—only, this time, to his concepts of 'scientific revolutions' and 'extraordinary science'. This is followed by a discussion of the concept of 'technological discontinuity', which is explored and illustrated alongside the technology S-curve model that we introduced in **4.4**, and the notion of 'disruptive technologies'.

5.3.1 **Scientific revolutions and 'extraordinary science'**

In returning to the seminal work of Kuhn, we shift focus from the operation of 'scientific paradigms' and 'normal science', to:

- the circumstances that lead to the emergence of anomalies within a scientific discipline that the existing paradigm struggles to accommodate;
- the subsequent emergence of a period of 'extraordinary science'; and eventually to
- a 'scientific revolution'—that is, the replacement of the current scientific paradigm by a new one.

As we noted in **5.2.1** earlier, normal science does not aim to generate novelty as such, but to focus activity towards extending *'the scope and precision with which the paradigm can be applied'*. In doing so, scientists within a scientific discipline—whether astronomy, physics, or chemistry, for example—construct and refine increasingly elaborate and specialized equipment, skills, and vocabulary. This process of 'professionalization' leads to a narrowing of vision and to a considerable resistance to any change in the prevailing paradigm.[39] The irony is, however, that the more sophisticated the equipment and skills, the more likely that anomalies, if they arise, will be identified and lead to novelty.

Thus, Kuhn argues:

> Without the special apparatus that is constructed mainly for anticipated functions, the results that ultimately lead to novelty could not occur. And even when the apparatus exists, novelty ordinarily emerges for the man who, knowing *with* precision what he should expect, is able to recognise that something has gone wrong. Anomaly appears only against the background provided by the paradigm. The more precise and far-reaching that paradigm is, the more sensitive an indicator it provides of anomaly and hence of an occasion for paradigm change.[40]

In order to accommodate such anomalies—that is, unanticipated findings—it is common for ad hoc modification of the prevailing theories and laws to occur, in order to eliminate any perceived conflict with the current paradigm.[41] According to Lakatos, this is more likely to occur to 'peripheral' rather than 'core' theories and laws within the paradigm.[42]

Kuhn neatly summarizes the reluctance to dispense with the prevailing paradigm so easily, arguing that:

> As in manufacture so in science—retooling is an extravagance to be reserved for the occasion that demands it. The significance of crises is the indication they provide that an occasion for retooling has arrived.[43]

Over time, however, the persistent emergence of anomalies and the accumulation of adjustments to accommodate them can give rise to increasing complexity and discrepancies within the prevailing theories and laws of a scientific field, whilst reducing their accuracy.[44] As a result, scientists may begin to lose faith in the current paradigm and start to consider alternatives. Such a situation is likely to give rise to a crisis. An alternative hypothesis of the origin of such crises argues that a sustained period of growth leads to a kind of crescendo of activity, ending in an exhaustion of the ideas that had stimulated the growth.[45] Either way, a crisis is accompanied by a shift from a period of 'normal science' to one of 'extraordinary science'; this is characterized by a change in attitude to the existing paradigm that manifests itself in a willingness to be more experimental and less constrained by the paradigm, by open discontent, and in debate and disagreement over, as well as a proliferation of competing versions of, the 'core' theories and laws.[46]

An episode of extraordinary science is similar in character to the 'era of ferment' during the emergence of a new technology and, ultimately, leads to discontinuity. Kuhn refers to such discontinuity in science as a 'scientific revolution', where an existing scientific paradigm is replaced by a new one through:

a reconstruction of the field from new fundamentals, a reconstruction that changes some of the field's most elementary theoretical generalizations as well as many of its paradigm methods and applications.[47]

Following such a scientific revolution, a new paradigm emerges and a new period of 'normal science' unfolds.

Kuhn's observations and interpretations of the nature of scientific endeavour, and his model of the cycle of scientific progress, provide a useful analogy for the cycle of technological progress, to which we now turn.

5.3.2 Technological discontinuity

Interest in technological discontinuities and their impact on sectors and organizations has been of increasing interest among academics since the mid-1980s, at which time a number of seminal works were published, led by Abernathy and Clark's article entitled 'Mapping the winds of creative destruction'. Other important contributors have included Tushman and Anderson, and Foster, through his work on the technology S-curve.[48] More recently, Christensen has investigated the nature and impact of what he terms 'disruptive technologies'.[49]

Tushman and Rosenkopf define a technological discontinuity as:

> those rare, unpredictable innovations which advance a relevant technological frontier by an order of magnitude and which involve fundamentally different product or process design *and* that command a decisive cost, performance, or quality advantage over prior product forms.[50]

Instances of such technological discontinuities are evident in our everyday lives: recent examples include the emergence of digital cameras to replace traditional 'film-based' cameras, and the emergence of flat-screen displays to replace the bulky cathode-ray tubes in televisions and computer monitors.

In their detailed historical study of innovation in US cement manufacture, airline transportation, and minicomputers, Tushman and Anderson found that each technological discontinuity that they identified had a major impact on an important measure of cost or performance for that sector; in the daily production capacity for cement, seat-miles-per-year capacity for airlines, and central processing unit (CPU) speed, respectively.[51] Tushman et al. also note that:

> dominant designs and technological discontinuities are demarcation points between fundamentally different competitive arenas . . . technological discontinuities . . . that trigger periods of technological and competitive ferment . . . are closed with the emergence of an industry standard or dominant design.[52]

Given the dramatic impact that a technological discontinuity can have upon a sector and the organizations within it, the question arises as to whether or not there are any predictable patterns to the origins, process, and consequences of discontinuities. We turn now to address these questions.

Foster argues that the events that lead up to a technological discontinuity are triggered by a technology approaching its technical limit—that is, as the potential of the existing technology declines along with the returns from research and development (R&D) effort, there is a growing recognition by those within the sector of the need to search for alternative possibilities and a growing sense of opportunity from those outside of the sector. Such opportunities might be addressed by the novel combination of existing technologies, or through the transfer of technologies already present in unrelated applications or sectors. Technological discontinuities might also emerge out of a scientific discovery or advance.

A technological discontinuity, for Foster, is represented by the gap between the S-curve of an existing technology and the S-curve of a new technology—that is, the *'point when one technology replaces another'*.[53] This perspective is depicted in **Figure 5.3**. Nevertheless, Foster recognizes that the process of substitution of the old by the new may take several years, because a new technology may capture market share only very gradually.[54] This would appear to be the case with digital television (see **Illustration 4.4**).

For Tushman et al., however, a technological discontinuity is not the point at which an old technology is replaced by the new, but the point at which a new technology ruptures the prevailing 'era of incremental innovation' and spawns an 'era of ferment', where the old technology competes alongside variants of the new technology for market acceptance before being superseded.[55] This perspective is depicted in **Figure 5.4**.

In **Figure 5.3**, we see that, for a while, R&D effort is expended in progressing the performance of both the old technology (represented by the left-hand S-curve) and the new technology (represented by the right-hand S-curve). Initially, the performance of the emergent technology falls below that of the old technology—this is represented where the right-hand S-curve lies below the left-hand S-curve. But eventually the old technology reaches its technical limit and, consequently, the new technology, with a

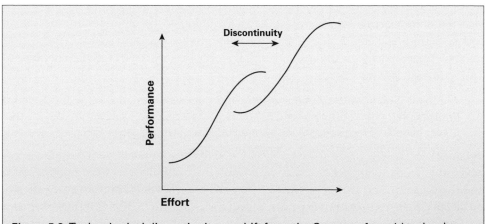

Figure 5.3 Technological discontinuity—a shift from the S-curve of an old technology (on the left) to the S-curve of a new technology (on the right) (Foster, 1986: 102)

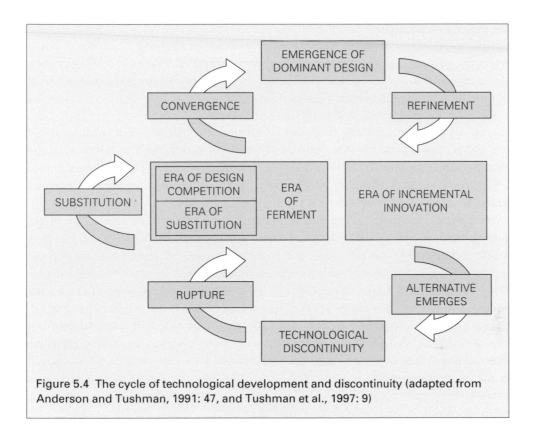

Figure 5.4 The cycle of technological development and discontinuity (adapted from Anderson and Tushman, 1991: 47, and Tushman et al., 1997: 9)

higher technical limit, begins to outperform the old technology. At this point, there is a shift—or a technological discontinuity—as the new technology replaces the old.

Turning now to the perspective of Tushman et al., captured in **Figure 5.4**, we see that a period of incremental innovation is disrupted by the emergence of a viable technological alternative. The arrival of a viable new technology creates a technological discontinuity by rupturing the prevailing technological trajectory and heralding a new 'era of ferment'. The era of ferment is comprised of two overlapping processes (both implicit in the technology S-curve). The first, termed the 'era of substitution', involves competition between the old and emergent technologies (in the S-curve model, this is represented by the two S-curves running alongside each other). It is not uncommon for existing technologies to have their performance improved markedly during this period of competition with the emergent technology, as organizations respond to the threat of new technology and new entrants to their sector.

The second, termed the 'era of design competition', begins following the substitution of the old technology by the new, and subsequent competition between design variants that each embody the new technology in a unique way (in the S-curve model, this is located within the lower portion of the right-hand S-curve). The era of design competition is embraced by the 'fluid' phase in the product-process cycle model, which involves the generation of product variety, and by the 'divergent design' phase of the dominant design model—both of which were discussed in detail in **Chapter 4** (see **4.5** and **4.6**, respectively). The era of design competition is brought to a close as and when convergence around one or a small number of designs gathers momentum, and a new period of dominant design and incremental innovation follows.[56]

The duration of the era of ferment can vary dramatically from sector to sector and is contingent upon a number of factors. With regard to the speed of substitution of the old technology by the new, this can depend, in part, upon the response of the incumbent firms within the sector—that is, on whether or not they choose to continue to invest in the old technology and defend their current position, or whether they choose to switch over to the new technology and embrace the discontinuity as an opportunity rather than a threat. The ease with which incumbent firms are able to adopt the new technology, of course, will depend upon the degree to which this requires new skills, competences, and resources. For Foster, the speed of substitution of the old technology is driven by the 'underlying economics'—that is, the cost–performance ratio: the quicker the new technology can be developed and refined to reduce costs whilst improving performance, vis-à-vis the old technology, the quicker the substitution process is likely to be.[57] Anderson and Tushman argue that it takes longer for a new dominant design to emerge when the discontinuity destroys rather than enhances existing know-how (this often includes the know-how of users and suppliers, as well as that of innovators).[58]

5.3.3 Competence-enhancing discontinuities versus competence-destroying discontinuities

In understanding the possible consequences of a technological discontinuity on existing firms within a sector, Abernathy and Clark make the important distinction between 'competence-destroying' and 'competence-enhancing' innovation, arguing that *'innovation is not a unified phenomenon: some innovations disrupt, destroy and make obsolete established competence; others refine and improve'*.[59]

Competence-destroying discontinuities *'significantly advance the technological frontier, but with a knowledge, skill, and competence base that is inconsistent with prior know-how'*,[60] and are at the heart of Schumpeter's theory of innovation, encapsulated by his notion of 'creative destruction'.[61] Such discontinuities are 'watershed' events within an industry. They are often initiated by new entrants, and are frequently accompanied by major shifts in the competitors and market leaders within the sector. These shifts occur because the competencies of existing players may not only be rendered obsolete, but may lock them into the existing technology; this has given rise to the term 'core rigidity', coined by Leonard-Barton, to which we will return in the next chapter.[62] On the other hand, new entrants to a sector are unconstrained by prior investments, existing technologies, or organizational inertia. Examples of competence-destroying product and process discontinuities include, respectively, the transistor that replaced vacuum tubes in computers (see **Illustration 5.3**), and the Pilkington float-glass process that replaced the Colburn drawing process in glass production.[63]

Illustration 5.3 Competence-destroying discontinuities—the case of the substitution of vacuum tubes by transistors in the computer industry

An often-cited example of a competence-destroying technological discontinuity is that of the replacement of vacuum tubes by transistors in the computer industry. Only two of the top ten vacuum tube producers survived the discontinuity: RCA and Philips.[64]

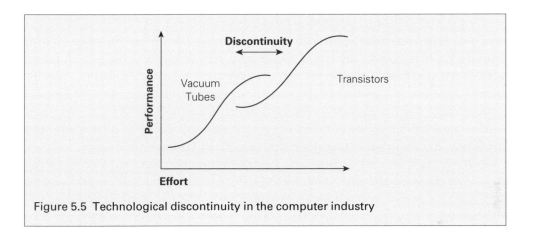

Figure 5.5 Technological discontinuity in the computer industry

In contrast, competence-enhancing discontinuities '*significantly advance the state of the art yet build on, or permit the transfer of, existing know-how and knowledge*', and are predominantly introduced by existing companies within the sector.[65] Such discontinuities help to conserve and reinforce the competences of existing players within the sector, and, as a result, can lead to the raising of barriers to entry and the lowering of the attractiveness of alternatives.[66] Examples of competence-enhancing product and process discontinuities include, respectively, the jet engine that replaced the propeller engine on aircraft and process control through computerization in industries such as cement production.[67]

Importantly, research by Tushman and Anderson found that such competence-enhancing discontinuities were more common than competence-destroying discontinuities.[68] Having made the distinction between competence-destroying and competence-enhancing discontinuities, however, it is also important to determine those competences that might be destroyed or enhanced, because, as Afuah argues:

> Technological change may not obsolete all of an incumbent's capabilities . . . if a discontinuity obsoletes technological capabilities leaving market ones intact, incumbents can have an advantage if such market capabilities are important and difficult to establish. Similarly, if supplier-focused capabilities are left intact in industries where supplier relations are important, incumbents may also have an advantage. Thus, a firm's ability to recognize just which of its capabilities will be rendered obsolete by the arrival of a technological discontinuity . . . while taking advantage of those capabilities that are not impacted by the technology can also be an asset.[69]

In relation to market capabilities, Afuah is referring to strengths in reputation, brand, distribution, and marketing and advertising, for example, whilst supplier capabilities embrace strengths built up around trust, negotiation, and supply-chain integration. To this should be added production capabilities. Anderson and Tushman argue that such capabilities, or complementary assets, are more likely to remain intact following a process, rather than a product, discontinuity, because product innovation generally has a greater impact on the value chain, often requiring changes in supplier and distribution channels.[70]

Given that many new entrants to a sector following a technological discontinuity are small entrepreneurial firms, established firms have many capabilities and competences at their disposal that set them apart. Such capabilities can enable the survival of established firms during technological discontinuities. Indeed, recent studies in the biotechnology and cellular telecommunications sectors, among others, have highlighted the importance and prevalence of symbiotic cooperative arrangements between new entrants and incumbents.[71] There is also increasing evidence that established firms across a range of sectors, including services, are forming alliances in response to a technological discontinuity in order to build competence and capability in the new technology.[72]

So far, we have assumed that technological discontinuities are brought about by the emergence of a new viable alternative. But technological discontinuities themselves might be triggered by non-technological discontinuities, such as major changes in the cultural, economic, political, or legal context. Consider the possible impact of the escalating price of oil, of the responses to the threat of climate change, and of the widespread deregulation in the energy and transportation sectors around the world. **Illustration 5.4** provides a good example of a non-technological discontinuity bringing about a technological discontinuity.

Illustration 5.4 The incandescent light bulb to be banned in Australia—the role of government and legislation in creating discontinuity

The traditional incandescent light bulb in widespread use today is actually based on late nineteenth-century designs from the likes of Thomas Edison and Joseph Swan. Despite its poor technical performance vis-à-vis more recent alternatives, such as the compact fluorescent light bulb (which uses about 25 per cent of the energy, and lasts between six and ten times longer), or light-emitting-diodes (LEDs), the incandescent light bulb continues to dominate the market for household lighting. Its longevity has been in good part due to the low costs of production and its wide availability in a range of sizes, shapes, and wattages. Alternatives remain more expensive to purchase, although cheaper to power, and concerns persist in relation to the compact fluorescent light bulb, for example, which contains a small amount of toxic mercury and, in its 'single-skinned' version, emits ultraviolet light at levels considered inappropriate for desk lighting.

But all of this is about to change. The Australian government recently announced its intention to ban the incandescent light bulb, and other governments, such as that of Canada, are likely to follow suit in the near future. The ban is designed to reduce Australia's greenhouse emissions by some 800,000 tonnes per year by 2012.[73] But the ban is also likely to yield renewed investment and innovation in lighting alternatives.

It is important also to consider non-technological and non-competitive responses to technological discontinuities once they emerge. Consider here the developments in biotechnology and the use of stem cells or genetically modified (GM) crops. Technological discontinuities may also be triggered by radical shifts in user preferences: what Tripas terms 'preference discontinuities'.[74]

5.4 Technological transitions and transformations

In the previous section, our attention was focused upon the prevalence, impact, and process of technological discontinuities. This discussion was in stark contrast to the preceding section, which centred on technological continuity. Although there are dissenting voices in the academic literature, it is easy to be left with the general impression that technological discontinuities necessarily involve rapid and seismic shifts in technologies and sectors; this is reinforced by the vocabulary employed, such as 'destruction', 'rupture', 'ferment', and 'turbulent'.

In this section, we shift the emphasis again, this time to the middle ground, to explore the notions of transition and transformation in relation to technological change.

5.4.1 Transitions and transformations in technology

Commentators tend to focus on the 'dramatic events' associated with the substitution of one technology by another. As Levinthal argues, however, *'discontinuities are generally not the product of singular events'*, but follow lengthy periods of development and gestation of a new technology.[75] This is perhaps not surprising, because, as Constant observes:

> the way we construct most evolutionary narratives, especially those about technology, typically effaces all that is stable, unchanging, or ordinary, and emphasizes almost exclusively that which is radical, unusual, innovative, or unique.[76]

Such a focus is likely, however, to lead to an exaggeration of the speed of technological change. This is the point that Levinthal is conveying through his apparently paradoxical phrase, *'the slow pace of rapid technological change'*.[77]

We also noted earlier that the development and adoption of a new technology within a sector is shaped and constrained by the prevailing technological regime. Technological regimes, as Smith et al. argue, *'have a tendency to exclude options and thereby introduce stability'*,[78] and are, in turn, shaped and constrained by the broader socio-technical landscape in which they are embedded; changes in socio-technical landscapes occur very slowly and often over decades.[79]

It has also been argued that the focus on the substitution of one technology by another gives rise to an overemphasis on discontinuity over continuity. Looking backwards from a substitution event, the origins of a new technology can often be traced back to the existing technology; only the new technology has evolved along an increasingly divergent pathway. Levinthal employs the biological metaphors of 'mutation' and 'speciation' to make this distinction: mutation gives rise to quantum and discontinuous changes in evolution, whilst speciation, which involves the *'separation of reproductive activity'* from the mainstream population, may give rise to 'genetic drift', especially when located within a distinct 'selection environment'. The notion of mutation has largely been abandoned in evolutionary biology in favour of the notion of speciation.

Levinthal argues that a similar shift in perspective provides useful insight into our interpretation of technological change.[80] New technology might also evolve from existing technologies through 'fusion' or through 'fragmentation'. Fusion occurs where two distinct technologies are combined to create a new hybrid technology.[81] A good example of this process is provided by the computed axial tomography (CAT) scanner, which resulted from the merging of X-ray technology with data-processing technology to produce 3D images of the body. Fragmentation occurs as a technology evolves to address a range of specialized applications and begins to evolve along a number of increasingly divergent trajectories. A good example of this process is provided by the evolution of telegraphy to address a wide range of applications (see **Figure 5.6**).

Speciation may occur when an existing technology within a sector is applied to a niche application within that sector. Initially, the modification of the technology to meet the specific requirements of the niche may be very minor—but if there exists selection criteria within the niche that are distinct from the mainstream of the sector, perhaps around functionality or price sensitivity, and there are resources available within the niche to develop the technology, then this can trigger a new and divergent trajectory for the technology. Ultimately, this divergent development process might lead to a distinct technology that is able to compete successfully with the prevailing technology and is thus able to 'invade' other niches, or even the mainstream applications within the sector.

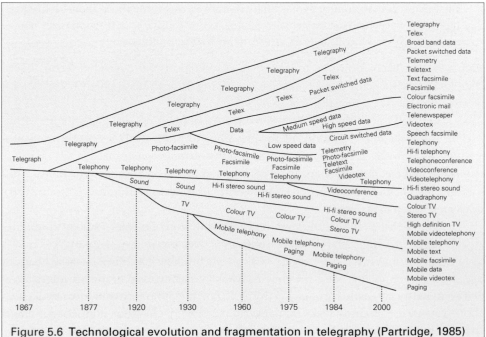

Figure 5.6 Technological evolution and fragmentation in telegraphy (Partridge, 1985)

> **Illustration 5.5 World's faster computer built with Sony PlayStation components**
>
> In June 2008, a new computing milestone was reported: an IBM supercomputer, codenamed 'Roadrunner', which ran at speeds equivalent to one thousand trillion calculations per second. This is twice as fast as the previous supercomputer. It is to be used to monitor the US nuclear stockpile, and for research into astronomy and climate change, for example. Perhaps what is most intriguing about the new supercomputer, however, is that it employs a 'hybrid' design, incorporating conventional supercomputer processors, as well as processors designed for the Sony PlayStation 3 (PS3). The PS3 processor, having been designed especially for the particular demands of the computer games niche, has evolved in a distinctive way from the mainstream computer sector.[82]

In taking the longer view and shifting emphasis away from the substitution event of one technology over another, we are able to highlight the importance of transitions and transformation in technologies, whether through speciation, fusion, or fragmentation, and reveal a more gradual, slower pace to technological change than is often portrayed.

5.4.2 Transitions and transformations in technological regimes

Technological regimes also undergo transitions and transformations. The origins and pathways of such transitions is an area of emerging interest and debate among academics. A cluster of academics within the Netherlands have been at the forefront of this debate, including Frank Geels, Johan Schot, and René Kemp.[83]

A key feature of this literature is the adoption of what is termed a 'multiple-level perspective' (MLP), which distinguishes between the socio-technical landscape, the technological regime, and niches. Under this perspective, pressure for a regime to change can come from 'above', following changes in the socio-technical landscape in which the regime is embedded, from 'below', where technology developed within a niche breaks through into the mainstream, or from 'within', arising from recognition within the sector that change is necessary. In addition, pressure can arise from 'outside', from organizations within other sectors which see new opportunities arising from within a sector. These pressures are illustrated in **Figure 5.7**. The destablization of the prevailing technological regime as a result of changes in the socio-technical landscape provides a window of opportunity for the emergence of alternative technologies and regimes.[84]

Taking each of these pressures in turn, we can summarize as follows.

- *Pressure from above*—changes within the socio-technical landscape, perhaps shifts in public opinion concerning climate change, can bring about conflict with the prevailing technological regime within sectors such as energy and automobiles, for example. Where these changes are accompanied by the mobilization and exertion of influence, such as through pressure groups, political parties, and groupings of influential scientists, then the technological regime can come under intense pressure to respond and adapt, or risk becoming obsolete. Where the shift in the socio-technical landscape is dramatic and rapid, then the prevailing technological regime may not be capable of adjustment. In such instances, a vacuum emerges that draws in outsiders with a range of possible alternatives and the prevailing technological regime is replaced.

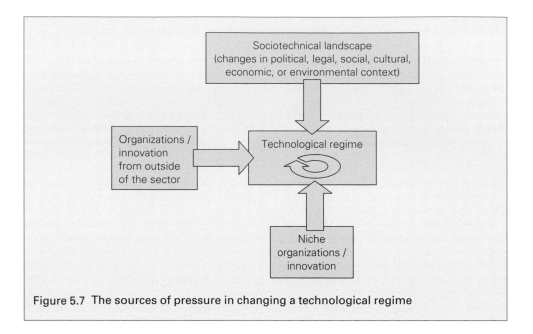

Figure 5.7 **The sources of pressure in changing a technological regime**

- *Pressure from below*—where innovation within a niche has led to the development of a distinct and viable alternative to the prevailing technological solutions within the mainstream of a sector, through functionality, price, or performance, advantages, for example, then this can create pressure on the mainstream technological regime to adapt to, or absorb, the technological regime within the niche, or be replaced by it. A good example of the process of absorption is provided in **Illustration 5.5**.

- *Pressure from within*—where there is recognition within a sector that the prevailing technological regime is reaching its limits in relation to providing future improvements in performance, for example, or is seen to be unable to respond adequately to expected changes in market demands or competition from alternatives, then this is likely to yield reflection and adaptation of the technological regime from within.

- *Pressure from outside*—as alternative technologies grow and evolve outside of a sector, there may be increasing recognition from 'outsiders' of the possible opportunities within that sector. These sectors may initially appear to be unrelated, but, over time, can create a genuine threat to the prevailing technological regime within a sector. A good example of this is provided by the shift to digital technology for cameras and video recorders. The external threat may encourage adjustments in the prevailing technological regime, but where the threat is substantial, and the new technology competence destroying and at odds with the current regime, then ultimately the current technological regime is likely to be replaced.

Whether or not a regime responds or succumbs to the pressures outlined above depends upon a number of factors, including the strength and coherence of the pressure exerted on the regime, and the adaptive capacity of its membership to adjust and adapt. Recalling our earlier discussion concerning scientific paradigms and revolutions, adaptation of a regime may be seen to be easiest where it only requires change to peripheral rules, but becomes more problematic when it necessitates change of the core rules. Nevertheless, Smith et al. argue that, *'Over time, we would expect more adaptive regimes to succeed and those with less adaptive capacity to be subsumed or substituted'.*[85]

For Poel, however, the manner and degree to which regimes are transformed varies between sectors and is dependent upon the prevailing pattern of innovation within the sector. Poel distinguishes between 'supplier-dependent', 'R&D-dependent', 'user-driven', and 'mission-orientated' patterns of innovation (based on the typology of Pavitt that we discussed earlier), and concludes that: *'The user-driven innovation pattern is relatively more constraining for processes of transformation than the other three patterns. The R&D dependent innovation pattern is the most enabling.'*[86]

5.5 Long waves and business cycles

Interest in long waves or cycles of economic activity has largely fallen within two periods: the first from the 1920s to 1950s, following the seminal work of the Russian economist Nikolai Kondratiev;[87] and the second, during the 1970s and 1980s. During these two periods, debate occurred within and between four research schools, which linked economic long waves or business cycles to levels of capital investment, clusters of innovations, crises in capitalism, and major wars, respectively.[88] Goldstein argues that *'interest in both rounds reflected contemporary problems in the world economy for which the long wave seemed to give a plausible explanation'.*[89] Whilst the intensity of the debate has subsided since its peak during the 1980s, long-wave theory continues to be of interest and relevance, for example, in attempts to project the likely impacts of ICT, such as the Internet, and of biotechnology and nanotechnology.[90]

We are interested here in the research that has attempted to determine the extent and nature of the relationship between economic cycles and innovative activity. This embraces the seminal work of economists such as Joseph Schumpeter and Simon Kuznets from the 1930s and 1940s,[91] and that of Gerhard Mensch, Christopher Freeman, and Carlota Perez, among others, during the 1970s and 1980s.[92] Important questions that this work has attempted to address include whether or not innovations occur in waves or clusters, and, if they do, whether these are synchronous with long waves in economic activity and where such innovation clusters fall in the economic cycle. Debate has also centred heavily on what such findings mean with regard to the nature of the relationship between economic activity and innovative activity, and the mediating role of the broader socio-economic and institutional system in which this activity is embedded. Given the detailed empirical evidence and arguments employed, we can provide here only an overview of the debate.[93]

5.5.1 The relationship between economic and innovation cycles

We start by providing a brief introduction to the work of Kondratiev. His major contribution revolves around the empirical evidence that he collated in order to demonstrate

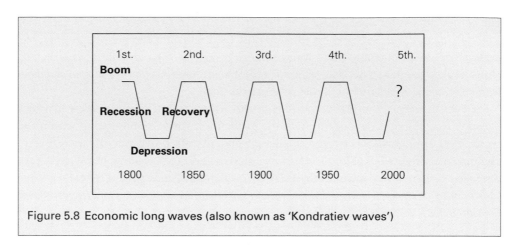

Figure 5.8 Economic long waves (also known as 'Kondratiev waves')

the existence of long-term business cycles of approximately fifty years in duration (i.e. longitudinal data from the UK, USA, Germany, and France, across a range of economic variables, such as commodity prices, wages, interest rates, and trade). Each business cycle was found to be characterized by four distinct phases: boom, recession, depression, and recovery. **Figure 5.8** provides a stylized depiction of these waves of economic activity and the individual phases of a single wave.

Kondratiev theorized that the pattern and duration of these business cycles was attributable to the periodic wearing out of capital equipment and its subsequent replacement in concentrated periods of large scale reinvestment during the upswing (i.e. recovery phase) of the economy. He also argued that important inventions and discoveries were concentrated in periods of recession, and were applied during the upswing of the economy in the jostling for market position.[94]

For Schumpeter, innovation was *'the outstanding fact in the economic history of capitalist society'*. Thus, in building on the work of Kondratiev, his aim was to develop a theory of business cycles that centred primarily upon innovation and the notion of 'lead sectors'. He theorized that radical innovations during a recession give rise to technological revolutions and the emergence of new lead sectors, which replace previously dominant sectors; this results in what we have referred to earlier as 'creative destruction'. It is argued that these new lead sectors drive the expansion of the economy during the upswing, led by entrepreneurs seeking to profit from vigorous exploitation of the emerging technology and followed by a 'bandwagon' of imitators. As a technology matures and is faced by diminishing returns, however, taking the economy toward recession, they too would be replaced by new technologies and new lead sectors. This is sometimes referred to as the 'depression trigger' hypothesis for the development and diffusion of radical innovation. He thus argued that each economic upswing was largely associated with the dissemination of one or more radical technologies, attributing the first long wave to steam power

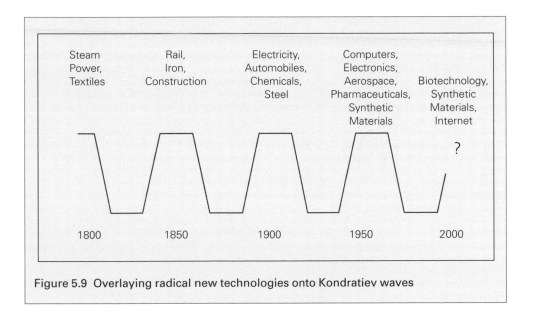

Figure 5.9 Overlaying radical new technologies onto Kondratiev waves

and textiles, the second to the railway boom, iron, and construction, and the third to electricity, chemicals, and automobiles; following Schumpeter's lead, these long waves are often referred to as 'Kondratiev waves'.[95] **Figure 5.9** illustrates the attribution of radical new technologies to the various Kondratiev waves.

Although Kuznets was not convinced by Schumpeter's notion of innovation clusters, subsequent work by Mensch found evidence of 'swarms of technical innovations' and 'long dry spells', with the innovative 'surges' or 'impulses' associated with the downswing of the economic long wave (i.e. the recession phase). These swarms or clusters of innovation included both radical innovations that brought about the emergence of new sectors, as well as major improvement innovations that rejuvenated existing sectors. Broadly speaking, the life cycle stages (i.e. growth, maturity, and decline) of these new innovations can be mapped onto the subsequent phases of the business cycle (i.e. recovery, boom, and recession). The depression phase then sees a shift in investment to alternatives, which triggers a new surge in innovation.[96]

In support of Schumpeter's notion of lead sectors, Freeman et al. argued that the upswing (i.e. recovery phase) in the business cycle was accompanied by a burst of growth across a small number of major new sectors and technologies. But in contrast to both Schumpeter and Mensch, they contended that depressions hinder, rather than stimulate, radical innovation, attributing the clustering of innovation to scientific and technological breakthroughs, and to periods of strong demand associated with the burst of growth in the upswing. This is sometimes referred to as the 'prosperity-pull' hypothesis for the development and diffusion of radical innovation. As such, they saw an important role for governments in stimulating the economy during depressions.[97]

Other research, such as that by Duijn, also found evidence of radical innovation occurring during economic upswings.[98] In contrast, however, Kleinknecht found evidence of product improvement and process innovations clustering on the upswing, and radical product innovations on the downswing, leading him to conclude that their was *'weak support for the prosperity-pull hypothesis . . . but . . . strong support for the depression-trigger hypothesis'*.[99] Thus, empirical studies have found varying, and sometimes conflicting, evidence of the degree of clustering of innovation, of the timing of such clustering in relation to the phases of the business cycle, and of the overall synchronicity of innovation and economic cycles. In part, this reflects differences in both sample and method.

For Perez, the above debate focused too narrowly on the relationship between the economic cycle and the innovation cycle. In developing her argument, she made the clear distinction between *'technological advance in terms of knowledge and inventions'*, which she viewed as *'a relatively autonomous process'*, and innovation, embracing the *'application and diffusion of specific techniques in the productive sphere'*, which she argued *'is very much determined by social conditions and economic profit decisions'*.[100] In this perspective, the application and diffusion of new technologies and innovations can be strongly inhibited, or facilitated, by the prevailing socio-institutional framework. Thus, she argued:

> Kondratiev's *long waves* are not a strictly economic phenomenon, but rather the manifestation, measurable in economic terms, of the harmonious or disharmonious behaviour of the *total* socioeconomic and institutional system (on the national and international levels).

From this perspective, 'harmonious behaviour', which occurs where there is a complementarity between the *'dynamics of the economic subsystem and related dynamics of the socio-institutional framework'*, results in an economic upswing, whereas 'disharmonious behaviour' results from a breakdown in this complementarity and leads to a 'structural crisis'—that is, a recession.[101]

Perez also contended that each upswing in the Kondratiev cycle is accompanied and facilitated by a new 'technological style'[102] or 'techno-economic paradigm',[103] defined as 'best productive common sense', which relates to the broad tendencies of the economy as a whole rather than an individual sector. Her concepts of 'technological style' and 'techno-economic paradigm' were much broader than those of her fellow economists of the time, such as the 'technological paradigm' concept of Dosi and the 'technological regime' concept of Nelson and Winter, but much closer to those articulated more recently by sociologists, such as the 'socio-technical landscape' concept of Geels (refer back to **Table 5.2**).[104]

The contribution of Perez is important, because it creates a link between national systems of innovation and economic long-wave theory. We will return to these broader concepts in **Chapter 11**.

5.6 Concluding comments

In this chapter, we have sought to build upon our discussion in **Chapter 4** of the models that have been developed to map the trajectory of a technology over its life cycle. Our discussion here has focused on explaining why technology might follow such trajectories, rather than develop in a more ad hoc manner, and how and why discontinuities may arise. In doing so, we drew upon the philosophy of science, through the work of Kuhn, evolutionary economics, through the work of economists such as Nelson and Winter, and the sociology of technology, through the work of sociologists such as Rip, Kemp, and Geels. Each of these disciplines have made valuable contributions to our understanding of how technology develops and progresses over time. In **Chapter 11**, we will return to develop the discussion around the shaping of innovation and technology within broader systems, through concepts such as national and regional 'systems of innovation'.

An appreciation of the patterns of innovation and technological progress, along with an understanding of why such patterns occur, is important for managers and policy-makers alike, who are seeking to comprehend and structure the past, and attempting to predict the future of their industry and regions. This is key to the development of appropriate capabilities and strategies. Such models also provide us with *'a language and a facility for talking about and directing technology'*.[105] Christensen et al. go further, arguing that:

> the only way to look into the future is to use these sorts of theories, because conclusive data is only available about the past . . . the best way to make accurate sense of the present, and the best way to look into the future, is through the lens of theory. Good theory provides a robust way to understand important developments, even when data is limited. And theory is even *more* helpful when there is an abundance of data.[106]

Thus, both this chapter and the last are seen as important precursors for our discussion of innovation strategy in **Chapter 6**.

CASE STUDY 5.1 INNOVATION AND DISCONTINUITY IN THE SERVICE SECTOR— REUTERS AND THE PROVISION OF FINANCIAL INFORMATION

Introduction

Reuters was founded in London in 1851 by Paul Reuter, a German immigrant. He provided a news and stock price information service, employing telegraphy technology and carrier pigeons—a very innovative service for its time. Reuters quickly established a reputation for speed, accuracy, and integrity. Over the years, the company's financial services have progressed from those focused on the provision of raw data, for example, concerning commodity and stock prices, to information, within which data was combined and integrated, and more recently, to analysis, in which such data and information is interpreted; each require very different competences. Today, the company is one of the world's leading news and information providers, collating data from some 160 stock exchanges and on 40,000 firms around the world. It has some 16,000 staff, 200 news bureaus, and had an annual turnover of over US$5bn in 2007. In 2008, Thomson bought the company for around US$16bn, to form Thomson Reuters, a major global player in the supply of financial information.

Employing emerging technology to enhance and develop service provision

Throughout its history, Reuters has sought to employ the latest technology to enhance the speed and scope of the dissemination of financial and market information, and to develop its range of services. By the early 1920s, for example, Reuters was employing radio to transmit news internationally, and exchange rate information by Morse code to Europe. By the mid-1960s, it was leading the way in transmitting market quotations via computer. In 1970, Reuters introduced technology that allowed the display of stock and commodity prices, and a decade later, it was the first to introduce technology that allowed financial traders to conclude trade via computers. By the mid-1990s, Reuters was utilizing television to provide live coverage of unfolding events on the computer screens of these financial traders.[107]

The arrival of the Internet

The rapid development of the Internet during the 1990s created unprecedented challenges to Reuters. In particular, it facilitated the emergence of countless new rivals, which were able to provide financial services much more cheaply than Reuters. It became clear to the company that its well-established and highly successful business model was increasingly at odds with such developments. It responded in 1997, by earmarking some £50m for the establishment of a new division based in Geneva to develop Internet-related services. During the late 1990s, the company also undertook a number of acquisitions and forged alliances with a range of technology and financial service companies, in order to gain access to the knowledge and competence necessary to operate in the new age of the Internet.

Questions

1. Apply the reverse product-process cycle to the case of Reuters. How useful do you think this model is in explaining innovation in financial information services?

2. How useful is the notion of 'discontinuity' for understanding the development of financial information services over the last 150 years?

FURTHER READING

1. For a good overview of the origins, processes, and pathways of transitions and transformations in technological regimes, see Smith et al. (2005) and Geels and Schot (2007).

2. See Devezas et al. (2005) for an application of long-wave theory to the development and diffusion of the Internet.

NOTES

[1] Kuhn (1962; 1996).

[2] Kuhn employs the term 'puzzles' to refer to problems with solutions (1962: 36–7).

[3] Kuhn (1996: 23–4).

[4] Crane (1972: 95).

[5] Lakatos (1978).

[6] Feyerabend (1974; 1975); Popper (1974).

[7] Slowik (1998).

8 Rosenberg (1969); Nelson and Winter (1977); Dosi (1982).

9 Dosi (1982: 153).

10 Ende and Kemp (1999: 835–7).

11 Ende and Kemp (1999); Poel (2003).

12 Ende and Kemp (1999).

13 Clark and Staunton (1989: 108–10).

14 Ende and Kemp (1999: 849).

15 Kemp et al. (1998: 182).

16 Nelson and Winter (1977: 56).

17 Dosi (1982: 152).

18 Data from 'Moore's Law timeline', available online at <http://www.intel.com> (accessed 7 September 2008).

19 Geels (2002: 1259).

20 Kemp et al. (1998: 182).

21 Geels and Schot (2007).

22 MacKenzie (1992).

23 Nelson and Winter (1977: 56–60).

24 Pavitt (1984).

25 Clark and Staunton (1989: 109).

26 Clark and Staunton (1989: 109).

27 Sundbo (1994; 1997; 2000); Gallouj and Gallouj (2000); Gallouj (2000).

28 Sundbo (2000).

29 Gallouj and Gallouj (2000: 36).

30 Gallouj and Gallouj (2000: 36).

31 Sundbo (2000: 119).

32 Schumpeter (1942: 83–4).

33 Drucker (1969).

34 Foster (1986: 48, 115).

35 Aberathy and Clark (1985: 14).

36 Anderson and Tushman (1991: 49).

37 Tushman and Anderson (1986); Anderson and Tushman (1991).

38 Aberathy and Clark (1985); Tushman and Anderson (1986); Rothaermel (2000).

39 Kuhn (1996: 64).

40 Kuhn (1996: 65).

41 Kuhn (1996: 78).

42 Lakatos (1978).

43 Kuhn (1996: 76).

44 Kuhn (1996: 68).

45 Kroeber (1957).

46 Kuhn (1996: 91).

47 Kuhn (1996: 85).

48 Abernathy and Clark (1985); Foster (1986); Tushman and Anderson (1986).

49 Christensen (1997; 2001a; 2001b).

50 Tushman and Rosenkopf (1992: 318).

51 Tushman and Anderson (1986).

52 Tushman et al. (1997: 5–7).

53 Foster (1986: 102).

54 Foster (1986: 160–1).

55 Tushman et al. (1997: 9).

56 Anderson and Tushman (1991); Tushman et al. (1997).

57 Foster (1986: 182).

58 Anderson and Tushman (1991: 49).

59 Abernathy and Clark (1985: 4).

60 Tushman and Anderson (1986: 460).

61 Schumpeter (1939; 1942).

62 Leonard-Barton (1992).

63 Anderson and Tushman (1991).

64 Foster (1986: 132–4).

65 Tushman and Anderson (1986: 460).

66 Abernathy and Clark (1985: 6).

67 Anderson and Tushman (1991).

68 Tushman and Anderson (1986).

69 Afuah (1998: 148–9).

70 Anderson and Tushman (1991: 48–9).

71 Ehrnberg and Sjöberg (1995); Tripsas (1997); Rothaermel (2000).

72 Lambe and Spekman (1997).

73 'Australia pulls plug on old bulbs', 20 February 2007, available online at <http://news.bbc.co.uk>, accessed 27 September 2008.

74 Tripsas (2008).

75 Levinthal (1998: 218, 222).

76 Constant (2002: 1253).

77 Taken from the title of a journal article—Levinthal (1998).

78 Smith et al. (2005: 1508).

79 Geels and Schot (2007: 400).

80 Levinthal (1998: 218), who cites Strickberger (1996) regarding perspective within evolutionary biology.

81 Kodama (1992); Levinthal (1998).

82 Fildes (2008).

83 Geels (2002); Poel (2003); Smith et al. (2005); Geels and Schot (2007).

84 Ende and Kemp (1999); Geels (2002); Smith et al. (2005); Geels and Schot (2007).

85 Smith et al. (2005: 1496).

86 Poel (2003: 66).

[87] Kondratiev (1925).

[88] Goldstein (1988).

[89] Goldstein (1988: 21).

[90] Linstone (2002); Silverberg (2002); Dewick et al. (2004); Castellacci (2006); Devezas et al. (2005; 2008).

[91] Schumpeter (1939); Kuznets (1940).

[92] Duijn (1981; 1983); Kleinknecht (1981); Clark et al. (1981); Freeman et al. (1982); Perez (1983).

[93] See from Goldstein (1988: chs 1–3), for an interesting overview of the data and debate.

[94] Kondratiev (1925); Rothwell and Zegveld (1985); Goldstein (1988).

[95] Schumpeter (1939); Rothwell and Zegveld (1985); Goldstein (1988).

[96] Mensch (1979); Rothwell and Zegveld (1985); Goldstein (1988).

[97] Clark et al. (1981); Freeman et al. (1982); Goldstein (1988).

[98] Duijn (1981).

[99] Kleinknecht (1981: 303).

[100] Perez (1985: 442).

[101] Perez (1983: 358).

[102] Perez (1983).

[103] Perez (1985).

[104] Dosi (1982); Geels (2002), respectively.

[105] Foster (1986).

[106] Christensen et al. (2004: xx–xxi).

[107] <http://www.thomsonreuters.com>. Read (1999) provides a detailed history of Reuters' exploitation of technology.

6

Innovation Strategies

Chapter overview

Learning objectives

This chapter will enable the reader to:

- appreciate the paradoxical nature of strategy and outline the various tensions that are present in formulating innovation strategies;

- outline the range of strategic decisions relevant to the management of innovation;

- assess the degree to which innovation strategies can, or should be, 'planned' or 'emergent', and the extent to which organizations are able to shape their environment and/or reinvent themselves in response to a changing environment;

- define concepts such as 'intangible assets', 'relational assets', 'core competence', and 'dynamic capabilities', and recognize their importance for the strategic management of innovation; and

- compare and contrast the various 'timing-to-market' innovation strategies open to organizations, and outline the assets, competencies, and capabilities required to undertake each strategy effectively.

6.1 Introduction

It is not our intention in this chapter to provide an overview of the strategic management literature; this is covered comprehensively in a large number and wide variety of strategic management texts.[1] We will, however, draw upon those areas of strategic management that are of particular relevance to the management and shaping of innovation. For example, in light of our discussion of trajectories and paradigms in **Chapters 4** and **5**, we will assess the balance between choice and inevitability in strategy formulation. We will also draw upon the work relating to 'intangible assets', 'core competencies', and 'dynamic capabilities'. In line with the network perspective introduced in **Chapter 3** and employed throughout this text, we adopt an extended version of the 'resource-based view' (RBV) to recognize the importance of 'relational' assets. We discuss the various 'timing to market' strategies (e.g. 'first to market' or 'late to market') that an organization might adopt, and the associated resources and competencies that each entail.

We start, however, by returning to the theme of paradox, introduced in **Chapter 2**. One strategic management text that stands out in this regard is *Strategy: Process, Content, Context*, by Bob De Wit and Ron Meyer, which adopts the notion of paradox as a central theme of strategic management. They argue:

> At the heart of every set of strategic issues, a fundamental tension between apparent opposites can be identified . . . Each pair of opposites . . . seem to be inconsistent, or even incompatible, with one another . . . If firms are competing, they are cooperating. If firms must comply to the industry context, they have no choice. Yet, although these opposites confront strategists with conflicting pressures, strategists must somehow deal with them simultaneously.[2]

This perspective fits well with our orientation to the understanding and managing of innovation.

By viewing strategy as inherently paradoxical, De Wit and Meyer outline a series of tensions, many of which are highly relevant to our discussion of innovation strategy and are outlined in **Table 6.1**. We will incorporate these paradoxes, both explicitly and implicitly, in our discussion of innovation strategy; they will also permeate subsequent chapters. In particular, the 'control–chaos' paradox is at the heart of our discussion of organizational structure (**Chapter 7**), the management of innovation and knowledge

Table 6.1 The paradoxical nature of strategy (adapted from de Wit and Meyer, 2004)

The strategic tension	Relates to
Logic vs creativity	Strategic thinking—concerns the tension between rationality and logic, on the one hand, and creativity and imagination, on the other, in relation to strategic analysis and formulation.
Planned vs emergent	Strategic formulation—concerns the tension between the degree to which strategy should be planned and 'deliberate' as opposed to 'emergent'.
Environment-led vs resource-led	Strategic drivers—concerns the degree to which strategy should be led or driven by a consideration of the opportunities in the environment or opportunities that arise from a consideration of an organization's resources and competencies.
Constraint vs choice	Strategic opportunity—related to the above tension, this concerns the degree to which strategic options are viewed as being constrained by an organization's environment, or can be generated by an organization's ability to modify its environment to create new opportunities. This reflects the tension between the environmental 'determinism' and 'shaping' perspectives.
Revolution vs evolution	Strategic change—concerns the degree to which an organization should embrace a 'continuous' or 'evolutionary' approach to change, as opposed to a 'discontinuous' or 'revolutionary' approach.
Competition vs cooperation	Strategic resources—concerns the degree to which an organization perceives the relationship with other organizations as essentially one based on competition or cooperation. Cooperative arrangements with other organizations, even direct competitors, can make available important 'relational' assets, but bring risks associated with the disclosure of propriety knowledge.
Globalization vs localization	Strategic location—concerns the degree to which innovation activities should be centralized or decentralized, and whether innovations should be generic or adapted to local market needs.
Control vs chaos	Strategic organization—concerns the degree to which organizational leaders and managers can or should control the activities and direction of the organization. This relates to the 'planned–emergent' strategy tension and to the appropriate mode of organising for an organization.
Profitability vs responsibility	Strategic objectives—this raises the tension that exists in relation to organizational 'purpose' and the interests the organization should serve. Increasingly, organizations are recognizing the need to serve a broader range of stakeholders.

processes (**Chapter 8**), and the role of informality and serendipity (**Chapter 9**); the 'competition–cooperation' paradox is central to our review of the sources of innovation (**Chapter 10**); and finally, the 'profitability–responsibility' paradox is a theme that emerges in our discussion of the social and political 'shaping' of innovation and technology (**Chapter 11**).

6.2 The innovation strategy-making process

In the following sections, we explore three areas. The first concerns the nature and scope of strategic decisions that are of particular relevance to the strategic management of innovation. The second centres on the degree to which innovation strategy is pre-planned or emergent, especially in relation to incremental versus discontinuous or radical innovation. And finally, we explore the notions of 'choice' and 'determinism' in the strategy-making process.

6.2.1 Strategic decisions in the management of innovation

Nonaka and Takeuchi, well known for their work on knowledge creation within organizations, argue that '*The most critical element of corporate strategy is to conceptualize a vision about what kind of knowledge should be developed and to operationalize it*'; they term this strategic vision 'organizational intention'.[3] Similarly, Leonard, who has written extensively on sustaining innovation, stresses the need for managers to articulate the organization's 'strategic intent', '*since employees require guidance to be productive*'.[4] But in establishing the strategic vision for an organization, managers must consider the 'substance' of strategy along a number of interrelated dimensions, principal among which are:[5]

- the selection of the technologies in which the organization will specialize;
- the proximity to the state-of-the-art in these selected technologies;
- the relative emphasis on basic research, applied research, and development;
- the degree to which technology will be developed internally or sourced externally;
- aggregate levels of investment in research and development (R&D);
- the degree to which R&D will be centralized or decentralized;
- the selection of the mechanisms for protecting R&D investments (e.g. patents);
- the selection of alternative routes for appropriating the benefits of innovation (e.g. licensing, outsourcing, and joint ventures).

Innovation is not only a strategic issue for young or high-technology industries: Grant, for example, argues that:

> the quest for differentiation in mature industries requires innovation . . . However . . .
> the limited opportunities for technology-based advantage creates impetus for innovation in marketing, product design, customer service, and organization.[6]

Nevertheless, sustaining innovative activity is a particular problem faced by mature companies.

Dougherty and Hardy suggest that for a mature organization to develop the capacity for sustained innovation, it must:[7]

- make resources available for innovative activity, to counter the prevailing practice in many organizations of supporting established activities;

- develop collaborative structures and processes to enable creative problem solving and the diffusion of innovations across the organization; and

- incorporate innovation as a meaningful element of strategy, through drawing upon existing competencies and assets, and, where necessary, through the reframing of strategy to reconnect it with innovation.

Baden Fuller and Stopford are much bolder in their assertions, arguing that maturity is a state of mind rather than a state of an organization or sector, and that, as such, every organization has the potential to rejuvenate and to transform its business environment. For them, rejuvenation is clearly a strategic endeavour.[8]

6.2.2 Planned versus emergent innovation strategy

There has been a long-running debate concerning the degree to which organizational strategy should be pre-planned or allowed to emerge in response to the dynamism exhibited by many sectors. Some of the best-known work in this area was undertaken by Henry Mintzberg and James Quinn during the late 1970s and the 1980s. This seminal work remains salient today. Mintzberg, for example, distinguished between 'deliberate *strategies, where intentions that existed previously were realized, from* emergent *strategies, where patterns developed in the absence of intentions, or despite them*'.[9] From this perspective, strategies were seen to emerge from a combination of proactive (planned) and reactive (emergent) strategic decision making. This was much the same observation as that made by Quinn, who noted that '*Real strategy* evolves *as internal decisions and external events flow together*'.[10] For Quinn, the ability to respond and react '*is essential in dealing with the many "unknowables" in the environment. Successful organizations actively create flexibility*'; it was not a matter of 'muddling through', but of logical, incremental, strategic decision making.[11] He thus argued that:

> Strategy deals with the unknowable, not the uncertain. It involves forces of such great number, strength, and combinatory powers that one cannot predict events in a probabilistic sense. Hence logic dictates that one proceed flexibly and experimentally from broad concepts toward specific commitments, making the latter concrete as late as possible in order to narrow the bands of uncertainty and to benefit from the best available information. This is a process of 'logical incrementalism' . . . [it] is not 'muddling' . . . It is conscious, purposeful, proactive, good management.[12]

In the approaches of both Mintzberg and Quinn, we see an attempt to embrace the paradoxes inherent in strategy formulation; theirs can be seen to be a 'both/and' approach as opposed to a 'either/or' approach.

This orientation seems particularly relevant for the strategic management of innovation, especially in relation to discontinuous or radical innovation. McDermott and O'Connor, for example, note that a critical issue for the management of radical innovation is *'how the opportunity should be leveraged . . . There are typically many choices and many unknowns . . . Deriving the best business model takes time and experimentation'*. Furthermore, in hi-tech sectors, ideas and opportunities often arise and emerge from lower levels in the organization and must percolate up, rather than down, the organizational hierarchy. Burgelman and Sayles contend that:

> Understandably we think of strategy formulation as top management work . . . But in the high-technology world, strategy often revolves around the innovation activities of relatively low-level technical and business people. To be sure, their decisions will require ratification by top management. Nevertheless . . . the reality is that those closer to the emerging technology will seek to define the business opportunity.[13]

Reid and Brentani forward the concept of the 'strategic web' to suggest how emergent radical ideas may be captured by the formal strategy-making process and 'meshed' with the stated strategy of the organization. They define a 'strategic web' as *'a porous and adaptable outline within which an organization has decided to play strategically in terms of markets, applications, technologies, and products'*.[14]

6.2.3 Determinism versus choice in innovation strategy

We start this discussion by disentangling two distinct issues relating to strategic choice. The first concerns the degree to which the strategic choices of an organization are constrained by the environment in which it operates—that is, the extent of 'environmental determinism'. The second concerns the degree to which such strategic choices are constrained by 'predetermined mindsets' of those determining strategy within an organization. This might be through, for example, the influence of prevailing technological paradigms (refer back to **5.2.2**), 'core rigidities',[15] or strategic 'recipes'.[16] Richard Whittington refers to this form of choice constraint as 'action determinism'. Action determinism occurs where *'actions are selected according to in-built preference and information processing systems'*.[17] Thus, strategic choice can be seen to face both external and internal forms of constraint.

We adopt a position not dissimilar to that of Clark, who argues that *'Contexts contain constraints as well as zones of manoeuvre'*[18]—that is, whilst environments are often seen to shape the strategic orientation of organizations, there is also scope for organizations to shape their environment. Organizations that bring to market discontinuous innovations, for example, have the possibility of radically altering their environments. As you may recall from **Chapter 5**, discontinuous innovation ruptures the prevailing 'era of incremental innovation' within an industry and spawns an 'era of ferment' (see **5.3.2**). Our position also recognizes the constraining influence of the prevailing technological paradigm or socio-technical regime within an industry, as well as the constraining influence of the broader socio-technical landscape, but nevertheless entertains the possibility that organizations can shape both 'regimes' and, to a lesser extent, 'landscapes' (see **5.4.2**). And because, as we noted in **5.2.3**, *'regimes result in technological trajectories'*,[19] then organizations can, in certain circumstances, ultimately shape technological trajectories.

In an interesting cross-sector study of the potential influence of different owner-ship structures on innovation strategy in US firms, Hoskisson and colleagues found that 'conflicting voices' existed between pension fund managers, *'who appeared to avoid invest-ing in firms that pursue external innovation through acquisition and prefer internal innovation'*, and professional investment fund managers, who appear to:

> prefer external innovation and are not predisposed to firms emphasizing internal innovation. They prefer more immediate returns . . . Investments in R&D are treated as an expense that reduces short-term returns.[20]

With regard to company directors, their results indicate that 'inside' directors (i.e. employees) favoured internal innovation and *'when inside directors are dominant, firms focus on internal innovation'*. In contrast, 'outside' directors (i.e. independent/non-employees), especially those with equity, preferred external innovation. They argue that this is important because of the trend towards the increasing dominance of 'outside' directors.[21] This research indicates one way in which 'action determinism' might influence strategic choice in relation to innovation.

In the strategic management literature, two competing perspectives stand out in rela-tion to strategy formulation: the 'environment-led' and 'resource-led' approaches. In the environment-led, or 'outside-in', perspective, the starting point in determining strategy is the analysis of the market, in order to identify attractive opportunities and the appropriate positioning of the organization within that market vis-à-vis competitors, customers, and suppliers, for example. This approach is typified by the work of Michael Porter.[22] Whilst this orientation 'favours' environment over resource, it does not dismiss the possibility that organizations may be capable of shaping their environments.

In stark contrast, in the resource-led, or 'inside-out', approach, strategy is determined following an analysis of the organization's resources and competencies in order to iden-tify those that are unique and difficult to imitate. Having identified the organization's strengths, strategy concerns the identification, adaptation, or creation of a market. This orientation, commonly referred to as the 'resource-based view' (RBV), is typified by the work of Gary Hamel and Coimbatore Prahalad, in relation to 'core competencies',[23] and David Teece and Gary Pisano, with regard to 'dynamic capabilities'.[24] From this latter perspective, Teece notes in relation to strategic choice that:

> the dynamic capabilities framework recognizes that the business enterprise is shaped but not necessarily trapped by its past. Management can make big differences through investment choice and other decisions. Enterprises can even shape their ecosystem . . . Managers really do have the potential to set technological and market trajectories, particularly early on in the development of a market . . . Indeed, the enterprise and its environment frequently coevolve.[25]

In the subsequent sections of this chapter, we draw upon the RBV; Grant notes that this approach emphasizes the uniqueness of each organization's collection of resources and competencies, and the importance of exploiting this uniqueness in establishing compet-itive advantage.[26]

We turn now to a discussion of the nature and variety of organizational resources and competencies, and their role in innovative activity.

6.3 Organizational resources and competencies

In the strategy and innovation literatures, the terms 'resource', 'asset', and 'endowment', are often employed interchangeably,[27] as are those of 'capability' and 'competence'.[28] Organizational competencies have been defined as an *'ability to perform an activity'*,[29] and *'firm-specific accumulations of expertise resulting from previous investments and from learning-by-doing'*.[30] Organizational endowments, resources, or firm-specific assets, such as brands, patents, reputation, skilled workers, and distribution networks, for example, do not in themselves create value, but must be combined and integrated to create organizational capabilities.[31] Grant illustrates this point by arguing that:

> On their own, Ford's engineers, designers, labs, studios, and IT resources are of limited value. Together, they can provide the new product development capability needed to create new models [of automobile].[32]

6.3.1 Intangible resources and assets

There has been increasing interest since the early 1990s in the 'intangible' resources or assets of organizations. This is, in large part, because of the recognition that such resources are a key driver of innovation and organizational value.[33] The notion of 'tangible' resources typically refers to those assets that are visible or can be touched, such as buildings, machinery, land, money, and inventories of raw materials and finished goods. In contrast, 'intangible' resources refers to those assets that are, for the most part, 'invisible', such as brand and reputation, know-how, and the social networks of employees.

De Wit and Meyer distinguish between tangible and intangible resources by referring to them as the 'hardware' and 'software' of the organization, respectively. For Hall, the concept of intangible resources embraces:

> the intellectual property rights of patents, trademarks, copyright and registered design; through contracts; trade secrets; public knowledge such as scientific works; to the people dependent, or subjective resources of know-how; networks; organizational culture, and the reputation of product and company.[34]

An important distinction often made between tangible and intangible assets is that whilst the former can be bought and sold, the latter can only be accumulated—that is, they result from *'adhering to a consistent set of policies over a period of time'*, such as those concerning employee training or reputation building.[35] Such a process of resource accumulation implies that intangible assets are 'inherently inimitable' (i.e. defy imitation), because *'would-be imitators need to replicate the entire accumulation path to achieve the same resource position'*.[36] It is perhaps not surprising, then, that Hall argues that such intangible resources are the 'feedstock' of the capabilities that allow an organization to achieve sustainable competitive advantage. Importantly, the notion of resource accumulation also implies a greater role for managers in the formation of organizational resources.

Tacit knowledge is often considered to be one of the most valuable intangible resources for the creation of sustainable competitive advantage;[37] its strategic significance lies in *'its potential to be valuable, rare, and hard for rivals to imitate'*.[38] But herein also lies a dilemma: for organizations to exploit the value of its tacit knowledge, it must find ways in which it can be transferred and replicated around the organization;[39] yet, for this to be achieved, some degree of codification of the tacit knowledge is required,[40] which, in turn, reduces the inimitability of the resource, allowing it to be more easily acquired by rivals. In overcoming this dilemma, Rivkin argues that organizations must find ways in which to balance the requirement for codifying tacit knowledge, so that it can be disseminated and employed widely within the organization, with the desire to maintain the tacitness of its knowledge, to reduce its acquisition and imitation by rival organizations.[41]

Grant contends that:

> For most companies, intangible resources contribute much more than do tangible resources to total asset value. Yet, in relation to company financial statements, intangible resources remain largely invisible.[42]

But such is the importance of intangible resources to organizations that increasing numbers are attempting to identify and value these assets, and to record them alongside tangible assets in their financial accounts. In this regard, the definition of 'intangible' assets provided by the International Accounting Standards Committee (IASC) is instructive:

> An identifiable nonmonetary asset without physical substance. An asset is a resource that is controlled by the enterprise as a result of past events (for example, purchase or self-creation) and from which future economic benefits (inflows of cash or other assets) are expected. Thus, the three critical attributes of an intangible asset are: identifiability, control (power to obtain benefits from the asset), [and] future economic benefits (such as revenues or reduced future costs).[43]

The IASC provides a number of examples of intangible asset, many of which are relevant to the strategic management of innovation, such as patents, copyrights, licences, and customer and supplier relationships. Nevertheless, despite this recognition of the importance of intangible assets, and the attempts to measure and report such intangible resources, much further work is still required in relation to the development of theory, definitions, and methods.[44]

Over the last decade or so, much effort has gone into the categorization of the various intangible resources accumulated and nurtured by organizations. This work is important because, as Subramaniam and Youndt note, it offers the *'means to parsimoniously synthesize the approaches by which knowledge is accumulated and used in organizations'*.[45] Of particular relevance to the study of innovation has been the identification of the various 'knowledge resources' of an organization. Collectively, these knowledge resources are increasingly referred to as 'intellectual' capital.[46] Although intellectual capital is defined in various ways and composed of different elements,[47] it may be usefully broken down into three components:

- 'human' capital;
- 'organizational' capital; and
- 'social', or 'relational', capital.

In the subsequent section, we discuss these various components of intellectual capital.

6.3.2 The components of intellectual capital

'Human' capital may be defined as *'the knowledge, skills, and abilities residing with and utilized by individuals'*; 'organizational' capital as *'the institutionalized knowledge and codified experience residing within and utilized through databases, patents, manuals, structures, systems, and processes'*;[48] and finally, 'social' capital, which we introduced in **3.4.3**, as:

> the sum of the actual and potential resources embedded within, available through, and derived from the network of relations possessed by an individual or social unit. Social capital thus comprises both the network and the assets that may be mobilized through that network.[49]

Social capital is sometimes referred to as 'relational' capital. Social capital embraces both intra- and inter-organizational connections, as well as informal (e.g. social networks) and formal (e.g. joint ventures and supplier agreements) relationships.

In **Chapter 3**, we also distinguished between three dimensions of social capital:

- the 'structural', referring to the configuration of the network;
- the 'cognitive', referring to the shared meanings and understanding between network members; and
- the 'relational', regarding trust, obligations, and norms between network members.[50]

Each of these aspects was explored in relation to its impact on the innovative capacity of the organization in **Chapter 3** (see 3.5.1–3.5.3).

Whilst it is useful to distinguish between these distinctive components of intellectual capital, it is also worth noting, as Subramaniam and Youndt do, that:

> the various aspects of intellectual capital are not always found in organizations in neat, separate packages. For example, individual knowledge (human capital) often becomes codified and institutionalized (organizational capital) and is transferred and leveraged in groups and networks (social capital).[51]

In a cross-sector study of around a hundred US organizations concerning the relationship between the different components of intellectual capital and innovative capability, it was found that organizational capital was strongly related to incremental innovative capability, and that human capital was strongly related to radical innovative capability, but only where it was accompanied by high levels of social capital. In explaining this latter finding, the researchers argue that *'unless individual knowledge is networked, shared, and chaneled through relationships, it provides little benefit to organizations in terms of innovative capability'*.[52]

The research yields two important findings for our understanding of the innovative capacity of organizations: firstly, that *'the value of human capital is inextricably tied to social capital'*, such that, to leverage existing human capital, *'it may be imperative for organizations to invest in the development of social capital to provide the necessary conduits for their core knowledge workers to network and share their expertise'*; and secondly, that social capital *'appears to be the bedrock of innovative capabilities . . . and* [fundamental] *for gaining the flexibility to selectively use these capabilities to meet market or competitive exigencies* [i.e. urgent demands]'.[53] In this regard, social capital is likely to be vital for the development of 'dynamic' capabilities—a concept to which we will return shortly.

In a complementary study, this time of US hi-tech new venture firms, the researcher found no relationship between the stock of human capital in the top management team and innovativeness, but, interestingly, did find a relationship between the diversity of human capital in the top management team and innovativeness.[54] Hayton explains this finding by noting the *'different perspectives that result from a heterogeneous group'*, and that *'diverse human capital also implies diverse sources of social capital and access to a broader range of social and professional networks from which new ideas can be acquired'*.[55]

Each of the various components of intellectual capital also require unique forms of investment, with:

> human capital requiring the hiring, training, and retaining of employees; organizational capital requiring the development of knowledge storage devices and structured recurrent practices; and social capital requiring the development of norms that facilitate interactions, relationships, and collaboration.[56]

Such 'investments' will be addressed in **Part III**, as our discussion shifts to an exploration of the ways in which organizations and processes can be organized to promote knowledge, creativity, and innovation. For example, in **Chapters 7** and **9**, we discuss modes of organizing that encourage the formation of boundary-spanning social networks, and in **Chapter 8**, we focus on processes and practices for the transfer and creation of knowledge.

6.3.3 Relational resources

De Wit and Meyer define relational resources as *'all of the means available to the firm derived from the firm's interaction with its environment'*.[57] From our earlier discussion of the prevalence and importance of sectoral networks and regional clusters to the innovative capacity of organizations (see **3.6.2** and **3.6.3**), and of the contribution of social capital to innovative capability noted above, it is clear that relational assets are of central concern for the strategic management of innovation.

But Lavie argues that:

> traditional perspectives on competitive advantage, such as the resource-based view . . . have envisioned firms as independent entities. Consequently, these perspectives have provided only a partial account of firm performance in view of the accumulated evidence of the proliferation and significance of interfirm alliances in recent years.[58]

As a result, such perspectives *'overlook the important fact that the (dis)advantages of an individual firm are often linked to the (dis)advantages of the network of relationships in which the firm is embedded'*.[59]

Dyer and Singh highlight a number of ways in which an organization can gain advantages from relational assets; these include the creation of specialized relation-specific assets, which result from the bringing together of existing assets from the collaborating organizations to create new assets that can be mobilized and drawn upon, through inter-organizational knowledge-sharing routines that generate learning within the organization, and through the leveraging of complementary resources of a partner organization. The degree to which an organization can benefit from such relational assets depends, however, on the extent to which it possesses 'absorptive' capacity—that is, the ability to *'recognize the value of new, external knowledge, assimilate it, and apply it to commercial ends'*.[60]

Illustration 6.1 Coping with technological discontinuity through the building of relational resources—the case of EMI and the recorded music industry

In 2000, the global market for recorded music was worth around £28bn; it was a market dominated by five major players—EMI, Universal, Sony Music Entertainment, Warner, and BMG—with a combined share of around 80 per cent. Whilst the introduction of the compact disc (CD) in the late 1970s had greatly improved the fortunes of the sector, by the late 1990s, with the emergence of MP3 and file-sharing services offered by companies such as Napster, the transfer of digital music through the Internet was beginning to present a very tangible threat to the music industry. The 'majors' were slow to respond to both the immediate threats and huge future potential of the Internet. But during 1999 and 2000, the industry witnessed a flurry of strategic alliances, licensing agreements, and technology partnerships as a response to this discontinuity in the distribution of recorded music.

In 1999, for example, EMI established arrangements with Internet music retailers, with music-maker.com, a site that allowed consumers to customize their CD downloads, Preview Systems, which was involved in technology for managing digital rights, and Liquid Audio, which was enlisted to help to digitize EMI's back catalogue. During 2000, EMI established a further seventeen major business relationships, including those with Microsoft, Roxio (which specialized in CD-burning technology), and Gigamedia Taiwan (which provided the opportunity to develop Internet retailing in the Far East). All of these arrangements provided EMI with an array of complementary and important relational resources to help it to secure its future in the new digital world.[61]

6.3.4 Slack resources and innovativeness

There has been a long-running debate concerning the relationship between 'slack' resources, and organizational performance and innovativeness. Slack resources may be defined as *'the pool of resources in an organization that is in excess of the minimum necessary to produce a given level of organizational output'*.[62] On the one hand, it is argued that slack resources relax internal controls, and encourage experimentation and risk taking through the redirection of resources toward projects with higher uncertainty, resulting in a positive impact on innovativeness.[63] It is also argued that slack resources allow an organization to adapt in complex and competitive environments.[64] In contrast, slack resources may buffer the organization from external shocks, and bring about complacency and misplaced optimism.[65] Furthermore, there is also evidence to suggest that organizations with less slack are able to leverage their more limited resources more efficiently and effectively.[66] Indeed, Baker and Nelson found, from their study of a sample of 'resource-constrained' firms:

that Levi-Strauss's concept of bricolage—making do with what is at hand—explained many of the behaviors we observed in small firms that were able to create something from nothing by exploiting physical, social, or institutional inputs that other firms rejected or ignored.[67]

To a certain extent, this apparent contradiction in the evidence concerning the impact of slack resources is tempered by evidence that suggests that slack resources have an inverse 'U-shaped' impact on innovation—that is, too little slack resource reduces experimentation, whilst too much slack reduces the discipline in the selection of projects, such that more poor ideas are pursued than promising ideas.[68]

In another study, involving a large sample of US technology firms, it was found that it was not the amount, but the nature, of the slack resource that mattered; in contrast to 'low-discretion' slack, 'high-discretion' slack was found to be positively related to the economic performance of the firm. George argues that this finding suggests '*that more slack is better for appeasing coalitions, experimentation, and risk taking*'.[69]

6.3.5 Core competencies and dynamic capabilities

So far, our discussion has focused on the various forms of resources, assets, or capital, and the degree to which they are associated with innovative activity. Grant argues, however, that '*resources are not very productive on their own*':[70] they need to be integrated and mobilized to create organizational competence or capability. Yet the relationship between resources and competence is complex and indirect. For Hamel and Prahalad, the key is the manner in which organizations leverage their resources; they suggest that resources can be leveraged through concentrating resources around a few clearly defined goals, accumulating resources, complementing and blending existing resources, conserving resources by employing them across various markets and products, and recovering resources by accelerating the speed by which investments in resources generate returns.[71]

We now turn our attention to the nature, formation, and role of organizational competencies and capabilities in relation to innovative activity; as we noted earlier, the terms 'competence' and 'capability' are generally employed interchangeably. The seminal work in this area was undertaken during the early 1990s, principally through the contributions of Gary Hamel and Coimbatore Prahalad, concerning 'core competencies',[72] and David Teece and Gary Pisano, with regard to 'dynamic capabilities'.[73]

For Prahalad and Hamel, '*competencies are the collective learning in the organization, especially how to coordinate diverse production skills and integrate multiple streams of technologies*'.[74] Similarly, Grant defines organizational capability as '*a firm's capacity for undertaking a particular productive activity*'.[75] Prahalad and Hamel also note that competencies do '*not diminish with use*', but are '*enhanced as they are applied and shared*'. Nevertheless, they warn that '*they still need to be nurtured and protected*'.[76]

Competencies may be expressed in relation to their functional area or value-chain activities and include, for example, capabilities in product design, brand management, the identification of market trends, speed of distribution, and radical product development. Competencies also range from lower-level capabilities, which reside around specific tasks, to higher-level capabilities, which are located at the functional or cross-functional level.[77] It is important, however, to view competencies not in isolation, but in combination: Danneels, for example, concludes, from his longitudinal study of the process of leveraging technological competence, that *'a competence to serve current customers and a lack of a marketing competence to search for new ones inhibits the leveraging of technological competence'*.[78]

Given the breadth and hierarchical nature of organizational competencies, it is not surprising that not all competencies provide competitive advantage. In this regard, Leonard usefully distinguishes between:

- 'core' capabilities—that is, those that *'constitute a competitive advantage for a firm'*;
- 'supplemental' capabilities—that is, those that add value to the core capabilities of the organization, but have the potential to be imitated; and
- 'enabling' capabilities—that is, those that are *'necessary but not sufficient in themselves to competitively distinguish a company'*.[79]

Prahalad and Hamel also distinguish core competencies from other competencies. They argue that for a capability to be considered a core competence, it must pass at least three tests:

> First, a core competence provides potential access to a wide variety of markets . . . Second, a core competence should make a significant contribution to the perceived customer benefits of the end product . . . Finally, a core competence should be difficult for competitors to imitate.[80]

The notion of 'dynamic capabilities' is subtly different from the concept of competence or capability, reflecting the distinctive *'ability to integrate, build, and reconfigure internal and external competencies to address rapidly changing environments'*.[81] The emphasis here, then, is on an organization's capacity to interact with, and adapt to or modify, its environment.

Teece disaggregates dynamic capabilities:

> into the capacity (1) to sense and shape opportunities and threats, (2) to seize opportunities, and (3) to maintain competitiveness through enhancing, combining, protecting, and when necessary, reconfiguring the business enterprise's intangible and tangible assets.[82]

The importance of these three capacities will 'wax and wane' in relative importance. Miyazaki, for example, argues that, at the early stage of an organization entering a specific field of endeavour, competence building is focused on knowledge acquisition to develop and deepen its comprehension of that field. At a later stage, once knowledge has been accumulated, competence building shifts to the development and bringing to market of innovations based on this knowledge.

The central importance of 'integrative' capacity to product development has been highlighted in a number of studies—that is, the ability of an organization to integrate both internal and external knowledge during the innovation process.[83] The concept of 'integrative' capacity should be viewed alongside the concept of 'absorptive' capacity discussed above. Such capacities are clearly of special interest and relevance to the strategic management of innovation.

6.4 Innovation strategies—'timing to market'

There have been a number of attempts to classify the different strategies adopted by organizations in relation to the 'timing to market' of their innovations relative to others in their sector. This is sometimes referred to as innovation 'posture' to distinguish this element of innovation strategy from other aspects. A key distinction is made between those organizations that aim to lead the sector and enter the market first (i.e. those adopting a 'first to market', or 'leadership', innovation strategy), and those that are content to follow once the market has grown large enough to enable economies of scale to be achieved (i.e. those adopting a 'late to market', 'cost minimization', or 'followership' innovation strategy). Although these innovation strategies were largely developed in relation to technological innovation, they clearly have relevance within the service sector, for example.

6.4.1 The strategic options—from leadership to followership

In **Table 6.2**, we outline the better-known classifications, or typologies, of innovation strategy. The different categories of one classification do not always map neatly onto those of another, but we have nevertheless attempted to align these different typologies.

In developing their typology, Maidique and Patch note that the individual strategies that they identify are not 'collectively exhaustive', but are *'simply intended to crystallize the concept of strategy in a way that is relevant in a technology-intensive environment'*.[84] Similarly, Freeman and Soete argue that:

> Any classification of strategies by types is necessarily somewhat arbitrary and does violence to the infinite variety of circumstances in the real world. The use of such ideal types may nevertheless be useful for purposes of conceptualization.[85]

Maidique and Patch also argue that individual innovation strategies are not mutually exclusive—that is, that an organization might adopt different innovation strategies in different markets at the same time, or different innovation strategies in the same market at different times. Given the different organizational structures, competencies, and resource required for each of the various innovation strategies, this requires organizations to manage in 'dual mode' or shift between modes over time (see **7.6** for a further discussion of this issue).

We will now provide a brief overview of each of the three main innovation strategies:

- 'first to market';
- 'second to market'; and
- 'late to market'.

The 'first to market', or leadership, offensive strategy

The adoption of a 'first to market' strategy requires an organization to be undertaking research and development close to the state-of-the-art in order to stay ahead of its competitors. This knowledge-intensive strategy necessitates substantial investments in

Table 6.2 Classifications of 'timing to market' innovation strategies

Classification Maidique and Patch (1988)	Porter (1985: 181–91)	Freeman and Soete (1997: 265–85)	Miles and Snow (1978)	Strategic aim of strategy
'First to market' or 'leader' strategy	Leadership strategy	Offensive strategy	Prospector strategy	To be first into the market with an innovation in order to gain first-mover advantages, such as monopoly profits.
'Second to market' or 'fast follower' strategy		Defensive strategy	Analyser strategy	To learn from the mistakes of the first-mover and enter the market with an improved innovation in the early stages of the life cycle.
'Late to market' or 'cost minimization' strategy	Followership strategy	Imitative strategy	Defender strategy	To enter the market later in the life cycle, once demand has grown sufficiently to allow significant economies of scale to be achieved. The aim is to gain cost advantage over competitors.
'Market segmentation' or 'specialist' strategy		Opportunist strategy		To innovate for a particular application/niche. May occur at various stages of the life cycle.
		Dependent strategy		To accept a subordinate role to its (often much larger) customers, who initiate the innovation process and provide the technical specifications.
		Traditional strategy	Reactor strategy	To continue much as before. Little emphasis on innovation and little capability to innovate.

applied R&D, to enable the identification of technological alternatives and to build the foundations of knowledge of a new technology in order to push it along the early phase of the technology S-curve. First-movers also require competencies in developing primary demand within a market.

Such organizations are those that are most likely to create technological discontinuities and initiate new technological trajectories, and, as such, they face high risks, but also high potential rewards. Pioneers, then, must take a long-term perspective and accept high risks, and emphasize flexibility over efficiency. Whilst the adoption of a first-to-market innovation posture and the allocation of appropriate resources might increase the likelihood of an opportunity presenting itself to an organization, Lieberman and Montgomery argue that, ultimately, *'opportunities for first-movership are by no means controlled by the firm alone'*.[86]

Illustration 6.2 eBay—benefiting from first-mover advantage

Established in 1995, eBay provides an example of a successful first mover. Today, eBay is the world's largest online auction site and a major brand. In 2007, the company achieved revenues of US$7.7bn (up from US$0.7bn in 2001), and traded goods with a total value of US$60bn. Its model is simple, but effective: the company is unburdened by stock, warehousing, and distribution; it simply provides a virtual space and a set of software tools. But a simple business model is simple to imitate, and plenty of companies have entered the market at different stages, including Yahoo! and Amazon.

So how was eBay able to build and sustain its first-mover advantage? Having attracted a critical mass of buyers and sellers, which is important in this market, eBay was able to benefit from the advantages associated with 'network externalities' (see **4.6.2**)—that is, traders now naturally gravitate to eBay, because this is where the buyers are clustered; a cycle that is very difficult for competitors to break. This, combined with a strong brand and an interface that is easy to use, has enabled eBay to capture a large share of the market.

The 'second to market', or 'fast follower', defensive strategy

Organizations adopting a 'second to market', or 'fast follower', innovation strategy must also be operating close to the state-of-the-art. As such, they must also have competencies in advanced R&D and be pursuing a knowledge-intensive strategy. Fast-followers do not seek to be first, nor do they wish to be left too far behind, and thus they must be responsive and adaptable—this requires the organization to combine elements of flexibility and efficiency, and often the rapid mobilization of financial capital. Although the risks and uncertainties are lower than for the first-mover, the fast-follower must be able to learn from the mistakes of those that proceed them, and be capable of stimulating secondary demand within the market.

Illustration 6.3 Google—the dominance of a market by a late-mover

Today, Google is by far the dominant Internet search engine. Figures from industry analysts, such as Hitwise, Neilsen/Net Ratings, and Net Applications, reveal that Google now accounts for some 60–65 per cent of search engine hits in the USA, and nearly 80 per cent worldwide. In contrast, the percentage of hits by Yahoo! has declined to around 20 per cent in the USA and 13 per cent worldwide. These are staggering figures given that Google was, in fact, a late mover into a fairly established market—one in which Yahoo! was the major player.

The explanation for this turnaround, in large part, concerns the expertise and software that lie behind the competing search engines. The focus of Yahoo! was almost the opposite to that of Google. Yahoo! concentrated its energy in creating an extensive directory of web links, but licensed the search engine technology from others, including AltaVista and Google. Google, on the other hand, developed its search engine technology in-house; this search engine does not employ a directory, but scours the whole of the Internet for content.

The 'late to market', or 'followership', imitative strategy

Organizations adopting a 'late to market', or 'followership', innovation strategy do not aspire to keeping up with the state-of-the-art technology. Indeed, they will tend to adopt established technologies, often through licensing. Such organizations require competencies in process development, and an emphasis on efficiency and control. Given that followers seek to compete on cost, they require access to large amounts of capital to establish large-scale production for economies of scale, and a focus on minimizing distribution costs.

6.4.2 First-mover advantages versus late-mover advantages

There has also been much debate over the last twenty-five years or so concerning the relative advantages and disadvantages of being first to market. Porter argues that the adoption of a leadership or followership strategy by an organization is based on an assessment of three factors: first-mover advantages, first-mover disadvantages, and the degree to which a lead can be sustained in the market.[87]

In the following discussion, we explore the mechanisms through which either first-mover or late-mover advantages are achieved. To a large extent, first-mover disadvantages and late-mover advantages are two sides of the same coin.

First-mover advantages

Initially, first-movers, or pioneers, are able to reap 'excess' or 'monopoly' profits. Subsequent first-mover advantages are derived from three primary sources: technological leadership, the pre-emption of assets, and buyer switching costs.[88] Lieberman and Montgomery argue that, although such *'first-mover advantages dissipate over time . . . [they] are enhanced by longer lead times before competitive entry'*.[89]

Technological leadership enables pioneers to move along the 'learning' or 'experience' curve ahead of competitors, allowing them to generate a cost advantage. This cost advantage can be sustained as long as the pioneer is able to restrict the diffusion of its learning beyond the organization and maintain market share.[90] Technological leadership also allows the pioneer to patent emerging technology ahead of competitors. Such pre-emptive patenting activity can not only protect R&D investments, but can also deter rivals or raise formidable barriers to entry. Lieberman and Montgomery argue, however, that *'patents confer only weak protection, are easy to "invent around", or have transitory value given the pace of technological change'*.[91] Pioneers also have the opportunity to develop complementary organizational innovations that are less 'visible' to outsiders, often slower to diffuse externally, and, as a result, may offer more durable first-mover advantage.[92]

Pioneers may be able to gain control of important assets through the pre-emptive acquisition, or 'capture', of scarce resources, such as skilled labour, specialist suppliers, and distribution channels, as well as natural resources. The pre-emptive investment in plant and equipment, however, whilst possibly deterring smaller entrants through the threat of potential cost and price cuts, does not appear to be important or effective in practice.[93]

Finally, first-movers have the opportunity to create switching costs for customers, such as those relating to a buyer's financial 'investment' in a product, the accumulation of product-specific learning (as exemplified by the continued dominance of the QWERTY keyboard—see **Illustration 1.5**), and contracts or incentives that 'lock in' the buyer. There is also evidence to suggest that first-movers are able to create reputation and brand loyalty in an uncertain, emerging market, especially where it is possible to influence perceptions about which attributes of a product or service are important.[94] All of these factors can create additional costs to followers in their attempts to attract customers from a pioneer organization, even where the late entrant has a superior, or a lower-priced, 'me-too' offering, although such factors appear to be more important for consumer rather than industrial markets.[95] Advantages may be enhanced and sustained by first-movers where they are able to influence or set industry standards (see our discussion of the emergence of a dominant design in **4.6**).[96]

Late-mover advantages versus first-mover disadvantages

First-movers may, however, also face a number of disadvantages, some of which can provide potential benefits that may be accrued by followers. First-movers often face substantial pioneering costs, such as those related to gaining regulatory approval, educating buyers, and developing infrastructure and complementary innovations.[97] In contrast, late entrants have the possibility of 'freeriding' off such market and technological investments made by first-movers. Furthermore, in advancing market and technological knowledge, first-mover investments help to reduce market and technological uncertainty, and thus the risk faced by late-movers.

Followers also have the opportunity to learn from the successes and mistakes of pioneers.[98] Such learning can have a dramatic impact on the relative survival rate of early followers as compared to pioneers in 'really new' (as opposed to 'incrementally new') product markets, in which first-movers may struggle to overcome market resistance and build primary demand.[99] In some cases, mistakes may damage the reputation and brand of the first-mover organization, such as those concerning poor reliability or safety, for example. Late entrants might also be better placed to take advantage of, and adapt to, subsequent changes in the market or technology; well-established first-mover organizations may be financially and psychologically committed to past investments in technology and equipment, for example, as well as past successes. The combination of these factors might undermine the initial advantages enjoyed by first-movers.[100]

As a market matures and a dominant design emerges, this generally signals a shift to price competition. In such circumstances, large organizations with the ability to establish large-scale facilities and benefit from economies of scale may enter the market late and usurp the pioneer organization. Furthermore, as we saw in **Illustration 4.3** concerning the development of the bicycle, despite the presence of a dominant design, opportunities may still emerge for late entrants through the 'stretching' of the design (e.g. through new and improved materials, components, or accessories), or the introduction of new 'design families'. More substantive changes in the market or technology might also benefit late entrants who are not encumbered by earlier investments.

Table 6.3 Leaders, followers, and commercial success in emerging markets (Grant, 2002: 347)

Product	Innovator	Follower	The winner
Jet airliner	De Haviland (Comet)	Boeing (707)	Follower
Float glass	Pilkington	Corning	Leader
X-ray scanner	EMI	General Electric	Follower
Office PC	Xerox	IBM	Follower
VCRs	Ampex/Sony	Matsushita	Follower
Diet cola	R. C. Cola	Coca-Cola	Follower
Instant camera	Polaroid	Kodak	Leader
Pocket calculator	Bowmar	Texas Instruments	Follower
Microwave oven	Raytheon	Samsung	Follower
Plain-paper copier	Xerox	Canon	Not clear
Fibre-optic cable	Corning	Many companies	Leader
Videogames console	Atari	Sony/Nintendo/Microsoft	Followers
Disposable diaper	Proctor and Gamble	Kimberley-Clark	Leader
Web browser	Netscape	Microsoft	Follower
MP3 players	Diamond Multimedia	Apple/Sony/others	Followers

Comparing the commercial success of leaders and followers

There are plenty of good examples of both sustained first-mover advantage, such as eBay (see **Illustration 6.2**), and of market dominance by a follower, such as Google (see **Illustration 6.3**). Perhaps not surprisingly, empirical evidence is also mixed. Some studies have indicated that pioneering is, on average, 'marginally unprofitable', whilst other research has revealed enduring market share for pioneers who survive. **Table 6.3** highlights the winner (i.e. leader or follower) for a sample of radical product innovations over the last few decades. Interestingly, in the majority of cases from this sample, the ultimate 'winner' was the follower, rather than the innovator or pioneer in the emerging market, with the pioneer 'winning' in only four of the fifteen industries.

In another study, Rosenbaum found a higher proportion of industries dominated by the first-mover (i.e. four out of ten), although this still represents a minority.[101]

So how might we explain why a first-mover innovation strategy is successful in some instances and not others? The evidence suggests that the advantages of pioneering are contingent upon the resources, capabilities, and behaviour of both pioneer and follower organizations, and upon the technological and sectoral context. In this regard, Lieberman and Montgomery argue that '*Pioneering may prove advantageous to some firms in some circumstances, but it is not necessarily a superior strategy for all entrants*'.[102] In their study of late entrants in the household electrical equipment sector, for example, Shamsie et al. concluded that whilst:

> late movers do face considerable hurdles in trying to penetrate the market . . . [their] performance . . . is much more dependent on the resource pool that they can draw upon . . . [and is] driven by the quality, price, and innovativeness of their products relative to those offered by their competitors.[103]

Similarly, Shankar et al., in their study of pharmaceutical products, found that '*innovative late movers grew faster than pioneers, slowed the growth* [of] *pioneers, and reduced the effectiveness of pioneers' marketing efforts*'; this was not the case for the non-innovative late movers in their study.[104]

Others have argued that first-mover advantages are much more closely associated with the degree to which an innovation can be protected by patents or copyright and the potential to establish a standard or dominant design in the market.[105] A study of survival rates of pioneers and early followers by Min et al. is also instructive. They found that, in 'really new' product markets, only 23 per cent of pioneers survived twelve years, as compared to 39 per cent of early followers, whereas in relation to 'incrementally new' product markets, the survival rate of pioneers rose to 66 per cent, whilst that of early followers remained steady at 38 per cent.[106]

Furthermore, the relative advantages of pioneering largely depends on timescale. On this point, Lieberman and Montgomery contend that:

> Pioneering firms can enjoy significant first-mover advantages but be less profitable than later entrants when viewed over an extended period . . . this raises the possibility that there may be both first-mover and late-mover advantages in a given market . . . [depending] on the point in time that the market is observed.[107]

This position is supported by an extensive survey undertaken by Boulding and Christen, of the revenue and cost performance of a large sample of consumer and industrial business units over a period of some fifty years. They found that whilst pioneers, on the whole, benefited from a profit advantage over followers in the initial years, these benefits eroded over time, such that after about a decade (ten years for consumer markets and twelve years for industrial markets), these advantages had been completely dissipated. Ultimately, the analysis by Boulding and Christen revealed that '*pioneers were substantially less profitable than followers over the long-run*'.[108]

6.4.3 Sustaining leadership

Typically, first-mover advantages are only sustained through proactive and careful protection and nurturing. Indeed, Porter notes that '*first-mover advantages can be dissipated through aggressive spending by later entrants unless the first mover invests to capitalize on them*'.[109] The degree to which a first-mover can sustain its leadership position within a market is influenced by a number of factors. Porter, for example, highlights the importance of maintaining control over the underlying technology. He notes that where technology is sourced externally, then the organization should seek to build alliances or to establish exclusive arrangements. In relation to in-house technology, Porter suggests that patenting, human resource policies aimed at retaining key employees, and vertical integration are possible ways in which the rate of knowledge diffusion or 'leakage' to competitors can be reduced.[110] Lieberman and Montgomery also suggest developing '*designs that are deliberately difficult to reverse engineer*'.[111]

Sustaining a leadership position, however, requires more than simply protecting the underlying technology; it also necessitates continual product innovation (see

Illustration 6.4) and the expansion of production capacity, especially where the market is growing rapidly.[112]

Illustration 6.4 Sustaining technological leadership through incremental innovation—the case of Sony and the Sony Walkman

Sony's policy of continual incremental product development of the Sony Walkman allowed it to maintain its dominant position in the personal stereo market for around twenty years. Table 6.4 shows that Sony was first to market, with a whole series of incremental innovations in the decade following the launch of the original Sony Walkman in 1979.

Table 6.4 *'Sony's technological leadership in personal portable stereos'* (Sanderson and Uzmeri, 1995: 770)

Feature	Firm	Date	Imitated?
First 'Walkman'	Sony	79	Y
AM/FM stereo radio	Sony	80	Y
Stereo recording	Sony/Aiwa	80–81	Y
FM tuner cassette	Toshiba	80–81	Y
Autoreverse	Sony	81–82	Y
FM headphone radio	Sony	81–82	Y
Dolby	Sony/Aiwa	82	Y
Shortwave tuner	Sony	83	
Remote control	Aiwa	83	Y
Separate speakers	Aiwa	83	Y
Water resistance	Sony	83–84	
Graphic equalizer	Sony	85	Y
Solar-powered	Sony	86	
Radio presets	Panasonic	86	Y
Dual cassette	Sony	86	
TV audio band	Sony	86–87	Y
Digital tuning	Panasonic	86–87	Y
Child's model	Sony	87	
Enhanced bass	Sony	88	Y

Lieberman and Montgomery argue, however, that:

> Ultimately, the sustainability of a first-mover advantage depends upon the initial resources captured by the pioneer, plus the resources and capabilities subsequently developed, relative to the quality of resources and capabilities held by later entrants.[113]

Porter, meanwhile, notes: *'Where the first mover does not have adequate resources, the first early mover with resources can often be the firm to gain the benefits of first-mover advantages.'*[114]

6.5 Concluding comments

In this chapter, we have sought to provide an overview of some of the key issues for the strategic management of innovation. We began by highlighting the paradoxical nature of strategy—this is a theme that permeates both this and other chapters. We also discussed the scope and process of strategic decision making in relation to innovation management, and indicated the degree to which organizations have choice in the strategies that they adopt and pursue.

The heart of the chapter focused on two main themes. The first theme concerned the nature and variety of resources and competences available to the organization, and their contribution to innovative capacity. Of particular interest are the intangible and relational resources that comprise the 'intellectual' capital of the organization, and the notion of 'dynamic capabilities'. This discussion, informed by the RBV, provides a useful bridge between the proceeding chapters and those that follow.

The second theme related to the 'timing to market' innovation strategies. We explored the advantages and disadvantages of pioneer and follower innovation strategies, and drew upon empirical evidence to highlight the comparative commercial success of each. It is clear from this empirical evidence that, despite the common perception of first-mover advantage, followers often benefit from substantial late-mover advantages and have usurped first-movers in a wide array of markets.

We now turn our attention to the management of the innovation and knowledge creation processes within organizations.

CASE STUDY 6.1 WHAT NEXT AFTER SEMICONDUCTORS? THE STRATEGIC RESPONSE TO ALTERNATIVE FUTURES IN THE COMPUTER INDUSTRY

Introduction

In earlier chapters, we have discussed the development of the microprocessor (see **Illustrations 4.1**, **4.2** and **5.2**). We noted that, over the last forty years, dramatic increases in the speed and density of microprocessors have been achieved through a combination of major product and process innovations. The exponential progress in the density of microprocessors is illustrated in **Figure 5.2**, in relation to Intel's series of semiconductors. Such a trajectory of technical progress was predicted by Gordon Moore forty years ago.[115]

There appears to be little consensus as to when the technology might reach its maturity phase. In large part, this is because technological advances are redefining its technical limits. Nevertheless, such exponential improvements in progress are not unsustainable in the medium to long term. Furthermore, in recent years, radical technological alternatives are emerging that show the potential for improving the technical performance of computers by orders of magnitude.

Unconventional computing alternatives

Researchers in universities and laboratories around the world are working on an array of highly novel and unconventional alternatives to the silicon-based electronics technology that is currently employed by the computer industry. Although many of these alternatives are still at the theoretical or conceptual phase, and are thus many years away from possible commercial exploitation, they offer huge potential in relation to computation power. Furthermore, many of these alternatives operate at the molecular level and offer the possibility of the development of minute devices.

A number of approaches are emerging under what may be termed 'chemical computing'. These exploit the reaction of different chemical 'soups' to different inputs. Changes in the shape of molecules, or the manner in which a wave travels through the liquid, can then be observed.

Another set of approaches may be grouped under the label 'biomolecular computing'. These perform computation through the manipulation of synthetic strands of DNA.

A further set of possibilities is emerging from the exploitation of quantum mechanics, known as 'quantum computing'.[116]

Questions

1. To what extent do you think the future directions for computing are competence destroying or competence enhancing for organizations currently operating in the computer industry? How might computer firms shape their innovation strategies in light of such potential impacts?

2. Given the wide array of possible technological futures for the computer industry, and the technological and commercial uncertainties that persist around each, how might a computer firm best position itself in terms of resources and competencies?

FURTHER READING

1. For a discussion of paradox in relation to leveraging knowledge, see Coff et al. (2006), and concerning the RBV, see Lado et al. (2006).

2. A useful overview of the literature concerning strategic choice is provided by Child (1997).

3. For a discussion of 'relational assets' with respect to competitive advantage, see Dyer and Singh (1998), and as an extension to the RBV, see Lavie (2006).

4. In relation to 'intangible assets', see Hall (1992); on 'social' and 'intellectual' capital, see Nahapiet and Ghoshal (1998); on 'core competence', Prahalad and Hamel (1990); on 'dynamic capabilities', Teece (2007); and on 'core rigidity', see Leonard-Barton (1992).

5. For a discussion of first-mover and late-mover advantages and disadvantages, see Lieberman and Montgomery (1988), Shankar et al. (1998) and Min et al. (2006). Lieberman and Montgomery (1998) also provide a summary of a range of 'timing to market' studies.

6. Lastly, Schilling (2003) provides an interesting case study of 'technological leapfrogging' in the US videogame console industry.

NOTES

[1] For example, Mintzberg et al. (2002); Grant (2002); de Wit and Meyer (2004); Thompson et al. (2007).

[2] De Wit and Meyer (2004: 13).

[3] Nonaka and Takeuchi (1995: 74).

[4] Leonard (1995: 139–40).

[5] Maidique and Patch (1988: 241); Grant (2002: 340–2).

[6] Grant (2002: 373).

[7] Dougherty and Hardy (1996).

[8] Baden Fuller and Stopford (1992).

[9] Mintzberg (1987: 13).

[10] Quinn and Voyer (1998: 103).

[11] Quinn and Voyer (1998: 108).

[12] Quinn (1988b: 104).

[13] Burgelman and Sayles (1986: 31).

[14] Reid and Brentani (2004: 181).

[15] Leonard-Barton (1992); Leonard (1995).

[16] Dosi (1982); Spender (1989); Geels (2002).

[17] Whittington (1988: 524).

[18] Clark (2000: 297).

[19] Geels (2002: 1259).

[20] Hoskisson et al. (2002: 710).

[21] Hoskisson et al. (2002: 711–12).

[22] Porter (1980; 1985; 1990).

[23] Prahalad and Hamel (1990); Hamel and Prahalad (1994).

[24] Teece and Pisano (1994); Teece et al. (1997); Teece (2007).

[25] Teece (2007: 1341).

[26] Grant (2002: 137).

[27] De Wit and Meyer (2004: 336).

[28] Grant (2002: 145).

[29] Afuah (1998: 384).

[30] Coombs (1996: 345).

[31] Afuah (1998: 54); Grant (2002: 139).

[32] Grant (2002: 139).

[33] Quinn (1992); Coombs (1996); Nahapiet and Ghoshal (1998); Canto and González (1999); Bounfour (2003).

[34] Hall (1992: 135).

[35] Dierickx and Cool (1989: 1506).

[36] Knott et al. (2003: 192).

[37] Barney (1991); Freeman (1991); Nonaka and Takeuchi (1995).

[38] Coff et al. (2006: 452).

[39] Nonaka and Takeuchi (1995); Tsai (2001).

[40] Zander and Kogut (1995).

[41] Rivkin (2001).

[42] Grant (2002: 141).

[43] International Accounting Standards Committee (1998: 38.8).

[44] Leitner (2005).

[45] Subramaniam and Youndt (2005: 451).

[46] Nahapiet and Ghoshal (1998); Youndt et al. (2004).

[47] For example, see Table 1 in Hayton (2005: 139).

48 Subramaniam and Youndt (2005: 451).

49 Nahapiet and Ghoshal (1998: 243).

50 Nahapiet and Ghoshal (1998: 243, 251).

51 Subramaniam and Youndt (2005: 452).

52 Subramaniam and Youndt (2005: 459).

53 Subramaniam and Youndt (2005: 459).

54 Hayton (2005: 149).

55 Hayton (2005).

56 Subramaniam and Youndt (2005: 452).

57 De Wit and Meyer (2004: 336).

58 Lavie (2006: 638).

59 Dyer and Singh (1998: 660).

60 Cohen and Levinthal (1990: 128).

61 Wollard (2001).

62 Nohria and Gulati (1996: 1246).

63 Bromiley (1991); Nohria and Gulati (1996).

64 Levinthal (1997).

65 Thompson (1967).

66 Baker and Nelson (2005); Katila and Shane (2005).

67 Baker and Nelson (2005: 329).

68 Nohria and Gulati (1996: 1260).

69 George (2005: 672).

70 Grant (2002: 144).

71 Hamel and Prahalad (1994).

72 Prahalad and Hamel (1990); Hamel and Prahalad (1994).

73 Teece and Pisano (1994); Teece et al. (1997); Teece (2007).

74 Prahalad and Hamel (1990: 82).

75 Grant (2002: 145).

76 Prahalad and Hamel (1990: 82).

77 Grant (2002: 147).

78 Danneels (2007: 532).

79 Leonard (1995: 4).

80 Prahalad and Hamel (1990: 83–4).

81 Teece et al. (1997: 516).

82 Teece (2007: 1319).

83 Henderson (1994); Iansiti and Clark (1994).

84 Maidique and Patch (1988: 239).

85 Freeman and Soete (1997: 265).

86 Lieberman and Montgomery (1988: 54).

87 Porter (1985: 182).

88 Porter (1985: 186–9); Lieberman and Montgomery (1988).

89 Lieberman and Montgomery (1988: 1121).

90 Porter (1985: 187); Lieberman and Montgomery (1988).

91 Lieberman and Montgomery (1988: 43).

92 Porter (1985: 187); Teece (1980).

93 Lieberman (1987).

94 Porter (1985: 186–7); Lieberman and Montgomery (1988).

95 Robinson and Fornell (1985); Robinson (1988).

96 Porter (1985: 187); Lieberman and Montgomery (1988).

97 Porter (1985: 189).

98 Maidique and Patch (1988); Boulding and Christen (2001).

99 Min et al. (2006).

100 Porter (1985: 182); Lieberman and Montgomery (1988).

101 Rosenbaum (1998).

102 Lieberman and Montgomery (1988: 54).

103 Shamsie et al. (2004: 81).

104 Shankar et al. (1998: 67).

105 Grant (2002: 346).

106 Min et al. (2006).

107 Lieberman and Montgomery (1988: 51–2).

108 Boulding and Christen (2001: 21).

109 Porter (1985: 188–9).

110 Porter (1985: 185–6).

111 Lieberman and Montgomery (1988: 54).

112 Lieberman and Montgomery (1988: 54).

113 Lieberman and Montgomery (1998: 1113).

114 Porter (1985: 188–9).

115 Moore (1965).

116 Based on news item 'Future directions in computing', 13 November 2007, available online at <http://news.bbc.co.uk> (accessed on 25 October 2008).

Part III

The Management of Innovation Within Organizations

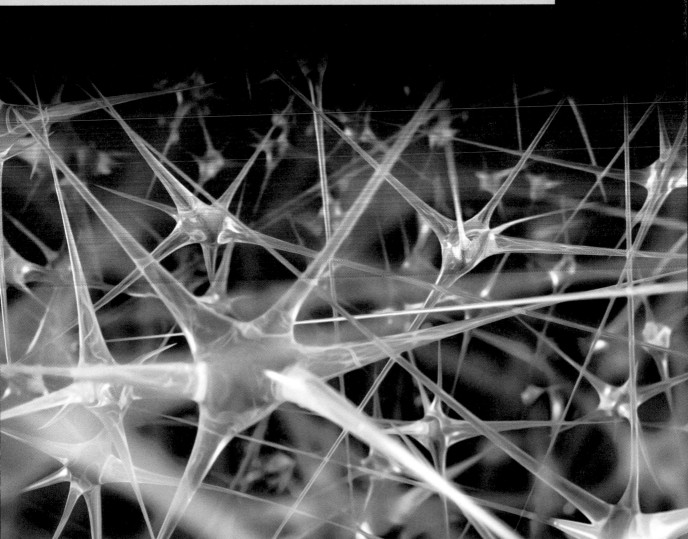

Introduction

For Burns and Stalker (1961: 22): *'If . . . the greatest invention of the nineteenth century was the invention of the method of invention, the task of the succeeding century has been to organize inventiveness.'* Burns and Stalker provide an interesting account of the attempts by industrial organizations through the nineteenth century and the first half of the twentieth century to absorb and manage inventive and innovative activity—in their words, *'to digest the thing they had swallowed'* (1961: 22–36). A key challenge of organizations in the latter part of the twentieth century was the integration of inventive and innovative activity of research and development (R&D) departments with the activities of other organizational functions, such as marketing and production, and with the strategy of the organization. More recently—and in particular since the 1980s—organizations have sought to find effective ways to connect to the marketplace and the external science and technology base, by building organizational structures that are increasingly permeable, flexible, and embedded within their environments.

Central problems in the management of innovation

Part III focuses on the management of innovation within organizations—on the way in which successful organizations have managed to 'digest' the activities of the inventor and the entrepreneur. As in other parts of the book, Part III is split into three chapters: **Chapter 7** centres on outlining the characteristics of organization structure and culture that are conducive to innovation—what may be termed the 'formal organization'; in contrast, **Chapter 8** shifts the focus to process, looking at both the innovation process, and more broadly, at the knowledge creation and transfer process within organizations; and finally, **Chapter 9** centres on the role of social networks in innovative activity—what may be termed the 'informal organization' or 'social organization.'

It is important to reiterate here that, whilst material related to specific themes—such as strategy, structure, and process—have been separated into chapters, these themes are inextricably interrelated and intertwined. Whilst writing this book, we have made a number of explicit links between the material in the various chapters; whilst reading the text, it is equally as important that you attempt to make your own connections, drawing from your own experience and exposure to innovation within organizations.

Van de Ven (1988) provides an interesting set of questions that help to frame the issues surrounding structure, culture, and process, in relation to the management of innovation, in a novel and open manner. The starting point is the contention that general managers, such as chief executive officers (CEOs), deal with a rather different and less well-understood set of problems in managing innovation from those dealt with by functional managers, such as those within the R&D, or marketing and sales, departments of an organization.

In attempting to develop a broader perspective on innovation, seen as more useful for general managers, Van de Ven suggests four 'central problems in the management of innovation' worthy of investigation, as follows:

1. *The structural problem of managing part-whole relationships*—this refers to the problem of managing the complexity and interdependence of the multitude of micro-activities and interactions that occur during the development of an individual innovation. This problem will be addressed in both **Chapter 7**, in particular with regard to the concern for building and managing formal links between different subunits (e.g. functions or divisions) of the organization, and in **Chapter 9**, in relation to building informal bridges and linkages.

2. *The strategic problem of institutional leadership*—that is, the problem of creating an intra- and extra-organizational 'infrastructure' that is conducive to innovation. This issue will be dealt with in **Chapter 7**, in relation to organizational structure and culture, and throughout **Part IV** of the book, with regard to extra-organizational infrastructure.

3. *The human problem of managing attention*—this is referring to the problem of encouraging individuals in organizations to focus on new ideas, needs, and opportunities. This issue will be discussed in both **Chapter 7**, with regard to culture and the development of an innovative 'climate', and **Chapter 9**, in relation to, for example, 'innovation as an emotional process'.

4. *The process problem of managing ideas into 'good currency'*—of particular importance here is the management of the social and political dynamics associated during the various stages of the innovation process as it moves towards implementation and diffusion. This issue will be addressed in **Chapter 8**, in relation to the management of the stages of the innovation process, and in **Chapter 9**, concerning the management and facilitation of informality in the innovation process.

What emerges from this and other categorizations of the overarching problems and issues surrounding the management of innovation (for example, Dougherty and Hardy, 1996), is the interrelatedness of the themes of strategy, structure, culture, and process, but also the role that power plays in, for example, the allocation of resources, the framing of strategy, and the direction of innovative effort and problem solving. Indeed, power is a theme that threads both implicitly and explicitly throughout the book—in particular, the power to 'shape' innovation.

7

Organizing for Innovation—Organization Structure and Culture

Chapter overview

Learning objectives

This chapter will enable the reader to:

*   identify the range of organizational features—including, structure, bureaucracy, culture, size, profit-orientation, spatial distribution, and physical layout—that impact the innovative capacity of an organization;

*   appreciate the appropriateness of alternative modes of organizing in relation to the external environment and the innovation posture of an organization;

- outline the key features of the 'organic'/'mechanistic' and 'integrative'/ 'segmentalist' modes of organizing, and their impact on the innovative capacity of an organization or organizational unit;

- understand the notion of the 'ambidextrous' organization and identify the circumstances in which an organization will need to be managed in 'dual mode';

- indicate the factors impacting the extent and nature of 'strategic linking' required by an organization, and the range of 'strategic linking' options available; and

- appreciate the relationship and tensions between workspace design and innovative activity.

7.1 Introduction

In 1966, Tom Burns noted in his preface to the second edition of the now classic 1961 text by Burns and Stalker *The Management of Innovation*, that:

> I am more than ever impressed with the extraordinary gap that exists between the perceptiveness, intellectual grasp, and technical competence of the people who work in industrial concerns, and the cumbrous, primitive, and belittling nature of the administrative structures by which they direct their efforts, and of the constraints they see fit to impose on their thinking and liberty of action.[1]

Despite the immense interest in the 'management of successful innovation' since the 1960s, by practitioners, consultants, policymakers, and academics alike, it is noteworthy to see that, thirty years on, Dougherty and Hardy found that the creative and innovative potential of individuals was still thwarted within many organizations:

> Our findings suggest that even when product innovation did occur, it did so in spite of organizational systems and only because of the unstinting efforts of individuals who bucked the system. Only in one case [the study embraced a total of forty cases of innovation within 15 U.S. companies from across four business sectors] . . . were resources really available, collaborative processes in place, and strategic meaning and value attached to innovation, in anything like the ideal organizational configuration recommended in the literature.[2]

Thus, a key challenge for many organizations remains the adoption of a mode of organizing that encourages innovation and creativity, allows flexibility and agility to respond in a timely manner to changes in the environment, and facilitates integration and coordination both internally, between subunits, and externally, with relevant stakeholders. The findings of empirical work since the 1960s have provided organizations with a great deal of guidance as to the nature of the modes of organizing that can either encourage or hinder innovation and creativity within organizations. It is the aim of this chapter to review this literature.

7.2 Understanding 'organization'

It is important, before we proceed, to ensure that we have a common understanding of 'organization' and its key components. For Richard Daft, a well-known writer on management, organizations *'are social entities that are goal-directed deliberately structured activity systems with an identifiable boundary'*.[3]

Looking at the components of this definition in turn:

- 'social entity'—organizations are composed of individuals and groups who interact with each other;

- 'goal-directed'—organizations exist for a purpose, which, translated into goals, provides direction and coherence to its activities;

- 'deliberately structured activity systems'—organizations consciously group activities into roles, groups, departments, and divisions; and

- 'identifiable boundary'—organizations must be distinct from their environment, and this is, in part, achieved through distinguishing between organizational members and non-members.

A further definition is provided by Hall:

> An organization is a collectivity with a relatively identifiable boundary, a normative order (rules), ranks of authority (hierarchy), communications system, and membership coordinating systems (procedures); this collectivity exists, on a relatively continuous basis in an environment, and engages in activities that are usually related to a set of goals; the activities have outcomes for organizational members, the organization itself, and for society.[4]

In this definition, Hall is more explicit about the organizational components of rules, hierarchy, and procedures, and he incorporates the additional elements of:

- time—the organization is seen as existing over a period of time;

- environment—indicating a relationship between the organization and its external context; and

- outcomes—for a variety of stakeholders, not only senior managers and shareholders.

Furthermore, Hall also raises a degree of ambiguity in relation to the notions of 'boundary' and 'goals', which resonates particularly well with our discussion regarding the informal or social organization in **Chapter 9**. This ambiguity is useful in helping us reflect upon the many assumptions that we 'take for granted' about organizations.

Jaffee outlines three such key assumptions:

- that organizational boundaries allow internal processes to be distinguished from external processes;

- that roles, tasks, and functions within organizations can be bounded and distinguished in a meaningful and distinct manner; and

- that *'organisations are instruments designed to accomplish definable goals'*.[5]

For some, these assumptions are increasingly 'fragile', because they appear increasingly at odds with contemporary and emerging forms of organization—sometimes termed 'post-modernist' forms of organization.

Jaffee outlines a range of features associated with the 'post-modernist' organization; a number of these characteristics are of particular relevance to this text (see **Table 7.1**), because they reflect practices often associated with innovative organizations. You may recognize some or all of these features from your own work experience.

Table 7.1 Some key features of the post-modernist organization (adapted from Jaffee, 2001: 284–6)

1. The emergence of the virtual organization	The traditional notion of an organization as a physical entity that is located in a fixed place becomes obsolete and, as a result, it becomes increasingly difficult to determine when individuals 'enter' or 'leave' the organization.
2. The rise of organizational networks and alliances	The use of a variety of forms of organizational cooperation and collaboration increasingly blur the boundaries of organizations, making it *increasingly difficult to determine where one organization "ends" and another "begins"'*.
3. The practice of managing rather than eliminating uncertainty	The acceptance by managers of chaos rather than order, and of multiple, less clearly defined, and often paradoxical goals.
4. Flexibility and variety in working arrangements	The emergence of working arrangements such as flexitime, temporary contracts, telecommuting, and outsourcing are eroding the traditional distinctions between work, home, and leisure.
5. The transient nature of the roles and careers of individuals	The increasing practice of organizing work around projects produces *a continuous shifting of roles and relationships for individuals within organisations*, which is reinforced by the emerging practices of 'multitasking', 'multiskilling', and multiple careers. As the frequency with which individuals within organizations change positions, roles, tasks, and projects increases, so formal structures and job descriptions become less relevant.
6. The demise of personal office space	The emergence of practices such as 'hot-desking' underline the transient nature of an individual's presence in an organization.

7.3 Modes of organizing and their relationship to context, strategy, and innovative capacity

Whilst this chapter will focus predominately on the ways in which organizations organize —through their structure, culture, systems, etc.—to support and promote innovative and creative activity, it is also important to bear in mind that an organization's mode of organizing must be congruent with its external environment and the innovation 'posture' that it adopts within this environment.

It is useful, therefore, to begin our discussion by looking at a number of different 'influences' on our understanding of the 'appropriateness' of alternative 'modes of organizing', as follows:

- *The relationship between the mode of organizing and external environment*—key work in this area includes a number of classic studies from the 1960s, such as *The Management of Innovation* by Burns and Stalker,[6] to which we will return later in this chapter, and *Organization and Environment* by Lawrence and Lorsch.[7] In this latter work, for example, Lawrence and Lorsch found that the greater the certainty and stability in the environment in which an organization operates (as defined by levels of competition, changes in product innovation, and the predictability of the supply and demand for inputs and outputs), the greater the formalization and rigidity of its organizational structure—that is, an organization's structure is contingent upon the nature of its environment; in certain or stable environments, organizations have the opportunity to develop a highly formalized and rigid set of practices, routines, and structures, whilst in uncertain or unstable environments, organizations are forced continually to adapt, requiring a less formalized and more flexible structure. This 'contingency theory' view implies that its environment, to a large extent, determines an organization's structure. A useful concept linked to this perspective is that of 'requisite variety';[8] this argues that, for an organization to adapt and survive in its environment, it must reflect the variation and complexity in this environment with that of its internal structure. This body of work represents what is sometimes termed an 'environmental determinism' approach—that is, the structure of the organization is determined by the nature of its environment.

  Other academics have attributed greater agency to organizations, by noting their ability to modify their environment, or to choose an environment that matches their structure. On this latter point, Mintzberg argues that *'organizations can select their situations in accordance with their structural designs, just as much as they can select their designs in accordance with their situations'.*[9] Highlighting the agency of individuals, Child argues that the design of organizational structures is *'an essentially political process in which constraints and opportunities are functions of the power exercised by decision makers in the light of ideological values'.*[10] In recent years, there has been an increasing emphasis in innovation studies on the importance of the development of extra-organizational linkages to embed an organization in its environment; this is highlighted by the 'interactive' model of innovation,[11] as discussed in **3.2.2**.

- *The relationship between the mode of organizing and innovation posture*—this link has already been discussed in the previous chapter, but it is worth noting again the link between innovation posture (i.e. innovation strategy, as defined by 'timing to market') and organizational structure, outlined in work such as that by Miles and Snow,[12] and Maidique and Patch.[13] Here, for example, we see an explicit link made between the 'first to market' (Maidique and Patch) and 'prospector' (Miles and Snow) strategies, and the emphasis on organizational flexibility over efficiency.

In contrast, we see an explicit link made between the 'late to market' or 'cost minimization' (Maidique and Patch) and 'defender' (Miles and Snow) strategies, and the emphasis on the need for efficiency, hierarchical control, and rigidly enforced procedures.

- *The relationship between the mode of organizing and internal innovative capacity*—this link has also been an area of great interest to academics and practitioners alike, and is informed by the above text by Burns and Stalker, and, for example, the classic 1980s innovation text *The Change Masters*, by Kanter.[14] Both of these books highlight the important impact that organizational structure has upon the innovative capacity of an organization, advocating more open, flexible, integrative structures to support innovation and creativity. Both of these texts will be discussed in detail in this chapter. A number of studies in the area of innovation studies have also focused on specific intra-organizational interfaces, for example, between the research and development (R&D) and marketing functions.[15] Another area of interest here is the role and contribution of social networks and boundary-spanning to innovative activity within organizations—this is core theme of **Chapter 9**.

7.4 Organizational boundaries

The establishment of an organization implies that a distinction has been made between members and non-members—that is, some individuals are admitted to participate in the activities of the organization, whilst others are excluded.[16] Modern organizations typically define organizational members contractually. Aldrich defines organizations as '*goal-directed, boundary-maintaining, activity-systems*';[17] maintaining this boundary, by distinguishing between members and non-members, involves the establishment of an authority empowered to admit some and exclude others. The ability of the organization to control its boundaries is critical for the maintenance of its autonomy.[18]

Lincoln, however, questions the need for separate theories, terminology, and research techniques for examining inter- and intra-organizational structures and processes, arguing that:

> If [organizational] boundaries are hard, objective facts, then links across them might be thought to have a very different quality from links within them. But if boundaries are vague, permeable and shifting, perhaps there is no reason to treat inter-organizational relations separately.[19]

In such an approach, an inter-organizational tie would simply be seen as just another relationship that happens to span what the observer views as the organizational boundary.

The term 'boundary-spanning' is used to refer to those linkages that transcend the boundaries of a given organization, whilst the term 'boundary-spanner' refers to those organizational members who are involved in boundary-spanning activity. Unlike boundary-crossing (eg. via hiring), boundary-spanning activity involves the flow of information and knowledge to and from the organization, rather than the flow of personnel. The degree and form of boundary-spanning activity varies according to a variety of organizational and environmental factors that are often inextricably linked.[20]

A number of studies from the early 1960s to the early 1970s were able to show a positive relationship between a higher dependence on other organizations for resources (excluding a parent organization), and a lower formalization and standardization of organizational structure.[21] Thus, one would expect a greater incidence of informal boundary-spanning activity for organizations that are more dependent on other organizations for resources. This view is supported by Aldrich who argues that:

> As inter-organisational dependence increases, organisations tend to display a more flexible and open structure, characterized by less formal and standardized procedures, greater decentralisation of decision-making, and decreased impersonality of relationships.[22]

With the proliferation of scientific and technical specialities,[23] noted earlier, the likelihood of inter-organizational resource dependence has notably increased, such that Prahalad and Hamel contend that even those with the largest technological resources can develop only a narrow range of core competencies.[24]

In addition, as noted earlier, research has also indicated that organizational structure is influenced by environmental uncertainty and that the most salient feature of organizational environments today is their rate of change.[25] These studies suggest that environmental uncertainty is likely to increase the incidence of informal boundary-spanning activity. Indeed, Aldrich argues that:

> As organisations are confronted with increasing environmental uncertainty due to instability, heterogeneity, or turbulence, more flexibility is observed in their structures and activities . . . Under such conditions, there is some evidence that organisations tend to be more effective if they are more decentralised and specialised, and less formalised and standardised.[26]

Furthermore, Aldrich also argues that '*boundary-spanning roles are expected to proliferate when organizations are in concentrated, heterogeneous, unstable, and lean environments*'.[27]

Internal and external boundary-spanning communication plays an important part in aiding, for example, coordination, integration, flexibility, responsiveness, and novelty within innovative organizations (see **Illustration 7.1**). By 'internal' boundary-spanning, we are referring to interaction across internal boundaries, such as those bounding project teams, functions, and divisions; by 'external' boundary-spanning, we are referring to interaction across external boundaries, such as the organizational boundary itself, as well as sectoral, regional, and national boundaries.

Illustration 7.1 Cancer Research UK—boundary-spanning and the development of new cancer drugs

Once we've worked out the structure of a protein, the very exciting thing is sitting down with colleagues from other disciplines to examine how, if it is significant in cancer, we could try to fix it. By putting all our perspectives together we hope to accelerate the process of producing new, more effective drugs for cancer patients. For example, only a hundred yards away from where I sit is the breast cancer clinic. I know the people there who are involved in trials of new treatments, Whatever I'm doing in my laboratory, I can link it right the way through to patients in the clinic over there. And that's so important.

(Cancer Research UK Scientist)[28]

The extent to which organizational members are able to interact across such boundaries is influenced by a whole range of organizational features, such as:

- *formal structure (i.e. organizational chart)*—in its narrowest sense, this represents the formal linkages within the organization, indicating the chain of command, reporting lines, and channels of communication; this is sometimes referred to as the 'formal organization'. Formal structures that emphasize vertical linkages up the chain of command, rather than horizontal linkages between the subunits of the organization, hinder boundary-spanning communication;

- *informal structure (i.e. social networks)*—this represents the informal and social linkages within the organization. These are not prescribed or defined by management, but emerge over time through social interaction; this is sometimes referred to as the 'informal organization'. Informal structures can facilitate boundary-spanning communication;

- *level of bureaucracy*—the degree to which an employee's behaviour with regard to acting, deciding, and communicating is prescribed by management, through defined and documented rules, procedures, tasks, and roles. High levels of bureaucracy can mitigate against spontaneity, serendipity, creativity, and informal boundary-spanning communication;

- *organizational culture*—the set of shared meanings, beliefs, values, and assumptions, held by individuals, or groups of individuals, within the organization. Organizational cultures that encourage sharing, informality, communication, interaction, and experimentation, for example, can foster boundary-spanning communication;

- *organization size*—large organizations, with several hundred and possibly many thousands of employees, are often organized into subunits. These subunits may be based on functional activities (e.g. marketing or production), or on product groupings, for example. Such grouping of employees can create barriers within the organization, particularly where subunits are differentiated by localized activities, 'languages', and subcultures. Where barriers emerge, this can hinder boundary-spanning communication;

- *organization type (e.g. profit versus non-profit)*—profit-orientated organizations are more likely to discourage or restrain external boundary-spanning communication in an attempt to prevent the 'leakage' of information and knowledge that it may view as having a 'market value'. Allen terms the restraining of external interactions of employees 'enforced localism';[29]

- *physical work environment*—building layout and design can have a profound impact on the boundary-spanning interaction and communication patterns of individuals within an organization, particularly where functionally and culturally distinct subunits are further separated by walls and floors. For Curtis et al.: '*We are all powerfully affected by the space in which we work. Good* [workspace] *design can promote knowledge sharing and transfer.*'[30] We explore this issue further at the end of the chapter, in **7.8**; and

- *physical location*—empirical evidence indicates that there is an inverse relationship between spatial proximity and communication—that is, the greater the physical distance between two individuals, the lower the frequency of their communication. Technologies, such as the email and telephone, only partly help to overcome this problem, particularly between individuals within functionally and culturally distinct subunits. Thus, boundary-spanning communication can be hindered by the spatial distribution of subunits.

The core focus of the remainder of this chapter centres on alternative modes of organizing and their impact on the innovative capacity of an organization. We use the term 'mode of organizing' to embrace the combined components of the formal organizational structure, the level of bureaucracy, and the organizational culture.

7.5 Organizing for innovation

In this section, we focus on the work of a number of widely cited and influential writers from the interrelated areas of innovation and knowledge management; each can provide us with insights into how organizations might best organize to foster innovative activity, and knowledge creation and sharing.

7.5.1 Burns and Stalker—'mechanistic' versus 'organic' modes of organizing

One of the most cited texts with regard to organizing for innovation is *The Management of Innovation* by Burns and Stalker. The 'mechanistic' and 'organic' modes of organizing detailed within it (in **Chapter 6**) have proved incredibly resilient since publication in 1961—the third edition was published in 1994. The work was based on an in-depth qualitative study of a sample of Scottish and English organizations, predominantly from the electronics sector.

Perhaps less well known and referenced is the rich detail within the text regarding 'informal organization' and 'organizational politics' (sometimes termed 'micropolitics') within technology-orientated organizations—themes that have as much resonance within organizations forty-five years on. The role of informal organization and the impact of organizational politics on the management and shaping of innovation are discussed further in **Chapter 9**; here we will focus on outlining the 'mechanistic' and 'organic' modes of organizing, and their impact upon the innovative capacity of an organization.

We discussed the relationship between the mode of organizing and the nature of the external environment in **7.3**. For Burns and Stalker, the mechanistic mode '*appeared to be appropriate to an enterprize operating under relatively stable conditions*', whilst the organic mode '*appeared to be required for conditions of change*'.[31] It is for this reason that the organic mode of organizing is strongly associated with innovative and adaptive organizations. Burns and Stalker go on to emphasize that:

> Both types represent a 'rational' form of organisation, in that they may both, in our experience, be explicitly and deliberately created and maintained to exploit the human resources of a concern in the most efficient manner feasible in the circumstances of the concern . . . We have endeavoured to stress the appropriateness of each system to its

own specific set of conditions . . . In particular, nothing in our experience justifies the assumption that mechanistic systems should be superseded by organic in conditions of stability. The beginning of administrative wisdom is the awareness that there is no one optimum type of management system.[32]

The two modes 'represent a polarity' and 'not a dichotomy'—that is, they define 'ideal types' at the two ends of a spectrum. Indeed, Burns and Stalker noted that, '*In practice most concerns show features of both, but approximate wholly or in part more closely to one or the other ideal type as they adjust to conditions of either stability or of change*'.[33] **Table 7.2** outlines the key features of the 'mechanistic' and 'organic' modes of organizing. When looking at each of the individual characteristics that constitute these two modes, think how each of these map onto your own experience of a particular organization, and assess where you might locate the organization on a spectrum between the two ideal types.

A number of overarching features of these two modes of organizing emerge from **Table 7.2**. The 'mechanistic' mode can be seen to be characterized by hierarchy, bureaucracy, and role specialization. In discussing the mechanistic mode, Burns and Stalker stress that:

> the individual 'works on his own', functionally isolated; . . . He works at a job which is in a sense artificially abstracted from the realities of the situation the concern is dealing with, the accountant 'dealing with the costs side', the works manager 'pushing production' . . . the rest of the organization becomes part of the problem situation the individual has to deal with in order to perform successfully; i.e., the difficulties and problems arising from work or information which has been handed over the 'responsibility barrier' between two jobs or departments are regarded as 'really' the responsibility of the person from whom they were received.[34]

The emphasis on 'functional isolation' and the 'handing of work over the responsibility barrier', is very reminiscent of the linear models of innovation (i.e. 'technology-push' and 'market-pull') discussed in **3.2.1**. These organizational features are not considered to be conducive to successful innovation.

In contrast, the 'organic' mode can be seen to be characterized by network structures, integration, boundary-spanning, and role fluidity. Indeed, for Burns and Stalker, under the 'organic' mode: '*The individual's job ceases to be self-contained; the only way in which 'his' job can be done is by his participating continually with others in the solution of problems . . .*'[35] The more 'organic' the system of management employed by an organization, the more difficult it becomes to distinguish the 'formal organization' from the 'informal organization' and the greater the reliance on organizational culture, rather than hierarchy, to ensure cooperation of organizational members. As Spender and Kessler stress, however, it '*does not lead to "flat" organizations, with individuals treated as equals*', but rather to those that are highly structured and stratified, often in relation to expertise, for example.[36] Furthermore, they argue that the '*informal structures must always be subordinated to the formal . . .* [such] *that the organization's purposes remain dominant*'.[37] Once again, if we refer back to **Chapter 3**, we see that the 'organic' mode of organizing and the 'coupling', or 'interactive', models of innovation (see **3.2.2**) send very similar messages in relation to the importance of boundary-spanning interaction for the fostering of innovative activity.

Table 7.2 The characteristics of mechanistic and organic modes of organizing (adapted from Burns and Stalker, 1961: 120–2)

'Mechanistic'	'Organic'
1. The problems and tasks facing the organization are broken down into specialist, functional tasks and distributed around the organization accordingly.	1. Specialist knowledge within the organization, wherever it resides, is seen as a potential contributor to the overall problems and tasks of the organization.
2. The tasks of individuals are largely abstracted from the broader goals of the organization, such that their accomplishment is seen as a means to an end in itself, rather than contributing to the ends of the organization.	2. The tasks of individuals are 'located' in the broader context of the organization.
3. Tasks attached to specific functional roles are defined precisely by superiors.	3. Tasks are subject to continual redefinition through interaction with others.
4. The responsibility of an individual is closely linked to his or her assigned functional role (and its associated tasks).	4. The commitment and responsibility of an individual is seen as stretching beyond specific tasks or roles.
5. A hierarchical structure of control, authority, and communication exists.	5. A network structure of control, authority, and communication exists.
6. Knowledge is seen as concentrated at the top of the hierarchy.	6. The dispersed nature of knowledge within the organization is recognized.
7. 'Vertical' interaction and communication within the organization predominates (i.e. between superior and subordinate).	7. 'Lateral' or 'horizontal' interaction and communication within the organization predominates (e.g. between functions).
8. There is insistence on loyalty to the organization and obedience to superiors.	8. Commitment to the goals and progress of the organization is seen as more important than loyalty or obedience.
9. Greater importance and prestige is attributed to internally specific (local), rather than externally relevant (cosmopolitan), knowledge, experience, and skills.	9. Greater importance and prestige is attributed to externally relevant (cosmopolitan), rather than internally specific (local), knowledge, experience, and skills.

Burns and Stalker also highlight differences between these two modes in the manner by which change is 'absorbed': under the organic mode, *changes from any direction were regarded as what they manifestly were—circumstances which affected every part of the firm and everybody's job, in some way*, whilst under the mechanistic mode:

the response to change was usually to create a new group, or to reconstitute the existing structure, or to expand an existing group which would be largely responsible for meeting the new situation, and so not disrupt the existing organization.

Rather than isolate and minimize the organizational response to environmental change, typical of mechanistic modes of organizing, organizations operating under organic principles adapt, in an attempt to absorb and embrace such change.[38]

The organic mode of organizing is reliant upon a shared set of values and a strong culture. This is far from easy to achieve: in her study of a sample of major US companies, Kanter, for example, found that providing employees with a *common focus within a shifting structure, especially when the company got too big to transmit its philosophy by osmosis* was problematic.[39] She goes on to cite the case of Hewlett-Packard and the 'HP way'—a case that has long been part of corporate folklore—as one mechanism through which successful and innovative companies were developing shared values and strong cultures. The 'HP way'—a set of long-held, distinctive, and articulated corporate values, centred on trust, respect, integrity, communication, teamwork, and innovation (see **Illustration 7.2**)—is reinforced by other mechanisms, such as a careful recruitment process and excellent employee benefits.

Illustration 7.2 Hewlett-Packard and the 'HP way'

Hewlett-Packard was founded in 1939. It is a highly successful and innovative US computer technology company. In 2005, it had an annual turnover of US$86bn. Today, the company is perhaps best known for consumer products, such as its LaserJet printers, its personal computers (PCs), and its digital cameras.[40]

The 'HP way' is a set of distinctive and articulated corporate values, the origins of which can be traced to the early years of the organization and to its founders—Bill Hewlett and Dave Packard. These values were first formalized in 1957 when the company was floated on the stock market.

For many, the 'HP way' has been at the heart of the commercial and technological success of the company. In summary, these values are:[41]

1. trust and respect for individuals;
2. a focus on high levels of achievement and contribution;
3. the conduct of business with uncompromising integrity;
4. the achievement of common objectives through teamwork;
5. the encouragement of flexibility and innovation.

These values have been reinforced by 'careful selection procedures', *formal training in its management philosophy in order to ensure transmission and continuity*', and *virtually permanent employment* [less so since the 1990s] *with excellent benefits to keep turnover low and thus retain a stable workforce*'.[42]

Whilst these values have remained incredibly resilient, it has been argued by some that the 'HP way' has come under increasing pressure since the late 1980s, from an increasingly competitive computer market and since its large-scale merger with Compaq in 2002.[43]

A further example is provided by LG Corporation, the South Korean electronics, telecommunications, and chemicals group (see **Illustration 7.3**).

Illustration 7.3 The LG Corporation and the 'LG way'

LG is a South Korean company operating in the electronics, telecommunications, and chemicals sectors. Founded in 1947, LG is today a major player in many of its markets. This highly successful and innovative company achieved sales of 94 trillion Korean Won in 2007 (i.e. US$80bn), and employs 160,000 people in its operations across the globe.

The 'LG way' is an articulation of the company's core beliefs, values, and aspirations '*that guides the thoughts and actions of LG people*'. At the heart of the LG way is the practice of 'Jeong-Do' management (translated as 'right way'), with its twin concerns of creating value for its customers through service and innovation, and respecting the 'human dignity' of its employees. For LG, this respect for 'human dignity' is achieved through valuing 'the ingenuity and autonomy' of its employees, the striving '*to maximize the capabilities of its people, and helping them realize their potential on the job*', and is '*reinforced through LG's everyday management practice of rewarding its people based on capabilities and performance*'.[44]

7.5.2 Kanter—'segmentalist' versus 'integrative' modes of organizing

The research underpinning Kanter's *The Change Masters* embraces a detailed study of ten US companies—six in new hi-tech sectors, such as computers and electronics (including Hewlett-Packard, Polaroid, and General Electric), and four in mature sectors, such as automobiles, telecommunications, and insurance (including General Motors)—and a total of 115 innovations developed by these companies. A core objective of the research was '*to learn in detail about exactly how people came to be connected to productive and innovative actions*'.[45]

From her detailed analysis of these organizations and innovation projects, Kanter reveals two distinctive modes of organizing: these she terms: 'segmentalist' and 'integrative'. They are not dissimilar to the 'mechanistic' and 'organic' modes, respectively, as identified by Burns and Stalker, and they too are viewed as 'ideal types' at two ends of a spectrum. Nevertheless, Kanter's framework places unambiguously at 'centre stage' the core importance of 'integration' to innovation and entrepreneurial activity within organizations.

Kanter summarizes these two modes very effectively:

> I found that the entrepreneurial spirit producing innovation is associated with a particular way of approaching problems that I call 'integrative': the willingness to move beyond received wisdom, to combine ideas from unconnected sources, to embrace change as an opportunity to test limits. To see problems integratively is to see them as wholes, related to larger wholes, and thus challenging established practices—rather than walling off a piece of experience and preventing it from becoming touched or affected by any new experiences.[46]

> . . .

> The contrasting style of thought is anti-change-orientated and prevents innovation. I call it 'segmentalism' because it is concerned with compartmentalizing actions, events, and problems and keeping each piece isolated from the others. Segmentalist approaches see problems as narrowly as possible, independently of their context, independently of their connections to any other problems. Companies with segmentalist cultures are likely to have segmented structures: a large number of compartments walled off from one another—department from department, level above from level below . . . Even

innovation itself can become a specialty in segmentalist systems—something given to the R&D department to take care of so that no one else has to worry about it.[47]

Building on this foundation, Kanter goes on to identify a number of factors that either promote or hinder innovative activity. In segmentalist mode, Kanter argues that innovation is often hindered by top management, who regard new ideas from below with suspicion, who over-bureaucratize the approval process for proceeding with new projects, who focus on criticism rather than praise and treat problem identification as a sign of failure, who operate on a 'need to know' basis in relation to information diffusion, and who assume that only they really know their business. Thus, for Kanter, segmentalism stifles innovation by making it 'unattractive and difficult' for employees '*to take initiative to solve problems and develop innovative solutions*'.[48]

In contrast, Kanter argues that creating a 'culture of pride', and a 'culture of change' rather than of 'tradition', are important to promoting innovation and individual initiative. By 'culture of pride', Kanter is referring to an '*emotional and value commitment between person and organization*' that makes '*people feel they "belong" to a meaningful entity*'; a culture that she associates with progressive management practices that view employees as '*important (not just well treated)*' and '*capable*', and which is reinforced by market success.[49] The 'HP way' combines this 'culture of pride' and 'culture of change' very effectively.

Drawing heavily from the work of Kanter, Bate develops a useful cultural framework from his post-mortem of the now infamous British Rail 'Advanced Passenger Train' (APT) project (see **Illustration 7.4**). The framework that emerges from his analysis, incorporates a number of cultural dimensions, which he terms 'cultural-isms'. The four key dimensions of this framework are 'segmentalism', 'conservatism', 'isolatism', and 'elitism'; for each dimension, Bate also outlines a number of 'switch-ons' (factors that promote innovation) and 'switch-offs' (factors that stifle innovation)[50]. We summarize the key elements of this framework in **Table 7.3**.

Illustration 7.4 British Rail and the APT project—the impact of organizational culture on the success of a radical innovation project

The Advanced Passenger Train (APT) is an interesting case study of innovation, not least because it '*will probably go down in history as one of the most successful innovations ever to have failed*'.[51] The APT was a bold and ambitious vision of a '*new generation of high performance train*', initiated in 1967; by 1985, the project had been 'deferred indefinitely' following a series of humiliating and very public trials in the UK. For Bate:

> it was within the management culture that the seeds of failure lay . . . this culture negatively influenced the APT programme in almost every dimension of its being, from beginning to end: strategy, concepts, attitudes, structures, systems, processes and responses.[52]

This culture is analysed by Bate under the various 'culturalisms' outlined in **Table 7.3**.

- '*Segmentalism*'—there was a long history of rivalry between the different departments and professional groupings within British Rail, encouraged by the heavy emphasis on the '*bureaucratic principle of organizational differentiation*'. Of particular importance was the rivalry between 'traditional' railway engineers, who saw the APT as 'totally impractical' and wasteful of scarce resources, and research scientists, who viewed the project in relation to technological possibilities and of long-terms goals. Another key struggle was between the 'innovators' and the 'book-keepers', who held the 'purse strings'.

- *'Conservatism'*—the conservative British Rail culture wanted 'continuity not change' and 'evolutionary rather than radical innovation'. The APT project threatened both of these well-embedded cultural positions.
- *'Isolatism'*—commercial and technical judgements were made by the APT project team with little openness to ideas or inputs from outside, such as professional marketers and designers from other industries.
- *'Elitism'*—British Rail was weighed down with a labyrinth of hierarchical reporting procedures and protocols to observe. Projects such as the APT were required to pass through slow and cumbersome review processes, and, ultimately, needed permission, legitimation, and sponsorship from top management.

Table 7.3 The 'cultural-isms' impacting an organization's innovativeness (adapted from Bate, 1994)

The 'cultural-isms'	The 'switch-ons' and 'switch-offs'
'Segmentalism'	*Switch-ons*—a culture that encourages networking, boundary-spanning, integration, and flexibility.
	Switch-offs—a culture that encourages functional boundaries and the compartmentalization of tasks and actions.
'Conservatism'	*Switch-ons*—a risk-taking and non-conformist culture.
	Switch-offs—a culture that supports continuity rather than change.
'Isolatism'	*Switch-ons*—an open culture that encourages interaction with external sources of expertise.
	Switch-offs—a closed culture that undervalues external expertise, and is dogged by a 'not invented here' syndrome.
'Elitism'	*Switch-ons*—a culture that encourages two-way vertical communication, supported by a flattening of the hierarchy.
	Switch-offs—a culture that limits vertical communication and mobility, supported by protocol, reporting lines, and hierarchy.

7.5.3 Nonaka and Takeuchi—managing the knowledge creating organization

We now turn to the work of two Japanese academics, Ikujiro Nonaka and Hirotaka Takeuchi, who are well known for their book *The Knowledge-Creating Company*, published in 1995. The text is based on research conducted within a sample of twenty highly successful and innovative companies (mostly Japanese), including NEC, Mazda, Mitsubishi, Fujitsu, Nissan, Matsushita, and Fuji Xerox. This work has been particularly influential in promoting the debate around the processes of knowledge creation and knowledge transfer within organizations. We will look at these processes in **Chapter 8**. Here, we draw upon the broader organizational context that Nonaka and Takeuchi emphasize in relation to developing the *'capability of a company as a whole to create new knowledge, disseminate it throughout the organization, and embody it in products, services, and systems'*.[53]

Nonaka and Takeuchi introduce five enabling 'conditions' that they see as key to providing the appropriate context for knowledge creation within an organization: 'organizational intention'; 'autonomy'; 'fluctuation' and 'creative chaos'; 'information redundancy'; and 'requisite variety'.[54] They also discuss a sixth factor, which we will term the 'inside–outside connection'. We will deal with each of these in turn.

- *'Organizational intention'*—this refers to the importance of creating and communicating a strategic vision about the kind of knowledge that should be developed within the organization. Nonaka and Takeuchi argue that, because a given technology might be employed across various product divisions in a large company, such as communications, computers, and digital equipment, this vision might be best abstracted and communicated in relation to 'core' or 'base' technologies, such as 'pattern recognition' or 'image processing'. For Nonaka and Takeuchi, without organizational intention, it would be impossible for individuals and groups within the organization *'to judge the value of information or knowledge'* that they have created or received.

- *'Autonomy'*—Nonaka and Takeuchi also stress the importance of allowing organizational members 'to act autonomously', in order to motivate them to create new knowledge, and to increase the possibility of serendipity and unexpected connections and opportunities arising in the organization. They see the 'self-organizing', cross-functional team as providing a particularly fruitful context in which individuals can act autonomously.

- *'Fluctuation'* and *'creative chaos'*—'fluctuation' occurs when organizational members are allowed to respond and adapt to ambiguous signals from their environment. Nonaka and Takeuchi argue that this can lead to *'a 'breakdown' of routines, habits, or cognitive frameworks'*, forcing individuals to question and reconsider their assumptions, and that it ultimately fosters the process of knowledge creation. 'Creative chaos' refers to the intentional chaos created by organizational leaders by evoking a 'sense of crisis'. This is intended to create 'tension' within the organization, in order to focus the minds of individuals on problem definition and crisis resolution. They warn, however, that *'the benefits of "creative chaos" can only be realized when organizational members have the ability to reflect upon their actions. Without reflection, fluctuation leads to "destructive chaos"'*.[55]

- *'Information redundancy'*—this refers to the 'intentional overlapping' of information within different parts of an organization. Nonaka and Takeuchi highlight a number of ways in which such information redundancy can be built into an organization, for example, through the 'fuzzy' division of labour between different departments engaged in cross-functional projects, through the establishment of competing teams undertaking the same project, and through the 'strategic rotation' of organizational members between different functions and divisions. But whilst Nonaka and Takeuchi argue that redundant information can help to create unusual and unpredictable interactions and communication paths within an organization, it also leads to an increase in the amount of information that needs to be processed and may result in 'information overload'.

- *'Requisite variety'*—we discussed this concept briefly towards the start of this chapter: to recap, it posits that, for an organization to adapt and survive in its environment, it must reflect the variation and complexity of this environment within its internal structure.[56] Nonaka and Takeuchi argue that *'Developing a flat and flexible organizational structure in which the different units are interlinked with an information network is one way to deal with the complexity of the environment'*.[57]

- *'Inside–outside connection'*—this refers to the importance to innovation and knowledge creation of the connection and interaction between an organization and its environment. This enabling condition is centrally important to Nonaka and Takeuchi, who argue that:

> What is unique about the way Japanese companies bring about continuous innovation is the linkage between the outside and the inside. Knowledge that is accumulated from the outside is shared widely within the organization, stored as part of the company's knowledge base, and utilized by those engaged in developing new technologies and products. A conversion of some sort takes place . . . from outside to inside and back outside again in the form of new products, services, or systems.[58]

We will look much more closely at the nature, role, and importance, of external sources to the innovation process in **Part IV**—within **Chapter 10**, in particular.

There is a noticeable overlap between the organizational features that underpin the enabling 'conditions' of Nonaka and Takeuchi, and those of the 'organic' and 'integrative' modes of organizing (for example, 'a flat and flexible' structure, cross-functional teams, and informal communication networks). Nevertheless, their perspective tends to focus on broader and more abstract constructs, such as 'creative chaos', 'information redundancy', and 'requite variety', and thus complements the earlier work of Burns and Stalker, and of Kanter.

7.5.4 'Structured formality' versus 'unstructured chaos'—a difficult balancing act?

One of the key themes to emerge from the work of Burns and Stalker, Kanter, and others is the negative impact that bureaucracy and managerial control can have on the creativity and innovativeness of individuals and teams. Yet the 'organic' and 'integrative' modes of organizing should not be interpreted as proposing 'unstructured chaos', free of bureaucracy and managerial control. For Spender and Kessler:

> Nor is the organic organization 'free-floating' and without direction. It remains a purposive tool in the hands of those who hold power, control the resources and shape the organization's cultural system.[59]

For Kanter:

> True 'freedom' is not the absence of structure—letting the employees go off and do whatever they want—but rather a clear structure which enables people to work within established boundaries in an autonomous and creative way. It is important to establish for people, from the beginning, the ground rules and boundary conditions under which they are working: what can they decide, what can't they decide? . . . The fewer the constraints given a team, the more time will be spent defining its structure rather than carrying out its task.[60]

Thus, some degree of bureaucracy and managerial control is regarded as desirable in relation to the management of innovation, to promote, for example, the congruence between team and organizational-level goals, the coordination of disparate activities, and the allocation of scarce resources between competing projects. Indeed, from his study of large, innovative US enterprises, Quinn observed:

> Neither structured formality nor unstructured chaos works well alone. Innovative companies seem to evolve a sophisticated approach to 'managed chaos', which recognizes the realities of how major technological innovations evolve, and harnesses this process to corporate needs.[61]

Whilst a number of studies have concluded that individuals and teams are more creative when they have relatively high autonomy in their day-to-day work,[62] too little 'structure' can also create relatively high ambiguity in roles and tasks, and this may lead to loss of focus, frustration, and anxiety. For example, Kanter argues that *'Without structure, groups often flounder unproductively, and the members then conclude they are merely wasting time'*,[63] whilst Burns and Stalker observed that:

> The organic form, by departing from the familiar clarity and fixity of the hierarchic structure, is often experienced by the individual manager as an uneasy, embarrassed, or chronically anxious quest for knowledge about what he should be doing, or what is expected of him, and similar apprehensiveness about what others are doing.[64]

The challenge for managers thus becomes one of finding a balance between formality and informality, and between control and chaos.

7.6 **The ambidextrous organization—managing in 'dual mode'**

As noted in **Chapter 2**, the 'ambidextrous' organization is one that has the ability to switch between alternative modes of organizing,[65] or to operate simultaneously in dual modes of organizing[66]—in this instance, the mechanistic and organic modes.

Why might an organization require such a capability? There are a number of such circumstances that are important to a more subtle understanding of the management of innovation:

- a change in the certainty or stability in the external environment of the organization, for example, through the maturing of an existing technology or discontinuous innovation;

- variation in the organizational requirements of different innovation activities during the various stages of the innovation process;
- variation in the organizational requirements of different functional activities within the organization; and
- to enable the effective management of portfolios of products and services that are at different stages of maturity.

Whilst the first two of the circumstances listed above require the capability to oscillate between alternative modes of organizing, the latter two require the capability to manage simultaneously alternative, and conflicting, modes of organizing—that is, to manage in 'dual mode'. We will discuss these each in turn. Following this discussion, we will look at the implications of moving between alternative modes of organizing and operating in dual modes, because, as Spender and Kessler contend: '*As soon as we set up the opposition of the mechanistic and the organic within the firm, we must offer a theory of its resolution.*'[67]

7.6.1 A change in the certainty or stability of the external environment

We noted earlier in this chapter the relationship between the certainty and stability of the external environment within which an organization operates, and the formality and flexibility of its internal structure. This 'contingency theory' perspective argues that there is 'no one best' way in which to organize; rather, organizations must 'match' their structure to their environment.[68] From this perspective, it follows that, as the external environment changes, then so too must the internal structure of the organization to ensure a continued match. Indeed, Burns and Stalker argue that '*the relation of one form to the other* [organic to mechanistic] *is elastic, so that a concern oscillating between relative stability and relative change may also oscillate between the two forms*'.[69]

Returning to the life cycle models discussed in **Chapter 4**, these indicate that we would expect, for example, increasing certainty and stability in an organization's environment to occur as a key technology within that environment moves along the technology life cycle towards maturity; the associated product-process cycle model, indicates a shift from innovation to efficiency. Both of these models highlight a need for organizations to migrate from organic to mechanistic modes of organizing as a sector matures. In contrast, discontinuous innovation—a shift onto a new technology S-curve—flags the need to return to more organic modes of organizing as uncertainty and instability return to the external environment.

7.6.2 Variation during the innovation process

Despite the volume and variety of literature concerning the management of the various stages of the innovation process—which will be discussed in much greater detail in **Chapter 8**—there is remarkable coherence around the proposition that the earlier stages of the innovation process are inhibited by mechanistic modes of organizing, because they require, for example, conditions that allow free thinking and the free flow of information.

In contrast, the latter stages of the innovation process are viewed as being more suscept-ible to mechanistic modes of organizing, because a reduction in uncertainty surrounding the innovation is accompanied by greater technical definition and goal orientation of the project. Thus, this literature indicates that there needs to be a shift in management style from organic to mechanistic during the course of an innovation project.[70]

Spender and Kessler[71] argue, however, that this literature looks at the innovation process too narrowly, from within the project, rather than from within the organization in which it is embedded. Drawing upon the work of Quinn,[72] they highlight the significant role played by formal planning activities—associated with a mechanistic mode of organizing—at the very early stages of the innovation process and in the 'back-ground' during the life of the project. To resolve this incongruence in the literature, Spender and Kessler propose a model of the innovation process that is broader—see **Figure 7.1**. It involves the 'nesting' of the innovation project (and its narrower set of innovation stages) within the 'enveloping bureaucratic system' of the 'host' organization. The project is initiated following a period of planning by the organization and, later, the output of the project is embedded back into the organization during implementation. They argue that project initiation can only occur when the project is 'released, legiti-mated, and resourced' by the 'host' organization. This process of 'bureaucratic release' is accompanied by the adoption of an organic mode of organizing within the project, whilst the embedding of the output of the project back into the organization is accompanied by a process of 'bureaucratic capture', which involves a reassertion of the mechanistic mode of organizing over the innovation process. Quinn alludes to this process of 'bureaucratic capture' by noting that:

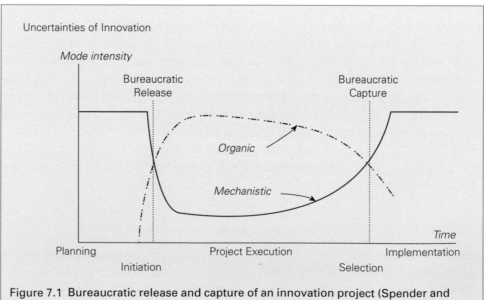

Figure 7.1 **Bureaucratic release and capture of an innovation project** (Spender and Kessler, 1995: 47)

> As the [innovation project] . . . approaches ultimate exploitation, uncertainty decreases, and the costs escalate. This is where formal planning, using the full array of program planning, economic evaluation, and progress monitoring techniques can pay high dividends.[73]

Whether we adopt the narrower ('project-centred') or broader ('project-nested') notions of the innovation process, it is clear that the management of the innovation process involves shifts in orientation between the organic and mechanistic modes of organizing.

7.6.3 Variation in functional requirements

A number of writers since the late 1960s have argued that because organizations incorporate activities with varying organizational requirements, they must embrace simultaneously both organic and mechanistic modes of organizing. Thompson, for example, contrasted 'efficiency-seeking' activities, which require a mechanistic system of management, with 'uncertainty-resolving' activities, which require an organic system of management: the former he associated with the stable, 'productive core' of an organization; the latter with its uncertainty-absorbing periphery—a 'buffer zone' from the external environment.[74] This perspective is illustrated in **Figure 7.2**, and is contrasted with the 'either/or' approach of Burns and Stalker.

The distinction between efficiency-seeking activities and uncertainty-absorbing activities is one that can, to a certain degree, be overlaid on different functions or subunits within the organization: the former characterizing the predominant focus of manufacturing; the latter, that of R&D. Indeed, within the classic study that we noted earlier, Lawrence and Lorsch revealed that subunits within the same organization face varying degrees of environmental uncertainty or stability. They argued that the greater the variation in the environmental uncertainty or instability faced by organizational subunits, the

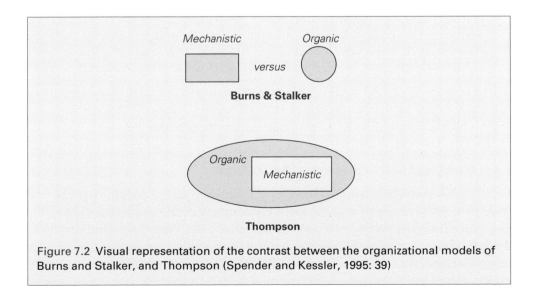

Figure 7.2 Visual representation of the contrast between the organizational models of Burns and Stalker, and Thompson (Spender and Kessler, 1995: 39)

greater their 'differentiation' with regard to their internal structure. R&D subunits, for example, were found to face a more uncertain environment than did the sales subunits, with manufacturing subunits facing the least uncertain environment of the three. From a contingency theory perspective, they argued that the mode of organizing of a subunit must be congruent with its respective external environment. Thus, one would expect R&D departments to have an organic mode of organizing and manufacturing departments a mechanistic mode of organizing.[75]

The notion that different activities or subunits within the same organization might require different and conflicting modes of organizing raises issues with regard to their integration and coordination—we address this key issue shortly.

7.6.4 Variation inherent within portfolios of products or services

It has been argued that sustained competitive advantage is derived from the ability of managers to build and manage organizations that can simultaneously develop incremental and discontinuous innovation.[76] Thus, successful organizations—other than perhaps newly established entrepreneurial firms—would be expected to have portfolios of products or services that are of varying maturity. Furthermore, in large organizations with multiple divisions, the balance in maturity of portfolios of products or services might vary greatly between individual divisions. But to nurture both incremental and discontinuous innovation requires an organization to operate dual modes of organizing, because they require fundamentally different organizational settings: broadly speaking, the mechanistic mode for incremental innovation, and the organic mode for discontinuous or radical innovation.

Tushman et al. outline the contrasting organizational contexts for incremental and discontinuous innovation as follows:

> Continuous, incremental improvement in both the product and associated processes . . . requires organizations with relatively structured roles and responsibilities, centralised procedures, efficiency-orientated cultures, highly engineered work processes . . . In dramatic contrast to incremental innovation, discontinuous innovation emerges from entrepreneurial, skunkworks types of organizations. These entrepreneurial units are relatively small, have loose, decentralized structures, experimental cultures, loose, jumbled work processes, strong entrepreneurial and technical competences.[77]

The implication is that for organizations to sustain innovation successfully over time, they must find ways to not only manage in 'dual mode', but to manage the potential conflict between organizational subunits operating in these different modes.

7.6.5 Managing a shift in the mode of organizing

A shift in the mode of organizing, whether at the organizational level in response to a change in the environment or at the project level in relation to the stage of the innovation process, requires managerial skills in recognizing the appropriate timing of that shift and in managing that shift itself—essentially a process of 'change management'. At the project level, the concepts of 'bureaucratic release' and 'bureaucratic capture' provide a neat way of framing the managerial process of shifting between modes of organizing.

At the organizational level, however, there has been much debate in recent years as to the extent to which organizations can actually adapt to their environment by altering their mode of organizing, or indeed, 'oscillate' between alternative modes. For some, such organizational change is seen as relatively unproblematic, whilst for others, organizations are characterized by 'structural inertia'[78] and 'cultural inertia',[79] and there is empirical evidence that would suggest that 'transitions' by organizations are, at a minimum, problematic.[80] Concepts such as 'core rigidity'[81] and 'zones of manoeuvre',[82] as discussed in **Chapter 6**, help to explain why such change might be difficult for organizations. But as Clark and Staunton note: '*The revelation of slow, uneven transition has not halted the plethora of best-selling studies of fast, almost frictionless change.*'[83]

Perhaps equally as difficult as managing a shift in the mode of organizing of an organization is managing an organization simultaneously in 'dual mode'. Senior management teams within such organizations need to have the heterogeneity and competencies to handle dual modes of organizing (i.e. mechanistic and organic) and dual modes of innovating (i.e. incremental and discontinuous). Furthermore, such ambidextrous organizations have 'built-in contradictions', are 'inherently unstable' and 'internally inconsistent', and, if not managed proactively, can foster internal conflict and politics. Conflict might occur, for example, between well-established, cash-generating, mechanistic subunits, generating today's incremental innovation, and young, entrepreneurial, cash-absorbing subunits, generating tomorrow's discontinuous innovation. Given that the 'power, resources, and traditions' of an organization are generally located within the established subunits, young, entrepreneurial subunits are often at risk of sabotage.[84]

In managing such 'differentiated' organizations, Tushman et al. argue that:

> the senior management team must provide clear roles and responsibilities for the contrasting units and be a force for integration and balance . . . tensions can be managed by a senior team that can . . . articulate a clear, common vision within which these tensions make sense.[85]

7.7 Integrating and coordinating innovative activity within organizations

A core characteristic of both the 'organic' and 'integrative' modes of organizing discussed above is the facilitation of horizontal communication between teams, functions, and divisions, within an organization, to promote innovation, coordination, and integration. Such horizontal communication can be achieved through an array of formal integrative mechanisms (discussed below) and informal integrative mechanisms (discussed in **Chapter 9**).

Nadler and Tushman refer to this set of formal integrative mechanisms as 'strategic links' due to their strategic importance to the organization, arguing that:

> Strategic linking issues follow directly from strategic grouping choices. Strategic grouping focuses resources by product, market, discipline, or geography. This grouping

of resources puts some resources together *and* splits other resources . . . Strategic linking involves choosing formal structures that link units that have been split during strategic grouping. Once strategic grouping decisions have been made, the next step is to coordinate, or link, the units so that the firm can operate as an integrated whole.[86]

The following sections address a number of questions relating to strategic linking:

- what factors influence the extent of strategic linking required by an organization?
- what formal strategic linking mechanisms are available to the organization?
- what factors influence the appropriateness of adapting a particular linking mechanism?

7.7.1 Identifying linkage requirements between organizational subunits

There are a number of factors that impact the extent of formal strategic linking required by an organization to integrate its subunits effectively. One such factor is the variation or 'differentiation' in the internal structure or mode of organizing of the various subunits within the organization.

For Lawrence and Lorsch, integrative mechanisms are an important counterbalance to the 'centrifugal forces' of differentiation between subunits. In their cross-sector study, they found that the most highly differentiated of the 'high-performing' organizations had established the most elaborate set of formal integrating mechanisms, including permanent cross-functional teams and 'integrative departments'. Thus, the greater the variation in the internal structure or mode of organizing between the subunits of an organization, the greater the need for formal strategic linking.[87]

A second factor is the degree of task interdependence between the various subunits of the organization. Nadler and Tushman provide a useful distinction between three forms of task interdependence, and highlight their respective linkage requirements:[88]

- '*pooled interdependence*'—where organizational subunits are essentially independent of each other except for the sharing of a range of organizational resources: for example, the divisions of a diversified firm operating in completely different product or market niches. Such organizations have low task interdependence between subunits, and therefore, have minimal coordination and linkage requirements;
- '*sequential interdependence*'—where there is task interdependence between organizational subunits by virtue of a process that flows sequentially from one subunit to another: for example, an organization that incorporates divisions that extract raw materials (e.g. oil), others that process these same raw materials (e.g. polymers), and still others that produce a variety of end products from these polymers (e.g. construction materials). Because subunits in such a flow of work through the organization depend on the work undertaken by subunits earlier in the process, close attention must be paid to coordination and timing in order that the flow of work remains both smooth and efficient; and

- *'reciprocal interdependence'*—where different subunits must work together to complete a task: for example, in the design and development of an innovative new product, where there may not only be task interdependence between functions such as R&D, production, and marketing, but also between different divisions, where the innovation requires the integration of knowledge and technology from more than one division. In instances of reciprocal interdependence, no single subunit can accomplish its task without the active contribution of other subunits, and this thus imposes substantial coordination and linking requirements in relation to problem solving.

Thus, the greater the task interdependence between organizational subunits, the greater the need for coordination and joint problem solving, and therefore the greater the need for strategic linking.

7.7.2 Options for linking organizational subunits

Having highlighted some of the factors impacting the extent of strategic linking required by an organization, the next step is to outline the range of formal integrative mechanisms available for strategic linking. Nadler and Tushman argue that '*The more complex the degree of work/task interdependence, the more complex the formal linkage devices must be to handle work-related uncertainty*'.[89] They go on to describe a variety of formal strategic linking mechanisms.

Hierarchy

This is the simplest form of structural linking, where coordination between individuals at the same level is accomplished through their common boss. Whilst, through 'focused, sustained, and consistent' action by the linking manager, this can be an effective linking mechanism, it is limited by the inherent cognitive and information-processing capacity of individuals, which can quickly lead to overload. Nevertheless, this is viewed as '*one of the most pervasive formal linking mechanisms*'.[90]

Liaison roles

This is a formal boundary-spanning role assigned to individuals within the organization '*that serve as information conduits and initiators of problem-solving endeavours deep in the organization . . .* [although] *they rarely have authority to back up their positions*'.[91] Burns and Stalker argue that one of the benefits of the liaison role is they '*offer the possibility of retaining clear definition of function and of lines of command and responsibility*'.[92] To be effective, though, such roles should be assigned to those who are capable of understanding and translating the contrasting 'languages' and 'coding schemes' of different functions, specialisms, and divisions within an organization.[93] This role is generally undertaken alongside other activities and is rarely a full-time responsibility.

Allen, however, argues against formalizing such liaison or boundary-spanning roles, which he believes '*seems unnecessary and could even prove undesirable*', favouring that '*recognition be afforded on a private, informal basis*'.[94]

We will return to the role of informal boundary-spanning in **Chapter 9**.

Cross-unit groups

These groups are made up of task-relevant representatives from different subunits within the organization, for example, from different business functions or different technology-orientated divisions. Whilst such groups may be focused on particular products, projects, markets, or clients, for example, and they may be permanent, temporary, or ad hoc, their common objective is to bring relevant expertise together to deal with a specific and joint task. In comparison to the liaison role, cross-unit groups *'provide a more extensive forum for information exchange, for coordination, and for the resolution of conflict between work units'*.[95]

Integrator roles or departments

The integrator role is responsible for taking a general management perspective and in aiding multiple or cross-unit work groups to accomplish a joint task, such as the development of a complex new product, process, or service. A key task of the integrator role or department is to adjudicate where differences and conflicts arise within, and between, cross-unit groups. Product managers, brand managers, regional managers, and account managers are all examples of the integrator role.

Matrix structures

Matrix structures involve dual chains of command, combining functional and product orientations, for example, to improve coordination and balance power between subunits. For subordinates, this entails having two bosses, and the inherent tensions and micro-politics that this brings; power issues are thus made more salient in this structure.[96]

Nadler and Tushman argue that *'When it is important to give equal attention to several critical contingencies and when information-processing demands are substantial, matrix structures are appropriate'*.[97] But they also warn that:

> Matrix structures are very complex. They require dual systems, roles, controls, and rewards . . . Further, matrix managers must deal with the difficulties of sharing a common subordinate, while the common subordinate must face off against two bosses . . . Given its complexity and inherent instability, a matrix structure should be reserved for situations in which no other linking alternative is workable.[98]

Nevertheless, Kanter notes the important opportunities that such a structure provides for building informal working relationships that cut across functional and subunit boundaries within the organization, and for coalition building for when individuals need help in mobilizing support or resources to complete a task or project.[99] The 'matrix structure' would be considered the most appropriate linking mechanism where there was a high degree of task interdependence and thus a need for greater organizational integration.

This is not an exhaustive list of formal linking mechanisms, but it does provide a good overview of the range of structural possibilities for integration. It is also worth noting that these linking mechanisms are not mutually exclusive. Furthermore, organizations are not static entities, and thus, as they change and adapt, the extent and nature of task interdependence between subunits will shift over time. It is also likely that changes in task interdependence will be accompanied by changes in the appropriateness of a given strategic linking mechanism.[100]

7.7.3 **Evaluating the linking mechanisms**

The five linking mechanisms discussed above can be evaluated along the following dimensions:

- *Cost/resource*—the resources absorbed, and thus the costs incurred, are greater the more extensive and complex the linking mechanism. 'Matrix structures' involve more individuals, and require more resource to accommodate dual sets of structures, systems, and procedures than other linking mechanisms. At the other extreme, 'hierarchy' and 'liaison roles', involve a smaller number of individuals, but in more intensive coordination roles. For Nadler and Tushman:

  > The essence of making linking decisions is to choose those formal linking mechanisms that most effectively handle work-related interdependence. Using overly complex linking mechanisms will be too costly and inefficient, whilst using too simple linking mechanisms will not get the work done.[101]

- *Dependence on informal organization*—the greater the complexity and the greater the work-related interdependence that accompanies the linking mechanism, the greater the reliance on the informal organization for its effective operation. Thus, whilst linking mechanisms based on 'hierarchy' rely largely on the formal organization, cross-group units, integrator roles, and matrix structures depend far more on the informal organization to resolve conflict and ambiguity brought about by greater interdependence. Indeed, Nadler and Tushman contend that '*without an informal organization that deals openly with conflict, that has collaborative norms and values, and that can deal with the complexities of dual-boss relations, matrix organizations will not work*'.[102]

- *Information-processing capacity*—simple linking mechanisms, such as 'hierarchy' and 'liaison roles', have limited information-processing capacity, because they rely on a relatively small number of individuals to link organizational subunits. In contrast, 'matrix structures' can draw upon more resources and perspectives, and are thus capable of dealing more effectively with more information and greater uncertainty. Thus, Nadler and Tushman argue that:

  > Managers must choose those sets of linking mechanisms that match the information-processing demands of their unit's work interdependence. A mismatch between information-processing requirements and the choice of linking devices will be associated with relatively poor coordination and lower organization performance.[103]

7.8 **The physical work environment—designing the work environment to foster innovative activity**

Most of the discussion in this chapter has focused on the ways in which organizational structure and culture either foster or constrain the innovativeness of an organization. This is not surprising, given that these issues have been a major preoccupation of academic work on innovation since the 1960s. One area in which there has been relatively little academic research, however, despite its potential impact on the innovation process and knowledge work, concerns the physical work environment. This has been partly rectified by a recent flurry of academic interest.[104] Nevertheless, many decisions relating to the work environment of knowledge workers continue to be made with very little consideration or understanding of the implications for work performance.[105] Curtis et al. find this odd, noting that '*When a production line in a factory is designed, an immense amount of effort and thought goes into the flow of materials in the production process*'; they ask: '*How many organizations give the same degree of thought to the flow of knowledge?*'[106]

In this chapter, we have written of the importance of organic and integrative modes of organizing to innovation, yet buildings and workplace design can constrain the very boundary-spanning interactions and informal networking behaviours that these organizational forms are intended to promote. Furthermore, Curtis et al. argue that '*echoes of the old structures and ways of working are frozen into the architecture*'.[107] Thus, it is not surprising that Skyrme has observed that '*Many conventional offices are not conducive to knowledge sharing. They have enclosed offices or cubicles that impede communication, or have poorly designed open areas that are distracting*',[108] whilst Curtis et al. argue that '*the physical space we work in is often poorly adapted to the task of capturing, organizing and exploiting knowledge*'.[109]

Designing 'effective' buildings and work environments requires a deep understanding of the kinds of behaviours that the organization wishes to encourage. It also requires a deep understanding of the kind of work environments that suit different types of tasks, and an appreciation of the different responses and reactions that different types of worker might have to different types of workspace. Despite the relative paucity of academic research, there are, nevertheless, an increasing number of 'good' examples of the successful interweaving of organizational design and architectural design, such as the British Airways headquarters 'Waterside' site near London Heathrow, and the British Telecom headquarters in central London. Common in such designs is the inclusion of a variety of different types of workspace and social space. The Waterside building, for example, designed by Norwegian Niels Torp, incorporates six buildings connected by a covered central 'street'. The central street, with shops and cafes, is a place where people can meet informally and serendipitously. Elsewhere, there are quieter, more private places.[110]

Such examples are increasingly providing architects and workspace designers with a set of design principles. Even so, a number of dilemmas arise when designing buildings and workspaces, such as the tensions between the following:

- *Open versus closed workspaces*—whilst closed workspaces can aid concentration and reflection, important in such tasks as software development, open workspaces can promote project coordination and decision making, and enable employees to build social networks that can foster trust and 'comfort' in seeking and providing information and assistance. Open workspaces can also promote boundary-spanning and flexibility, because:

  > eliminating walls and panels reduces the clear lines between individuals and groups, creating a more permeable set of boundaries that can ebb and flow as groups change and evolve over time.[111]

But open workspaces, if poorly designed, can be distracting,[112] and the need for 'quiet space' is often overlooked.[113] The issue is not whether open or closed spaces are most appropriate, because both serve useful purposes, but determining the right balance between open and closed workspaces. Furthermore, there are many types of open workspace, for example, high or low-panelled cubicles for individuals, clusters of low-panelled cubicles for teams, separated off from other clusters by high panels, or completely open areas.[114] Overall, however, Becker and Sims argue that open workspaces are most effective *'when communication and interaction are critical elements of the work process'*, and for them, *'Few jobs or professions don't apply'*.[115]

- *Workspaces versus social spaces*—informal and boundary-spanning interactions, important to the innovation process, are facilitated by the provision of social spaces, such as cafes and other communal zones. Such social spaces may be located between different functions, for example, to help nurture cross-functional inter-actions, or away from workspaces, to encourage more open interactions. Often these social spaces are networked, to allow for Internet and intranet connection. Such social spaces are increasingly common in innovative organizations.

- *Person-specific versus task-specific workspaces*—traditionally, employees have a fixed workspace in which they undertake all or most of their tasks; however, most roles in organizations incorporate a variety of different tasks, each potentially requiring different types of workspace. Software engineers, for example, typically associated with spending much of their time working alone on coding, actually spend up to a third of their time interacting or collaborating with others.[116] Emergent trends include the increasing use of 'hot-desking', facilitated by advances in computing and telecommunications, and the provision of workspaces dedicated to particular tasks, such as team meetings, individual 'quiet work', and 'technology-intensive work' (for example, high-resolution graphics, scanning, etc.).

- *Stability versus mobility and flexibility*—in many sectors, an increasing proportion of employees spend much of their time 'on the road', meeting clients. One organiza-tional response is to assign whatever workspace is available to such workers on those occasions that they enter the office. This practice is sometimes referred to as 'hotelling'. Whilst this can be a cost-efficient way of allocating space, it does little to promote social capital and informal networks.[117]

- *Individual preference versus organizational effectiveness*—whilst knowledge workers prefer closed offices, they communicate much better in open offices.[118] Thus, for Becker and Sims, a key dilemma is *'whether one forces employees to occupy space they do not prefer . . . in the hope that they will adapt as they realize the benefits, but at the risk of alienating and demoralizing them'*.[119]

This section has sought to highlight the importance of the work environment to the innovation process and knowledge work, and to indicate the sorts of dilemmas that exist in attempting to design effective workspaces. But whilst the area is relatively underde-veloped academically, a set of design principles appear to be emerging from the projects of leading-edge architects and work space designers working alongside innovative companies.

7.9 **Concluding comments**

This chapter has sought to provide an overview of the key features of the internal context, such as structure, culture, physical layout, and the level of bureaucracy, which impact upon the innovative capacity of an organization. At the heart of this discussion is the degree to which the organizational context either promotes or hinders boundary-spanning across team, functional, and divisional boundaries. Boundary-spanning is presented as an important factor in promoting communication, flexibility, integration, and responsiveness, all of which are key to the processes of innovation, and knowledge creation and sharing. Indeed, boundary-spanning is a central feature of the seminal work on the management of innovation, and is core to the 'organic' and the 'integrative' modes of organizing.

One of the key themes to emerge from the work of Burns and Stalker, and of Kanter, among others, is the negative impact that bureaucracy and managerial control can have on the creativity and innovativeness of individuals and teams. Yet it is important to re-emphasize here that the 'organic' and 'integrative' modes of organizing should not be interpreted as proposing 'unstructured chaos', free of bureaucracy and managerial control. For Kanter:

> True 'freedom' is not the absence of structure—letting the employees go off and do whatever they want—but rather a clear structure which enables people to work within established boundaries in an autonomous and creative way.[120]

A second key concept discussed in this chapter is that of the 'ambidextrous' organization. Such an organization is one that has the ability to switch between alternative modes of organizing, or to operate simultaneously in dual modes of organizing—in this instance, the mechanistic and organic modes.

A number of circumstances were discussed that might require an organization to be ambidextrous:

- a change in the certainty or stability in the external environment of the organization, for example, through the maturing of an existing technology or discontinuous innovation;
- variation in the organizational requirements of different innovation activities during the various stages of the innovation process;
- variation in the organizational requirements of different functional activities within the organization; and
- to enable the effective management of portfolios of products and services that are at different stages of maturity.

Whilst the first two of the circumstances listed require the capability to oscillate between alternative modes of organizing, the latter two require the capability to manage simultaneously alternative, and conflicting, modes of organizing—that is, to manage in 'dual mode'.

In this chapter, we have focused on the ways in which managers can strive to establish an organizational context that is conducive to innovation. In the next chapter, we centre our discussion on the management of the innovation process, teams, and the knowledge creation and sharing processes. In the remaining chapter of **Part III**, the discussion moves away from the 'formal' management and organization of innovation, to centre on 'emergent' and 'informal' processes and structures; here, we will discuss the important, although often overlooked, role of social networks, informal learning, and serendipity, for example, in the innovation process.

CASE STUDY 7.1 SUN MICROSYSTEMS—MATCHING ORGANIZATION CULTURE AND THE PHYSICAL WORKING ENVIRONMENT

Introduction

Sun Microsystems is a highly innovative and commercially successful developer of computer hardware and software. Sun is a US company, with headquarters in California.[121] Formed in 1982, with only four employees, the company grew rapidly, reaching US$1bn in annual revenue by 1988. Following a highly successful floatation on the US stock market in 1986, Sun grew rapidly in market value and is currently ranked around 170th in the Fortune 500; by 2004, Sun had annual revenues of some US$11bn and was employing 35,000 people across a hundred countries.

The backbone of Sun's huge commercial success has been its highly regarded range of workstations, powered by its own 64-bit scalable processor architecture (SPARC) processors. But the company is also widely viewed as a global leader in network computing solutions, developing industry standards, such as, network file sharing (NFS) technology (first introduced in 1984); and Java (first introduced in 1995), which is present on over a billion devices around the world, from mobile phones to smart cards. Sun Microsystems UK has a turnover of around US$1bn and employs around 2,800 people.[122]

The 'Sun culture'

Since its inception, Sun Microsystems has fostered an entrepreneurial and progressive culture, which it believes is at the heart of its ability to attract and retain world-class technologists, to innovate, and ultimately, to succeed in a highly competitive and rapidly changing business environment. The 'Sun culture' embraces a non-corporate ethos and emphasizes, for example, collegiality, dynamism, innovation, optimism, responsiveness, intensity, and collaboration. What emerges is a highly interactive and informal culture, but one that is also driven and creative.

Company literature specifically highlights the importance of boundary-spanning interaction:

> Innovation within Sun comes from individuals pursuing their ideas. We deliberately hire individuals who march to the beat of their own drum. But we cannot innovate through individual effort alone. Our success is tied directly to our ability to work as a team, within and across operating companies and corporate functions, that are loosely coupled, yet highly aligned.

Such a progressive culture necessitates an equally progressive building design to promote, rather than inhibit, core features such as dynamism and collaboration.

Sun and flexible working

Since the mid-1990s, Sun has promoted more flexible work practices amongst its employees. This has been the result of a number of factors, such as coping with rapid growth at a time when most of its offices were filled to capacity, the need to contain the escalation of real-estate costs worldwide, and responding to changes in the marketplace that have required the greater use of cross-functional teams and increasing numbers of employees spending more of their time with customers.

The company now distinguishes between two types of employee:

- *assigned personnel*—those requiring a permanent workspace—that is, largely non-mobile, office-based administrative staff, such as those in the human resource and accounting functions;

- *unassigned personnel*—any employee who, due to his or her job function, does not require a permanent, designated desk. These employees are sometimes known as 'nomadics', or 'walkers'.

In this case study, we are particularly interested in the latter of these two types, i.e. the so-called 'unassigned personnel'.

Sun's experience in building flexible working environments

From Sun sites around the world, the company has learnt much in recent years from its experiences of designing flexible offices and of the realities of working in them. Sun now has a number of flexible offices, the first of which was a pilot scheme in Farnborough, in the UK; this was followed by sites in Phoenix, Southfield, Somerset, and Alpharetta, all in the USA. In 1998, construction of Guillemont Park, in the south east of the UK, began; it was designed to replace six Sun offices across the locality and result in Sun's largest flexible office worldwide.

Enabling the Sun culture—architectural responses

In designing a building to embrace rather than inhibit the Sun culture, the architects needed to embody a number of broad features within the physical design of the buildings and site: it had to be user-friendly, comfortable, visually stimulating, human in scale, and relate to the outdoors. Perhaps more importantly, it needed to provide spaces in which to think and focus, and informal places to meet, as well as places in which to have fun, outdoor spaces in which to gather, and a design that promoted interaction and reasons to cross boundaries. Such architectural responses are evident at the new Sun site at Guillemont Park.

Guillemont Park—a new working environment

Guillemont Park has a variety of different types of workspace, to allow individuals and teams to work in a flexible and interactive manner. Employees are able to sit anywhere in the building that best suits their needs on any particular day. This flexibility is facilitated by networking technology—the 'Sun Ray' hot-desking architecture—which allows employees to log on to the system from any desk in the building via a smartcard, connecting them to their files and applications located on a central server.

Workspace types within Guillemont Park include:

- *neighbourhoods*—office space is divided into areas called 'neighbourhoods'; these incorporate different groups of employees selling to a common marketplace, to enable them to work together and thus encourage an integrated solution for the customer. These neighbourhoods are unstructured and boundaryless, and employees are encouraged to work wherever their tasks are most easily and effectively completed. Each neighbourhood has its own 'neighbourhood manager';

- *cockpits*—an individual workspace within the open plan areas;
- *hot spaces*—'hot spacing' is the concept of using whatever unassigned workspace is available, whenever it is required, whether on an hour-by-hour or day-by-day basis. It usually applies to mobile staff who work away from the office for much of the time and do not need a personal, permanent workspace;
- *group assigned areas*—an area in which a specific group of employees has priority, but where individuals in the group 'hot-desk' within that area day by day. This type of space is commonly for groups that, as a whole, are based at the office for the majority of their work time; and
- *media stations*—these areas incorporate a range of multimedia, such as high-specification scanners, printers, etc.

Employee on-site facilities also include a gym, a 'well-being' centre, a concierge, retail outlets, and a restaurant. Guillemont Park is also set in a landscaped setting, consisting of trees, lakes and watercourses.

Questions

1. In what ways might the physical architecture of a traditional office building hinder creativity and innovation?

2. Drawing upon examples from the case study, in what ways might the physical architecture of an office building be designed to foster creativity and innovation?

3. Can you see any potential problems or issues that might arise with such an open and flexible working environment as that highlighted in the case?

CASE STUDY 7.2 W. L. GORE & ASSOCIATES—A UNIQUE CORPORATE CULTURE

Introduction

Gore & Associates is perhaps best known for GORE-TEX®—a high performance waterproof, windproof, and breathable fabric; it is to outdoor wear and leisure wear what the Intel micro-processor is to the PC. In fact, the Gore product range, '*designed to be the highest quality in their class and revolutionary in their effect*', provides innovative solutions in electronic and electrochemical materials, cables and cable assemblies, and medical and surgical products.

Despite its humble beginnings—originally set up in the basement workshop of the founder's house in the late 1950s—Gore & Associates is today a multinational enterprise with 6,000 employees (known as 'associates'), facilities in forty-five locations around the world, and annual sales worth around US$1.5bn. But what makes Gore & Associates stand out from other innovative, and technically and commercially successful, enterprises is its unique corporate culture and its mode of organizing.

The following is taken from the company's website:

> How we work sets us apart . . . Our founder, Bill Gore created a flat lattice organization. There are no chains of command nor pre-determined channels of communication. Instead, we communicate directly with each other and are accountable to fellow members of our multi-disciplined teams.[123]

Given the central importance of Bill Gore to both the technical direction and organizational form of the company, it is useful to reveal some of his background to understand their origin.

From polymer scientist to entrepreneur

During the 1940s, Bill Gore had worked as a polymer scientist in the research laboratories at DuPont. He had been part of a 'taskforce' of twenty scientists who had been charged with developing new products based on a polymer patented by DuPont in 1937 (polytetraflouroethylene, or PTFE). At around this time, an alternative polymer was developed by another team within DuPont and Gore's group was disbanded.

But Gore continued to work on applications of PTFE in his basement workshop during the evening, weekends, and holidays for another year or so. During this time, he developed a method of coating electrical computer wires with PTFE. DuPont was not interested in his invention, and so Gore decided to leave DuPont and found his own company. Due to the positive experiences that Gore had had whilst working in a taskforce at DuPont, he decided to organize his own enterprise around the taskforce concept rather than along traditional organizational structures—that is, what he terms the 'flat lattice' organization.

A novel mode of organizing

There are a number of key features that characterize the mode of organizing at Gore & Associates:

- *a 'flat' hierarchy*—this refers to the decentralized nature of decision making 'involving those closest to a project' within Gore, and highlights the lack of a chain of command that is present in most other organizations; 'associates' (as opposed to employees) are guided by their sponsor (rather than directed by a boss);

- *a 'lattice' organizational structure*—this refers to the complex structure of the social networks that are fostered by the open culture within Gore; there is no formal organizational chart representing predetermined channels of communication that are also typical of most other organizations. For the lattice organization to be effective and efficient, however, everyone must know each other, and this limits the size of a division's workforce to a maximum of 200 associates;

- *'self-organizing'*—activities are organized by *'voluntary commitments and universal agreement on the objectives'*. What emerges is a dynamic, self-organizing system of innovation, in which *'teams organize around opportunities and leaders emerge'*. This allows Gore & Associates to be very responsive to technological and market opportunities; and

- *'guiding principles'*—associates are guided by four core principles, which centre around fairness to others, freedom to 'grow' and help others to 'grow', freedom to decide and monitor one's own commitments, and the need to consult with others where one's actions might impact the reputation of the company.

Not only has the Gore corporate culture proved to be a successful formula for innovation, it has placed Gore & Associates high in the rankings of the 'best places to work' in the USA, the UK, Germany, and Italy.

Questions

1. How well does the Gore & Associates mode of organizing fit with the prescriptions of academics such as Burns and Stalker, Kanter, and Nonaka and Takeuchi?

2. Given the technical and commercial success of Gore & Associates, does its corporate culture represent a panacea to organizing for innovation?

3. What potential problems might arise from the Gore & Associates style of organizing, for example, in relation to decision making and conflict resolution?

FURTHER READING

1. Given the importance of the work of Burns and Stalker (1994), and Kanter (1985), to our current conceptions of organizing for innovation, it would be useful to review these texts.

2. For a discussion of the organizational challenges and options for the management of innovation within multinational companies, see the seminal work of Ghoshal and Bartlett (1988a; 1988b), and Barlett and Ghoshal (1998).

3. For a discussion concerning the 'ambidextrous' organization, see Tushman and O'Reilly (1996) or Tushman et al. (1997), and for a debate around 'ambidexterity' and 'punctuated equilibrium' as alternative organizational responses to undertaking 'exploration' and/or 'exploitation' activities, see Gupta et al. (2006).

4. For an interesting discussion of recent research concerning innovation, creativity, and workspace design, see the 2007 special issue—49(2)—of the *California Management Review*, and in particular, articles by Allen (2007), Chan et al. (2007), and Elsbach and Bechky (2007).

NOTES

[1] Preface to Burns and Stalker (1966: xxxv).

[2] Dougherty and Hardy (1996: 1144–5).

[3] Daft (2008: 9).

[4] Hall (1999: 30).

[5] Jaffee (2001: 281).

[6] Burns and Stalker (1961; 1966; 1994).

[7] Lawrence and Lorsch (1967).

[8] Ashby (1960).

[9] Mintzberg (1988: 277).

[10] Child (1972: 6).

[11] Rothwell (1983); Rothwell and Zegveld (1985).

[12] Miles and Snow (1978).

[13] Maidique and Patch (1988).

[14] Kanter (1985).

[15] Gupta and Wilemon (1988); Moenaert et al. (1994a; 1994b).

[16] Weber (1947).

[17] Aldrich (1979: 4).

[18] Aldrich (1979).

[19] Lincoln (1982: 27).

[20] Aldrich (1979).

[21] Pugh et al. (1969); Hasenfeld (1972); Haas and Drabek (1973).

[22] Aldrich (1979: 122).

[23] Badaracco (1991).

[24] Prahalad and Hamel (1990).

[25] Dill (1958); Lawrence and Lorsch (1967); Duncan (1972).

[26] Aldrich (1979: 124).

[27] Aldrich (1979: 263).

[28] Cancer Research UK (2004: 3).

[29] Allen (1977).

[30] Curtis et al. (2002: 29).

[31] Burns and Stalker (1994: 5).

[32] Burns and Stalker (1994: 119–25).

[33] Burns and Stalker (1994: 126).

[34] Burns and Stalker (1994: 123–4).

[35] Burns and Stalker (1994: 125).

[36] Spender and Kessler (1995: 37).

[37] Spender and Kessler (1995: 40).

[38] Burns and Stalker (1994: 8).

[39] Kanter (1985: 134).

[40] See <http://www.hp.com>.

[41] See <http://www.hp.com>, or <http://www.hpalumni.org>—a website for HP alumni.

[42] Kanter (1985: 134).

[43] Dong (2002).

[44] See <http://www.lg.co.kr>.

[45] Kanter (1985: 27).

[46] Kanter (1985: 27).

[47] Kanter (1985: 28).

[48] Kanter (1985: 101).

[49] Kanter (1985: 149).

[50] Bate (1994).

[51] Bate (1994: 102).

[52] Bate (1994: 106).

[53] Nonaka and Takeuchi (1995: 3).

[54] Nonaka and Takeuchi (1995: 73–83).

[55] Nonoka and Takeuchi (1995: 79).

[56] Ashby (1960).

[57] Nonaka and Takeuchi (1995: 83).

[58] Nonaka and Takeuchi (1995: 6).

[59] Spender and Kessler (1995: 37).

[60] Kanter (1985: 248).

[61] Quinn (1988a: 123).

[62] Bailyn (1985); West (1987).

[63] Kanter (1985: 248).

[64] Burns and Stalker (1994: 122–3).

[65] Duncan (1976).

[66] Tushman et al. (1997).

[67] Spender and Kessler (1995: 40).

[68] Burns and Stalker (1961; 1966; 1994); Lawrence and Lorsch (1967).

[69] Burns and Stalker (1994: 122).

[70] Hellgren and Stjernberg (1995); Nobelius and Trygg (2002).

[71] Spender and Kessler (1995).

[72] Quinn (1988a).

[73] Quinn (1988a: 133).

[74] Thompson (1967).

[75] Lawrence and Lorsch (1967).

[76] Tushman et al. (1997).

[77] Tushman et al. (1997: 14).

[78] Hannan and Freeman (1984); Baum (1996).

[79] Tushman and O'Reilly (1996).

[80] Miller and Friesen (1984).

[81] Leonard (1995).

[82] Clark (2000).

[83] Clark and Staunton (1989: 40).

[84] Tushman et al. (1997).

[85] Tushman et al. (1997: 6).

[86] Nadler and Tushman (1988: 471).

[87] Lawrence and Lorsch (1967).

[88] Nadler and Tushman (1988).

[89] Nadler and Tushman (1988: 471–2).

[90] Nadler and Tushman (1988: 475).

[91] Nadler and Tushman (1988: 476).

[92] Burns and Stalker (1994: 170).

[93] Tushman and Katz (1980).

[94] Allen (1977: 61).

[95] Nadler and Tushman (1988: 477).

[96] Kanter (1985).

[97] Nadler and Tushman (1988: 479).

[98] Nadler and Tushman (1988: 480–1).

[99] Kanter (1985: 168).

[100] Nadler and Tushman (1988).

[101] Nadler and Tushman (1988: 481).

[102] Nadler and Tushman (1988: 482).

[103] Nadler and Tushman (1988: 483).

[104] Allen (2007); Allen and Henn (2007); Chan et al. (2007); Elsbach and Bechky (2007); Toker and Gray (2007).

[105] Davenport (2005).

[106] Curtis et al. (2002: 26).

[107] Curtis et al. (2002: 26).

[108] Skyrme (2000: 185).

[109] Curtis et al. (2002: 26).

[110] See <http://www.nielstorp.no>.

[111] Becker and Sims (2000: 39).

[112] Skyrme (2000).

[113] Davenport (2005).

[114] Becker and Sims (2000: 9).

[115] Becker and Sims (2000: 11).

[116] Becker and Sims (2000).

[117] Davenport (2005).

[118] Davenport (2005).

[119] Becker and Sims (2000: 49).

[120] Kanter (1985: 248).

[121] See <http://www.sun.com>.

[122] See <http://www.sun.co.uk>.

[123] See <http://www.gore.com>.

8

Managing the Innovation Process

Chapter overview

Learning outcomes

This chapter will enable the reader to:

- outline the key stages and activities of the innovation/new product development (NPD) process, and appreciate the degree to which these are generic or vary across different sectors and between different types of innovation;

- identify the key 'go/no-go' evaluation criteria employed during the life of the innovation/NPD process;

- recognize the range of factors that impact the creative and innovative performance of teams;

- understand the key stages of the Nonaka and Takeuchi model of knowledge creation and sharing within organizations; and

- provide a critique of the key assumptions underlying the 'mainstream' literature on the innovation/NPD process, and of the knowledge creation and transfer processes.

8.1 Introduction

There is a high proportion of 'failure' and 'dropout' as innovations progress from original idea to market launch (studies reveal failure rates ranging from 30 per cent to as high as 95 per cent),[1] which is a sobering fact for both project and senior managers. Given this failure rate, and the associated development and marketing costs to the innovating organization, it is not surprising that a major area of investigation in innovation studies has been centred on revealing 'success factors'. Indeed, this concern was at the heart of many of the earliest studies of innovation, such as *Project SAPPHO* and *Wealth From Knowledge*.[2] We have already spoken of the importance of external networking and the sourcing of external inputs in the innovation process (**Chapters 1** and **3**), and of an appropriate strategy (**Chapter 6**) and organizational context (**Chapter 7**). This chapter focuses on the innovation process itself, which is often referred to as the 'new product development' (NPD) process—two terms that will be used interchangeably throughout this chapter.

In discussing the innovation process, we draw upon a range of literatures, including those of marketing, work and social psychology, knowledge management, and innovation studies, to provide an overview of three core areas of interest to the management of innovation within organizations:

- the management of the innovation process, from idea generation through to implementation and diffusion;
- the management of innovation teams; and
- the management of knowledge creation and knowledge transfer.

Key questions that we will address in this chapter include:

- how do successful innovators generate innovative ideas and select 'winners' rather than 'losers'?
- how are successful innovation projects managed?
- how are successful innovation teams managed?
- how do successful innovators generate and share new knowledge?

There are many and varied reasons why an innovation might fail. For example, a lack of top management support leading to under-resourcing, a lack of cross-functional interaction leading to expensive designs or delays in commercialization, or the development and launching of innovations with minimal or unclear advantage over existing offerings, leading to poor market performance and slow diffusion. Many of the causes of failure can, however, be reduced dramatically by a more robust management of the innovation process, such as through the systematic and iterative use of idea screening techniques to 'kill off' weak ideas as early on as possible, the use of cross-functional teams, the employment of sophisticated techniques for idea generation and problem solving, and the structured involvement of users in the testing of concepts and prototypes to identify 'problems' that can be corrected prior to market launch.

Indeed, for Tidd et al.:

> The real test of innovation success is not a one-off success in the short-term but sustained growth and through continuous invention and adaptation. It is relatively simple to succeed once with a lucky combination of new ideas and receptive market at the right time—but it is quite another thing to repeat the performance consistently . . . In our terms, success relates to the overall innovation process . . .[3]

In this chapter, we draw upon a range of studies to outline the important routines and techniques employed in the innovation process of successful 'serial' innovators—these do not in themselves ensure success, but they can help to reduce the probability of failure. But it is important to remember that these 'lessons' of successful innovation are necessarily generalized in nature. There is some evidence to suggest that success factors are not universal; rather, they are likely to be contingent upon the complexity and novelty of the innovation, and to vary to some extent between sector.[4] Our discussion takes a middle path, by highlighting a number of such major variations, whilst at the same time seeking to provide a clear and coherent overview of the innovation process.

This chapter can be broadly divided into three distinct sections: the first reviews the various ways in which the innovation process may be modeled (**8.2**) and outlines the generic stages, activities, techniques, and decision criteria employed during the innovation process (**8.3**); the second focuses on the management of innovation teams (**8.4**); and the third centres on the management of the knowledge creation and transfer processes (**8.5**).

8.2 An overview of the innovation process

8.2.1 The innovation process and organizational routines

An 'organizational routine' is a sequence of actions that has been 'learnt' through experience. This experience is stored in the 'memory' of the organization within formal structures, procedures, and processes, as well as in informal structures and conventions, such that it can be evoked given the appropriate stimuli. The knowledge relating to organizational routines is highly tacit, yet the embedded nature of routines allows them to exist independently of specific individuals. Furthermore, routines are both 'firm-specific' and 'context-specific'. The triggering of a routine requires the recognition of the internal or external stimuli and the initiation of the 'unfolding' of the appropriate sequence of actions with which it is associated. In this sense, it involves a process of problem recognition, rather than problem solving.

First introduced by Cyert and March in the early 1960s,[5] the concept of organizational routines has proved to be insightful and resilient, continuing to inform debate on the behaviour of organizations.[6] The DNA metaphor is often invoked in relation to the embedded nature of organizational routines,[7] but, as Clark cogently argues, '*routines are neither replicated or replicators*', because they are subject to adaptation and reinterpretation.[8]

In relation to the management of innovation, organizations may draw upon a range of organizational routines regarding the scanning of the environment for opportunity, the assimilation of knowledge and technology from external sources, and the monitoring and controlling of the innovation process, for example. But to reinforce the point made in the above paragraph, such routines are both 'firm-specific' and 'context-specific'. Thus, Tidd et al. argue:

> Successful innovation management routines are not easy to acquire. Because they represent what a firm has learned over time, through a process of trial and error, they tend to be firm specific. Whilst it may be possible to identify the kinds of things which 3M, Toyota, Hewlett-Packard or others have learned to do, simply copying them will not work. Instead each firm has to find its own way of doing things—in other words, developing its own particular routines.[9]

Nevertheless, research concerning the management of the innovation process provides 'helpful clues' as to the identification and nature of the key routines that distinguish between success and failure. During this chapter, we will explore the set of integrated routines that constitute the NPD process of successful innovators.

Illustration 8.1 Building idea screening routines through experience in a small and innovative software company

We all have a pretty good understanding of the kind of characteristics that an appropriate idea would have. And that comes from doing this for several years. We've gradually built experience up. So if a new idea comes up, we can almost instantly know whether it looks like a fit or not . . . so ideas get filtered very quickly now. In the old days, we had lots of ideas, and we didn't know well enough what was good and what was bad, and we wandered too much . . . Today we have an appreciation of a combination of what we're good at, and what the market would need. Having far too many ideas floating around, that's not a problem. It's more a matter of deciding which ones to go for.

(Founder of Intelligent Applications Ltd)[10]

8.2.2 Models of the innovation process

The innovation process has been conceptualized and represented in many different ways since the 1950s, with major contributions from the marketing and operations management literatures. In these literatures, in particular, the innovation process is generally referred to as the 'new product development' (NPD) process, and, as the term suggests, it has largely focused on the development of new products rather than services. Nevertheless, these models are generic in nature, and thus have relevance to both service and process innovation.

A useful typology for categorizing these approaches is provided by Mike Saren.[11] From this typology, the most common categories of approach across which you will come are as follows:

- *'Department stage' models*—these are typical of the early representations of the NPD process where individual stages of the process broadly map onto functional departments, such as development, manufacture, marketing, and sales. These are exemplified by the linear models of innovation (i.e. the technology-push and market-pull models) discussed towards the beginning of **Chapter 3** (see **Figure 3.1**). In such models, innovation activities within each department are distinct and separate, so that the innovation is passed 'over the wall' from one department to the next. The lack of boundary-spanning communication in such an over-the-wall process is often found to lead to problems in functional coordination and frequently to delays. Delays can arise, for example, when difficulties in manufacturing a product design are only picked up when the design is passed from the product development department to the manufacturing department.

- *'Activity stage' models*—a common approach for representing the NPD process is to relate individual stages to the activities of the process, such as idea generation, idea screening, and concept testing, for example. Early versions of this approach tended to perpetuate the 'over the wall' mentality of the department stage models, but nevertheless began to represent the process as iterative, with the inclusion of 'feedback loops' between individual stages. Later versions of this approach, from the late 1980s, began to emphasize activities run in parallel—otherwise referred to as 'simultaneous engineering' or 'concurrent engineering'.

- *'Decision stage' models*—another common approach is to represent the NPD process as a series of decisions, such as 'is the idea worth considering?', 'does the idea match the strategy and resources of the organization?', and 'have we developed a technically and commercially viable product?'. Thus, the NPD process incorporates a series of 'go/no-go' or 'drop' decision points.

- *'Network' models*—this approach began to emerge in the 1980s. Early on, it is represented by the 'coupling' model of innovation, and later, by the 'interactive' model of innovation (refer to discussion towards the beginning of **Chapter 3** and in **Figure 3.2**). These models highlight the interaction, integration, and coordination of activities and actors involved in the innovation process, and emphasize the importance of both intra- (internal) and inter- (external) organizational relationships and networks in the innovation process.[12]

Despite the efficacy of the network model of innovation in capturing the complexity of the innovation process, the model most commonly used and cited in the literature would appear to be a combination of the activity stage model and the decision stage model;[13] this represents a rational model of the innovation process. Whilst such a model is a fairly simple representation of the NPD process, it does serve to highlight a number of issues that are in need of discussion:

- the activities that constitute the innovation process;

- the nature and concept of decision points or stage gates, through which the idea must pass, and the associated selection criteria; and

- the ways in which the NPD process may be accelerated.

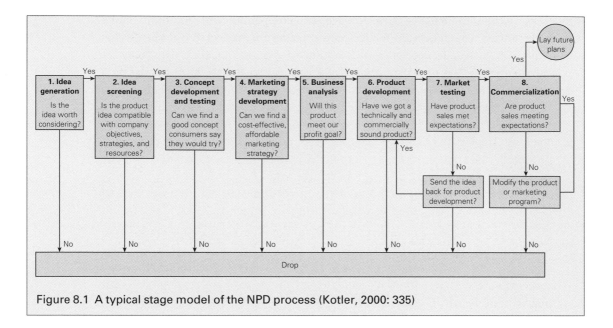

Figure 8.1 A typical stage model of the NPD process (Kotler, 2000: 335)

8.2.3 A rational stage model of the innovation process

In the combined 'activity-decision' stage model, the NPD process is conceptualized and represented as a linear series of activity stages, with feedback loops between stages. Each stage is connected to the next by a decision point, or 'stage gate', through which the process must pass before moving forward. Although the number and labelling of the activity stages and stage gates may vary between the various NPD models of this type, there is reasonable similarity between them—generally being made up of around six–eight activity stages, and separating many of the activities as in **Figure 8.1**. But the largest differences occur between those stage models that have emerged from different perspectives. From the marketing perspective, for example, stage models often include 'marketing strategy development' and a greater emphasis on commercialization (as in **Figure 8.1**). In contrast, and perhaps not surprising, from a design or engineering perspective, there is a greater emphasis on the core development activities.[14] **Figure 8.1** is based on a widely available example developed by Kotler and adopts a marketing perspective.[15]

In this approach, the overall NPD process may be visualized as a 'development funnel', in which the original pool of ideas is repeatedly screened and filtered at each stage, in a gradual process of uncertainty reduction until one or a small number of innovations emerge to be launched into the marketplace.[16] This gradual process of screening and filtering is important, because, as Quinn contends, '*innovation is probabilistic*'—that is, it is difficult to be certain early on in the NPD process which of the pool of ideas is likely to be most successful.

In his study of a sample of large, innovative US enterprises, Quinn found that the most innovative firms consciously initiated several competing prototype projects to proceed in parallel during the product development stage. The competition between projects was seen to yield motivational benefits for the project teams and informational benefits for senior managers in the selection process. 'Shoot-outs' between competing approaches in such organizations might not occur until advanced prototypes have been developed.[17] Regarding the use of parallel projects, Quinn argues that:

> Although anathema to many who worry about presumed *efficiencies* in R&D, greater *effectiveness* in choosing a right solution can easily outweigh 'duplication costs'. This is likely to be the case when the market genuinely rewards higher performance or when large volumes justify increased technical or cost sophistication. Under these conditions competing approaches can both improve probabilities of success and decrease development times . . .[18]

Kotler proposes that, as an idea moves through the stages of the NPD process, management should constantly revise its estimate of the overall probability of the success of the idea using the following formula:

> overall probability of success = probability of technical completion × probability of commercialization given technical completion × probability of economic success given commercialization

Through monitoring the overall probability of success of an idea throughout the NPD process, management can be alerted to a movement in this probability. Where the overall probability of success of an idea falls below an accepted threshold within the organization, management must actively consider the possibility of terminating the project.[19]

As a project progresses through the different stages of the NPD process, it imposes different requirements on the organization's management and resources. Typically, increasing amounts of resource, such as skilled labour and finance, are required. Also, as the emphasis shifts between technical, market, and financial considerations, so the balance in functional resource will shift. For example, in the development of a new product, one would expect the dominant functional resource during the business analysis, product development, and market testing stages to shift between the finance, R&D, and marketing departments, respectively. Research has also indicated that whilst the ability to shape and influence the design of an innovation declines rapidly during the product development phase, the involvement of senior managers in the innovation process is often heavily weighted to the latter stages, such as commercialization, and for manufactured products, the 'tooling up' of production, where the possibilities for shaping the innovation are much reduced and/or expensive to implement.[20]

Much of the emphasis in the NPD literature has been on understanding the process and success factors in the development of products rather than services. What research on service innovation that has been undertaken suggests that the development process is largely informal or ad hoc in nature.[21] Indeed, from their in-depth, cross-sector study of the US service industry, embracing consulting, retailing, financial services, and hospitality, Martin and Horne found that around 70 per cent of their sample of 217 service firms did not have a 'new service development' strategy, and around 60 per cent did not have a formal development process. Furthermore, and in stark contrast to empirical evidence concerning the new product development process, neither the absence of a development strategy nor the informality of the innovation process had a significant impact on the success or failure of the resulting service innovation.[22]

One way in which they attempt to explain some of these differences is in relation to the customizable nature of services:

> a customized new service is by definition never really out of the developmental stage. It is always being modified. The entire product development process is geared towards being able to mass-produce economical orders of nearly identical goods, an assembly line mentality. As services move away from that logic, perhaps the entire service development process becomes more entrepreneurial.[23]

But Martin and Horne did reveal some similarities in the development of successful new products and services: firstly, the fit of the new offering with the existing businesses of the organization, and secondly, the involvement of users and customers in the innovation process. Thus, whilst there are significant differences in the process of developing a successful new product as compared to that of a successful new service, there are also some noteworthy similarities.[24]

8.2.4 'Stage' gates and 'go/no-go' decision criteria

In **Figure 8.1**, each of the NPD process stages is concluded by a 'yes' ('continue', or 'go') or 'no' ('drop', or 'no-go') decision point—as noted above, these are commonly referred to as 'stage gates'.[25] Increasingly, stage models of the innovation process can be seen to incorporate such stage gates, although many representations and studies of the NPD process continue to neglect these 'go/no-go' decision points.[26] The stage-gate approach allows for the systematic screening, monitoring, and progression of the NPD process, and, as such, improves the management of risk through the life of the project. A recent study of best practice revealed that nearly 60 per cent of organizations employ a stage-gate approach in their NPD process.[27]

The decision to continue or drop (i.e. stop) the development of a new product or service is based on an evaluation of the performance of the project at each of the stage gates. Such decisions are generally made by either middle or senior management. Each of the stage gates requires the collection of specific information that allows the project to be evaluated against a set of criteria that are specific and appropriate to that stage gate.[28] Whilst there has been much research that has focused on identifying the range of evaluation criteria employed during the life of a project, studies have tended not to focus on the criteria and their weightings utilized at specific stage gates.

A recent study by Hart et al. has, however, thrown some light on the evaluation criteria and their weightings, utilized at specific stage gates. For example, in their study of project evaluation criteria from a sample of 166 managers in Dutch and UK manufacturing companies, they found that:

1. 'technical feasibility' and 'product uniqueness' criteria were important in the idea generation and concept testing stage gates;

2. 'market acceptance' criteria were important throughout the NPD process, but particularly at the commercialization stage gate;

3. 'product performance' criteria were especially important at the product development and market testing stage gates; and

4. 'financial' criteria were important at the business analysis and commercialization stage gates.

Hart et al. also found little discernable difference in the usage of evaluation criteria in relation to market share, organization size, innovation posture, or between the Dutch and UK organizations in their sample.[29]

Nevertheless, other research has indicated that the overall importance of different evaluation criteria may vary in some sectors. For example, in their cross-national study of the grocery sector, Stagg et al. found that evaluation criteria concerning branding, distribution, and promotions were of greater importance as compared to earlier studies.[30] This is perhaps not surprising given the relative importance of these factors to the grocery sector as compared to other sectors.

Selecting the 'right' criteria or metric with which to measure performance at each stage gate is an important precursor to project managers making the 'right' decisions. And making the 'right' decisions is increasingly important, given the rising costs of developing new products and services, and the detrimental impact of committing organizational resources to the 'wrong' project.

8.2.5 'Fuzzy' gates

Whilst the stage-gate approach to the NPD process enables greater management control over the progression of a project, it can have a number of negative impacts, for example:

- it can slow the progression of a project, and hence increase the time to market for the new product or service. This is because the stage-gate approach requires projects to wait at a gate until all of the required tasks have been completed and thus prevents the overlapping of activities from different stages;

- it can reduce the smoothness of progression of a project, through the need to wait at stage gates; and

- it can create an overemphasis on evaluating projects and teams in relation to the completion of individual stages of the NPD process, which can lead to a loss of sight of the more holistic and higher-level criteria. This is a key problem in the management of what Van de Ven refers to as 'part–whole' relations—that is, *'impeccable micro-logic often creates macro nonsense, and visa versa'*.[31]

One solution to these problems has been the introduction of 'fuzzy gates', which allow 'conditional go' decisions. These provide a greater degree of fluidity within the NPD process by allowing the overlapping of stages. This is sometimes referred to as a 'third-generation' NPD process[32]—the first generation being the introduction of the stage model itself (during the 1960s), and the second generation involving the addition of stage gates (during the 1980s).

Table 8.1 The range of evaluation criteria employed by organizations during the NPD process (adapted and summarized from Stagg et al., 2002: 463–70)

Evaluation criteria category	Examples of specific criteria for evaluating innovation
Product differential advantage	1. It is unique to the marketplace. 2. It will be superior to existing offerings. 3. It is patentable.
Product promotion	1. It will have a clear 'unique selling proposition'.
Product characteristics	1. Its specifications are clear. 2. It has a variety of applications.
Corporate synergy	1. It will fit the organization's current businesses. 2. It will employ existing skills and resources. 3. It will have senior management support.
Trade synergy	1. It will be easy to get 'trade' support.
Nature of the market	1. It will have few competitors. 2. It will have a mass market. 3. It will be subject to legislation.
Competitive and market intelligence	1. We understand the market for this innovation. 2. We know how competitors will react.
Financial potential	1. ROI potential is high. 2. Sales growth potential is high. 3. Market share potential is high. 4. It will offer high profit margins. 5. It has a high probability of success.
Market strategy	1. It will enhance the organization's reputation. 2. It will yield a new range of offerings. 3. Its launch will make it difficult for competitors.
Product branding	1. An existing brand could be employed.
Market acceptance	1. It will satisfy identified customer needs. 2. It will be consistent with customer values.

A further advantage of the fuzzy-gate approach is outlined by Jones, who argues that by:

> being more outcome rather than task focused . . . [fuzzy gates have permitted] organiza-
> tions to build prioritization models and, through a shift in authority away from phase
> reviewers towards programme managers, enabled projects to move through the process
> with more flexibility than was possible using more rigid stage gates.[33]

There are, however, disadvantages as well as advantages in organizations adopting the fuzzy-gate approach: for example, mistakes being made due to the skipping of steps, inefficiencies arising from variations in the flexibility of different stages, and the increased complexity of resource allocation.[34]

8.2.6 The 'fuzzy front end' of the NPD process

The majority of the research on the NPD process has focused on the activities and decisions that occur at, or following, the point at which an innovation project has been established—that is, the period leading up to, and following, the 'bureaucratic release' of a project.[35] As a result, there is a paucity of research centred on the embryonic stages of the innovation process—what is sometimes termed the 'fuzzy front end'. More specifically, the 'fuzzy front end' refers to the very earliest phases of the innovation process, prior to the establishment and formalization of the innovation project and team. Reid and Brentani describe this initial phase as roughly denoting '*all time and activity spent on an idea prior to the first official group meeting to discuss it*' and embracing the '*process of identifying, understanding, and acting on emerging patterns in the environment*'.[36]

The activities in the fuzzy front end can be broadly grouped into 'early' and 'late' activities. 'Early' activities include problem and opportunity recognition and structuring, and information gathering and exploration, whilst 'late' activities include initial idea generation, concept development, and pre-screening. This embryonic stage of the innovation process is seen as disproportionately important, because the fuzzy front-end activities and decisions can:

- determine the direction or path of the subsequent development process;[37]

- provide the greatest time savings at the least expense of any stage of the innovation process;[38] and

- improve the eventual market performance of the resulting innovation.[39]

But there are fundamental differences during the fuzzy front end for incremental and discontinuous innovations.

With regard to incremental innovation, Reid and Brentani argue that '*organizations are aware of and are involved in the NPD process from the project's beginning . . . structured problems or opportunities typically are laid out at the organizational level and are directed to individuals*'.[40] This fits with the process outlined by Spender and Kessler, who argued that the organizational-level planning stage was the first phase of the innovation process, leading on to the formal initiation (i.e. 'bureaucratic release') of the innovation project (refer back to **7.6.2**).[41] In contrast, Reid and Brentani propose that, during the fuzzy front end of discontinuous innovation:

> information typically is unstructured and is brought into the organization by individuals without such activity being explicitly directed by other persons in the organization. Venture groups in organizations may be directed generically to 'find something new'; however, the problems or opportunities are not identified or structured by the organization . . . As such, the role of individuals [during this phase] takes on heightened importance . . .[42]

Thus, with discontinuous innovation, the fuzzy front end can be seen to proceed the organizational-level planning stage noted above.

Despite the importance of discontinuous innovation to organizations and the importance of the fuzzy front end to discontinuous innovation, however, there remains a great deal of uncertainty among both academics and practitioners regarding the activities and decisions of the 'fuzzy front end' of the innovation process.[43] Some insights have been provided, for example, in relation to communication at the R&D–marketing interface;[44] this aspect is discussed in **Chapters 7** and **9**.

8.2.7 Accelerating the NPD process—increasing speed to market

So far in this chapter, we have discussed the importance of 'good' management of the innovation process, through the adoption of a formalized NPD process incorporating stage gates, as well as the importance of 'good' decision making, through the use of appropriate evaluation criteria and techniques. Research has, however, also highlighted the importance of the 'timely introduction' to the market for the success of new products and services.[45]

There are a number of drivers behind the need to accelerate the NPD process, such as the emergence of increasingly competitive markets and the shortening of product life cycles, as well as the need for organizations to recuperate the escalating costs of innovation. Thus, the 'speed to market', sometimes referred to as 'time to market', is also a key objective in the management of innovation. The timely introduction of an innovation requires not only a focus on reducing the time to develop a product or service, but also a focus on reducing the time to launch and roll out an innovation.[46]

There are a number of ways in which the NPD process may be accelerated, for example:[47]

- encouraging the role of the product champion;
- where appropriate, conducting activities in parallel;
- discouraging the spreading of resources thinly across too many projects;
- the decentralization of decision making, to encourage team commitment and ownership, whilst reducing excessive external interference;
- better integration of suppliers with the NPD process and innovation team (we discuss this issue in **10.5**);
- the early involvement of (potential) customers in the NPD process (we discuss this issue in **10.3**); and
- employing enabling technologies, such as computer-aided design (CAD) and stereolithography, to allow early visualization and rapid prototyping of the innovation.

But there is some evidence to suggest that the factors that enable an acceleration of the NPD process are, to some extent, contingent upon the nature of the project.[48] For example, in their exploratory study of the 'speed to market' of product innovation in ten large US companies, Kessler and Chakrabarti found a number of differences, as well as similarities, in the factors that impacted the speed of development of radical versus incremental innovation. With regard to the differences, they found:

> Under some circumstances, project managers should clearly specify the product concept (radical project) and at other times allow for some vagueness (incremental project). In some cases it is functional to appoint a variety of low-level champions (radical project) and in other cases to appoint fewer but more powerful champions (incremental project). Sometimes assigning a project leader higher (radical project) or lower (incremental project) in the organizational hierarchy will speed up projects. Sometimes it is faster to co-locate teams and test products frequently (radical projects); other times it may be faster to spread out teams and limit testing (incremental project).[49]

In providing an international dimension, Wong notes that the timely launch and roll out of new innovation across multiple countries is a particular problem for organizations:

> The need to meet diverse customer and subsidiaries' needs and expectations in different countries presents a substantial challenge for product managers. Not least, product, technical and marketing criteria may have to be adapted to meet each country market's requirements prior to commercialization on an international scale. In addition, product managers have to manage and co-ordinate NPD activities among headquarters and subsidiaries and agents in different country markets.[50]

Recent research concerning the international roll-out of new innovations has identified a number of important drivers that impact their timeliness. These drivers include appropriate levels of marketing and technical resource, project integration and proficiency, and strong coordination between the headquarters and the subsidiaries or agents in the target countries.[51]

8.3 The stages of the innovation process

We now turn our attention to a more detailed discussion of the individual stages of the innovation process.

8.3.1 Idea generation

The generation of high-quality ideas is critical to the NPD process, because the effectiveness of subsequent stages is limited by the potential promise of the initial ideas. In many case studies of successful invention and innovation, the origin of the idea is attributed to serendipity—that is, a chance meeting, a chance conversation, a chance result from an experiment, or a chance connection of otherwise disparate information: a somewhat mysterious process, little short of alchemy.

Illustration 8.2 highlights the importance of extensive social networks and serendipity to the idea-generation process.

Illustration 8.2 'Casting the net widely'—serendipity and the emergence of ideas and opportunities from networks

Networking is absolutely vital. In all our successful innovations it's been a combination of networking and serendipity . . . We actually have huge networks . . . and we've had just a few cases where two nodes have basically come together, and something has popped out of it . . . it's unpredictable where out of the network the opportunity arises. So you've just got to cast the net and not be too judgemental, and see what happens . . . *The trick to being in the right place at the right time is to be in a lot of places!*

(Founder of Intelligent Applications Ltd)[52]

In contrast, a number of researchers have argued that idea generation is best served by a 'continuous and systematic' search process, rather than a reliance on serendipity, and guided by the strategy of the organization[53]—that is, a system that is connected, proactive, and organized, rather than fragmented, reactive, and ad hoc. That is not to say that idea generation is an activity that should be limited to a small number of people within the organization; on the contrary, ideas can originate from anywhere both within and outside of the organization, and thus what is needed is a way of capturing and directing these ideas. Indeed, Sowrey argues that:

> instead of individual development decisions being taken on an opportunistic *ad hoc* basis . . . a system is required in which ideas are generated through an organized network with a central collection point. This network should cover both inside and outside the company and utilize a variety of idea collection methods and techniques.[54]

Although it is possible that an idea may be presented 'fully formed' to an organization, perhaps by an employee, a customer, a supplier, or an independent inventor, typically, the idea-generation stage will involve a number of distinctive tasks, such as:

- scanning of the environment for both market and technology 'signals';
- the collection of relevant market and technical data and information;
- data analysis; and
- the generation of ideas for potential new products, services, or processes.

Thus, although creativity is at the heart of the idea-generation process, data collection and data analysis are also important components. Indeed, environmental scanning for a combination of market and technology signals is vital (this links into our discussion of the 'interactive' model of innovation in **Chapter 3**). Furthermore, given the principle of requisite variety (noted in **Chapter 7**), such scanning is better achieved by making the activity a responsibility of all employees within the innovative unit, rather than relying on a few boundary-spanners.[55]

Table 8.2 provides an overview of the breadth of techniques that can be employed to generate ideas in a more systematic manner. These techniques can be grouped according to the orientation from which they draw inspiration, for example, the marketplace, existing products and services, and competitors.

In contrast to research concerning product innovation, there is empirical evidence to indicate that service organizations rely heavily on 'competitive imitation' and 'customer canvassing' during the idea-generation stage, rather than on sophisticated idea-generation

Table 8.2 An overview of the breadth of idea-generation techniques (adapted and summarized from Sowrey, 1987)

Orientation of technique	Examples of techniques
Market-orientated	1. *Repertory grid*—involves the identification of 'constructs' employed by users to distinguish between products. Commonly used constructs can be mapped to identify gaps and new possibilities. 2. *Segmentation analysis*—focuses on identifying distinct market segments by clustering the market through a variety of different user characteristics. Identifying distinctive market segments can aid the process of generating distinctive ideas to service such segments. 3. *Activity analysis*—involves the study of tasks and activities undertaken by users, and user attitudes in undertaking these tasks. This helps to identify user needs, which can initiate the search for new products and services to fill these identified needs.
Product-orientated	1. *Attribute listing*—involves the systematic listing of the attributes of all components of an existing product or service. Each attribute of each component is then taken in turn, reconceived of in a variety of ways, and the changes related back to the overall product or service. 2. *Forced relationships*—relies on the creation of an artificial or forced link between two or more unrelated products, services, or ideas as the starting point for the generation of new ideas.
Creativity-orientated	1. *Brainstorming*—a common technique employed to generate a large quantity of ideas within groups. Creativity is stimulated by the 'chain reaction' of ideas through interaction within the group, the encouragement of 'wild' ideas, and the deferment of judgement. 2. *Lateral thinking*—ideas are arrived at through the attempt to break free from current ways of thinking, by restructuring existing patterns of information in the mind, and by exploring alternative and novel ways of viewing a situation, problem, or object, for example.
Competitor-orientated	1. *Benchmarking*—involves learning and idea generation through the structured comparison of the products, services, and processes of an organization with those of its competitors, or perhaps those outside the sector that are 'world class' in relation to a particular activity. 2. *Reverse engineering*—a method of gaining in-depth information of the offerings of competitors, such as detailed technical specifications. Typically, this involves the 'de-assembly' of the products of competitors. This can allow imitation or improvement through adaptation.
Employee-orientated	1. *Company suggestion schemes*—increasingly employed to elicit ideas from employees across the organization. Typically, such schemes are formalized, ongoing, and involve rewards for those whose ideas are implemented. 2. *Job rotation*—the movement or seconding of employees from one function to another, or one subunit to another, is an innovative way to bring about fresh and innovative connections between existing staff, practices and innovations.
Technology-orientated	1. *Delphi model*—this involves the surveying of the opinion of relevant experts with regard to future development areas with the most potential. There may be several, increasingly focused, iterations of the survey in order to ascertain whether there is convergence of opinion. 2. *Theoretical limits test*—experts are invited to forecast the furthest limits to which a technology could be potentially extended. Possible applications are then sought and considered.

techniques, such as the repertory grid and attribute listing.[56] Furthermore, in their cross-sector study of 217 US service organizations, Martin and Horne found that '*most participant managers said they were unfamiliar with these more sophisticated techniques*'.[57]

8.3.2 Idea screening

Just as the idea-generation stage requires a 'central collection point' for ideas, the idea-screening stage requires a 'central evaluation point', which might be at either the organizational or subunit level. Kotler proposes the regular meeting of an 'idea committee', whose role it is to review the range of ideas with which they are presented.

In undertaking this review, he advocates a series of filters for reducing the original pool of ideas down into the most promising. The first filter involves the categorization by the committee of each idea as 'promising', 'marginal', or 'reject'. In the second filter, promising ideas are researched by individual committee members, who subsequently report back to the idea committee. The third filter involves a 'full-scale' screening of the remaining promising ideas, culminating in the selection of those ideas that should progress to the next stage of the NPD process.[58] The screening out of weak ideas as early as possible is important, because costs rise substantially as the idea progresses through the NPD process. Furthermore, progressing with weak ideas diverts resources away from other, more promising, ideas. Despite this, research indicates that organizations often perform poorly in relation to killing off weak ideas.[59]

The 'full-scale' screening of an idea may involve the use of one or more techniques. Where more than one technique is employed, the resulting evaluations may be weighted differently according to organizational preferences and priorities. Idea-screening techniques include, for example:

- *'frontier' models*—the idea is evaluated in relation to its closeness to the 'frontier' of existing offerings in the marketplace: the closer to the 'frontier', the higher the idea will be rated;

- *risk analysis*—the idea is evaluated in relation to a series of different risk categories, such as the technical and cost risks associated with the development process, and the financial and reputational risks following commercialization;

- *'portfolio' models*—rather than evaluating the idea in isolation, it is evaluated alongside the existing portfolio of offerings by the organization: for example, in relation to novelty, risk, and target market segment. In using such a model, the organization will have a preference regarding the balance of novelty, risk, and the target markets of its offerings. In this way, high-novelty and high-risk ideas, for example, may be accepted where the portfolio of offerings is able to absorb the downsides of the potential failure of such ideas; these ideas might otherwise be screened out if evaluated in isolation;

- *customer evaluation*—users are approached to assess the degree to which the idea would meet potential and existing user needs (in **10.3**, we will discuss in much greater detail the important role played by users in the NPD process);

- *'scoring' and 'checklist' models*—such models often incorporate a wide range of criteria against which the idea is assessed, for example, in relation to its synergy with current corporate strategy and resources, the range of possible applications, the nature of the competition within the marketplace, and the potential for return on investment (ROI), market share, and profit margins.

8.3.3 Concept development and testing

Tidd et al. argue that *'innovation is more than simply coming up with good ideas, it is the process of growing them into practical use'*.[60] This leads us on to the next stage of the NPD process—concept development—which involves the translation and elaboration of an idea (e.g. a 'soft' drink with added vitamins) into a concept that is expressed in terms that are meaningful to the end user (e.g. a health supplement for children, a health supplement for busy professionals, *or* an energy boost for the sporty). Kotler argues that this is important, because *'consumers do not buy product ideas; they buy product concepts'*.[61] At this stage, the idea may be translated into a number of different possible concepts (as in the above example), through assessing who will use the product or service, what its primary benefits will be, and the location and context in which it will be used or consumed. The translation of the idea into a concept aids the identification of the market segment, and hence the evaluation and positioning of the concept against current market offerings.

Having established one or more concepts, the objective of concept testing is to present these new concepts to users (or potential users) from the appropriate target market. For incremental innovations, a representative sample is likely to be the most appropriate, whilst for major or discontinuous innovations, it is probably more appropriate for the sample to focus on 'lead users' (i.e. those that recognize new needs or adopt new offerings earlier than the majority in the marketplace—refer to the discussion in **4.7**, concerning the diffusion curve and different 'adopter categories', and **10.3.5**, regarding 'lead users'). The concept may be presented to the user sample either as a physical prototype or in a more abstract form. But as Kotler argues, *'the more the tested concepts resemble the final product or experience, the more dependable concept testing is'*.[62]

A popular tool used in concept development and testing is 'conjoint analysis'.[63] This is a method for assessing the utility that users attach to different combinations of the attributes of a hypothetical product or service, by asking users to rank different hypothetical 'offers'. The results of this analysis can allow product developers to identify the most appealing combination of attributes for a concept. This technique has a wide range of commercial uses.[64] It is important to note, however, that this tool was developed for identifying and evaluating variations of existing innovations, and is therefore of much more limited use for identifying or developing high-novelty innovations or applications.[65]

8.3.4 Marketing strategy development

Following the testing of the concept, the next stage is to produce a preliminary marketing strategy. This is essentially a provisional plan for the launch and commercialization of the concept into the marketplace.

Kotler outlines a number of key components of such a plan:[66]

- a description of the size, structure, and consumption behaviour of the target market;
- the planned positioning of the concept relative to existing offerings in the market;
- the initial marketing budget and distribution strategy;
- the medium-term goals in relation to sales, market share, and profit;
- the 'marketing mix' strategy over time; and
- the longer-term goals in relation to sales, market share, and profit;

The further articulation of the positioning of the concept and the target market required of this stage is an important precursor to the subsequent stages of business analysis and product development. Indeed, research has revealed that the most important criteria for success lies in the uniqueness and superiority of the innovation: those with high product advantage through these criteria were found to succeed in 98 per cent of cases, compared to 58 per cent in instances of moderate advantage, and only 18 per cent in cases of minimal advantage. A well-defined product concept prior to development, and the careful definition and assessment of the target market, product requirements, and product benefits, have also been shown to be important.[67]

8.3.5 Business analysis

The business analysis stage involves the preparation of cost, revenue, and profit projections for each of the concepts that have survived the earlier filtering processes. Costs need to be estimated for the development, launch, and ongoing marketing of a concept, as well as the production or delivery costs of the final product or service. Annual or quarterly sales revenues need to be estimated. Having established cost and revenue estimates, a cash-flow analysis can be undertaken. With these figures, it is also possible to estimate profit levels, the payback period (i.e. the time that it will take to cover the costs of the development of the innovation), and the rate of ROI.

The time period covered by these projections will depend on the 'scale' of the project. For incremental innovations with relatively low development costs (e.g. a new type of chocolate bar), projections may only be required for a two to three-year period. For discontinuous innovation with high development costs (e.g. a hydrogen-powered automobile engine), projections may be required for much longer periods, in order to absorb the high development costs and account for the possibility of the slow diffusion of the innovation. There exist a number of management accounting tools, such as 'sensitivity analysis', that allow for a more sophisticated approach to appraising the financial viability of an idea—but these are beyond the scope of this text.

This stage is equally as important for public sector and not-for-profit organizations, although the process is more likely to focus on the costs of development and provision in relation to the impact and diffusion of the innovation.

The purpose of the business analysis stage is to evaluate the attractiveness of one or more ideas, and hence inform the decision-making process. The organization may wish to finance only one project, or only those that meet a certain threshold with regard to

ROI, for example. Either way, it can help an organization to prioritize the allocation of its limited resources. Thus, at this stage, it is likely that further ideas will be screened out.

8.3.6 Product development

If the concept passes the business analysis stage, then it moves to the heart of the NPD process—the translation of the concept into a new product, process, or service. This stage usually involves a huge increase in investment in the project, over and above that at earlier stages in the NPD process.

Product development involves a number of interrelated activities, as follows.

- *The translation of customer requirements into technical specifications*—the specification needs to indicate the functionality, performance, and styling requirements, and work within agreed cost and production constraints. The detail and complexity of the specification will vary with the nature of the new product, process, or service, with the specifications for new motor vehicles or aircraft, for example, running into many volumes. One technique that has been employed by a number of leading Japanese (e.g. Toyota) and US firms (e.g. Ford and AT&T) is 'quality function deployment' (QFD). This technique provides a framework for the systematic evaluation of customer requirements, allowing for the measurement of trade-offs and costs of different customer requirements, and for the identification of product improvement or differentiation opportunities. Another key role of QFD is the systematic translation (i.e. deployment) of customer requirements into specifications and language that is meaningful to engineers. QFD encourages communication between the marketing, R&D, and production functions, and provides a useful tool for structuring the process of translating customer requirements into technical specifications.[68] On the downside, QFD requires the collation of a large amount of market and technical information, and research has indicated that only 25 per cent of projects employing QFD have enjoyed quantifiable benefits.[69] Furthermore, QFD is perhaps best suited to the development of incremental improvements of existing offerings rather than to discontinuous innovation.

- *The translation of technical specifications into a product design*—design can be seen as a distinct activity within the product development stage, in which the technical specification is 'given specific physical form'.[70] For technical products, design incorporates the configuration of components, the selection of materials, and the overall form and visual order (i.e. the aesthetic). The design process will often involve the building of scale or full-sized models; such models may be used to test the impact of air or water currents on the integrity of the design, for example, although the 'scaling up' of scale models can often highlight unexpected design errors.[71] The design process has been substantially enabled in recent years by major advances in computer simulation and modelling that allow sophisticated and rapid testing of stress and temperature, for example, on alternative designs.

- *Problem solving and the search for solutions*—the processes of product design and development are intertwined and iterative. The process of design will constantly raise questions with regard to technical feasibility, which may highlight the need for problem solving to develop new solutions or lead to the search for alternative existing solutions. Product designers and development engineers must work closely together: the selection of appropriate materials by designers, for example, will require engineers to provide information regarding the strength, flexibility, and corrosive resistance of these materials. Designers must also work closely with production engineers, because the final design can have a substantial impact on production quality, costs, and times. As **Illustration 8.3** highlights, this stage might involve the generation of many possible solutions from which the innovation team must select.[72]

Illustration 8.3 The development of draught Guinness in a can

The 'in-can system' (sometimes called a 'widget') was launched by Guinness in 1988 after four years of development. It has been hugely successful and has subsequently been mimicked or licensed by many other brewing companies. The innovation allowed Guinness served from a can to have the same creamy head as draught Guinness. The impetus for the development had been the growing trend for drinking beer at home and the fact that, at the time, the vast majority of draught Guinness drinkers did not drink the take-home version of the product—Guinness Extra—which did not have the distinctive creamy head.

Research to solve this problem had been 'bubbling' in Ireland for some time. A system had been developed that involved syringing the drink once opened. Although launched in Ireland, it was felt inappropriate for the English market.

The brief for the innovation project included the following:

- the product must be in a can;
- the process must not be unusual to the consumer; and
- the product must be recognized by the consumer as draught Guinness.

Over a hundred techniques were developed and tested in-house. These were narrowed down to only a few of the best ideas, before can manufacturers in the UK and USA were approached. Finally, the 'in-can system' solution was selected.[73]

- *Prototype development and testing*—there are wide variations in what designers and developers may consider to be a prototype, from two-dimensional sketches or three-dimensional computer models (e.g. CAD or virtual reality), right through to fully functional working models. More important than providing a precise definition of what constitutes a prototype is the highlighting of the key roles of a prototype. For Wheelwright and Clark, '*prototyping can be an important vehicle for cross-functional discussion, problem-solving, and integration*'.[74] Leonard concurs, arguing that prototypes are '*one of the most neutral and yet provocative mechanisms for shared problem-solving*'. Indeed, she sees prototypes as important 'boundary-spanning objects', encouraging communication across functional boundaries, between project teams and senior managers, and with users and suppliers.[75] Given this role, the prototype should not be seen as an end in itself; frequently rapid prototyping and the development of 'quick and dirty' prototypes are sufficient. **Illustration 8.4** provides an example of a highly successful innovation that adopted such an approach. Highly refined prototypes might close down new ideas and communication.[76] The testing of prototypes within the organization is often referred to as 'alpha testing', whilst testing by users is often referred to as 'beta testing'. A recent study of fifty projects within twenty-five companies found that prototyping was more appropriate for high-novelty rather than low-novelty innovations.[77] Nevertheless, prototyping is an important tool for speeding up the innovation process, encouraging cross-functional interaction, and eliciting high-quality feedback from users. Indeed, Peters argues: *'Effective prototyping may be the most valuable "core competence" an innovative organization can hope to have.'*[78]

Illustration 8.4 Rapid prototyping and the development of the Netscape Navigator 3.0

After only six weeks into the development project, the team released a prototype version of Navigator 3.0 on an internal website for testing by their colleagues. The 'quick and dirty' prototype captured the essence, although not all of the functionality, of the new version of the software and was able to generate meaningful feedback very early on in the life of the innovation project. Following the internal release of a second prototype and the ironing out of any glitches, a third prototype was released into the public domain around ten weeks into the development. This, and subsequent, prototype versions of the software allowed Netscape to react to feedback from users through much of the NPD process.[79]

The product development stage may involve one or more iterations of the above activities. The more complex and novel the innovation, the more complex this stage is likely to be.

8.3.7 Market testing

The market testing stage involves the introduction of the innovation into an 'authentic' market or application setting, and will thus require both the branding and 'packaging' of the new product, process, or service. Market testing is an important mechanism for obtaining detailed market information, serving to gauge the size of the market, and to ascertain the reaction of users and distributors to the 'consumption' and 'handling' of the innovation. The techniques during this phase of the NPD process are beyond the scope of this text, but there are, however, many introductory marketing texts that can provide you with an overview of such techniques and direct you towards further readings.[80]

8.3.8 Commercialization

There are a number of key manufacturing, distribution, and marketing decisions that need to be made during the commercialization stage of the NPD process. With respect to marketing, decisions are required in relation to 'when' (the timing of the launch), 'where' (the geographical scope of the launch), 'to whom' (target market), and 'how' (the launch strategy). Marketing costs are a major cost during this phase: the marketing costs for new food products, for example, can typically represent 57 per cent of first-year sales.[81] Production set-up costs can also be substantial for new manufactured products, particularly those involving the need for new equipment and buildings.

The marketing, sales, manufacturing, and distribution activities involved in the initial launch and ongoing commercialization of an innovation are, however, also beyond the scope of this text. Again, there are a multitude of introductory marketing texts and introductory operations management texts that can provide you with an overview of these activities, and direct you towards further readings.[82]

The commercialization stage of the innovation is of interest here for a number of specific reasons:

- for evaluating the 'success' of the innovation, for example, through financial criteria (e.g. sales turnover, profit, ROI, payback period), market criteria (e.g. speed of diffusion, market share), and strategic criteria (e.g. competitive advantage)—refer back to the earlier discussion on 'assessing the success of an innovation' in **1.5**;

- for gaining an appreciation of the patterns of diffusion and adoption of an innovation—refer back to the discussion on the 'diffusion curve' in **4.7**;

- extensive research on the role of the user in the innovation process highlights the importance of 'multiple and continuous' interaction with users[83]—for a comprehensive discussion of the 'role of the user in the innovation process', see **10.3**;

- linked to the above point, innovation is viewed as an ongoing, iterative process, where post-commercialization feedback from the marketplace, for example, can trigger and inform the process of re-innovation. This feedback might not only include ideas for new functionality, for example, but might highlight novel ways in which the innovation has been adapted or employed in unforeseen applications—refer back to the discussion on 're-innovation' in **1.6**, and also the discussion on the diffusion curve concerning 'adoption and adaptation' in **4.7**.

With regard to innovations that are adopted internally, read 'implementation' for 'commercialization', and with respect to public sector and not-for-profit organizations (e.g. government departments, universities, and charities), read 'diffusion' for 'commercialization'.

8.3.9 Post-project evaluation

An important, although often overlooked, stage of the innovation process is the post-project audit or review—this is a 'post-mortem' of the process itself. Such an evaluation of what went well and what went badly is an important learning routine within successful innovative organizations, regardless of whether the project was a success or failure.

A post-project audit provides a formal structure for team members to reflect upon the processes, activities, and decisions throughout the life of the innovation project. It is important that the audit involves all team members to promote a multi-perspective evaluation, and adopts an objective, non-judgemental, blame-free forum in which mistakes and problems can be revealed, rather than concealed. The degree to which project objectives were met is an important measure of the success of a project, and thus the evaluation process is aided where the original project objectives were formalized and documented.

Whilst it is important to capture and codify such lessons from in-house innovation projects, however, what really matters is whether these lessons systematically feed into the routines and processes of future projects.[84]

8.3.10 **Reflections on the management of innovation**

A partial understanding of the innovation process—for example, of seeing the development of an idea in isolation rather than as part of a wider system, of viewing innovation as involving only 'breakthroughs' (i.e. discontinuous innovation), whilst ignoring the vast potential of incremental innovation, and of seeing innovation as only being initiated internally whilst rejecting ideas from outside the organization—can be damaging to the management of innovation within an organization.[85] Simple models of innovation are likely to perpetuate such partial understandings. Indeed, there are a number of criticisms that might be leveled at the 'departmental', 'activity', and 'decision' stage representations of the innovation process. Perhaps, most obvious is the oversimplification of what is often a messy, complex, iterative, and non-linear process.

Quinn, for example, observes that innovation:

> is a process that rarely can be planned and controlled with the kind of analytical certainty managers associate with other operations. Instead, technology tends to advance in a bubbling, intuitive, tumultuous process—more akin to a fermentation vat than a production line.[86]

This is particularly true of the 'fuzzy front end' of the innovation process, which is very difficult to 'capture' or standardize.[87] That is not to say that the NPD process cannot be managed; rather, project managers must learn to cope with the messiness, ambiguity, and complexity common to many innovation projects.

Through case-study research, Van de Ven et al. have explored in detail the limitations of such simplistic models of innovation. What emerges from this research is the notion of innovation as a dynamic process or 'journey' during which ideas proliferate and diverge, success criteria change, the innovating unit is restructured through external interference or unexpected events, and during which cycles of optimism and pessimism highlight the emotional rollercoaster that often characterizes the innovation process.[88] We will return to the metaphor of 'innovation as a journey' in **Chapter 9**.

In the preceding sections of this chapter, the NPD process has been discussed from what may be termed a rational 'project management' perspective—that is, the concern has been on highlighting, for example, the key stages of the innovation process, and the appropriate techniques and 'go/no-go' criteria associated with these stages, in order to facilitate the smoother running of the project and increase the probability of a successful outcome. But whilst a common perspective within the NPD literature, such an orientation over-rationalizes the decision-making processes within innovative activity. Indeed, Thomas argues that:

> Despite penetrating criticism and a paucity of empirical support, researchers still find it difficult to relinquish 'rational actor' models of organizations and organizational decision-making. This reluctance can be explained, in part, by demands for parsimony that encourage researchers to build into their theoretical models reductionist assumptions about human behavior and motivation.[89]

The adoption of an 'interactive' or 'network' perspective of the innovation process helps us to move away from such a rational model of innovative activity. Indeed, viewing innovation as a social and political process highlights quite different issues with regard to project management. Through such a 'lens', the process of screening ideas, for example, is no longer seen as simply a rational choice of selecting the 'best' idea through the use of appropriate selection techniques and criteria, but also as a social and political choice enacted through the mobilization of coalitions and alliances, and the application of power and influence.

In order to develop a 'fuller' appreciation of the innovation process, in **Chapter 9**, we will view the innovation process from a number of alternative orientations:

- 'innovation as a social process'—see **9.4.1**;

- 'innovation as a political process'—see **9.4.2**;

- 'innovation as an emotional process'—see **9.4.3**.

8.4 Managing innovation teams

Task-orientated teams and groups are an important component of organizations. It is not surprising, then, that there has been an abundance of research on teams and groups over the last fifty years that has sought to understand, for example, the factors affecting group formation, group cohesion, group decision making, and group effectiveness. For an overview of this research, see any introductory text on organizational behaviour or management.[90]

Our particular focus in this section concerns the factors impacting the creativity and innovativeness of teams and groups. Key researchers in this area include, for example, Michael West of Aston Business School, in the UK, who has written extensively on innovation within teams, and Teresa Amabile, who has written widely on the environmental determinants of individual creativity.[91] The following is not an exhaustive list, but it does highlight a number of the key dimensions of teams that impact their innovativeness and creativity.

8.4.1 The encouragement of creativity and new ideas

Amabile et al. argue that encouragement operates at three levels: organizational, supervisory, and team or work group. Furthermore, they argue that it is the perception that individuals and teams have of such encouragement that really impacts their creativity and innovativeness.[92] We have already discussed the organizational level of encouragement in **Chapter 7**, with regard to the 'organizational climate' for innovation: here, we indicated that the broad encouragement within an organization of risk taking and idea generation, the fair and supportive evaluation of new ideas, as well as the rewarding and recognition of creativity, all played a positive role in encouraging creativity. At the supervisory level, research has shown that goal clarity, open interactions between supervisor and team members, and supervisory support for the ideas of the team play an important part in motivating innovative behaviour within the team. Finally, at the level of the team, creativity can be encouraged, for example, through mutual openness of team members and the constructive challenging of new ideas, and a shared commitment to the project; this level is sometimes referred to as 'team climate'.

8.4.2 External demands

Whilst there is evidence to suggest that creativity is impacted negatively by external demands and threats, research has also indicated that work pressure can promote innovativeness.[93] In attempting to explain such contradictory evidence, Amabile et al. differentiate between 'workload pressure' and 'challenge'—the former having a negative impact, especially where it is perceived by the team as being imposed upon them, and the latter a positive impact on creativity—arguing that 'some degree of pressure could have a positive influence if it is perceived as arising from the urgent, intellectually challenging nature of the problem itself'.[94] Sustained high demands on individuals can, however, lead to apathy or learned helplessness.[95]

8.4.3 Team leadership

Team leaders can play a number of important roles, such as ensuring the clarity of team goals and encouraging the commitment of team members to such goals, managing conflicts of interest and personality within the team, facilitating the constructive interaction of competing perspectives in diverse teams, and providing support for creativity and innovation. Indeed, a number of these roles attempt to address the 'dilemmas of teamwork', which can otherwise '*drive a wedge between individuals and the team*'.[96] Team leaders can also play important 'championing' and 'boundary' roles, such as negotiating with senior management for additional resource, coordinating with other teams, and buffering the team from external pressures and interference. Research has also indicated that individuals are more creative when they have a sense of ownership and perceive themselves to have freedom in undertaking the tasks that they are allocated; this would indicate that leading innovation teams requires a 'light touch' style of management.[97]

8.4.4 Team membership

There is a great deal of research that highlights the positive contribution that diversity in professional background, knowledge, skills, and abilities of team members can bring to the innovative performance of a team.[98] Research has also indicated the importance to team performance of the presence of a range and balance of roles within the team.

Perhaps the work best known in this area is that of Belbin, who identified a number of key team roles, including: 'innovator' (to generate new ideas), 'monitor-evaluator' (to ensure critical reflection), 'team worker' (to help to build social cohesion and relationships within the team), 'shaper' (to direct the team), and 'completer' (to focus on output). For these roles to be enacted successfully, there needs to be a good match between the personality and behavioural preferences of the incumbents of these roles.[99]

Team diversity in relation to functional background is also an important contributor to successful innovation—we have already spoken of the important integrating role played by cross-functional teams within innovative organizations (see **7.7**). In their study of forty-five cross-functional innovation teams in five hi-tech companies, Ancona and Caldwell found that the recruitment of a new team member often led to a dramatic increase in communication with the functional area from which they were drawn.[100] The importance of the permeability of team boundaries and the openness of team networks is discussed in more detail below.

Overall, team diversity increases the pool of task-related skills, information, and perspectives at the disposal of the team, and can allow for more comprehensive and creative idea generation and decision making. But diversity within teams also requires supervision and management, because, as West argues, '*diversity of background may also be a hindrance, if group processes do not facilitate integration of team members' perspectives*'; goal clarity is one way in which such integration might be facilitated.[101]

8.4.5 Team resources

The level of resourcing of a project has a direct bearing on the ability of a team to perform, through the provision of appropriately skilled staff and finance for equipment and travel, for example. Of particular interest here is research that has suggested that the creativity of a team is directly related to the resources that it is allocated.[102] The perceptions of team members of the adequacy of resources can also impact their feelings about the intrinsic value of the project in which they are involved.[103] Thus, negative perceptions of resourcing may lead to low morale and motivation within the team.

We will discuss in more detail in the next chapter (see **9.5.1**) the important role played by 'product champions' in securing resources for specific projects.

8.4.6 Team task characteristics

Research has also indicated that the nature of the task undertaken by a team can impact the 'task orientation' of its membership. In summarizing the literature, West notes that the intrinsic motivation of a task is positively influenced by task characteristics, such as task 'completeness' (i.e. where the team has a complete task to perform), the variety of demands, the opportunities for learning and for social interaction, task significance, and the possibilities for developing the task.[104]

8.4.7 Team permeability and openness

Although research has long highlighted the value of external networking of project teams to the NPD process, there nevertheless exists a tension between the effectiveness and originality of a team's communications. On the one hand, efficient and effective communication is more likely to occur within a densely connected team. Such cohesive ties facilitate trust and cooperation amongst individuals, and minimize internal conflict, allowing the team to focus on issues that maximize consensus.[105] On the other hand, a number of studies have found that, as teams and groups become increasingly cohesive, they can also become increasingly inward-looking. This can have a negative impact on the creativity and innovativeness of the team.

For example, Rogers and Kincaid argue that:

> ingrown communication patterns . . . discourage the exchange of new information with the environment . . . lack openness . . . [and] may simply facilitate the pooling of ignorance among the individual members.[106]

In their investigation of the 'not invented here (NIH) syndrome', Katz and Allen found that performance increased in stable project teams of up to one and a half years' tenure remained steady for a time, but noticeably declined after five years. They argue that this is because stable project teams become *'increasingly cohesive over time and begin to separate themselves from external sources of technical information and influence by communicating less frequently with professional colleagues outside their teams'*, such that it begins to become *'increasingly complacent about outside events and technology developments'* and *'to believe it possesses a monopoly of knowledge of its field'*.[107] For Boissevain, *'it is no coincidence that criticism often comes from outsiders . . . Their social and geographic distance gives them perspective, it also ensures that they are less vulnerable to counter-pressures'*.[108]

The above research conclusions resonant well with the findings of Allen in his study of communication patterns in an R&D laboratory (discussed in **Chapter 3**). This indicated that:

> despite the hopes of brainstorming enthusiasts and other proponents of group approaches to problem-solving, the level of interaction within the project group shows no relation to problem-solving performance.[109]

Thus, it is important that team boundaries remain 'permeable'.

In summary, successful innovation teams are more likely to be those that combine a dense set of internal linkages, which facilitate trust, consensus, and efficient and effective internal team communication, with an open network linking the team members to a variety of external sources, and which expose the team to in-flows of new ideas and information.[110]

8.5 Managing knowledge creation and knowledge transfer

Since the early 1990s, a number of prominent writers on management and organization, such as Drucker and Quinn, have argued that knowledge now represents the key organizational resource for sustained competitive advantage.[111] It is not surprising, then, that the last decade has seen a burgeoning literature concerning the management of knowledge within organizations.[112] Many of the themes that have emerged in this area—such as 'the nature and types of knowledge', 'knowledge generation', 'knowledge codification', 'knowledge transfer', and 'the politics of knowledge'—are clearly also of relevance to the management of innovation. Indeed, many of these themes have already been touched upon either implicitly or explicitly in earlier chapters: for example, 'the nature of knowledge' in **Chapter 1**, 'knowledge as an organizational resource' in **Chapter 6**, and 'knowledge transfer' in **Chapters 3** and **7**. We also noted earlier (see **3.6.1**), national differences in knowledge transfer behaviour.[113]

It is not the intention to summarize the knowledge management literature in this text, not least because of the breadth and depth of this literature; rather, we seek to illustrate the connections and overlaps between knowledge management and innovation management. Thus, for an excellent overview of the key themes of knowledge management see *Knowledge Management in Organizations: A Critical Introduction* by Donald Hislop of Longhborough University Business School, *Managing Knowledge Work* by Sue Newell, Maxine Robertson, Harry Scarbrough, and Jacky Swan, all members of the IKON Group at Warwick Business School, or *Managing Knowledge* by Stephen Little and Tim Ray of the Open Business School. From a practitioner perspective, among the most influential writers in the area of knowledge management are Thomas Davenport and Larry Prusak: their text *Working Knowledge* has been widely read by managers and is also worth a read.[114]

Depending on one's starting point or perspective, knowledge creation and knowledge transfer can either be seen as narrow activities that contribute directly to individual innovation projects, or a much broader organizational activity, which initiates, nourishes, and surrounds these individual innovation projects. Both of these orientations are of relevance to the management of the innovation process. Specific and directed knowledge creation and transfer activities are at the heart of each of the various stages of the NPD process discussed earlier in this chapter: for example, in serving the idea generation, concept development, and product development phases. More ephemeral and emergent knowledge creation and transfer activities are important during the 'fuzzy front end' of the innovation process, but also contribute serendipitously and informally throughout the life of an innovation project.

The essence of this broader orientation, which embraces the dynamic and emergent nature of knowledge creation and transfer activities within organizations, is captured by Leonard in her well-known book *Wellsprings of Knowledge*:

> knowledge does not appear all at once. Rather, knowledge accumulates slowly, over-time, shaped and channeled into certain directions through the nudging of hundreds of daily managerial decisions . . . knowledge reservoirs in organizations are not static pools but wellsprings, constantly replenished with streams of new ideas and constituting an ever-flowing source of corporate renewal.[115]

The processes and mechanisms by which such 'knowledge reservoirs' trigger and sustain innovative activity are not, however, particularly well understood.

In the next subsection, we will discuss the frequently cited work of Nonaka and Takeuchi regarding knowledge creation and sharing within organizations, but, before we do so, it is worth reflecting on a few broad observations of the 'mainstream' knowledge management literature.

- It is 'overwhelmingly optimistic', and rarely attempts to address seriously the problems and barriers to knowledge creation, knowledge codification, and knowledge transfer within organizations.[116]

- It is dominated by a 'knowledge supply' perspective, rather than a 'knowledge demand' perspective, where the focus is on the user and uses of knowledge.[117]

- It has often been information technology (IT) or tools-driven, leading to an over-emphasis on knowledge capture and codification, and the role of IT in knowledge transfer, and an underemphasis on the complexities of organizational processes.[118]

- It has largely treated 'tacit knowledge' as knowledge 'not yet articulated' rather than knowledge that *cannot* be articulated.[119]

Keep these points in mind when reading the next subsection—we will return to these issues in due course.

8.5.1 The Nonaka and Takeuchi model of knowledge creation

Perhaps the most cited writers in the area of knowledge management are two Japanese academics, Ikujiro Nonaka and Hirotaka Takeuchi, who are well known in both the academic and practitioner communities for their book *The Knowledge-Creating Company*, which was published in 1995. This text has been particularly influential in raising the profile of tacit knowledge within business schools and organizations in the West, as well as reinvigorating the debate concerning the processes of knowledge creation and knowledge transfer.

The work is based on in-depth interviews with managers from a sample of approximately twenty companies (mostly Japanese), including NEC, Mazda, Mitsubishi, Fujitsu, Nissan, Matsushita, and Fuji Xerox. From this research, Nonaka and Takeuchi argue that, at the heart of the difference between Japanese and Western organizations, is the way in which they understand knowledge. They contend that Western managers '*view knowledge as necessarily "explicit"—something formal and systematic*', whilst Japanese managers, it is argued:

> recognize that the knowledge expressed in words and numbers represents only the tip of the iceberg. They view knowledge as being primarily 'tacit'—something not easily visible and expressible . . . highly personal and hard to formalize, making it difficult to communicate or to share with others.[120]

In this text, Nonaka and Takeuchi attempt to describe and model the processes of inter-action between tacit and explicit knowledge—how it is captured and transferred within the organization.

Their model consists of four 'modes of knowledge conversion'. Iterations of this process lead to a 'knowledge spiral', as new knowledge is created, captured, and transferred. For Nonaka and Takeuchi, '*organizational knowledge creation is the key to the ways in which Japanese companies innovate*'.[121]

We will now briefly outline these processes:

Modes of knowledge conversion and the knowledge spiral

The four 'modes of knowledge conversion' are sometimes collectively referred to as the Socialization–Externalization–Combination–Internalization (SECI) model. This model is based on the assumption that knowledge is created through the interaction between tacit and explicit knowledge (see **Figure 8.2**). Furthermore, the knowledge conversion process is viewed as a social process—that is, it is enacted through interaction between individuals.

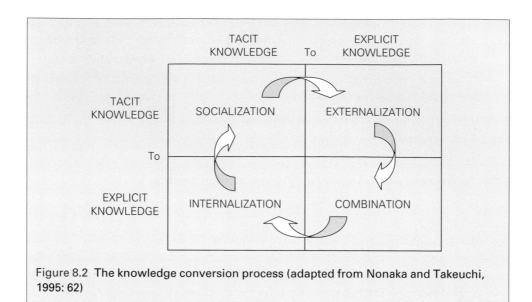

Figure 8.2 The knowledge conversion process (adapted from Nonaka and Takeuchi, 1995: 62)

It is worth noting here that Nonaka and Takeuchi distinguish between what they term 'cognitive' and 'technical' dimensions of tacit knowledge: the former *'consists of schemata, mental models, beliefs, and perceptions so ingrained that we take them for granted . . . these implicit models shape the way we see the world around us'*, whilst the latter *'encompasses the kind of informal and hard-to-pin-down skills or crafts captured in the term "know-how"'*.[122] In making this distinction, Nonaka and Takeuchi argue that *'the articulation of tacit mental models . . . is a key factor in creating new knowledge'*.[123]

We turn now to a brief overview of the four modes of knowledge conversion.

- *Socialization*—this conversion mode involves the transfer of tacit knowledge through the sharing of experiences, and the creation of shared mental models and technical skills—this is referred to as 'sympathized knowledge'. The tacit knowledge need not be articulated; indeed, it is more likely that it is transferred through observation, imitation, and practice. The master–apprentice relationship is often employed to explain this mode of knowledge conversion. Socialization is often used as a way of sharing tacit knowledge between product developers and customers, before, during, and after the innovation process. But for tacit knowledge to be communicated widely within the organization, it must be converted into words and numbers that are readily comprehensible to anybody within the organization—which leads us to the next mode of knowledge conversion, i.e. 'externalization'.

- *Externalization*—this conversion mode involves the creation of explicit knowledge from existing tacit knowledge through dialogue and collective reflection. The process is frequently driven by the use of metaphors and analogies as a way of attempting to articulate the tacit knowledge into explicit concepts—referred to as 'conceptual knowledge'. This process is often used in the concept development stage of the NPD process—see **8.3.3**. Once tacit knowledge is rendered explicit, it is capable of being communicated more widely and rapidly around the organization, allowing for the possibility of combining the knowledge with other existing explicit knowledge. This leads us to the 'combination' mode of knowledge conversion.

- *Combination*—this conversion mode involves the exchanging, combining, and reconfiguring of existing explicit knowledge, through a whole range of media from face-to-face meetings, to the use of the Internet and intranet, to create new explicit knowledge—this is referred to as 'systemic knowledge'. As we have noted in earlier chapters, boundary-spanning networks can bring together novel combinations of knowledge. Nonaka and Takeuchi describe this as *'a process of systemizing concepts into a knowledge system'*.[124] Such a process often occurs in the idea-generation stage of the innovation process and in problem solving during the product development stage of the innovation process. The application or use of this combined explicit knowledge leads us to the final mode of knowledge conversion, i.e. 'internalization'.

- *Internalization*—this conversion mode involves the internalization of explicit knowledge by individuals. Their 're-experiencing' of the explicit knowledge through 'learning by doing', leads to the creation of new tacit knowledge—referred to as 'operational knowledge'. A similar process occurs during the prototype development and testing phases of the innovation process (see **8.3.6**).

Illustration 8.5 Knowledge conversion using virtual reality

An innovative application of virtual reality (VR) has been in the area of sports training—this also serves as a useful example for demonstrating the knowledge conversion process itself. Typically, training in sport occurs through 'socialization'—a process not dissimilar to that of the craft apprenticeship. Through the use of sensors attached to the body and respective equipment, however, movement and orientation in response to a series of situations (e.g. in tennis, this might include a range of serves, volleys, and backhands) can be employed to capture the tacit knowledge of top sportsmen and women: Andre Agassi, the US tennis player, has been involved in such a project.

At one level, the 'motion capture' technology is simply capturing data; at another, it is argued that it actually captures the essence of an expert's skill by recording his or her responses to a range of situations. This is equivalent to the process of 'externalization'. An aspiring sportsperson might have the opportunity to replay and learn from the movements and responses of a number of top sportsmen and women in his or her sport—this could be considered a process of 'combination'. By using the same technology, the trainee can then, through 'trial and error', gain new knowledge by reviewing the mechanics of their own actions—a process of 'internalization'.[125]

The above four modes of knowledge conversion represent one iteration of the knowledge creation process. The subsequent socialization of tacit knowledge created through the internalization phase leads to a new iteration or cycle of knowledge creation—this is referred to as the 'knowledge spiral'. Nonaka and Takeuchi describe the overall organizational knowledge creation process as follows:

> an organization can not create knowledge by itself. Tacit knowledge of individuals is the basis of organizational knowledge creation. The organization has to mobilize tacit knowledge created and accumulated at the individual level. The mobilized tacit

knowledge is 'amplified' through four modes of knowledge conversion and crystallized at higher . . . [organizational] levels. We call this the 'knowledge spiral' . . . Thus, organizational knowledge creation is a spiral process, starting at the individual level and moving up through expanding communities of interaction [group, department, division, organization, inter-organization].[126]

8.5.2 Reflections on the management of knowledge

Whilst the Nonaka and Takeuchi model of knowledge creation is, on the surface, fairly seductive and, as a result, has been widely cited, it embodies a number of assumptions concerning both the nature of knowledge and knowledge workers. Such assumptions, whilst implicit, are nevertheless commonplace in the mainstream knowledge management literature. It is important, therefore, to reflect upon these assumptions in order to gain an insight into the extent to which knowledge can actually be 'managed' within organizations.

Micro-politics, power, and ownership

Implicit in the Nonaka and Takeuchi model discussed above is the assumption that employees will be willing to share freely their knowledge with others within the organization. This is not something that can be taken for granted. Yet, without the willingness of individuals to share their knowledge, the processes of 'knowledge conversion' and of the 'knowledge spiral', as outlined by Nonaka and Takeuchi, cannot operate effectively or efficiently.

Linked to the willingness of employees to share their knowledge is the issue of ownership: who owns the knowledge of an employee—the employee or the employer? In concepts such as 'intellectual capital' (discussed in **Chapter 6**), the assumption is that organizations not only own the knowledge captured in patents, for example, but also the knowledge stored in the heads of their employees. For McInerney and LeFevre, however, the picture is not so clear: they argue that whilst '*all explicit material created on the job belongs to the organization and its storage in knowledge repositories is appropriate. Who owns the knowledge residing within employees is another question*'.[127] Furthermore, where management lay claim to the knowledge within their organization, this does not mean that they actually possess this knowledge.[128]

Given the importance attributed to knowledge—and tacit knowledge, in particular—not least by Nonaka and Takeuchi, such ambiguity over ownership further complicates the process of managing knowledge within organizations. Issues of knowledge ownership and knowledge sharing are thus inextricably intertwined with issues of power and micro-politics, as are knowledge management initiatives themselves. This later point is raised by Storey and Barnett, who argue that the:

> different meanings of, and approaches to, the possibilities for KM [knowledge management] lead, therefore, to potential micro-political battles over the ownership of KM initiatives. There is plenty of scope for turf wars given the problems of getting to grips with such an elusive phenomenon as 'knowledge'.[129]

The 'slippery' nature of knowledge

A second broad area of critique of the mainstream knowledge management literature concerns the assumptions around the nature of knowledge. For Blackler, '*Knowledge is multifaceted and complex, being both situated and abstract, implicit and explicit, distributed and individual, physical and mental, developing and static, verbal and encoded*',[130] whilst for McKinlay, '*The most sophisticated databases can capture "knowledge bytes" but they cannot appropriate the subtle and ephemeral social processes that constitute tacit knowledge*'.[131] Both of these quotes serve to highlight the 'slippery' nature of knowledge and warn us against adopting overly simplistic notions of knowledge.

The context-specific nature of much of the knowledge within organizations, for example, is largely underplayed or ignored within the mainstream literature. Indeed, although Nonaka and Takeuchi acknowledge that '*information and knowledge are context-specific and relational*', they also argue that the conversion of tacit knowledge into explicit knowledge allows it to be '*valuable to the company as a whole*', thus implying that it may flow freely around the organization from one context to another, such as between the different functions or divisions.[132] Hayes and Walsham contrast this dominant perspective in the knowledge management literature, which portrays knowledge as '*an entity that can be possessed and traded*', with the relational perspective, which views knowledge as being '*provisional and context bound . . . as residing in an evolving, continuously renewed set of relations of persons, their actions and the world*'.[133] This later perspective has major implications for knowledge codification and transfer within organizations.

A more fundamental critique of the mainstream literature is presented by Tsoukas, who argues that:

> Nonaka and Takeuchi's interpretation of tacit knowledge as knowledge-not-yet-articulated—knowledge awaiting for its 'translation' or 'conversion' into explicit knowledge—an interpretation that has been widely adopted by management studies, is erroneous: it ignores the essential ineffability of tacit knowledge, thus reducing it to what can be articulated. Tacit and explicit knowledge are not the two ends of a continuum but the two sides of the same coin: even the most explicit kind of knowledge is underlain by tacit knowledge. Tacit knowledge consists of a set of particulars of which we are subsidiarily aware as we focus on something else . . . The ineffability of tacit knowledge does not mean that we cannot discuss the skilled performances in which we are involved. We can—indeed, should—discuss them provided we stop insisting on 'converting' tacit knowledge . . . Tacit knowledge can not be 'captured', 'translated', or 'converted' but only displayed, manifested, in what we do.[134]

At the heart of this line of argument is the distinction between 'knowledge' and 'knowing': Cook and Brown express this as the distinction between '*what is* possessed in the head' as compared to '*what is* part of practice'.[135]

The work of Tsoukas, and Cook and Brown, draws heavily upon the seminal work of Polanyi from the 1960s.[136] Polanyi, for example, has argued that '*we can know more than we can tell*'. He referred to this as 'tacit knowing': the ability to do things in practice without the capacity to say how we perform them, such as riding a bicycle or recognizing a face.[137] Taking this cue, Cook and Brown contend that:

> We hold that knowledge is a tool of knowing, that knowing is an aspect of our inter-action with the social and physical world, and that the interplay of knowledge and knowing can generate new knowledge and new ways of knowing.[138]

The above discussion only touches upon the emerging debates in the area of knowledge management. Nevertheless, it serves to illustrate some of the complexities and barriers to managing knowledge. It also serves to emphasize the importance of social interaction, rather than knowledge capture, to the processes of knowledge creation and transfer within organizations

8.6 Concluding comments

In this chapter, we have sought to provide an overview of the innovation process—often referred to as the new product development (NPD) process in the marketing and operations management literature. We outlined a variety of ways in which the innovation process has been conceptualized, such as through the use of 'department stage' models, 'activity stage' models, 'decision stage' models, and 'network' models. We choose to employ a model that combines the activity stage and decision stage models to highlight the different activities, stages, and decisions that are broadly generic to the innovation process, and we introduced and discussed a model of knowledge conversion and transfer.

Whilst such conceptualizations provide a useful structure for a discussion of the innovation process, and knowledge creation and transfer processes, it is important to re-emphasize that they oversimplify what are often very messy, complex, iterative, and non-linear processes. Quinn, for example, observes that innovation:

> is a process that rarely can be planned and controlled with the kind of analytical certainty managers associate with other operations. Instead, technology tends to advance in a bubbling, intuitive, tumultuous process—more akin to a fermentation vat than a production line.[139]

Furthermore, such models are overly rational and partial: for example, they lack an appreciation of the social, political, and emotional dimensions of such processes. That is not to say that these processes cannot be managed; rather, project management and knowledge management are also, as a consequence, complex.

In the next chapter, we explore the role and importance of informality and serendipity during the innovation process, to provide a counter-perspective to the rational, planned, project management perspective of much of the mainstream NPD literature.

CASE STUDY 8.1 FKI BABCOCK ROBEY LTD AND THE DEVELOPMENT OF THE EURONOX INDUSTRIAL BOILER[140]

Introduction

The EURONOX range of industrial boilers was developed in the early 1990s and won the Queen's Award for Environmental Achievement in 1993. At the time, industrial boilers were notorious for emitting noxious elements of nitrogen and sulphurous oxides into the atmosphere. Although technologies had been developed to reduce these emissions, they had done little to cure inherently poor boiler design and high running costs. In contrast, the EURONOX boiler was developed to address both of these issues and, as a result, led the market in delivering an efficient low-emission boiler. The innovator, FKI Babcock Robey Ltd—a medium-sized engineering company—was established in 1988 following a series of mergers (it is now part of Wellman Robey).

The EURONOX project resulted directly from an explicit strategy to develop environmentally friendly products as a way of entering the highly competitive Continental European marketplace for industrial boilers. But this search for new markets was itself triggered by the declining UK boiler market on which the company relied heavily. As FKI's sales and marketing director says:

> EURONOX was developed principally to take us into new markets . . . Our motivation was solely to look at how we were going to take our company forward. We couldn't afford to sit around in a declining, stagnant marketplace. So we wanted to look at new market opportunities. And Europe was one. Europe was a challenge . . . if we were going to go into Europe we would first of all bump into countries that had much tighter emission regulations . . . It was most certainly a market-led development.

The strategy team and environmental scanning

A small 'strategy' team, headed by the sales and marketing director, was established to investigate the long-term views relating to emission controls in the UK and Europe, and the possible strategic and marketing options. At the start of the project, the strategy team spent several months looking at both German boiler manufacturers (their competitors) and German boiler users. This was critical to directing the development programme. The team collected as much product and market information as possible. It also evaluated every conceivable aspect relating to prevailing emission laws and those mooted for the future, as well as assessing customer awareness of environmental legislation.

The analysis of all of this information highlighted one fundamental market need not being addressed by boiler manufacturers: an environmentally friendly boiler that met legislative requirements, but which was also energy-efficient. In responding to environmental legislation, boiler manufacturers had modified existing boiler designs without too much concern for the resulting inefficiencies or for the customers' need to find ways to cover the additional capital expenditure.

> For six months we looked around Europe and quickly homed in on Germany, where legislation is relatively tight and low NOx combustion was a fairly well practised art. We spent nearly two and a half months in Germany looking at our German competitors. We were sometimes direct and sometimes very discrete . . . In those early discussions there was the possibility that we might even collaborate . . . We talked to the [German] companies . . . We managed to get all of their product literature. We visited lots of [their customer] sites without their knowledge, and two of their factories, with their knowledge! We saw many of them across the six-month period in exhibitions. So we were able to get to know quite a lot.

The technical team

A small cross-functional 'technical' team was then established during the spring of 1992. This team included representatives from engineering, quality, and procurement, and was led by the chief engineer. All project teams within the company have a sponsor from senior management to smooth the way for the team and provide a space for creativity. The sales and marketing director became the project sponsor. Rather than attempting to modify existing boiler designs, the team went back to the drawing board. As the chief engineer recalls:

So what we decided to do was go back to a clean sheet of paper . . . We decided to home in on the idea that if we could reduce the operating running cost, then that was an ingredient that would go towards paying for the increased capital cost for this NOx reduction. On the otherside, if we could actually engender into the design of the boiler some unique features in heat transfer, and we are a heat transfer specialist . . . we could actually increase the overall operating efficiency . . . Within six months we had actually developed a new boiler.

Collaboration and the development of EURONOX

During this design process, a decision was made to collaborate with Saacke Ltd—a manufacturer of burners and a key supplier of burners to FKI Babcock, which incorporated them into their boiler range. Right from the start, the collaboration was formalized: '*So we approached Saacke and got them to sign a ten-year confidentialty agreement. Then we exposed our ideas and from that point onwards we collaborated.*'

The technical teams from the two companies worked extremely closely during the development process, meeting monthly and sometimes weekly, and, as a result, a strong working relationship was established between the members of the two organizations. As the chief engineer says:

> We came together in a planned process. We had separate tasks designated. Clearly defined . . . These meetings . . . included quite a lot of sessions which were what we called 'open sessions', where we would literally sit around a table wondering what the next step would be and just throwing ideas in the air.

This partnership proved crucial to the technical development of EURONOX. A number of unique technical aspects to the EURONOX boiler were designed in collaboration, such as the gas recirculation system employed. The project team:

> spent an awful lot of time on the configuration of the boiler . . . so we looked at the mechanical constraints, the pressure constraints, but also what the thing looked like because we wanted it to look different and to really look like a boiler for the new millennium . . . So every conceivable element of design was gone through. We made no assumptions [from the past].

According to the sales and marketing director, this collaborative relationship was quite unique to the industry:

> It was quite a culture change in here, and it was also a culture change in Saacke, because it was the first time in the industry that a boiler-maker and burner-maker had actually put their heads together . . . and the two technical teams actually sat down and behaved as if they were one . . . For the first time a burner company had to think like a boiler maker, and a boiler maker, ourselves, had to think like a burner maker. So we had to have quite a lot of cross-fertilization . . . And it worked very well.

The two technical teams '*behaved as if they were one. You couldn't at the time discern who worked for Saacke and who worked for FKI. They were a very cohesive team*'.

The personalities involved also aided the success of the collaborative process:

> It very much worked because of the personalities in the two teams . . . We had been in a longstanding [company-level] relationship, and it has also been a relationship of stability of people within each of the companies . . . It's very important because when you have problems, the problems are dealt with without too much fuss. So the formality is there. The discipline is there. But then there is a close understanding of the needs of one another as companies and as individuals. And a very genuine desire to succeed, to solve problems on a day-to-day basis.

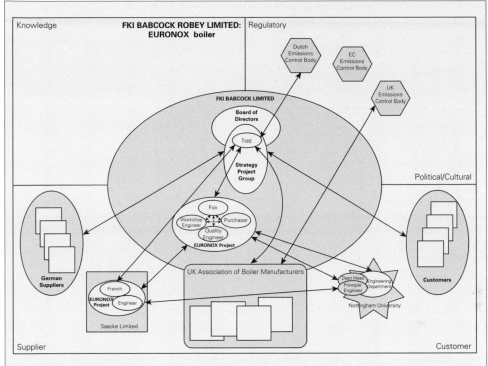

Figure 8.3 Actors and links mobilized in the development of the EURONOX industrial boiler (Steward et al., 1996)

Collaboration and the testing of the EURONOX prototype

To test the performance of the design, the EURONOX boiler needed to be fully implemented within a customer site: a 'guinea pig' customer was needed. The development team approached Nottingham University, which had made enquires with FKI Babcock regarding a new boiler a few months prior. The University was shown the design, liked what it saw, and agreed for the EURONOX boilers to be installed. There was a great deal of trust on both sides very early on and, as a result, the principal engineer and head of engineering at the University were asked to join the project team, which was still working on designing the boiler. Those from Nottingham University quickly built up a strong working relationship with those from the FKI–Saacke development team.

This relationship was crucial during the implementation of the boiler, with the technical team being given full and flexible access to the university site. It was the key element in the implementation phase of the project, because it allowed for the testing of the innovation *in situ* for the first time. This was especially important, because FKI Babcock had not built a full-scale working prototype prior to this installation. According to the sales and marketing director:

> They worked with us very closely on site [implementation]. And that is the area where they really started to have influence . . . They helped enormously . . . We spent three months at Nottingham . . . We worked very closely with the operators on site, the guys with their hands on the boilers. Making sure that any changes in operating characteristics were freely discussed, in case there was something that was in our design that had the possibility of degrading [performance].

The implementation proved very successful. Following the test implementation, the new boiler design was soon ready for commercialization and subsequently proved to be successful in the marketplace.

Questions

1. To what extent are the stages of the NPD process, discussed earlier in this chapter, visible in the development of the EURONOX boiler?

2. How useful is it to view the development of the EURONOX boiler as a 'social process', rather than simply as a series of stages or activities?

3. What aspects of the development team strike you as novel and/or conducive to creativity and innovativeness?

CASE STUDY 8.2 GLAXO PLC AND THE DEVELOPMENT OF ZANTAC

Introduction

Zantac (ranitidine)—a drug for the treatment of stomach ulcers—was introduced to the market by Glaxo in 1981. It was originally discovered in its R&D laboratory at Ware, in the UK, in 1976. At the time, Glaxo was a UK-based pharmaceutical company, but it has subsequently been involved in a series of mergers. Zantac became not only Glaxo's biggest ever commercial success, but also one of the best selling individual drugs ever introduced by the world's pharmaceutical industry. The success of the product was partly due to its technological innovativeness. Yet it was not a radical innovation in therapeutic terms, because it followed the lead set by the SmithKline-RIT drug Tagamet (cimetidine), which had been on the market since 1976. Zantac did, however, offer some modest advantages in terms of effectiveness and dosage.

Organizational innovations played a key role in transforming a significant, rather than spectacular, technological innovation into an enormous and unprecedented commercial success. These organizational innovations were concerned with two critical issues: the rapid 'speed' of commercial introduction following the initial discovery in the laboratory, and the wide 'scope' of early international market entry. These were achieved through radical changes in the organization of product development and of product marketing. At the heart of these changes was a shift in the organizing principle of development and marketing activities from 'sequentiality' to 'simultaneity'.

The incentives for the global exploitation of a pharmaceutical innovation

Pharmaceutical companies have strong incentives for the effective global exploitation of a product innovation. The general attractiveness of a larger market is given added significance for pharmaceutical companies because:

- there are large R&D costs associated with the introduction of a new chemical entity as a drug for human use;

- rights over exploitation rely almost exclusively on patent protection, which is of fixed duration;

- competitor products are often introduced by other research-based companies to challenge market share; and

- the market for a new drug is usually limited to a particular segment of the population within a market, but this segment is present in many other national markets.

These features have been present since the birth of the modern pharmaceutical industry in the 1940s and 1950s. But they were exacerbated by trends in pharmaceutical innovation during the 1960s and 1970s concerning:

- declining rates of drug discovery and innovation;
- the increasing length of time from discovery to market; and
- stronger public and regulatory requirements on safety and efficacy.

The combination of these emergent factors led to:

- higher uncertainties in the innovation process, with fewer new synthesized chemicals likely to reach the commercialization stage;
- larger costs for bringing an individual innovation to the market; and
- shortening of the effective period of patent protection after market entry, due to the lengthening of the period between patenting and market launch.

The impediments to the global exploitation of a pharmaceutical innovation

The attractions of the international marketplace are offset by a number of difficulties. Again, there are some specific factors affecting the pharmaceutical industry, in addition to the generic problems of operating at a distance in countries with different languages and cultures, such as that:

- disease patterns and perceptions vary between different countries;
- medical and scientific attitudes and practices vary in different countries;
- regulatory requirements usually operate at a national level and may differ or need to be based on clinical trials conducted in that country; and
- national governments may set conditions on the operation of multinational firms requiring, for example, local production facilities to be established in order to gain access to markets, etc.

For companies to take advantage of global opportunities and respond to such national variation requires a combination of decentralized autonomy and centralized coordination. The balance between these contrasting requirements may be struck in different ways through different organizational arrangements.

Glaxo's organizational innovation—the parallel model of drug development

The traditional model of drug innovation has been that of a 'pipeline', based on a linear sequence of activities. There are three main phases in this pipeline:

1. discovery;
2. development; and
3. marketing.

Each of these phases is further composed of a sequence of substages. For example, 'development' involves laboratory-based toxicological and pharmacological studies in animals, followed by clinical studies in humans at increasing levels of scale and complexity; 'marketing' involves initial entry into one national market, often the domestic market for convenience, a large market for the scale of initial returns, or a market with the speediest regulatory process for early market entry. Following initial market entry, other national markets are entered sequentially to establish a global market.

Glaxo's response to the increasingly difficult context for drug innovation in the mid-1970s was a radical departure from this linear pipeline model. Instead of the traditional sequential process, a shift was made toward conducting activities simultaneously wherever possible. This was expressed in two ways.

- *Time to market*—according to the then director of Glaxo's research laboratory, plans were drawn up whereby tests were undertaken '*almost simultaneously and not on a sequential basis . . . had we done our tests strictly one after the other we could have lost a year or more on development time*'. The adoption of this approach did, of course, carry a risk, because it was necessary to assume success in the different stages of scientific testing and development. The overlapping of phases meant that the time from the first chemical synthesis of ranitidine to the market launch of Zantac was five years and four months compared with a conventional timescale of seven to ten years. A further shift to simultaneity concerned the relationship between development and production. The secondary production controller of Glaxo Pharmaceuticals recalled: '*We got going on production facilities before the clinical trials or even the animal trials were finished.*' Thus, construction of a £25m purpose-built manufacturing plant in Singapore started well before Zantac had achieved regulatory clearance.

- *Geographical scope of clinical tests and market entry*—the other sphere in which simultaneity was adopted concerned the entry into different national markets. The Glaxo group chief executive stated:

  > We didn't develop in the UK first, then in the next country and so on. Rather we launched worldwide in as short a time as possible. We made sure that everyone throughout the world had what they needed in the way of local clinical data in order to get the fastest possible approval in their own market.

Before the Zantac project, Glaxo tended to use only the UK for its clinical trials, with few other countries involved. It now changed to a coordinated international programme involving twenty different countries. According to the then medical director of Glaxo Group Research, this was necessary to '*treat enough patients quickly enough . . . for international registration purposes*' and to obtain '*broad clinical experience of the product, because medical practices differ so widely between countries*'. The ability to coordinate simultaneous clinical trials and regulatory submissions in twenty different countries relied on the company having a subsidiary in each of these countries. In the traditional, mainly Commonwealth, markets of Glaxo, there were long-established subsidiaries; in newer markets—particularly in the USA, Germany, and Japan—there were new acquisitions. The effectiveness of this process is demonstrated by the fact that the international 'launch roll-out' was accomplished with astonishing speed: only three years elapsed between the first and last major launches.

The pioneering of 'co-marketing'

Glaxo also adopted a novel approach to the marketing of Zantac. The emphasis here was less on coordination, than on responsiveness to local conditions. The generation of clinical information in each of the different local settings was used as the springboard for the local marketing effort. By conducting clinical trials in all of the main potential markets, they were able to create demand for Zantac among opinion leaders before the product was launched. Advertising of Zantac was closely tailored to each local market, where local conditions were allowed to determine the local marketing 'message'. There was no UK-generated core campaign to be followed everywhere.

A global information database was, however, maintained by Glaxo Group Research, which contained all material published on Glaxo's products—this was accessible to all Glaxo staff throughout the world at any time.

The major innovation in the marketing of Zantac was the pioneering of 'co-marketing'. Traditionally, a pharmaceutical company would choose between alternative options for marketing its product in different national markets: if it had a weak foothold in the market, it would simply license its product to a well-established company in that country; if it had an established subsidiary, it would market it exclusively through its own company. Traditionally, Glaxo had relied on licensing in the USA, and selling through its own subsidiary companies in the Commonwealth and some European markets. The conventional model of the development of international business views the process as a sequence of stages from exports, through licensing, to joint ventures and wholly owned subsidiaries. The co-marketing concept introduced the radical notion that a company could pursue more than one of these options simultaneously.

Although the presence of subsidiaries in different countries had been important for Glaxo's coordination of its worldwide programme of clinical trials and product registration, it cleverly avoided being bound by this when it came to marketing. Instead, the company adopted a flexible and pluralistic approach to marketing arrangements in different countries. As well as marketing through its own subsidiary company in each market, it also made a simultaneous licensing or joint venture arrangement with a major player in that market. This is termed 'co-marketing'. The aim was to maximize speedy market penetration through the deployment of larger marketing resources than were available in-house. For example, in Italy, the product was also licensed to Menari; in France, it was licensed to Fournier; in Germany, it was also marketed through a joint venture company, Cascan. In each of these countries, the collaborating company sold the product under a different brand name. In the USA, the subsidiary made an unprecedented arrangement with Hoffman La Roche, under which both companies simply sold the same product under the same trademark in the same package.

The pioneering of co-marketing and the coordination of local subsidiaries in a wide variety of national markets required a new balance between autonomy and control within the organization. This required a reversal of the strong centralizing trend within the organization that had occurred during the 1950s and 1960s. The emerging power of ICT allowed such a change to occur through its ability to facilitate coordination and communication between relatively autonomous national subsidiaries.

The adoption of these organizational changes associated with the introduction of Zantac resulted in the transformation of Glaxo into the fastest growing and most profitable of the world's drug companies during the 1980s. It also established new standards for the speed and scope of product development and marketing through its embrace of simultaneity in place of sequentiality.

Questions

1. Outline the novel aspects of the NPD process for Zantac.
2. What are the risks and benefits associated with such a process?
3. What organizational structures do you think might support such a process and why?

FURTHER READING

1. For an alternative to the dominant 'pipeline' metaphor implicit in most of the NPD literature, see Van de Ven et al. (1999), who adopt the metaphor of the 'journey' to provide insight into the innovation process.

2. For a range of interesting case studies of invention and innovation, see Petroski (2006), who focuses on the important role of learning from failure in successful innovation, and Leinhard (2006), who traces the myriad of individual contributions in the emergence of new technologies.

3. For an overview and review of the emerging literature on 'virtual' teams, see Martins et al. (2004).

4. For an interesting discussion of recent research concerning the management of trade-offs and paradoxes in managing the innovation and knowledge creation processes within the 'cultural' industries, see the 2007 special issue of the *Journal of Organizational Behaviour*, and in particular, articles by Perretti and Negro (2007), concerning team composition and innovation in the making of Hollywood feature films, and Cohendet and Simon (2007), which discusses a case study of knowledge creation in a videogame company.

5. For a discussion of 'emotion', 'empathy', and 'care' in the knowledge creation and knowledge sharing processes, see Von Krogh (1998) and Von Krogh et al. (2000).

NOTES

[1] Tidd et al. (2001: 16)

[2] Achilladis et al. (1971); Langrish et al. (1972), respectively.

[3] Tidd et al. (2001: 49).

[4] Dvir et al. (1998); Tidd and Bodley (2002).

[5] Cyert and March (1963).

[6] Nelson and Winter (1982); Levitt and March (1988); Clark (2000).

[7] For example, Nelson and Winter (1982).

[8] Clark (2000: 98).

[9] Tidd et al. (2001: 48).

[10] Interview by Steve Conway with the founder of Intelligent Applications Ltd, a small, innovative, software company.

[11] Saren (1984).

[12] Rothwell and Zegveld (1985); Rothwell (2002).

[13] Cooper and Kleinschmidt (1986); Page (1993); Urban and Hauser (1993); Hart et al. (2003).

[14] Jones (2002: 47).

[15] Kotler (2000: 335).

[16] Wheelwright and Clark (1992).

[17] Quinn (1988a).

[18] Quinn (1988a: 129).

[19] Kotler (2000: 337).

[20] Roberts (1986).

[21] Rathmell (1974); Gronroos (1990); Martin and Horne (1993).

[22] Martin and Horne (1993).

[23] Martin and Horne (1993: 63).

[24] Martin and Horne (1993).

[25] Cooper (1993).

[26] Saren (1994).

27 Griffin (1997).

28 For detailed summary of criteria, see Cooper and Brentani (1984); Stagg et al. (2002).

29 Hart et al. (2003).

30 Stagg et al. (2002).

31 Van de Ven (1988: 112).

32 Cooper (1994).

33 Jones (2002: 89).

34 Crawford (1992).

35 Spender and Kessler (1995).

36 Reid and Brentani (2004: 171, 182).

37 Reid and Brentani (2004).

38 Smith and Reinersten (1991).

39 Cooper and Kleinschmidt (1995).

40 Reid and Brentani (2004: 170).

41 Spender and Kessler (1995).

42 Reid and Brentani (2004: 177).

43 Reid and Brentani (2004).

44 Moenaert et al. (1994a; 1994b).

45 Griffin (1993); Cooper and Kleinschmidt (1994).

46 Cooper and Kleinschmidt (1994); Chryssochoidos and Wong (1998).

47 Milson et al. (1992); Rothwell (1992); Towner (1994); Kessler and Chakrabarti (1999).

48 Kessler and Chakrabarti (1999).

49 Kessler and Chakrabarti (1999: 244).

50 Wong (2002: 121).

51 Chryssochoidos and Wong (1998).

52 Interview by Steve Conway with the founder of Intelligent Applications Ltd, a small, innovative, software company.

53 Drucker (1985); Sowrey (1987).

54 Sowrey (1987: 39, 53).

55 Weick (1969); Van de Ven (1988).

56 Martin and Horne (1992; 1993).

57 Martin and Horne (1993: 57).

58 Kotler (2000).

59 Cooper and Kleinschmidt (1990).

60 Tidd et al. (2001: 37).

61 Kotler (2000: 337).

62 Kotler (2000: 338).

63 For a detailed case study example of this approach, see Page and Rosenbaum (1987).

64 Wittink et al. (1994).

65 Tidd and Bodley (2002).

66 Kotler (2000).

[67] Cooper and Kleinschmidt (1990).

[68] Hauser and Clausing (1988).

[69] Griffin (1992).

[70] Roy (1986b).

[71] Oakley (1984).

[72] Ibid.

[73] Conway (1994).

[74] Wheelwright and Clark (1992: 149).

[75] Leonard (1995: 83).

[76] Schrage (1993).

[77] Tidd and Bodley (2002).

[78] Peters (1997: 96).

[79] Adapted from Iansiti and MacCormack (1997).

[80] For example, Kotler (2008).

[81] Kotler (2000).

[82] For example, Slack et al. (2006) and Kotler (2008).

[83] For example, Hippel (1976; 1977a; 1977b; 1977c; 1978a; 1978b; 1978c; 1988); Shaw (1985; 1993).

[84] Leonard (1995: 131–3); Tidd et al. (2001: 270–2).

[85] Tidd et al. (2001: 43–4).

[86] Quinn (1988a: 128).

[87] Hellgren and Stjernberg (1995); Nobelius and Trygg (2002).

[88] Van de Ven et al. (1989), Van de Ven et al. (1999).

[89] Thomas (1994: 238).

[90] For example, Buchanan and Huczynski (2006), or Mullins (2007).

[91] For useful overviews, see West (2002) and Amabile et al. (1996).

[92] Amabile et al. (1996).

[93] Claxton (1997).

[94] Amabile et al. (1996: 1161).

[95] Maier and Seligman (1976).

[96] Kanter (1985: 256).

[97] Bailyn (1985); West (1987).

[98] Andrews (1979); Ancona and Caldwell (1998); Paulus (2000); Joshi and Jackson (2003).

[99] Belbin (1981).

[100] Ancona and Caldwell (1992).

[101] West (2002: 363).

[102] Damanpour (1991); Kanter (1985).

[103] Amabile et al. (1996).

[104] West (2002).

[105] Gargiulo and Benassi (2000).

[106] Rogers and Kincaid (1981: 136).

[107] Katz and Allen (1982: 7).

[108] Boissevain (1974: 22).

[109] Allen (1977: 122).

[110] Conway (1997).

[111] Quinn (1992); Drucker (1993).

[112] For example, Leonard (1995); Nonaka and Takeuchi (1995); Davenport and Prusak (1998); Holden (2002).

[113] For example, see Inkpen and Pien (2006); Michailova and Hutchings (2006).

[114] Newell et al. (2002); Little and Ray (2005); Davenport and Prusak (1998).

[115] Leonard (1995: 3).

[116] Storey and Barnett (2000).

[117] Ibid.

[118] Scarbrough et al. (1999).

[119] Tsoukas (2003).

[120] Nonaka and Takeuchi (1995: 8).

[121] Nonaka and Takeuchi (1995: 3).

[122] Nonaka and Takeuchi (1995: 8).

[123] Nonaka and Takeuchi (1995: 60).

[124] Nonaka and Takeuchi (1995: 67).

[125] Adapted from Watts (2001).

[126] Nonaka and Takeuchi (1995: 72).

[127] McInerney and LeFevre (2000: 11).

[128] Alvesson (1993).

[129] Storey and Barnett (2000: 147).

[130] Blackler (1995).

[131] McKinlay (2000: 119–20).

[132] Nonaka and Takeuchi (1995: 59, 13).

[133] Hayes and Walsham (2000: 69).

[134] Tsoukas (2003: 425).

[135] Cook and Brown (1999: 382).

[136] Tsoukas (2003); Cook and Brown (1999)

[137] Polanyi (1966: 4).

[138] Cook and Brown (1999: 381).

[139] Quinn (1988a: 128).

[140] Steward et al. (1996).

9

Social Networks and Informality in the Innovation Process

Chapter overview

Learning outcomes

This chapter will enable the reader to:

*   outline the origin and characteristics of 'informal' or 'social' organization, and indicate how these 'emergent' structures overlay and interact with the 'formal' organization;

*   identify the role of informality and serendipity in the innovation process;

*   appreciate the new product development (NPD) process as a social, political, and emotional process;

- identify the role of 'product champions' and 'boundary-spanners' in the innovation process;

- apply the 'deviance-based' versus 'competence-based' perspectives to explain product championing and boundary-spanning behaviour; and

- outline the key features of the informal networks of scientists and engineers.

9.1 Introduction

In recent years, the mainstream innovation studies literature has almost universally espoused the virtues of informal networks.[1] Informal networks are seen as an important device for promoting communication, integration, flexibility, and novelty, within and between organizations. They are viewed as structures that supplement, complement and add value to the organization. A number of these authors have also highlighted the potential for tension or conflict between the activities of the 'informal' organization and the 'formal' organization. This tension is largely seen to manifest itself in the potential for dissonance in the goals and objectives of these two 'structures'.[2] The corollary of this position has, however, been that the utility of the informal organization can be harnessed if it is 'brought into line' with the goals and objectives of the formal organization.[3] In this sense, this latter perspective assumes that the informal organization is essentially a misdirected, but benign, creature.

Despite the long-held interest and recognition of the importance of informal networks to the innovation process, there has been relatively little in-depth research compared to other topics such as the new product development (NPD) process or alliances, and other forms of formal inter-organizational collaboration and cooperation. In explaining this paucity of research, Freeman argues that informal networks are *very hard to classify and measure*.[4] This, however, is beginning to change—in part, because of the recent popularization of social network analysis (SNA) by the likes of David Krackhardt and Rob Cross, and the resulting rise in interest among managers and practitioners.[5]

A key objective of this chapter is to provide a deeper understanding of informal or social organization, and to highlight the 'micro-social' and 'micro-political' dimensions of the innovation process. Within this text (but also more generally in the literature), the terms 'prescribed' or 'formal' network are used synonymously with the term 'formal' organization (i.e. the organizational chart), whilst the terms 'emergent', 'informal', or 'social' network, as well as 'informal' or 'social' organization, are also used synonymously.

It is worth noting here that informal networks have little regard for organizational boundaries, whether 'internal' boundaries, such as those between teams, functions, or divisions, or 'external' boundaries, between organizations or sectors, for example. Thus, when we are discussing informal organization, we are inherently discussing social structures that span organizational boundaries. In this regard, this chapter provides an important bridge between **Part III**, with its focus on the management of innovation within organizations, and **Part IV**, with its focus on the impact of organizational embeddedness and context in the shaping of innovation.

9.2 **Informal and social organization**

The recognition of the importance of informal or social networks within organizations can be traced at least as far back as the late 1930s.[6] For Blau and Scott, '*it is impossible to understand the nature of formal organization without investigating the networks of informal relations*',[7] whilst more recently, Krackhardt and Hanson have argued:

> Many executives invest considerable resources in restructuring their companies, drawing and redrawing organizational charts only to be disappointed by the results. That's because much of the real work of companies happens despite the formal organization . . . informal networks can cut through formal reporting procedures to jump start stalled initiatives and meet extraordinary deadlines.[8]

In contrast, the organization studies literature has long provided a more critical perspective; this recognizes the complex interplay and interweaving of the formal and informal organization.[9]

Indeed, others have highlighted the dysfunctional potential of informal networks. Krackhardt and Hanson, for example, argues that '*informal networks can just as easily sabotage companies' best laid plans by blocking communication and fomenting opposition to change*',[10] while Stacey contends that:

> the informal organisation is essentially destabilising. It exists, sometimes in place of the formal organisation and sometimes in competition with it . . . the informal . . . organisation is the way in which the formal organisation is changed and it is so changed in the most fundamental way by altering the existing paradigm.[11]

This is perhaps not surprising, because, as Burns and Flam note, '*participants in the organization bring into it external statuses, relationships, network and organizational ties, each with their own social rule system, which may or may not contradict the formal system*'.[12] Thus, there is a clear recognition in both the organization studies and innovation literatures of the existence of some form of 'shadow' structure that overlays and intertwines with the formal structure within organizations.

We turn now to look at the ways in which the formal and informal organizational structures may be distinguished.

Illustration 9.1 Mapping the informal organization using SNA—the case of Masterfoods USA

Social network analysis (SNA) used to be the preserve of academic research; this is no longer the case. Increasingly, organizations are employing social network tools and techniques to reveal the informal networks within and between their various organizational units. One example is provided by Masterfoods USA, a division of Mars Inc., famous for the Mars Bar, M&Ms, and Snickers. Masterfoods comprises three core businesses (confectionary/snacks, ready-made meals, and pet food), employs around 8,000 employees, and accounts for approximately 30 per cent of the revenue of Mars.

A couple of years ago, the organization decided to restructure in order to become more market-orientated. A key aspect of this reorganization was the decentralization of the research and development (R&D) group into the various business units. SNA was employed to understand the current R&D network, in order that important relationships within the R&D community could be maintained following restructuring, but also to identify ways in which the R&D network could be enhanced to improve collaboration between different parts of the organization. A number of specific recommendations emerged from the network analysis.

SNA also allowed the organization to assess the fragility of its R&D network by looking at the impact of removing a small number of highly connected individuals from the network.[13]

9.2.1 Comparing the informal and formal organization

Organizational charts and job descriptions generally reflect the formal structure or 'prescribed' network in a given organization. Such prescriptions are often guided by the missions and strategies of the organization, even though their explicitness may vary greatly between one organization and another.[14] In contrast, 'informal' or 'emergent' networks refer to the often covert and unsanctioned informal relations that emerge over and above such prescribed patterns of interaction.[15] The formal organizational structure is often referred to as the 'formal' organization, while the informal structure is frequently termed the 'informal' or the 'social organization'.

For Freeman:

> Behind every formal network, giving it the breath of life, are usually various informal networks . . . Personal relationships of trust and confidence (and sometimes fear and obligation) . . . For this reason cultural factors such as language, educational background, regional loyalties, shared ideologies and experiences, and even common leisure interests, continue to play an important part in networking.[16]

Blau and Scott posit that 'informal organizations develop in response to the opportunities created and problems posed by their environment, and the formal organization constitutes the immediate environment'.[17] This view is supported by Tichy, who argues that 'metaphorically, a prescribed organizational network provides pegs from which emergent networks hang' and that 'variations in a prescribed network therefore alter the emergent networks'.[18] But Schwartz and Jacobson argue that while 'to some extent these [informal] links . . . reflect the formal structure . . . the formal structure does not define the sociometric [informal] structure'.[19] **Figure 9.1** provides an illustration of the different structures that are typical of the formal and informal organization.

Formal and informal organization may be distinguished along a series of dimensions, including the origin and stability of the structure, the nature and flow of influence, and the basis for interaction of individuals in the network—as summarized in **Table 9.1**.

As discussed in **7.7**, the degree to which the prescribed formal network and the emergent informal or social networks overlap is dependent upon whether the organization is 'mechanistic' or 'organic'. The mechanistic organization uses bureaucratic principles to design, plan, and prescribe roles. Thus, the influence and information channels are likely to be highly prescribed. In addition, given high task certainty, individuals will have little discretion in their choice of work-related contacts. In contrast, organic organizations are

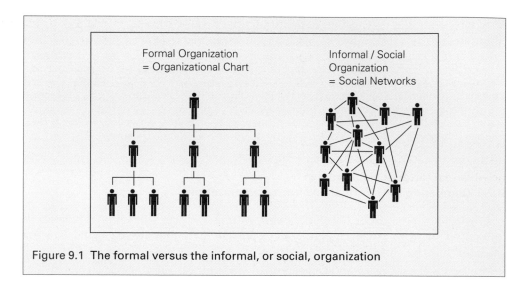

Figure 9.1 The formal versus the informal, or social, organization

Table 9.1 A comparison of the characteristics of informal and formal organization (adapted from Gray and Starke, 1984: 412)

	Formal organization	Informal organization
Structure: a) origin b) rationale c) stability	Prescribed Rational Stable	Emergent Emotional Dynamic
Influence: a) base b) type c) flow	Position Authority Top-down	Personality Power Bottom-up
Communication: a) channels b) networks	Formal channels Well defined and follow formal channels	'Grapevine' Poorly defined and cut across formal channels
Individuals included	Those indicated by formal position and role	Only those deemed 'acceptable'
Basis for interaction	Prescribed by functional duties and position	Spontaneous and personal characteristics

most effective when there is high task uncertainty. Because complex or highly variable tasks can not be pre-programmed, it is likely that informal relations arise to accomplish them.[20] Thus, Tichy et al. contend that:

> This implies that the emergent network of interaction will closely follow the prescribed network in a mechanistic setting . . . [while] in organic organisations, emergent networks may differ considerably from prescribed networks.[21]

It is apparent from a review of the organizational studies literature that variations exist in the meaning attached to the term 'informal' organization. Mouzelis distinguished between four categories of meaning:[22]

- 'informal' as deviation from the formal;
- 'informal' as irrelevant to organizational goals;
- 'informal' as unanticipated; and
- 'informal' as 'what really goes on in organizations'.

These categories are not mutually exclusive, although they do highlight different views of what is important about informal organization. Thus, whilst informal organization might be emergent and not officially prescribed by management, it does not necessarily mean that its activities are irrelevant to the organization's goals. Indeed, this chapter provides many instances in which this is not the case. Even where a social network within an organization may appear, at first sight, to be non-task-orientated—for example, a five-a-side football team that meets at lunch breaks—it nonetheless creates personal connections that may potentially be mobilized in the future for task-orientated activities.

9.2.2 Why do informal organization's emerge?

It is argued that the need for informal organization arises from two major failings of bureaucratic control: firstly, the subordination of individuality, and the alienating and demotivating nature of bureaucracy; secondly, the inability of the bureaucratic structure to handle environmental ambiguity and uncertainty. On the first point, Stacey contends that:

> People deal with the shortcoming of the bureaucratic system by colluding to operate a 'mock' bureaucracy (Gouldner 1964) and acting instead within an informal organisation that they set up themselves (Blauner 1964) . . . [paying] lip service to the rules but tacitly agree not to enforce them.[23]

On the second point, Tichy notes, for example, that 'social networks play important roles in business organizations', because 'unplanned structures . . . emerge because organizations are so complex that plans can never anticipate all contingencies'.[24] Stacey sees the informal organization as the mechanism that people employ to:

> deal with the highly complex, the ambiguous, the unpredictable, the inconsistent, the conflicting, the frustrating . . . They use it to satisfy social and motivational needs . . . and as the tool to promote innovation and change.[25]

Wolek and Griffith, however, have argued that the reliance on informal organization is *'somewhat troublesome, for it is the* formal *channels which seem to be much more amenable to control and institutional support'*, noting that informal networks are *'sometimes interpreted as a sign of both weakness and need for better formal systems'*.[26] Certainly, informal networks present a number of issues for managers in their attempts to steer an organization towards its formally defined goals and objectives. We will address the issue of goal congruence in **9.2.5**.

In considering the emergent trajectories of social networks within and between organizations, Kilduff and Tsai consider two distinct processes: one 'goal-directed' and the other 'serendipituous'. In the former, the network trajectory is 'energized' by a goal and leader, or leading group, and *'exhibits purposive and adaptive movement towards an envisioned end-state'*; in the latter, the network trajectory *'relies on processes of chance and opportunity'*, because it *'has no pre-existing goal around which members cluster'*.[27] Informal organization may emerge along either of these trajectories.

9.2.3 Revealing different types of informal network

So far, we have made the distinction between formal and informal, based on the degree to which the structure is sanctioned and prescribed by management. Alternatively, social structures within organizations may be distinguished by the nature of what flows through network linkages—that is, what is exchanged between individuals.

Tichy et al. distinguish between four types of network:[28]

- *friendship networks*—to reveal connections based on friendship;
- *influence networks*—to reveal power structures;
- *communication networks*—to reveal who is sharing information; and
- *economic networks*—to reveal the patterns of exchange of money and goods;

Krackhardt and Hanson employ slightly different categories to distinguish between the following three types of network:[29]

- *advice networks*—to reveal the prominent players in an organization on whom others depend to solve problems and to provide technical information;
- *trust networks*—to reveal the pattern of sharing with regard to delicate political information and support in a crisis; and
- *communication networks*—to reveal the employees who talk to each other on a regular basis.

In mapping out these various types of network in the design group of a California-based computer company, Krackhardt and Hanson found that major differences were revealed in the key individuals identified and the structures highlighted, between the formal, advice, communication, and trust networks. They argue that an analysis of the trust and advice networks, in particular, provided a clearer picture of the dynamics at work in the design group that they investigated.

Stephenson and Krebs employed a similar typology in their study of employee diversity in a large US organization. By mapping the prescribed, work, information, and support (mentoring) networks among a group of employees, and by distinguishing between gender and race, they were able to highlight the degree of isolation or integration of female and minority employees.[30]

9.2.4 Informal organization and boundary-spanning

A key characteristic of informal and social organization is their tendency to span organizational boundaries—team boundaries, functional boundaries, and even the organizational boundary itself. The importance of social or informal networks to innovative organizations is highlighted by Kreiner and Schultz, who argue that:

> In recent years, the traditional boundary activities bridging the company to its environment has been supplemented with a host of collaborative ventures . . . Most noticeable, probably, is . . . when it takes the form of strategic alliances and other formalised collaborative structures. Increasingly, we are becoming aware that much of such collaboration is also pursued along unpaved paths in the undergrowth of less formalised, personalised networks.[31]

Such boundary-spanning interaction is at the heart of the 'interactive model' of innovation (as discussed in **3.2.2**). The interactive model places great emphasis on the ability of innovative organizations to manage relationships across interfaces, both within the firm (between project groups, functional departments, and divisions), and externally (within and across industrial sectors, and between the science base and marketplace, for example). In particular, studies of successful technological innovation have highlighted the importance of the internal marketing and R&D interface,[32] and of interacting with external organizations and individuals, such as customers and suppliers.[33]

9.2.5 Tensions between the formal and informal organization

The informal organization has been shown to be a valuable mechanism through which 'fresh' ideas and information filter into the innovation process, and, as such, it represents an important 'intangible' organizational resource that is difficult for competitors to replicate. Although there is much evidence to highlight the efficacy of the informal organization in supplementing and complementing the activities of the formal organization, however, informal networks can be dysfunctional as well as functional. This is not surprising, because while management may set the parameters for informal exchange behaviour, '*what actually gets traded is determined by day-to-day interactions of engineers, marketers, and product developers*'.[34]

A key managerial concern involves the flow of information across the organizational boundary, because informal boundary-spanning activity not only provides for the sourcing and acquisition of external information and know-how, but can also result in 'information leakage'. Mansfield postulates that the rapid diffusion of technology via informal channels is one reason why many firms have difficulty in appropriating benefits from their innovations.[35] This view is supported by Carter, who argues that *'exchangers of information do incur costs. The cost to the trader . . . is not the loss of the information itself, but rather the* competitive back-lash'.[36]

With evidence from his study of informal know-how trading between competitors in the US steel mini-mill industry, Schrader argues that employees are likely to trade information in accordance with the economic interests of their firms, with factors such as friendship of secondary importance.[37]

In contrast, other research has indicated that informal 'trading' or sharing of information by employees across organizational boundaries may be explained by friendships or personal objectives.[38] Hippel, for example, argues that *'it is clear that the benefits to individuals actually engaged in the trading may differ from those of the firms which employ them. But they do not necessarily differ'.*[39] Similarly, Carter contends that:

> since knowhow traders must proceed with significant autonomy, some agency problems are likely . . . possible sources of conflict lie in the employee's loyalty to and aspirations in his broader professional community.[40]

But often employees are simply unable to make an *'accurate judgement with regard to business matters'* or do not possess the managerial information to enable well-informed decisions.[41]

9.2.6 Micro-politics and the political landscape within organizations

As noted earlier, the mainstream innovation studies literature tends to see the informal organization as misdirected, but essentially benign, and thus in need of better management and control. This is perhaps not surprising, because, as Knights and Murray note:

> The process of OP [organizational politics] and the vast array of individual and collective strategies it involves, is necessarily riven through with a central paradox. For a great deal of managerial practice constructs a reality of its own activity that denies the political quality of that practice.[42]

The mainstream organizational studies literature stresses this more critical perspective, locating 'politics' at the heart of organization. Thus, for Stacey, informal networks:

> are essentially political in nature; that is people handle conflicting interests through: persuasion and negotiation; implicit bargaining . . . and application of power that takes the form of influence rather than authority; that influence is derived from personal capability and the breadth of other network contacts . . . the informal group can often exert sanctions—the fear of rejection—that are stronger than those of the formal organisation.[43]

Just as the informal organization is in constant flux, so too is the nature and locus of conflict and 'challenge' within the organization, both in relation to the formal organization, and to other cliques and coalitions within the informal organization. Indeed, Burns and Flam argue that *'power relations among the actors or groups advocating different* [rule] *systems become critical, since these will in part decide which of several competing or contradictory social rule systems will prevail'*, and that *'informal status hierarchies and leadership patterns develop, countervailing and providing a basis of challenge to those formally designed and super-vised by management'.*[44]

For Pettigrew, the key to an understanding of the 'political landscape' of an organization is an appreciation of the access to, and control of, information, because these are essential sources of power.[45] The informal organization constitutes an important element of this landscape: as Pfeffer argues, *'clearly, the power that comes from information control . . . derives largely from one's position in both the formal and informal communication networks'.*[46] Research has highlighted a robust link between the centrality of an individual within the informal network and the power that he or she accrues within the organization.[47]

An important adjunct to network centrality is the 'accuracy' of an actor's perception of the informal organization, or what Freeman et al. term 'social intelligence'. They found that an actor's ability to accurately recall social structure was a function of whether he or she was a member of the core group or, a peripheral or transitory member[48]—that is, network centrality and social intelligence are closely interconnected.

9.3 Informality and the innovation process

Peters argues that:

> Unfortunately, most innovation management practice appears to be predicated on the implicit assumption that we can beat the sloppiness out of the process if only we'd get the plans tidier and the teams better organized. The role of experiments and skunkworks, the zeal of champions . . . is denigrated as an aid fit for only those who aren't smart enough to plan wisely.[49]

To a certain extent, this position is mirrored in much of the mainstream NPD literature discussed in **Chapter 8**, with its focus on project management, and process standardization and formalization. Yet, in stark contrast, the innovation studies literature is also replete with research concerning the importance of informality,[50] serendipity,[51] rule-breaking,[52] deviance,[53] and unconventionality,[54] to the processes of fruitful information seeking, discovery, creativity, and innovation.

The increasing interest in 'communities of practice' (CoPs) among many leading organizations highlights a recognition by managers of the importance of informal and emergent networks to the vitality of an organization, coupled with the desire to encourage and direct such informal interactions toward specific goal-orientated activities and problem solving.[55] (See **3.4.4** for a fuller discussion of this concept.)

9.3.1 Serendipity and spontaneity

In a study of office interactions, Kraut et al. found that many smaller decisions within projects arose from brief, informal, and spontaneous encounters, resulting in improved project coordination. Furthermore, such informal and unintended interactions were not only seen as equally as valuable as scheduled project meetings for the accomplishment of tasks, but were also viewed as more effective in building social bonds between co-workers than formal meetings.[56]

Serendipity—that is, the unexpected or accidental discovery of something of 'value'—is a particularly intriguing concept. It is an event or process of which we all have first-hand experience in various aspects of our work and social lives. On the one hand, it is integral to the processes of information seeking, discovery, creativity, and innovation; on the other, it is elusive, unpredictable, and seemingly even 'magical' in quality, and hence has figured little in current models of information seeking, for example.[57]

Whilst the notion of planning for serendipity might appear oxymoronic, there are certain conditions that might increase its likelihood. Fine and Deegan consider the importance of temporal, analytical, and relational factors:[58]

- *temporal*—that is, the importance of being in the 'right place at the right time';

- *analytical*—the importance of purposeful searching, and active learning and analysis. Classic examples here include the discovery of penicillin by Alexander Fleming, that of the 'vulcanization' process by Charles Goodyear, and that of the pasteurization process by Louis Pasteur, who famously observed '*Chance favours only the prepared mind*';

- *relational*—that is, unexpected discoveries that emerge from social connections and interactions. One might expect a more 'open' and 'diverse' social network with 'bridges' and 'weak' links to other social networks to be one that yields novel and serendipitous connections, ideas, and discoveries.[59]

These factors are not mutually exclusive and, for many successful innovators, they are intuitive. Returning to **Illustration 8.2**, for example, we see that the founder of the small entrepreneurial software company is implicitly referring to the importance of temporal (i.e. 'right place at the right time') and relational factors (i.e. unexpected discoveries that emerge from social connections and interactions) to serendipity.

9.3.2 Informal exchange and communication

Many studies have indicated the importance of informal boundary-spanning exchange and communication to the innovation process,[60] and particularly in relation to the transfer of tacit knowledge.[61] Indeed, informal information and knowledge sharing are at the very heart of the knowledge creation process and the resultant 'knowledge spiral' of Nonaka and Takeuchi (as discussed in **8.5.1**). In study of a sample of commercially and technically successful UK innovations, it was found that informal exchange activity was a valuable contributor throughout the innovation process, and, in particular, during the phases of idea generation, problem solving, and prototype testing.[62] Given the importance of informal interaction and exchange, it is not surprising, then, that organizational 'downsizing' has been criticized for breaking '*the network of informal relationships*' on which innovation depends.[63]

How might one explain such informal exchange activity? Firstly, it is worth noting that the informal exchange or communication of information or knowledge between individuals generally occurs within ongoing social relations[64] and with 'particular others'[65]—that is, individuals exchange information and knowledge with those whom they know, with whom a relationship or a 'bond' has been established, and thus between whom trust and obligations have emerged.

In **3.5.1**, we discussed three types of 'tie' between individuals that help us to appreciate why individuals might undertake such informal exchange activity:[66]

- *instrumental ties*—those through which mutually rewarding economic exchanges can be operationalized;

- *affective ties*—those through which satisfying emotional sentiments, such as friendship, can be evoked; and

- *moral ties*—those in which a code of fairness and reciprocity are the main binding forces. Such ties are often embedded within wider groupings and social structures, resulting in obligations, rights, and duties[67] or '*professional ethics of co-operation*'.[68]

Research into informal exchange in the innovation process indicates that all three of these types of tie have a role to play. Furthermore, these types of tie should not be seen as mutually exclusive.[69]

Reciprocity is also important to the ongoing exchange of information and knowledge between individuals, although such expectations are often implicit and exchanges do not have to occur concurrently. A lack of reciprocity during the early stages of a relationship may prevent the further development of the relationship, whilst the decline in reciprocity in an established relationship may lead to an exit from the relationship of one of the individuals, or inertia in exchange activity.[70]

9.3.3 Managing informal exchange and interactions

The unpredictable nature of the linkage and interaction patterns within informal networks provides a number of challenges for organizations. Carter argues that '*because knowhow trading is informal and* off-the-books *such trading is difficult for the firm to evaluate and to manage*'; Kreiner and Schultz, in their study of the Danish biotechnology sector, found that '*the norms governing the interaction seem to reside in the network itself rather than in any of the participating organizations*'; Mueller, meanwhile, sees informal networks as '*short-lived, self-camouflaging and adisciplinary. They are invisible, uncountable, unpollable, and may be active or inactive*', thus highlighting the transient and intangible nature of informal networks.[71]

Hamel et al. suggest measures to restrain informal boundary-spanning activity.[72] But Bouty sees such an approach as counterproductive, given the reciprocal nature of informal exchange activity, such that '*trying to stop outgoing flows would clearly and directly result in an equivalent drying up of resources flowing in*', and pointless, because '*the efficacy of classical tools such as hierarchical monitoring . . . is illusory*'.[73] Alternatively, it has been suggested that an organization could employ mechanisms to induce desirable information transfer behaviour, for example, through incentive schemes to motivate employees to act in the interests of the organization or through the wider diffusion internally of information regarding the goals of the organization, and of the market and competitors.[74] For Cross et al., however:

while research indicates ways managers can influence informal networks at both the individual and whole network levels, executives seem to do relatively little to assess and support critical, but often invisible, informal networks in organisations.[75]

The mismanagement or non-management of social capital is evident through social networks lost to attrition[76] or stunted in their development through the 'blind' use of virtual working.[77]

9.3.4 Informal and incidental learning

It is argued that one of the most significant developments in recent years in the area of human resource development has been the increasing focus on informal or 'on-the-job' learning, and the shift away from the traditional reliance on formal or 'off-the-job' learning, such as training courses and seminars.[78] This shift has, in part, been explained by the many limitations associated with formal learning, such as abstraction from the realities of the workplace, the difficulty of transferring learning to practice, and the lack of relevance.[79]

Marsick and Watkins define formal learning as *'typically institutionally sponsored, classroom-based, and highly structured'*; in contrast, they define informal learning as *'not typically classroom-based or highly structured'*, in which *'control of learning rests primarily with the learner'*, and *'predominantly experiential and non-institutional'* in nature. Informal learning might result from networking, mentoring, or coaching.[80] Recent research into knowledge and skills acquisition by software developers found that learning took place primarily 'on the job', either through interaction with colleagues or 'learning by doing'.[81] But whilst some studies concerning the contribution of mentoring and coaching to learning in teams reveal a positive picture,[82] others have proved more mixed.[83]

Incidental learning may be viewed as a subcategory of informal learning—one that is not planned or intentional, but which is the *'by-product of some other activity, such as task accomplishment, interpersonal interaction, sensing the organizational culture, trial-and-error experimentation, or even formal learning'*.[84] Incidental learning, then, might be the result of 'learning from mistakes' as well as 'learning from successes'. In relation to the former, Neumann found that learning might arise from 'errors of omission' (i.e. where something was not done), from 'substantive errors of commission' (i.e. where what was done was incorrect), from 'errors of process of commission' (i.e. where the way something was done was incorrect), or 'action errors of commission' (i.e. where nothing should have been done).[85]

Because informal learning is often unplanned and ad hoc, outcomes of such learning are inherently difficult to measure and assess.[86] Nevertheless, Marsick and Watkins argue that informal and incidental learning *'represent a neglected, but crucial, area of practice'*. They also contend that whilst informal and incidental learning may be difficult to organize and control, they can be enhanced through:[87]

- *proactivity*—an individual's preparedness to take initiative;
- *critical reflexivity*—an individual's willingness to surface and critique one's own assumptions and 'taken-for-granteds' (i.e. 'double-loop' learning); and

- *creativity*—an individual's ability and willingness to explore possibilities and to break out of preconceived patterns of thinking and behaviour.

All of these actions are, however, highly influenced by the organizational context in which individuals are employed, and, in particular, by the degree of autonomy and empowerment present in the organization. Thus, if organizations wish to encourage informal and incidental learning, they must support openness towards experimentation, reflection, and the translation of learning into practice.[88]

9.3.5 'Skunkworks'—operating outside of bureaucractic control

The original concept of 'skunkworks' referred to emergent groups of loosely structured researchers who worked on their own projects without the sanction or knowledge of senior management. Thus, 'skunkworks' were product development projects that worked outside of the constraints of the formal organizational rules, structures, and decision-making processes, and drew upon unofficially diverted resources.[89] **Illustration 9.2** captures this approach well, in the development of the novel and commercially successful Apple 'Mac'.

Illustration 9.2 The development of the Apple 'Mac' personal computer—success through a 'skunkworks' project

The Apple Mactintosh, introduced in January 1984, was the first affordable personal computer (PC) to employ a graphical user interface (GUI) and mouse. For many, the launch of the Apple 'Mac' represents a major milestone in the history of the PC. Yet despite its eventual impact, the Apple Mac emerged from very humble roots: as an underground 'skunkworks' project initiated in 1979 by a handful of engineers, it was not viewed as a major project within Apple until early 1982. In stark contrast, the 'Lisa' PC (the first to incorporate GUI), launched by Apple in January 1983, was a commercial failure, despite involving a large team of engineers and enjoying 'corporate sponsorship'.

For some, Lisa's failure lay in Apple's attempt at trying to be a 'grown-up organization', to be 'too corporate', through establishing large projects and hiring seasoned managers. All of this was at odds with the 'rebel spirit' that had served Apple so well since its foundation in 1976. Folklore has it, for example, that, in these early days a 'skull and crossbones' flag flew over the Apple building.

The Mactintosh project consisted of a small team of exceptional, passionate, and motivated engineers, who embraced the original spirit and vision of Apple. It is reported that, by late 1980, Steve Jobs, co-founder of Apple, had become alienated by the corporate ethos of the Lisa project, but increasingly drawn to the 'fervent hobbyist' atmosphere of the Mactintosh project. He became involved and gradually took over leadership of the project, directing it toward his own personal vision of democratizing computing.[90]

More recently, the concept is one that has been consciously adopted or appropriated by organizations, and thus today more often refers to a formally established and mandated project or division that operates outside of the 'normal' rules, structures, and decision-making processes of the organization. Examples of such mandated 'skunkworks' have been documented at Ford, IBM, Apple, and Ericsson.[91] As such, these constitute examples of organizations operating in 'dual mode' (see **7.6**). Either way, for Quinn: '*Skunkworks help eliminate bureaucratic delays, allow fast unfettered communication, and permit the quick turnarounds and decisions that permit rapid advance.*'[92]

9.4 Innovation as a social, political, and emotional process

In the introduction to **Part III**, we outlined the four 'central problems' in the management of innovation, as set out by Van de Ven. The third of these—that is, '*the human problem of managing attention*'—raises a number of issues that may, in part, be categorized as the 'management of emotion'. The fourth of these 'central problems' concerned the '*process problem of managing ideas into good currency*'—that is, to successful commercialization or implementation.

The social and political 'landscape' within an organization are important shapers of these processes. For Van de Ven:

> The social and political dynamics of innovation become paramount as one addresses the energy and commitment that are needed among collations of interest groups to develop the innovation . . . People become attached to ideas over time through a social-political process of pushing and riding their ideas into good currency.[93]

Over the next three sections, we build upon the foundations laid in the earlier text of this chapter, to provide an alternative view to the rational, linear, project management perspective of the innovation process that dominated in **Chapter 8**. In this chapter, we argue that whilst an understanding of the stage model of the innovation process is important, it conjures up the imagery of a 'pipeline', in which ideas are managed through a series of processes and an innovation emerges at the end. We believe a more appropriate metaphor for the innovation process is one of 'a journey', on which an idea is 'navigated' through a varied social and political landscape. Here, one can envisage many different routes that the innovation process can take and many different destinations that might be reached.[94]

We now consider some of the social, political, and emotional dimensions to the innovation process.

9.4.1 Innovation as a social process

Social interaction is at the heart of many of the activities and processes that are fundamental to the innovation process, such as idea generation, problem solving, creativity, learning, information and knowledge sharing, and knowledge creation. It is a theme that runs strongly throughout this text and, in particular, through the chapters that comprise **Parts III** and **IV**.

For Rogers and Kincaid:

> the information explosion . . . is often handled by the structuring of interpersonal network links by individuals. *Know-who* thus begins to replace *know-how* as one of the main determinants of individual effectiveness.[95]

More specifically, social relationships and interaction are often viewed as vitally important for the transfer of 'tacit' or 'uncodified' knowledge.[96] Indeed, for Allen, '*ideas have no real existence outside of the minds of men*', whilst Burns contends that '*the mechanism of technological transfer is one of agents* [people], *not agencies* [intermediaries]'.[97]

Furthermore, given the distributed nature of information and knowledge, social interaction—and, in particular, informal boundary-spanning communication—is also important for the collation of the multiplicity of inputs that is typical of the successful innovation process.[98] Indeed, for Bouty, '*a diversified network is essential for capturing varied resources*'.[99] Social networks are also valuable for developing a common understanding between collaborators and for coordinating their activities,[100] as well as for enhancing the possibilities for serendipity.[101]

9.4.2 Innovation as a political process

An understanding of 'micro-politics' within organizations is important to our understanding of the innovation process.[102] This is highlighted by a study of failure in NPD by Jones and Stevens, who found that:

> The various 'sectional interests' of groups and individuals becomes particularly apparent during the NPD [new product development] process. Reputations, and consequently career prospects, can be enhanced or ruined according to the success or failure of a new product or service. Therefore, any attempt to set out a framework for NPD must include explicit recognition of the role played by micro-politics.[103]

Returning to the classic Burns and Stalker study of 1961, discussed extensively in **Chapter 7**, we see that the text is littered with examples and references to internal politics and careerism. Indeed, they note that internal politics and careerism were 'visibly present' in each of the companies in their sample, arguing that this arose from the competing commitments of individuals to the organization, to their career, to various internal 'political' groups, and to a multiplicity of social relationships within the organization.

Burns and Stalker also reported that internal politics and careerism gave '*rise to intricate manoeuvres and counter-moves, all of them expressed through decisions, or in discussion about decisions, concerning the organization and the policies of the firm*', and that these '*generate, or contribute to, manifest inefficiencies of communication within the working organisation*' and '*affect the rational adaptation and exclusive devotion of the working organization to its task*'.[104] Similarly, with regard to knowledge sharing, Hislop notes:

> While power has not been adequately dealt with in the knowledge management literature there has been a growing acknowledgement that not only can people's attitudes to participate in knowledge activities be highly variable, but that interpersonal or inter-group conflict in knowledge processes is not uncommon. Those issues are . . . linked to power.[105]

Thus, the importance of 'micro-politics' to an understanding of the management of innovation and knowledge transfer has long been recognized, although as noted above and in **Chapter 8**, it has often been underplayed or ignored in the NPD and knowledge management literatures.

For Thomas, this is partly explained by the fact that politics and political action are '*essentially private, backstage behaviours*' that are 'not openly discussed'.[106] Thomas also argues, however, that where politics is recognized as impinging upon the innovation process, it is almost always treated as an outcome or an obstacle to its implementation, rather than a driver of innovative activity.[107] Yet for Kanter, 'innovative accomplishments' require the exercise of power to bring about the 'disruption of existing activities', the 'redirection of organizational energies', and *to mobilize people and resources to get some-thing nonroutine done*'. Organizational politics also occur within project groups and especially within cross-functional teams, where individuals often act as 'representatives' of their function.[108]

Despite these insights, much of the NPD literature portrays the ongoing screening and filtering of ideas and possibilities, from the idea screening stage right through to com-mercialization (the 'development funnel'), as a rational decision-making process aimed at identifying the most promising ideas and allocating limited or scarce organizational resources to them. Yet this process of resource allocation has shown to be influenced by the lobbying of 'project champions', who frequently hold senior positions within the organization or subunit, and by various interest groups, who mobilize around competing ideas. This process is neatly summarized by Van de Ven, who emphasizes *'the centrality of ideas as a lobbying point around which collective action mobilizes'*.[109] The 'political' can also operate at a much broader level, with regard to the competition between scientific and technical paradigms, for example; this theme was discussed in **Chapter 5**.

9.4.3 Innovation as an emotional process

Since the early 1990s, the study of emotion in the workplace has been an emerging area of interest in fields such as work psychology and organization studies.[110] For Fineman, organizations are 'emotional arenas': *'Emotions are within the texture of organizations. They work mistily within the human psyche, as well as obviously in the daily ephemera of organiza-tional life.'*[111]

Perhaps the key popularizer of this theme has been Daniel Goleman, who published *Emotional Intelligence* in 1995. This research has relevance to the NPD process, because the activities associated with creativity, learning, and innovation are inherently 'emotional'— that is, faced with the challenges of dispensing with the 'old' and embracing the 'new', individuals can be confronted with a mixture of positive emotions, such as, excitement, joy, and pleasure, and negative emotions, such as, anxiety, fear, and frustration.

The following discussion covers some of the relevant contexts in which the emotions of individuals need to be considered in relation to innovation, knowledge, and learning.

Learning and idea generation

The processes of learning and idea generation within project teams can cause great anxiety among individuals, because they may be required to expose gaps in their know-ledge, expose their ideas to scrutiny (and possible criticism or mockery), and dispense with past knowledge and deeply held assumptions. Innovation and creativity may also surface difficulties that an individual has with the very process of learning itself, giving rise to feelings of inadequacy. It is not surprising, then, that Fineman argues that the process of learning is *'inextricably emotional or of emotion'*.[112]

The potential for emotional responses during the learning process can be illustrated through a comparison of two simple models of learning: 'single-loop' learning and 'double-loop' learning.[113] In the single-loop model, learning occurs through the monitoring of the results of an action and changing that action in a reactive way where the desired results are not being achieved. In such a model, learning is likely to be incremental, because it does not challenge underlying mental models or assumptions.

In contrast, double-loop learning requires individuals to surface and challenge their underlying mental models and deeply embedded assumptions, through openly talking about what Argyris refers to as the 'undiscussables'.[114] Such a learning process is proactive and can lead to the generation of radical new ideas.

In comparing these two models of learning, Van de Ven argues:

> Single loop learning represents conventional monitoring activity . . . Because it does not question the criteria of evaluation, single loop learning leads to . . . organizational inertia . . . Double loop learning involves a change in the criteria of evaluation. Past experiences are called into question, new assumptions about the organization are raised, and significant changes . . . are believed possible. While double loop learning can lead to change, it can also lead to low trust, defensive behaviour, undiscussibles, and to bypass tactics.[115]

Negative emotional responses can potentially have a direct and detrimental impact on learning and creativity within the innovation process: for example, Warr and Downing found that high anxiety was associated with low motivation; Goleman notes that anxiety and depression can hinder learning, whilst excitement can encourage learning.[116]

This latter point may be explained by the work of Kuhl, who contends that individuals have limited information-processing resources and that the cognitive resources available for learning can be reduced by non-task activities, including activities associated with coping with stress and anxiety. Thus, the higher the level of anxiety, the greater an individual's cognitive resources are diverted towards their coping activities.[117] But the relationship is not simple: for example, Brown notes that '*Negative and positive emotions are not necessarily distinct—one can, for example, be pleasurably anxious or fearfully excited*', while Hochschild argues that a mismatch between prior expectations and perceived outcomes can create negative emotions, such as dissatisfaction, that simulate action and learning.[118] For Von Krogh, the 'active empathy' of colleagues can play an important role in reducing the frustrations of those attempting to learn complex or 'difficult to grasp' concepts and tasks. He also highlights the importance of 'courage' in experimentation and in the voicing of opinion or feedback.[119]

The overall stresses and frustrations of the NPD process

Idea screening, project termination, and project failure are all events in the life cycle of the NPD process that are likely to yield emotional responses from project members. As we noted earlier, individuals become attached to ideas and projects, not least because of the emotional and physical energy that they invest in bringing them into 'good currency', but also because the success of an idea becomes increasingly intertwined with the success of the project team and project champion. Thus, the emotional 'fallout' that occurs when an idea is screened out of contention, when a project is terminated, or when a project fails needs to be recognized and managed. For example, in his discussion of development 'shoot-outs' between competing project teams, Quinn argues: *'Properly rewarding and reintegrating losing teams is perhaps the most difficult and essential skill in managing competing projects.'*[120]

More broadly, with regard to major innovation within large organizations, Quinn argues that the innovation process can seem *'ambiguous, uncertain, frustrating, and debilitating beyond belief'*. Such emotions also need to be recognized and managed if they are not to have an adverse impact on the motivation of members of complex and long-running innovation projects. In this regard, Quinn highlights the *'essential role of a "determined champion" in overcoming the many soul-wrenching disappointments and setbacks major innovation always seems to encounter'*.[121]

Team climate

The level of anxiety that an individual may experience in relation to learning and idea generation is, to some extent, moderated by the level of support for innovation within the team and what is sometimes referred to as 'intragroup safety'—that is, *'the psychological or psychosocial safety group members feel in the presence of their fellow group members and especially during whole group interactions'*.[122] Thus, it may be argued that the greater the perception of support for innovation and intragroup safety among team members, the lower the level of anxiety, and the greater the potential for learning, creativity, and innovation within the team.

But politics and competition within teams can often create anxiety and frustration among team members. On this issue, Kanter argues that *'Of course, declaring people a "team" does not automatically make them one, nor does seeking decisions in which many people have a voice ensure that democratic procedures will prevail'*.[123]

Organizational context

With regard to the broader organizational context, we noted in **Chapter 7** that the organic mode of organizing associated with innovative organizations is characterized by a reduction in clarity and stability of the hierarchical structure. The resulting ambiguity can create angst among individuals. Burns and Stalker argue that:

> [organic modes of organizing are] often experienced by the individual manager as an uneasy, embarrassed, or chronically anxious quest for knowledge about what he should be doing, or what is expected of him, and similar apprehensiveness about what others are doing.[124]

Von Krogh highlights the importance of 'care' within organizations—that is, a culture that encourages mutual trust, active empathy, and leniency of judgement, for example. He also notes the negative impact on knowledge sharing and creation where low levels of 'care' exist within an organization.[125]

Implications of emotion for the management of innovation projects

The consequences of negative emotions to the outcomes of creativity, innovation, and learning highlight the need for project managers to be aware that anxiety, just as much as excitement, is an inherent part of the NPD process. It follows that an important element of the project manager's role is the monitoring and channelling of the emotions of individual team members, and, where necessary, intervening at the individual and team level.

Such a role requires a project manager to:

- be self-aware—that is, to know their own feelings;
- be able to sense, and have empathy for, the feelings of others;
- handle relationships and conflict effectively; and
- be able to channel negative and positive emotions.

These are essentially core components of what is sometimes referred to as 'emotional intelligence'.[126]

At the organizational level, Von Krogh argues that management must encourage 'care' through a range of mechanisms, such as mentoring programmes, embedding trust, openness, and courage within the stated values of the organization, training programmes in 'care-based' behaviour, and social events.[127] In broad terms, this ties in with the contention of Van de Ven that *'the management of attention must be concerned not only with triggering the action thresholds of organizational participants, but also of channeling that action toward constructive ends'*, and with the concept of 'emotional energy',[128] which Huy believes *'represents a largely unexploited, yet ready* [organizational] *resource'* that can be channelled to 'realize strategic stretch'.[129]

9.5 Networkers and networking in the innovation process—roles, competence, and behaviour

Research has identified a number of key roles played by individuals in the innovation process. The most important of these are the roles of 'product champion',[130] and 'gatekeepers' and 'boundary-spanners' (introduced in **Chapter 3**).[131] Importantly, these roles are most often informal and emergent. In the subsequent sections, we explore these roles, assess the extent to which they should be understood in terms of 'deviance' or 'competence', whether they can be formalized, and consider to what extent the networking skills critical to these roles can be taught or learnt.

9.5.1 The role of product champions

One of the key roles in the innovation process is that of the 'product champion'. Indeed, the importance of the championing role to successful innovation has long been recognized.[132] A champion is an individual who 'champions' a project through various organizational obstacles that may arise during the innovation process, often at personal risk. Furthermore, the champion is likely to use cross-functional personal networks in place of the formal hierarchy.

Schon, who first coined the term, observed that 'champions' provided *'the energy required to cope with the indifference and resistance that major technical change provokes'* and, in so doing, display *'persistence and courage of heroic quality'*.[133] These characteristics of persistence and courage are also highlighted by others: for example, Afuah argues that champions do *'all they can within their power to ensure the success of the innovation . . . inspiring others with their vision of the potential of the innovation'*, whilst Quinn notes the role of the champion in *'overcoming the many soul-wrenching disappointments and setbacks major innovation always seem to encounter'*.[134]

Champions also play an important role in the procurement of resources to support projects and ideas, by providing the bridge between different levels of management in the organization. In this regard, Burgelman notes that *'Resources can be obtained if technical feasibility is demonstrated, but demonstration itself requires resources. Product championing activities serve to break through this vicious circle'*.[135]

In his review of the literature on champions, Shane identifies six valuable championing roles:

- they provide innovative individuals with autonomy from bureaucracy to enable the development of creative solutions;
- they employ loose monitoring mechanisms to enable the creative use of resources by innovative individuals;
- they build organizational support among managers, through developing cross-functional coalitions;
- they establish mechanisms for supporting inclusion, consensus, and equality in the decision-making processes concerning an innovation;
- they protect the project team from interference from the organizational hierarchy; and
- they employ informal means in the persuasion of others within the organization to support the innovation.[136]

Despite the importance of such roles to the innovation process, Dougherty and Hardy raise a note of caution, arguing that:

> rather than celebrate lone champions and other individual heroes, we suggest, as did Schon (1963), that primary reliance on such personal power is inherently ineffective for sustained innovation. Such power is limited by the reach of individual networks, knowledge, and experience and is easily uprooted by downsizing, restructuring, and changes in senior managerial focus. Moreover, it is unavailable to young people and newcomers who lack experience with an organization, as well as to people in organizations that have no history of building informal systems to get around the formal ones.[137]

Much of the research concerning product champions has been conducted in either US or UK companies. But in a large cross-national survey of nearly 4,500 individuals, within a cross-sector sample of forty-three organizations across sixty-eight countries, Shane found widespread evidence of product championing. This research suggests, therefore, that the entrepreneurial 'spirit' of the product champion is one that is not 'culture-bound'.

9.5.2 The role of boundary-spanners and gatekeepers

The boundary-spanner, or gatekeeper, is an individual who facilitates communication across functional and organizational boundaries and between activities.[138] Through their informal and personal networks, boundary-spanners and gatekeepers provide access to information and innovative ideas from outside the organizational unit. These are crucial roles for sustaining innovation within organizations, because, as Nonaka and Takeuchi argue:

> What is unique about the way Japanese companies bring about continuous innovation is the linkage between the outside and the inside. Knowledge that is accumulated from the outside is shared widely within the organization, stored as part of the company's knowledge base, and utilized by those engaged in developing new technologies and products.[139]

By virtue of their strategic network position, gatekeepers or boundary-spanners are exposed to large amounts of potentially relevant information. They are also often 'reasonably senior' within their organization.[140] Studies have found that gatekeepers in R&D laboratories, for example, have a significantly greater readership of professional and scientific literature, and maintain longer-term relations with experts in a broader, more diverse range of fields outside of their immediate working environment than the average researcher.[141]

Furthermore, gatekeepers also provide the main line of defence against information overload through the communication channels of the organization, by acting as 'information filters'. Where there exists ambiguity and uncertainty in the information that they receive, gatekeepers act as 'uncertainty absorbers', drawing inferences from the perceived facts and transmitting only the inferred information. They may also generate or sustain internal variation by channeling information about external developments to relevant parts of the organization.[142]

With reference to this variety of roles, Macdonald and Williams argue that *'the gatekeeper's function is clearly quite distinct from that of the general employee bringing information into the firm'*.[143]

Illustration 9.3 Bridging the gap between businesses—the role of informal networks and gatekeepers in technical problem solving at ICI

ICI is a multinational chemicals company. It develops and manufactures specialty chemicals that are employed in a wide range of products, such as paints, fragrances, and foods. R&D within the organization is decentralized across its various businesses, which themselves are distributed worldwide. A small central technology function exists to *'oversee and guide R&D activity within the businesses and facilitate inter-business technology transfer'*.[144]

In an insightful study of the informal problem-solving network within ICI's research and development function, it was found that the majority of the technical communication and interactions were limited to within, rather than between, the individual businesses. Looking at the two networks illustrated in **Figure 9.2**, we see, in the upper network map, the critical role played by individuals within the 'corporate centre' in providing a bridge between the various businesses; in the lower network map, where the corporate centre links have been removed, we clearly see that direct links between the four businesses are very few in number and relate to a small number of boundary-spanners or gatekeepers.

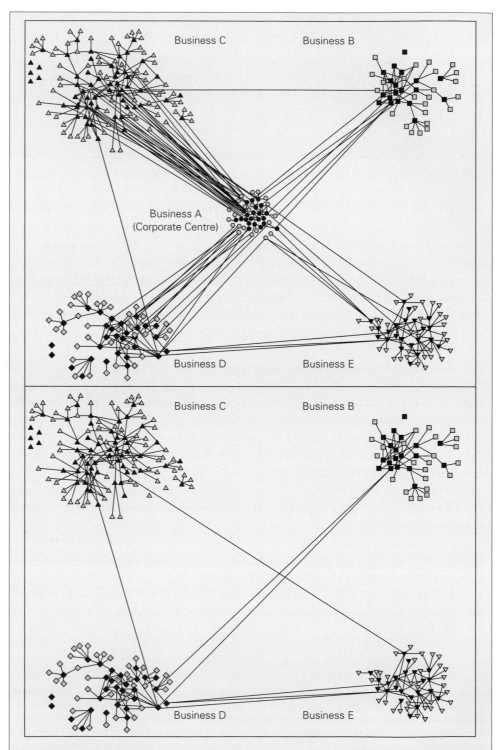

Figure 9.2 The informal problem-solving network within ICI's R&D function (Allen et al., 2007: 186)

Interestingly, the predominant intra-business (internal) pattern of informal communication and interaction occurs in spite of the formal organizational structures that have been implemented to encourage inter-business knowledge transfer and collaboration.

In attempting to explain this networking behaviour, the researchers highlight the geographical and cultural separation of the different businesses, which is seen as the result of a *'complex recent history of the company . . . of a number of acquisitions and the restructuring of operations in the late 1990s'*.[145]

Given the importance of informal boundary-spanning activity to the innovation process and the reliance on a relatively small number of specific individuals acting as boundary-spanners, the organization is, to some extent, vulnerable. Indeed, Lawton-Smith et al. argue that *'the downside of the key role which personal relationships play in collaborative ventures is over-dependence on certain individuals'*.[146] In his study of the communication patterns of engineers, however, Allen found that gatekeepers were easily recognized by the organization, with an overlap between guesses by management and the study data of around 90 per cent. In addressing this concern, Allen argues against formalizing the role of boundary-spanners, which he believes *'seems unnecessary and could even prove undesirable'*, favouring that *'recognition be afforded on a private, informal basis'*.[147]

9.5.3 Product championing and boundary-spanning—'competence' versus 'deviance'

Despite the importance of the championing role, only a small minority of individuals within an organization choose to become product champions.[148] The same is true of the boundary-spanner and gatekeeper roles.[149] Furthermore, these individuals risk their positions and reputation in the championing of ideas and the overcoming of organizational obstacles.[150]

So why do individuals become champions? One prominent explanation is that of social 'deviance', emphasizing the departure from customary norms, rules, and procedures commonplace in the adoption of such minority roles. This 'deviance' perspective focuses on the characteristics of individuals that cause them not to conform or comply with normal majority behaviour. In recent years, the personality or preferences of the individuals who act as product champions or gatekeepers has attracted attention in a similar way as has the psychological characteristics and traits of the individuals who become entrepreneurs.[151] Indeed, Shane argues that there is a strong parallel between the 'deviant' preferences and behaviour of champions with those of entrepreneurs.[152]

From a 'deviance' perspective, the management and rewarding of product championing becomes problematic. Indeed, Quinn argues that:

> Some of the most difficult problems for large organizations are: (1) how to tolerate and nurture the kinds of off-beat, driven personalities who tend to become champions; and (2) how to reward both champions and experts appropriately.

In fact, many innovative organizations recognize that champions generally work for more than monetary reward and have established alternative reward mechanisms, such as offering:

significant personal recognition, independence in research, appropriate power or visibility within the company or in the division exploiting the innovation, or that most cherished of technical incentives—the right to a major role in the next big innovation.[153]

But Aldrich and Zimmer criticize personality trait approaches to entrepreneurship as a failure to address the activity as a dynamic and relational process through which the individual succeeds in becoming *'embedded in networks of continuing social relations'*;[154] a very similar argument may be applied to the analysis of product champions and gatekeepers.

An alternative approach to understanding the adoption of such innovation roles is to view the behaviour as an expression of 'competence' rather than 'deviance'. While accepting that personality may well be important, the primary interest becomes that of identifying the distinctive skills and know-how needed to fulfil the networking roles required for innovation. Furthermore, Mintzberg argues that an emphasis on 'skills' is a more operational concept than intangible 'traits', because they can be directly related to behaviour.[155] The concept of competence is oriented toward practice; its overriding emphasis is on being 'able to perform', not to 'just know about'—or alternatively, it is *'not what a person knows but what a person can do'*.[156] There has, however, been some disagreement in the literature as to whether the competences of individuals can, or should, be assessed in relation to specific jobs or generic activities.[157] Of particular interest for the study of management of innovation is whether it is possible to map with more precision the ability to perform non-routine episodic networking roles that are neither attached to a specific function or occupation, nor carried out by all employees.

The concept of competence has also renewed attention to the relationship between the contrasting arenas of experience and education in the acquisition of skills and know-how. Thus, in adopting a competence perspective, two specific areas for investigation emerge: the nature of the particular and unusual networking competence exhibited by product champions and gatekeepers, and the routes through which it is learned or acquired.

9.5.4 Identifying networking competence

The 'core competencies' of the firm refers to the collective learning in the organization—especially in relation to the organization and coordination of activities—and the integration of technologies, in the delivery of 'value'. The concept encompasses many of the intangible resources of the organization that are difficult to replicate by other organizations and therefore are key to maintaining competitive advantage, such as organizational culture, know-how, and networks.[158] In addition, Prahalad and Hamel argue that *'core competence does not diminish with use. Unlike physical assets, which do deteriorate over time, competencies are enhanced as they are applied and shared'*.[159] Although this perspective acknowledges that individual competencies lie behind the competencies of the firm, Defillippi and Arthur argue that the broader dependence of these firm competencies on individual career behaviour also needs to be emphasized. They posit that *'competency accumulation through boundaryless careers can make a critical contribution to the unfolding competencies of firms and their host industries'*.[160]

Research on boundary-spanning roles in the innovation process has revealed some of the dimensions of competence required to perform such roles. For example, through their personal networks, both within and outside the organization, gatekeepers are exposed to large amounts of potentially relevant information. Their ability to absorb, assimilate, and transfer this information also allows them to act as information 'filters' and 'diffusers' to the relevant parts of the organization.[161] Central to the boundary-spanning role is the gatekeeper's ability to understand and communicate in a variety of 'languages' that build up around different disciplines and organizations.[162] Thus, the capability to translate 'contrasting coding schemes' or different 'languages' appears to be an important element of networking competence. Another attribute, associated with successful coupling between marketing and R&D functions, has been described as 'role flexibility'. This is defined as the ability to assume extra-functional tasks in the innovation process: the ability to step into different functional roles enables a better comprehension of the needs of other parties.[163]

Other studies of the R&D–marketing interface have emphasized the importance of credibility to cross-functional cooperation. Credibility in terms of communication depends on two aspects: 'information credibility', which relies on the quality of the information itself, and 'source credibility', which concerns the perceived character-istics and interpersonal skills of the information provider. These two aspects are highly correlated.[164]

Investigation of the gatekeeper role at the internal–external interface emphasizes the ability to engage in informal information exchange outside the organization, usually without encouragement from the employing organization itself. Thus, this role requires the skills to operate beyond the boundaries and beyond the control of the firm.[165] The different requirements of internal and external boundary-spanning are emph-asized by Stewart. The main challenge for interaction across internal boundaries is seen as obtaining cooperation from individuals within the organization over whom one has no formal control or authority. External contacts are considered to be dependent on skills in developing personal relationships, impression management, bargaining, discretion, and dealing with diversity.[166] Kotter provides a useful distinction between the activities of 'network building' and 'network using'. Building informal networks relies on a wide variety of 'face-to-face' methods, which include techniques of obliga-tion, identification, reputation, and manoeuvre, whilst using informal networks requires face-to-face skills of mobilization and influence. These interpersonal skills, along with information processing skills, are seen as central to the exercise of power among a network of people on whom an individual may depend, but over whom they have little direct control.[167]

By combining the innovation management and general management literatures, then, we are able to identify a number of individual-level competences that are important for the undertaking of the product champion, gatekeeper, and boundary-spanner roles. These individual-level competences include communication and translation skills, role flexibility, interpersonal skills, network building and nurturing skills, and the ability to 'mobilize' or 'leverage' a network.

9.5.5 Routes to building networking competence

Management education

Formal management education has been targeted as one route for the acquisition of networking competence. In the UK, there has been a large growth in the provision of both university management education and management development programmes, although the scale has consistently been viewed as below the level required to keep abreast of competitor nations.[168] A key advantage of this route is that it helps to challenge 'parochial beliefs' by bringing people together from very different business settings.[169] This is offset by the limits of traditional formal teaching methods and the need for new skill-oriented approaches, such as 'on-the-job' simulation or the creation of an 'appropriate climate' for the encouragement of entrepreneurial skills.[170]

The 'switchboard' effect of Master of Business Administration (MBA) degree education, through facilitating managerial mobility between organizations and functions, has been identified as a particular feature of UK business schools.[171] This view is supported by a study that compared the size and diversity of the intra- and inter-organizational social networks of a large sample of managers involved in the innovation process. Those with an MBA were found to have more extensive intra-functional and inter-organizational relationships—in large part, facilitated by greater job mobility and an appreciation of a wider set of functional roles and languages.[172]

In the UK, there have been direct suggestions that networking is a behavioural role based on skills and competences that can, and should, be taught in order to 'develop networking and networkers'.[173] The skills and competences highlighted include inter-personal skills of communication and judgement, information skills of data access, capture, and handling, and the ability to distinguish different types of networking, such as conscious versus intuitive, open-ended versus focused, and predatory versus reciprocal. There has also been official encouragement in the UK for training in innovation management to be a more prominent and widespread part of MBA programmes. The Economic and Social Research Council (ESRC) Innovation Training Materials initiative, for example, sought to progress this through the development of teaching resources. Such courses introduce the interactive network concepts of innovation studies.

Management experience

Management experience as a route to networking competence has also been considered in several studies, particularly with respect to career mobility. There is, however, a significant contrast in these studies between those that emphasize continuity and those that highlight change as a key to competence building.

Kotter identifies a career pattern of '*long tenure in one company and one industry*' in enabling individuals to become '*incredibly knowledgeable . . . and develop a large number of good working relationships*' as a 'key personal asset' for networking. Thus, effective networkers are not 'made' overnight, nor are they simply 'born', but are developed over many years. This career profile emphasizes stability and consistency. There is also a recognition that experience in a range of functions is important, but only if undertaken at an appropriate rate to facilitate growth of intellectual and interpersonal skills. This is seen as a responsibility of higher management, along with a direct role as 'coach' to assist in the process.[174]

Kanter strikes a different note. She sees mobility of people between jobs as a primary 'network facilitating condition'. As people move around, they take with them *'the potential to establish another information node and support base for a particular network in a different corner of the organisation'*. Constant and frequent moves, including lateral mobility, characterize the more integrative and innovative organizations. Although originally emphasizing employment security as a condition for high mobility, Kanter recognized in a subsequent study that attachment to one employer was declining and that mobility between organizations was increasing. In spite of the risks of anxiety and insecurity, the positive result was the encouragement of collaboration across functions, across business units, and across organizations.[175]

Movement between functions is seen as a distinctive US and UK 'generalist' career path. The merits are the development of *'dynamic outward looking strategic capabilities and broad inter-functional networks of communication that facilitate innovation, flexibility and coordination'*. The downside is seen as satisfaction *'in the glamour of change . . . not in the grind of implementation'*. From this perspective, executives are criticized for viewing themselves as *'competent since they are good at network socialising'*. In Germany, there is an opposite emphasis on functional specialization. Its strength is meticulous implementation, but its weakness is seen as strategic and organizational rigidity.[176]

With the decline of long-term attachment to an employer, the core competence of the organization is seen as becoming ever more dependent on individuals and their career paths. As a result, the 'boundaryless' career and 'portable' skills are receiving increased attention, along with 'layered' careers combining professional depth with managerial generalism.[177] There is also an increasing interest in new modes of learning involving mentoring, and links between the spheres of education and experience.[178]

9.6 The social networks of scientists and engineers

This section explores the similarities and differences in the nature and role of the social networks of scientists and engineers.

9.6.1 The social networks of scientists

One of the central approaches to the sociology of science during the late 1950s to early 1970s was the study of science as a social system—that is, science was treated not as a body of knowledge or as a set of methods and techniques, but as an organized social activity guided by a set of shared norms. The early work on the social organization of science focused on defining the principal norms governing scientific activity;[179] these writings were later to be considered as an idealized conception of science.[180] As a result of this, a number of writers began to argue that scientific behaviour should be viewed as pluralistic, rather than monolithic—that is, that the behaviour of scientists should be viewed as influenced by both their professional socialization and their institutional setting.[181]

From the mid-1960s, the emphasis began to shift towards a focus on the communication patterns of scientists and, through the adoption of a network perspective, an interest in the social structure of networks of scientists. Such studies repeatedly highlighted the crucial role of informal personal networks and the informal mechanisms through which these relationships were operationalized, to the science information system.[182] The findings of this early research continues to hold much resonance today.

This research also highlighted the importance of two types of subgroup: 'solidarity groups' and 'invisible colleges' (these were introduced in **3.6.1**). The 'solidarity group' consists of a group of collaborators that, under the leadership of one or two highly productive scientists, recruits and socializes new members, and maintains a sense of commitment to the area among existing members. The 'invisible college' is the apex or elite of a specialty, formed by the highly productive scientists who communicate with each other and transmit information informally across the whole field. In this way, the invisible college forms a communication network that links the various solidarity groups in a given speciality.[183] Indeed, Allen contends, '*it is probably safe to say that for university scientists this is a far more salient social system than is the university that employs them*'.[184]

In relation to the role of invisible colleges and solidarity groups within a given speciality, DeSolla Price argues that:

> [invisible colleges] effectively solve a communication crisis by reducing a large group to a small select one of the maximum size that can be handled by interpersonal relationships . . . [while the creation of solidarity groups] partly solves the problem of organising the lower-level scientists so that they can be directly related to the research life of the elite.[185]

DeSolla Price goes on to argue that one of the critical characteristics of the invisible college is its informal structure, and that:

> Possibly, if such groups were made legitimate, recognized and given newspaper-like broadsheet journals circulating to a few hundred individuals, this would spoil them, make them objects of envy or of high-handed administration and formality.[186]

But DeSolla Price also warns that an invisible college is self-perpetuating and has '*a built-in automatic feedback mechanism that works to increase their strength and power within science*'—that is, following our discussion on scientific paradigms in **5.2.1**, the invisible college reinforces the prevailing paradigm.

9.6.2 The social networks of engineers

The social organization and communication patterns of engineers and technologists has also long been an area of research interest, with the seminal work being undertaken between the mid-1960s and late 1970s. While this research also indicated the importance of emergent or informal social and communication networks to technology development,[187] Marquis and Allen argue that '*the communication patterns in the two areas of activity* [science and technology] *are not only largely independent of one another, but qualitatively different in their nature*'.[188]

The variations in the social organization of scientists and engineers can, in part, be explained by differences in the norms and values in science and technology. Unlike scientists who were able to organize freely and interact informally across organizational boundaries, Allen found that engineers *'are limited in forming invisible colleges by the imposition of organizational barriers'*. We noted in **3.6.1** that Allen referred to this phenomena as 'enforced localism'. This, it was argued, occurred because the vast majority of technologists and engineers were employed by profit-orientated organizations and were thus excluded from external informal communication in two ways: firstly, they must work on problems that are of interest to their employer; and secondly, they must refrain from early disclosure of research results in order to maintain the competitive advantage of the organization. Allen also noted that *'both of these constraints violate the rather strong scientific norms that underlie and form the basis of the invisible college'*.[189]

Debackere et al. view this notion of 'enforced localism' in technology as increasingly outdated.[190] For example, despite organizational constraints to informal communication across organizational boundaries, Hippel found informal know-how trading quite common in some industries, although essentially absent in others; Schrader also found extensive evidence of informal know-how trading between engineers from competing firms in his study of the US steel mini-mill industry. Rogers and Larson (and others) have highlighted the importance of friendship networks, job mobility, social focuses, and spatial proximity to the 'free-wheeling information exchange' between engineers in competing microelectronics companies in Silicon Valley; Kreiner and Schultz noted the importance of what they term 'bartering' or 'informal collaboration' between universities and industry in the Danish biotechnology sector. Finally, Hughes has clearly demonstrated that technologists do disseminate their knowledge via papers and patents.[191]

Furthermore, organizations are increasingly encouraging the formation and development of CoPs, in which engineers and technologists, for example, can coalesce and interact informally around technical specialities and issues, supported by knowledge sharing and communication technologies. Such CoPs cut across traditional organizational boundaries, and serve to supplement and complement the formal organizational structures.[192]

9.6.3 The blurring of the boundary between the social organization of science and technology

Debackere et al. contend that the boundary between what has often been referred to as the 'scientific community' and 'technological activity' is fading.[193] Indeed, it is commonplace in some sectors, such as pharmaceuticals, for university-based scientists and corporation-based scientists to network informally, and, as such, have overlapping networks.[194] This pattern has been reinforced by an increase in the incidence of private-sector funding of university research[195] and university 'spin-out' companies.[196]

As a result of this trend, Rappa and Debackere propose that the behaviour of academic researchers and industrial researchers should be studied at a more holistic level—that is, at the level of the *'technological community or R&D community'*.[197] Debackere et al. view such a community as consisting of *'scientists and engineers working towards the solution of an inter-related problem set, who are dispersed across both private and public sector organisations'*, through which *'information and knowledge flow quite freely'*. But they also note that such an R&D community *'allows for openness and secrecy at the same time'*.[198]

9.7 Concluding comments

In this chapter, we have focused on the role of social networks and informality in the innovation process. We have sought to highlight the important role of informal relationships, networks, interactions, and activities in the knowledge creation, knowledge sharing, and innovation processes of organizations. In so doing, we present the innovation process as a social, political, and emotional process, and as a 'journey', the path of which is shaped by informality and serendipity, as well as by formality and project management.

Throughout this chapter, as with the whole of **Part III**, we have focused on the internal structures, interactions, and processes that promote and inhibit innovation and creativity within the organization. Yet, as we have already alluded to several times, successful innovation rarely takes place in an organizational vacuum. Thus, in the remaining chapters (**Part IV**) we turn our attention to the ways in which innovative organizations interact with their environment during the innovation process. In **Chapter 10**, for example, we discuss the role played in the innovation process by a range of external sources, such as users, distributors, competitors, suppliers, universities, and independent inventors. These external actors can be both 'passive' suppliers of 'inputs' into the innovation process, as well as 'active' shapers of the innovation outputs.

CASE STUDY 9.1 THE DEVELOPMENT OF THE MRC-500 CONFOCAL MICROSCOPE—INFORMALITY AND RADICAL INNOVATION

Introduction

The development of the MRC-500 'confocal' microscope in the mid-1980s allowed the life sciences researcher to look into cells effectively for the first time. Whilst not the first confocal microscope to be developed, existing models were cumbersome, time-consuming to operate, and provided very dim images.

Interestingly, the working prototypes of the MRC-500 were not developed by the eventual commercializer of the innovation, Bio-Rad Ltd, nor by one of the number of university physics and engineering departments around the world, which had viewed the technique with great academic fascination since the mid-1970s. The innovation was, in fact, stimulated and developed principally by John White, a molecular biologist with a physics and engineering background, who was working at the Laboratory of Molecular Biology (LMB) in Cambridge. The LMB, funded by the Medical Research Council (MRC), is considered one of the most prestigious life science laboratories in the UK, if not the world.

Informality and the development of a radical innovation

The stimulus for the development came from the improvement in imaging required by the research project in which White was engaged at the time, although he quickly recognized the generality of the microscope's application and hence its commercial potential. White built the first prototype around 1985, essentially by adapting an existing scanning beam microscope; whilst not quite the same as the final commercialized product, it provided him with a good working model. At about this time, Brad Amos became involved in the development on a rather casual and informal basis. He was a cell biologist with an interest in optics, who was working as a contract researcher in the Zoology Department at Cambridge University. Although he had no official position at LMB, he found the laboratories there a more amenable place to work. Luckily for him, his wife worked in one of the LMB laboratories and he began to share her laboratory space.

White used the prototypes in his work throughout the development of the innovation. Equally as important, however, was that cell biologists from other projects within LMB also started to employ the prototype instrument in their research. Indeed, some of these colleagues began to pester White and Amos to make minor modifications, or to add features that neither had considered. In this sense, the LMB community provided a demanding and highly experienced user base before the microscope had even been commercialized.

From user innovation to commercial product

During 1986, White informally approached a number of companies to discuss the possibility of commercializing the confocal microscope that he had developed. After a number of unsuccessful attempts, he was approached by Andrew Dixon of Bio-Rad Ltd, who had heard of the prototype through the grapevine, and a formal collaboration was established around late 1986. What impressed Dixon almost immediately was not only seeing that the technology provided such good results, but also that it was being understood and adopted by real users.

Bio-Rad is US multinational, the management of which is technically competent, right up to the president who established the company. The management hierarchy is extremely flat and individual business units are given a great deal of freedom.

The MRC-500 project was subsequently funded by Bio-Rad, but the decision was made to continue the development of the microscope at LMB, where Dixon felt the energy and expertise for the product was located. As a result of the deal, Amos officially transferred to LMB, but while his wage was paid by Bio-Rad, he was accountable to the MRC. The development team grew to include three individuals from LMB, including White and Amos, and three individuals based at Bio-Rad in Hemel Hempstead, who were involved on the engineering and production side. In the event, the MRC-500 product launched had little resemblance to the original LMB prototype that had been 'lashed' together two years previously. For example, the prototype had a sort of analogue display and no control software, while the finished instrument had a computer display and was entirely software-controlled via a user interface.

Postscript

The MRC-500 confocal microscope has proved to be a commercial success, despite a fairly rapid response by the more established microscope manufacturers. Bio-Rad quickly built up a user base of several hundred, accounting for around 65 per cent of the total worldwide user base. In addition, the MRC-500 soon became the single largest selling product for Bio-Rad, representing 4–5 per cent of its total turnover. The MRC has also gained financially, receiving royalties on the product and further finance from Bio-Rad for future collaboration.

Questions

1. How important were informality and serendipity to the success of this innovation?
2. How might an organization encourage such informality and serendipity?

3. Do you see any management or organizational issues that might arise from such informality in the workplace?

FURTHER READING

1. For a discussion of how to reveal and 'map' social networks within organizations, see Krackhardt and Hanson (1993), Cross et al. (2002a), and Allen et al. (2007). The last of these, for example, provides a case study that employs SNA to compare and contrast the formal and informal knowledge networks within ICI (see **Illustration 9.3**).

2. For a discussion of the role and nature of micro-politics within the innovation process, see Swan and Scarbrough (2005), Jones and Stevens (1999), and Frost and Egri (2002), and in relation to project management, see Pinto (2000).

3. For a discussion of the role of serendipity in information retrieval, see Foster and Ford (2003).

4. For a recent overview and critique of the literature on contemporary careers, see Arthur (2008).

NOTES

[1] For example, Allen (1977); Kanter (1985); Freeman (1991); Kreiner and Schultz (1993).

[2] For example, Mansfield (1985); Hippel (1987); Carter (1989); Schrader (1991); Bouty (2000); Conway (1995).

[3] Schrader (1991); Bouty (2000).

[4] Freeman (1991).

[5] Krackhardt and Hanson (1993); Cross et al. (2002a); Cross et al. (2002b).

[6] Roethlisberger and Dickson (1939).

[7] Blau and Scott (1962: 6).

[8] Krackhardt and Hanson (1993: 104).

[9] For example, Blau and Scott (1962); Blauner (1964); Gouldner (1964); Pfeffer (1981); Burns and Flam (1987).

[10] Krackhardt and Hanson (1993: 104).

[11] Stacey (1996: 341).

[12] Burns and Flam (1987:214).

[13] Masterfoods USA SNA case study available online at <http://www.robcross.org> (accessed 30 September 2008).

[14] Chandler (1962).

[15] Roethlisberger and Dickson (1939); Mintzberg (1973); Tichy (1981); Monge and Eisenberg (1987).

[16] Freeman (1991: 503).

[17] Blau and Scott (1962: 6).

[18] Tichy (1981: 227).

[19] Schwartz and Jacobson (1977: 159).

[20] Burns and Stalker (1994).

[21] Tichy et al. (1979: 514).

[22] Mouzelis (1967).

[23] Stacey (1996: 340).

[24] Tichy (1981: 225).

[25] Stacey (1996: 341).

[26] Wolek and Griffith (1974: 411).

[27] Kilduff and Tsai (2003: 92).

[28] Tichy et al. (1979).

[29] Krackhardt and Hanson (1993: 111).

[30] Stephenson and Krebs (1993).

[31] Kreiner and Schultz (1993: 189).

[32] For example, Rothwell et al. (1974); Calantone and Cooper (1981); Bonnett (1986).

[33] For example, Langrish et al. (1972); Hippel (1988); Conway (1995).

[34] Hamel et al. (1989: 136).

[35] Mansfield (1985).

[36] Carter (1989: 158).

[37] Schrader (1991).

[38] Rogers (1982).

[39] Hippel (1987: 302).

[40] Carter (1989: 162).

[41] Bouty (2000: 63).

[42] Knights and Murray (1994: 31).

[43] Stacey (1996: 340).

[44] Burns and Flam (1987: 214, 233).

[45] Pettigrew (1973).

[46] Pfeffer (1981: 130).

[47] Laumann and Pappi (1976); Brass (1984).

[48] Freeman et al. (1988).

[49] Peters (1988: 138).

[50] Hippel (1988); Freeman (1991); Shaw (1993); Conway (1995).

[51] Conway (1997); Foster and Ford (2003); Eagle (2004).

[52] Olin and Wickenberg (2001).

[53] Shane (1994).

[54] Steiner (1995).

[55] Wenger (1998; 2000); Wenger and Synder (2000); Wenger et al. (2001).

[56] Kraut et al. (1990).

[57] Foster and Ford (2003).

[58] Fine and Deegan (1996).

[59] Conway (1997).

[60] For example, Allen (1977); Hippel (1977c; 1987; 1988); Kreiner and Schultz (1993); Shaw (1993); Conway (1995).

[61] Nonaka and Takeuchi (1995).

[62] Conway (1995).

[63] Dougherty and Bowman (1995).

[64] Granovetter (1985).

[65] Heimer (1992).

[66] Kanter (1972).

[67] Uehara (1990).

[68] Freeman (1991).

[69] Hippel (1987); Carter (1989); Schrader (1991); Freeman (1991); Conway (1995).

[70] Bouty (2000).

[71] Carter (1989: 155); Kreiner and Schultz (1993); Mueller (1986).

[72] Hamel et al. (1989).

[73] Bouty (2000: 63).

[74] Schrader (1991); Bouty (2000).

[75] Cross et al. (2002a: 26).

[76] Dess and Shaw (2001).

[77] Cross et al. (2002a).

[78] Boud and Garrick (1999); Coffield (2000); Tjepkema et al. (2002).

[79] Bryans and Smith (2000); Raelin (2000).

[80] Marsick and Watkins (1997: 12).

[81] Marks and Lockyer (2004).

[82] Mulec and Roth (2005).

[83] Borredon and Ingham (2005).

[84] Marsick and Watkins (1997: 12).

[85] Neumann (1988).

[86] Clarke (2004).

[87] Marsick and Watkins (1997: 3).

[88] Marsick and Watkins (1997); Bryans and Smith (2000).

[89] Peters (1988); Gwynne (1997).

[90] Levy (1994); Hertzfeld (2004)

[91] Single and Spurgeon (1996); Schrage (1999); Nobelius (2000).

[92] Quinn (1988a: 130).

[93] Van de Ven (1988: 105–6).

[94] Van de Ven et al. (1989); Van de Ven et al. (1999).

[95] Rogers and Kincaid (1981: 344).

[96] Nonaka and Takeuchi (1995).

[97] Allen (1977: 43); Burns (1969: 12).

[98] Burns and Stalker (1994); Allen (1977); Kanter (1985); Conway (1995).

[99] Bouty (2000: 63).

[100] Burns and Stalker (1994); Allen (1977); Kanter (1985); Conway (1995).

[101] Conway (1997); Jones and Conway (2004).

[102] Frost and Egri (2002).

[103] Jones and Stevens (1999: 175).

[104] Burns and Stalker (1966: xxix, 189, 101).

[105] Hislop (2005: 87).

[106] Thomas (1994: 238).

[107] Thomas (1994: 237).

[108] Kanter (1985: 212–13).

[109] Van de Ven (1988: 107).

[110] Fineman (1993; 2000); Ashkanasy et al. (2000); Lord et al. (2002).

[111] Fineman (1993: 1).

[112] Fineman (1997).

[113] Ashby (1952).

[114] Argyris (1990).

[115] Van de Ven (1988: 110–11).

[116] Warr and Downing (2000); Goleman (1995).

[117] Kuhl (1992).

[118] Brown (2000: 287); Hochschild (1983).

[119] Von Krogh (1998).

[120] Quinn (1988a: 129).

[121] Quinn (1988a: 131).

[122] West (1990, 2002).

[123] Kanter (1985: 260–1).

[124] Burns and Stalker (1994: 122–3).

[125] Von Krogh (1998).

[126] Goleman (1995).

[127] Von Krogh (1998: 143–5).

[128] Van de Ven (1988: 111).

[129] Huy (1999: 342).

[130] Schon (1963); Chakrabarti (1974); Shane (1994).

[131] Allen (1977); Aldrich and Herker (1977); Tushman and Katz (1980).

[132] Schon (1963); Rothwell et al. (1974); Peters and Waterman (1982); Kanter (1985).

[133] Schon (1963: 84).

[134] Afuah (1998: 38); Quinn (1988a: 131).

[135] Burgelman (2001: 694).

[136] Shane (1994).

[137] Dougherty and Hardy (1996: 1146).

[138] Allen (1977); Aldrich (1979).

[139] Nonaka and Takeuchi (1995: 6).

[140] Macdonald and Williams (1993b).

[141] Allen (1977); Tushman (1977); Tushman and Katz (1980).

[142] Aldrich (1979).

[143] Macdonald and Williams (1993b: 131).

[144] Allen et al. (2007: 184).

[145] Allen et al. (2007: 191).

[146] Lawton-Smith et al. (1991: 468).

[147] Allen (1977: 61).

[148] Howells and Higgins (1990).

[149] Allen (1977); Tushman and Katz (1980); Macdonald and Williams (1993b).

[150] Shane (1994); Afuah (1998).

[151] Collins and Moore (1970); McGrath and MacMillan (1992).

[152] Shane (1994).

[153] Quinn (1988a: 131–2).

[154] Aldrich and Zimmer (1986).

[155] Mintzberg (1973).

[156] Antonacopoulou (1993); Tate (1995).

[157] Boyatzis (1982).

[158] Prahalad and Hamel (1990); Grant (1991); Hall (1992).

[159] Prahalad and Hamel (1990: 82).

[160] Defillippi and Arthur (1994: 308).

[161] Aldrich (1979); Tushman and Katz (1980).

[162] Allen (1977); Tushman (1977).

[163] Moenaert et al. (1994a; 1994b).

[164] Gupta and Wilemon (1988).

[165] Macdonald and Williams (1993b).

[166] Stewart (1976).

[167] Kotter (1982).

[168] Constable and McCormick (1987); Storey (1994).

[169] Kotter (1982).

[170] Mintzberg (1973).

[171] Whitley et al. (1981).

[172] Steward and Conway (1997).

[173] Barrell and Image (1993).

[174] Kotter (1982).

[175] Kanter (1985; 1989b).

[176] Evans (1990).

[177] Defillippi and Arthur (1994); Arthur (2008).

[178] Megginson and Clutterbuck (1995); Marsick and Watkins (1997).

[179] Barber (1952).

[180] Barnes and Dolby (1971); Duncan (1974).

[181] Hill (1974); Mitroff (1974).

[182] DeSolla Price (1963); DeSolla Price and Beaver (1966); Lin et al. (1970); Crane (1972).

[183] DeSolla Price (1963); DeSolla Price and Beaver (1966); Crane (1972).

[184] Allen (1977: 137).

[185] DeSolla Price (1963: 85–90).

[186] DeSolla Price (1963: 85–6).

[187] Allen (1970; 1977); Frost and Whitley (1971).

[188] Marquis and Allen (1966: 1052).

[189] Allen (1977: 40–1).

[190] Debackere et al. (1994).

[191] Hippel (1987); Schrader (1991); Rogers and Larson (1984); Kreiner and Schultz (1993); Hughes (1987).

[192] Wenger (1998; 2000); Wenger and Synder (2000); Wenger et al. (2001).

[193] Debackere et al. (1994).

[194] Kreiner and Schultz (1993); Albertini and Butler (1995).

[195] Behrens and Gray (2001).

[196] Nicolaou and Birley (2003).

[197] Rappa and Debackere (1992).

[198] Debackere et al. (1994: 22, 32).

Part IV

The Impact of Context on the Management and 'Shaping' of Innovation

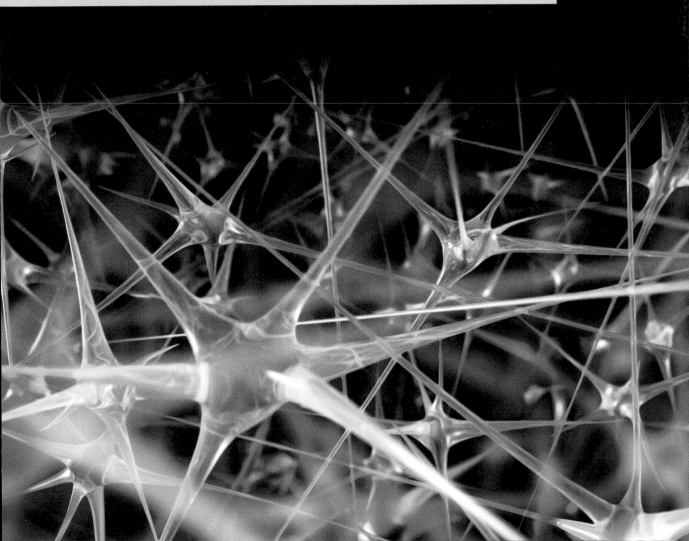

Introduction

In this fourth and final part, we shift our attention to an exploration of the ways in which organizations interact with their environments in relation to their innovative activities. Some of the background has already been set. For example, in **Chapter 1**, we noted the 'fragmented' and 'distributed' nature of knowledge, both within and beyond the boundaries of the organization, and highlighted the importance of external sources to the innovation process. In **Chapter 3**, we explored the nature and role of sectoral networks and regional clusters, as well as informal boundary-spanning social networks, and in **Chapter 6**, we employed an extended version of the resource-based view (RBV) to embrace such relational resources. In **Chapter 5**, meanwhile, we discussed the important role played by 'technological paradigms' and 'socio-technical landscapes', for example, in the constraining and shaping of technology trajectories and patterns of innovation.

In the following two chapters, we build on this foundation. In **Chapter 10**, we will focus on the external sources of innovation, highlighting the 'distributed' nature of innovation. In **Chapter 11**, we turn our attention to the broader context in which an organization is embedded, and adopt a 'systems' perspective of innovation to explore the ways in which context shapes innovation and the innovative activity of organizations.

The 'distributed' nature of innovative activity

Studies have consistently shown that new ideas leading to successful innovation seldom appear fully formed and articulated from a single source, and that the sources of innovation typically originate from a range of internal and external sources, and through a plurality of relationship types. Research concerning the sources of innovation takes a predominantly ego-centred approach—that is, the innovator is located at the centre of a set of relationships with users, suppliers, competitors, universities, and intermediaries, etc.—and this is the approach taken in **Chapter 10**. In this chapter, we provide an overview of the varied roles and contributions of such external sources during the innovation process, from idea generation through to adoption, adaptation, and re-innovation. We highlight the interesting occurrence of user and supplier innovation, in part, exemplified by the emergence of 'open-source' innovation; indeed, for Hippel and Krogh (2003: 210), *'open source software has emerged as a major cultural and economic phenomenon'*.

Innovation from a 'systems' perspective

In **Chapter 11**, we move away from the ego-centred approach of **Chapter 10**, where the innovating organization is presented centre stage, to a adopt a socio-centred perspective, where the innovator is embedded within a complex system. This builds on our discussion concerning sectoral networks and regional clusters in **Chapter 3**, and that of 'socio-technical landscapes', and 'technological styles', or 'techno-economic paradigms', in **Chapter 5**. These various concepts may be embraced within a 'systems' perspective.

Edquist (1997a: 14) defines a 'system of innovation' as incorporating '*all important economic, social, political, organizational, institutional and other factors that influence the development, diffusion and use of innovations*'. The perspective may be employed in relation to national systems of innovation (NSI), regional clusters or systems of innovation (RSI), or specific technological fields or sectors. We explore the role of such systems in supporting, shaping, and constraining innovation and innovative activity within organizations. We also introduce the theme of the 'transformative' potential of innovation (Steward, 2008): for example, how might innovation be mobilized and employed to address issues of climate change and environmental sustainability?

10

The Sources of Innovation—The Role of Users, Suppliers, and Competitors in the Innovation Process

Chapter overview

Learning objectives

This chapter will enable the reader to:

- gain an appreciation of the sources of innovation, and the role played by users, universities, suppliers, competitors, distributors, intermediaries, and consultants, in the development of new products and services;

- identify the mechanisms and relationships through which inputs from external sources flow into the innovation process;

- explain variations in the locus and variations of innovation; and

- understand the notions of 'open source' and 'open innovation', and recognize the issues and challenges that such phenomena present to innovating organizations.

10.1 **Introduction**

In our discussion of the innovation process in **Chapter 8**, we argued that innovation should not be viewed as resulting from a single idea, but from a bundle or ensemble of ideas, information, technology, codified knowledge and know-how, which may or may not be embodied within the new product, process, or service. It is also worth noting that new ideas seldom appear fully formed and articulated from a single source. Indeed, for Allen et al.:

> Bits and pieces of what eventually becomes a new idea arrive from a variety of sources . . . The individuals who introduce the new idea to the organisation, integrate these messages and in that way make their own creative contribution to the process.[1]

A key objective of this chapter is to reveal the origin of these 'bits and pieces'—that is, the sources of inputs to the innovation process.

There is strong empirical evidence to indicate that successful innovators employ a range of external sources (i.e. sources external to the innovating organization) during the innovation process, from users and suppliers, to competitors and universities. Much of the seminal work in this area was undertaken during the late 1960s and 1970s, when academics, policymakers, and managers alike were keen to identify the origins and sources of successful innovation. Subsequent studies have reinforced the key findings of this research. This body of empirical work has also shown how external sources can play a variety of important roles, such as stimulating the innovation process itself, highlighting key elements of functionality, form, and performance, generating technical solutions, and providing feedback on prototype or commercialized versions of the new product or service. In this chapter, we will make reference to both this seminal work and more recent studies.

In some sectors, such as scientific instruments, medical equipment, and commodities such as aluminium, it is fairly common for the innovation process to be dominated by either users or suppliers—that is, the 'locus of innovation' is either 'upstream' (material or component suppliers) or 'downstream' (users) of those organizations in the supply chain that might otherwise have been expected to dominate the innovation process. In this sense, what is of interest to this research is not so much the 'sources of inputs' to the innovation process, but the 'sources of innovation' itself. The seminal work in this area was undertaken by Eric von Hippel—professor of management of technology at the Sloan School of Management, MIT—during the 1970s and 1980s.[2] This chapter will outline the various explanations for this phenomenon and provide examples of those sectors where variations in the locus of innovation frequently occur. In many such instances, the 'user-innovators' and 'supplier-innovators' are not interested in commercializing their innovations, and thus it is important to explain why such organizations and individuals become involved in the innovation process in the first place, and the incentives that they may have in passing their innovations on to others to commercialize. Although it is important to note that sectors in which either users or suppliers dominate the innovation process are atypical, they provide a useful alternative perspective to force reflection on commonly held, and increasingly outdated, assumptions of the origin of ideas and innovation.

The realization of the importance of external sources of inputs to the innovation process requires an accompanying reorientation of the innovation activities of the organization to one that is outward-looking, not inward-looking, and one that is highly connected to, and not at 'arm's length' from, its immediate environment. It is important to note, however, that the external sourcing of ideas and inputs into the innovation process is not a trivial activity. It requires the organization to scan its environment proactively, and, in doing so, to build, nurture, and mobilize a diversity of relationships. It must also have the absorptive capacity to identify and assimilate 'useful' information, knowledge, and technology, and efficient and effective mechanisms for diffusing such information, knowledge, and technology within the organization, and, in particular, to those parts of the organization where it is of most value and relevance. Whilst this process can be facilitated by 'gatekeeper networks'[3] (see **3.6.1**), internal distribution, assimilation, and utilization is difficult and should not be taken for granted.[4] To this end, this chapter will focus not only on developing an understanding of the role and importance of external sources to the innovation process, but also on highlighting the nature of the mechanisms and relationships through which innovative organizations engage and interact with these external sources, and absorb external inputs into the innovation process.

10.2 **The sources of innovation**

The sources of inspiration for new products and services—their stimulus, their features and functionality, the solutions employed in their realization—are many, and often unexpected, such as those evident in **Illustration 10.1**. In this section, we will discuss the varied sources employed in the innovation process. In subsequent sections, we will take each major external source in turn, including users, universities, suppliers, competitors, distributors, intermediaries, and consultancies, to provide a deeper understanding of the manner in which each interacts with the innovating organization and contributes to the innovation process.

Illustration 10.1 The development of the festival 'blackout' tent—a collaboration between 'Blacks Leisure Group' and 'The Ministry of Sound'

The development of the 'blackout' tent for use at music festivals is a good example of how novel ideas can arise from unexpected sources. Blacks is the UK's leading outdoor leisure retailer. The Ministry of Sound is a well-known club and music brand in the UK; it has been at the centre of the UK music scene since the early 1990s and has been involved in numerous outdoor music festivals. The 'blackout' tent is the result of collaboration between these two organizations, drawing upon the unique experience and expertise of each. The tent is made of material that 'blacks out' outside light, allowing those within 'to recover in darkness' from the late nights and antics of the typical music festival, such as Glastonbury in the UK. Other key features include 'glow-in-the-dark' guy ropes, so that passers-by do not trip over the tent, and 'pop-up assembly', for ease and speed of construction.[5]

10.2.1 Seminal studies on the sources of innovation

The late 1960s and early 1970s saw a flourish of studies focused on revealing the sources of innovation. This research may be broadly divided into those studies the primary concern of which was to understand the relationship between basic science and technological innovation, and those that centred on attempting to reveal the relative importance of internal and external sources to the innovation process. With regard to the former stream of research, studies such as 'Project Hindsight'[6] and 'TRACES' in the USA,[7] for example, sought to trace back through time the scientific origins of a range of technological innovations (see **Figure 10.1** for an example of one of the cases investigated—the contraception pill).

In contrast, studies such as those by Myers and Marquis in the USA,[8] and 'Project SAPPHO', 'SAPPHO II', and *Wealth From Knowledge*, in the UK,[9] attempted to reveal the multiplicity of sources of ideas and inputs in the development of successful innovation. **Table 10.1** highlights the focus and key findings of this sample of seminal studies concerning the sources of innovation.

Three findings from this set of early studies are of particular interest. The first is that research from this period found little evidence to support the then commonly held belief that 'science feeds technology'. On the one hand, studies of the sources of ideas and inputs into the innovation process, such as *Wealth From Knowledge*, found a rather insignificant contribution from universities; on the other, studies such as 'Project Hindsight' and 'TRACES' had mixed success in identifying the scientific origins of a range of technological innovations. Although this might, in part, be explained by the often lengthy and complex path from 'laboratory to marketplace', these finding were reinforced by a contemporaneous study of the citation patterns in scientific and technology journals.[10] Such research led Allen, in the late 1970s, to conclude that:

> the activities of technology . . . while at various times in close harness with science, have developed for the most part independently. Science builds on prior science; technology builds on prior technology.[11]

This separation of science and technology was highlighted by the British government during the 1990s, for example; this gap between UK universities and businesses, it is

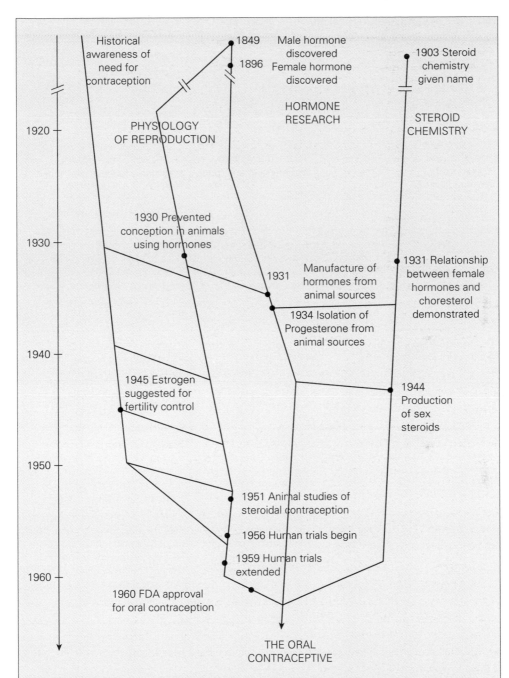

Figure 10.1 A historiograph showing the network of interactions in the development of the contraceptive pill (Illinois Institute of Technology, 1968)

Table 10.1 A sample of seminal studies concerning the 'sources of innovation'

Study		Focus	Key findings
Sherwin and Isenson (1967)	'Project Hindsight'	US attempt to trace technological advances back to their scientific origins.	In most cases studied, the trail 'ran cold' before reaching any activity that might be considered basic research.
Illinois Institute of Technology (1968) Battelle Memorial Institute (1973)	'TRACES'	US attempt to trace technological advances back to their scientific origins.	Did succeed in tracing the origins of six of the sample of technological innovations, but only after extending the time-horizon well beyond twenty years.
Jewkes et al. (1969)	*The Sources of Invention*	Study of 71 radical innovations of the twentieth century, e.g. radio, jet engine, penicillin, and xerography.	Two-thirds of the sample were developed by small firms or independent inventors rather than large firms, but a trend toward larger firms as the century progresses.
Myers and Marquis (1969)		US study of industrial innovations commercialized by a cross-sector sample of firms.	In approximately 25 per cent of innovations, the original idea originated outside of the innovating firm, and in about one third of cases, the external inputs expedited the solution.
Achilladelis et al. (1971) Rothwell et al. (1974)	'Project SAPPHO' and 'SAPPHO II'	Study of key factors in innovation process that discriminate between success and failure.	Successful innovation was aligned with better internal and external communication networks, and stronger understanding of user needs.
Langrish et al. (1972)	*Wealth From Knowledge*	UK study of winners of the Queen's Award for Technology.	Two-thirds of the important ideas employed in the development process were from sources external to the innovating firm; over half of these were from other firms and industries.

argued, has prevented a nation with a strong science base translating radical science into radical innovation.[12]

Illustration 10.2 The lag from discovery to commercialization—the case of magnetoresistance

The lag between the initial discovery of 'anisotropic magnetoresistance', by Kelvin in 1857, the discovery of 'giant magnetoresistance', by Fert and Gruenberg in 1988, and subsequent commercialization in products, such as miniaturized computer hard disks and iPods, is around 150 years.[13]

More recent research, however, has highlighted a much stronger link between science and technology, especially in sectors such as biotechnology, telecommunications, aerospace, and computer technologies,[14] and particularly in the USA.[15] A recent review of the UK government's science and innovation policies by Lord Sainsbury concluded that:

> The translation of university research into commercial goods and services has significantly increased in the past decade. The number of spin-off companies, the number of patents, the income from licensing agreements and the income from business consultancy have all increased. The performance of our universities in this area is now comparable with US universities. We are also seeing significant clusters of high-technology businesses growing up around our world-class research universities.[16]

For the UK, this, in part, reflects changes in the funding structure of universities, where industrial sponsorship and partnership are increasingly commonplace (as discussed in **9.6.3**). The UK government has also sought to bridge the gap between its scientific and technological communities, by encouraging and supporting interaction between universities and commercial organizations. Such policy initiatives are now widespread across the globe.[17]

Nevertheless, science and technology ought to be considered as 'loosely coupled', bridged by a diverse set of relationships between universities and industry, and characterized by a bidirectional flow of communication—that is, in some sectors, basic research is often 'use-inspired' or preceded by technological developments that require 'gap-filling' science. Indeed, in their study of university–industry interactions in applied research for microelectronics, Balconi and Laboranti found: '*On the whole, the research agenda of science is inspired by industry.*'[18] We will return to the role of universities in the innovation process later in the chapter (see **10.4**).

The second key finding highlighted by this body of research was the important role played by external sources in providing a multiplicity of inputs that shape the functionality of an innovation and the solutions employed in its realization. A diverse range of organizations and individuals, including users and consumers, universities, suppliers, competitors, distributors, and government research laboratories, have been found to play a role in the innovation process. Whilst the majority of these early studies, including those in **Table 10.1**, drew from a cross-sector sample and were interested in revealing the plurality of external sources involved in the innovation process, research from the late 1970s has increasingly focused on a single sector and often upon the role of a specific external source, such as users or suppliers. This latter research has provided a much deeper understanding of the origin of ideas, the impetus for innovation, and the way in which ideas and innovations are transferred from users and suppliers, for example, to innovative manufacturers and service organizations. Good examples of this latter body of research include studies of the role of the user in developing scientific instruments, medical equipment, and applications software, and the role of the supplier in developing industrial gas and thermoplastics process equipment.[19]

Studies since the late 1980s have only served to reinforce the importance of external sources to the innovation process, and the increasingly role played by users, universities, and competitors, whether through formal alliances or informal interactions.[20] For example, in their survey of over 4,500 firms in the Lake Constance region (i.e. the point of confluence of Germany, Austria, and Switzerland), Gemunden et al. conclude that *'Firms which do not supplement their internal resources and competence with complementary external resources and knowledge show a lower capability for realizing innovations'.*[21] In a major survey of the networking and innovation literature by Pittaway and colleagues, meanwhile, it was concluded that:

> the evidence shows that the innovation process, particularly complex and radical innovation processes, benefits from engagement with a diverse range of partners which allows for the integration of different knowledge bases, behaviours and habits of thought.[22]

Furthermore, a recent survey by IBM of CEOs found that only 17 per cent considered internal research and development (R&D) as the most significant source of innovative ideas. For Chesbrough, such patterns of innovative behaviour, where organizational boundaries are increasingly porous, are indicative of a new era of 'open innovation'.[23] We will explore this notion towards the end of the chapter (see **10.9**).

The third key finding relates to the origin of the stimulus of the innovation process itself. A review of this research indicates that between two-thirds and three-quarters of innovations are need-stimulated,[24] and that there is a positive correlation between need-stimulated innovation and commercial success.[25] An innovation may be said to be 'means-stimulated' or 'need-stimulated';[26] these terms are similar to the concepts of science or technology-push and market-pull, respectively, which were introduced in **Chapter 3**.

The above research, then, serves to reinforce the importance of the adoption of an 'interactive' or 'network' model of innovation. In the subsequent sections of this chapter, we will take an in-depth look at the role and importance of users, universities, suppliers, competitors, distributors, intermediaries, and consultancies in the innovation process.

10.2.2 Variations in the role and importance of external sources

Given the preponderance of studies in hi-tech sectors, such as information technology (IT) and biotechnology, one has to be careful not to overgeneralize the results of such research, and to miss subtle and not-so-subtle differences between organizations, sectors, and nations.

For example, a more sensitive review reveals that the importance of external sources to the innovation process has been found to be contingent upon a number of factors, such as the following.

- *Organizational size*—small and medium-sized enterprises (SMEs) tend to source a greater proportion of their inputs into the innovation process from external sources. This is, in part, explained by the need to supplement often very limited internal resources.[27] Indeed, research highlights the particularly important role that intermediaries and consultancies can play in the innovation process of SMEs— see **10.7.2** and **10.7.3**—through the brokering of knowledge, technology, relationships, and funding.

- *The nature of the innovation*—there tends to be greater external input into product innovation as opposed to process innovation,[28] with a study of German manufacturers, for example, suggesting that cooperation with users is more likely in product innovation and cooperation with suppliers more likely in process innovation.[29]

- *The radicalness or maturity of an innovation*—research has indicated that external sources—and, in particular, users—are of greater importance to mature products, characterized by incremental innovation. In such cases, users have often developed a sophisticated understanding of the performance, uses, and limitations of the innovation that is not evident at earlier phases, and are thus more capable of contributing to the innovation process.[30] In their review of the literature, Pittaway et al. noted that the evidence suggests an orientation toward suppliers and consultants for innovation new to a market (e.g. involving the transfer and adaptation of existing innovation), and universities for radical innovation.[31] Similarly, a recent study in Holland suggests that R&D cooperation with universities is linked to radical innovation, competitor cooperation to either radical or incremental innovation, and supplier cooperation to incremental innovation.[32]

- *Business sector*—there is clear evidence that hi-tech sectors, such as biotechnology, computers, and telecommunications, are characterized by dense networks of organizational and personal relationships, which are mobilized to source ideas and inputs into the innovation process, and, in some cases, for co-development. The utility of public sector research has been found to vary dramatically between sectors, being much more important in pharmaceutical and aerospace than plastics and fabricated metals.[33] Earlier studies have also found that external sources were less important for innovation in consumer goods[34] and in the services sector[35] than for industrial goods, although there is emerging evidence of increased networking in a range of service sectors, from banking to hotels.[36]

- *The stage of the innovation process*—external sources are employed more frequently in the initial idea-generation phase than in later problem-solving phases.[37] It has also been found shown that users are more often employed in the idea generation and field-testing stages, and universities, suppliers, and consultancies, for example, in the problem-solving stage of the innovation process.[38]

- *Nations*—in a recent survey of around 200 of the 'largest R&D performing' companies in the USA, Japan, and Europe, for example, a major variation was found in the degree to which companies collaborated with customers, with 52 per cent of Japanese companies reporting a high frequency of collaboration, as compared to 44 per cent of US firms and 38 per cent of European firms.[39] Studies have also indicated that a much stronger relationship has existed traditionally between universities and industry in the USA, as compared to Europe.[40]

Such variations highlight the value of the wide range of sector-specific surveys and case studies that have been undertaken, such as those within agriculture,[41] software,[42] aerospace,[43] scientific instruments,[44] medical equipment,[45] biotechnology,[46] automotives,[47] convenience stores,[48] and various outdoor leisure pursuits, such as mountain biking and kayaking,[49] as well as sector[50] and national comparisons.[51]

10.2.3 The channels and mechanisms for the transfer of external inputs into the innovation process

In addition to the sources of ideas and innovation, the literature also reveals a strong interest since the early 1960s in the 'mechanisms' and 'channels' for transferring external inputs into the innovation process. These can be usefully categorized along two axes: formal/informal, and passive/active. The notion of 'formal' channels would incorporate all those that have been formally sanctioned by managers, such as alliances and joint development projects with external organizations, i.e. planned and structured relationships and interactions. In contrast, 'informal' channels refer to the personal relationships and the emergent and unstructured interactions of individuals, perhaps through their attendance at conferences or general networking within a range of technical, professional, and user networks and communities. 'Active' channels or mechanisms of transfer refers to those that involve direct and multiple interactions between the innovating organization and external individuals or organizations, whilst 'passive' channels refer to those that do not involve direct interaction, e.g. textual material.[52] **Figure 10.2** organizes some of the more important mechanisms and channels for the sourcing of ideas and information into the above categories.

Whilst there is plenty of evidence to suggest a substantial rise in formal organizational-level relationships, such as strategic alliances,[53] there is also abundant research that highlights the importance of personal relationships and informal networking.[54] By

	Active	**Passive**
Formal	Alliances & Joint Ventures Co-Development Hiring of Staff Supply-Chain relationships Consultancy	Licensing Market Surveys Idea Competitions
Informal	Science and Engineering Networks Profession Networks (e.g. clinicians, librarians etc.) User Communities (e.g. sports and hobbies)	Journal articles, trade journals, and conference papers Internet searches Imitation, Counterfeiting, and Reverse Engineering

Figure 10.2 The diversity of mechanisms and channels for sourcing ideas

engaging with science and engineering networks, organizations are able to tap into a wide range of external expertise and research, whilst by interacting with profession networks (e.g. clinicians, librarians, designers), or user communities (e.g. software gamers, snowboarders, mountain bikers), organizations are able to determine emerging trends in needs, for example. Furthermore, Freeman argues that:

> Behind every formal network, giving it the breath of life, are usually various informal networks . . . Personal relationships of trust and confidence (and sometimes fear and obligation) are important both at the formal and informal level . . . For this reason cultural factors such as language, educational background, regional loyalties, shared ideologies and experiences, and even common leisure interests, continue to play an important part in networking.[55]

Innovative organizations also utilize a wide range of passive channels for sourcing ideas and information, from licensing and reverse engineering, to Internet searches, and scientific and trade journals.

Over the remainder of this chapter, we will explore the role played in the innovation process by a range of external sources, such as users, universities, suppliers, and competitors. Each innovator will employ a different balance in the mechanisms and channels illustrated in **Figure 10.2** to draw upon these different sources.

10.3 The role of the user in the innovation process

The user, or customer, was first identified as a key source of ideas and inputs to the innovation process in studies as early as the 1960s[56]—but it was the pioneering work of Eric von Hippel, from the mid-1970s, that appears to have caught the imagination of academics and practitioners alike.[57] Research on the role of the user continues to abound and appear in a wide range of management journals, such as those focusing on innovation studies, marketing, strategy, and operations management.[58] Governmental bodies have also highlighted the importance of the user to the innovation process. For example, the recent UK government policy paper on innovation—*Innovation Nation* (March 2008)—recognizes the important contribution and future potential of user involvement in the innovation process.[59]

10.3.1 From the user as a source of ideas to user innovation

One aspect that is striking about the literature is the wide variation in the nature of user involvement in the innovation process. A useful starting point, then, is to summarize this breadth along a spectrum, which may be banded in the following manner.

- *The user as a source of ideas*—it has long been recognized that an accurate understanding of user needs is near essential in developing commercially successful products.[60] To obtain such an understanding, the user is approached by the innovating organization to determine market needs or to elicit product feedback and new product ideas. In this sense, the role of the user is very limited and reactive, only 'speaking when spoken to'. This is perhaps the classic perception of the role of the user in the innovation process, but it greatly underplays their potential.[61]

- *Continuous interaction with user*—many studies have indicated the importance of 'multiple and continuous' interaction between the innovating organization and the user base, often on an informal basis, and at stages throughout the innovation process, from idea generation, specification setting, and prototype testing, to adoption and diffusion. Such interactions not only allow the innovator to reveal user needs and preferences, but may also offer opportunities for them to complement their own R&D expertise and efforts by plugging in to the technical strengths of their industrial customers.[62]

- *The user as co-developer*—users are formally integrated into the innovation process through, for example, invitation onto the development team by the innovator organization.[63] Novel forms of user co-development are starting to emerge: for example, in gaming software, innovators are organizing innovation within online consumer communities.[64] In other sectors, such as in medical equipment, surgeons were found to organize the co-development process, themselves, bringing together the various experts and manufacturers required to refine their prototypes and build commercial products.[65] Integrated toolkits are also being developed by innovators that allow their users to create and test designs for customized products or services, which can then be transferred seamlessly back to the innovator to deliver.[66]

- *The user as innovator*—driven by his or her own needs, the user determines the specification for the innovation, develops and tests fully functional prototypes, but often falls short of commercializing the innovation, instead approaching an organization upstream to fine tune the design and bring it to the marketplace.[67] Studies have revealed the high incidence of user innovation in a number of industries, such as scientific equipment, medical equipment, and semiconductor and electronic subassembly process technology,[68] and that it is prevalent among users of a wide range of software applications, from personal computer (PC) computer-aided design (CAD) systems, to library information systems and shipping documentation.[69] Recent studies have also highlighted the high incidence of user innovation among consumers in outdoor leisure and sports equipment, such as that used in climbing, mountain biking, and hiking;[70] innovation and free-revealing in many of these communities is similar to that found in the development and diffusion of 'open source software'.[71]

- *The user as entrepreneur*—having conceived and developed the innovation, the user takes the next step and commercializes the innovation. This is perhaps more prevalent than one might expect. In his development of a typology of technological entrepreneurs based on previous occupational background, Jones-Evans found that around 15 per cent had previously been end-users in the area of the innovation being exploited.[72] User entrepreneurs are perhaps more common in relation to consumer goods, because many successful serial entrepreneurs are keen observers of needs—their own as well as those around them—and adept at converting these into successful innovations. Good examples from the UK include the well-known case of James Dyson and his commercialization of the 'cyclone' bagless vacuum cleaner,[73] and the much less known case of Liz Paul and her commercialization of Menitest, a home testing device for meningitis.[74] Indeed, studies of user innovation in rodeo kayaking, and baby and juvenile products, such as car seats and high chairs, show a high incidence of commercialization by user innovators.[75] With regard to baby and juvenile products, 84 per cent of the 263 US firms sampled were founded by user-entrepreneurs—the vast majority by parents. User entrepreneurship is not, however, limited to industrial and consumer products: many charities, for example, are established by sufferers of illnesses, or their families, to provide innovative services for fellow sufferers and their carers. Such might be considered to be social entrepreneurship[76] by 'users'. **Illustration 10.3** provides such an example.

Illustration 10.3 The National Autistic Society (NAS)—social entrepreneurship

The National Autistic Society (NAS) was founded in the UK in 1962 by a group of parents who were seeking to ensure a better quality of life for their autistic children. Today, it is the leading UK charity supporting people—sufferers and family members alike—affected by autism. Like many such charities, this 'self-help', grass-roots society has evolved into a charity with a national presence, eighty branches, and 17,000 members, many of whom volunteer thousands of hours to help to support the activities of the organization. Today, NAS provides a wide range of innovative services in relation to training, advocacy, and support for sufferers and their families, including specialist residential and outreach support. The society also campaigns and lobbies on behalf of those impacted by autism.[77]

In the following subsections, we will explore in detail how innovating organizations are able to engage users in order to identify opportunities, capture ideas, obtain feedback on prototypes, and transfer innovations from the end-user. We will also explore the notion of 'lead users' and their important role in the innovation process.

10.3.2 Identifying customer needs

Involving users in the formative stages of an innovation is often crucial to its subsequent commercial success. In part, this is because there is often a notable mismatch in the perception of users and innovators with regard to the weighting of different product or service characteristics. For example, users are typically much more aware of the importance of the total 'life cycle' costs of products than innovators, who are often more concerned with purchase price. Information from users about needs is often bundled together with valuable design data, such as rough functional specifications, and implicit or explicit preferences with regard to the general type of solution to be adopted. Users can also play a vital role in establishing the optimum price/performance combination, which, in turn, helps establish the optimum design specification.[78] But innovators must be wary, becuase there are many ways that product planners can be misled by user information.[79]

In contrast to consumer markets, identifying customer needs in industrial markets can be more problematic and expensive. Not only are needs likely to vary more dramatically between industrial or professional users, but those needs are likely to change more rapidly. As a result, innovators must adopt different methodologies and techniques to identify customer needs between these two broad types of market. For example, whilst surveys and focus groups might work well with consumer products and services, and in particular for fast-moving consumer goods (FMCGs) such as groceries, multiple interactions between innovator and user organizations, or the hiring of engineers and scientists with user experience, work better for products and services intended for other organizations.[80] Increasingly, innovating organizations are also recognizing the potential of tapping into user (professional and consumer) communities.[81] Lead users, whether consumers or user organizations, can play a particularly important role in the early identification of user needs (see **Illustration 10.4**).[82] We will discuss the role and nature of lead users in more detail in **10.3.5**.

Illustration 10.4 Karrimor—drawing from the expertise of its users

Before being bought by Benetton in the late 1990s, Karrimor was a medium-sized family-owned company. It had been designing and manufacturing innovative outdoor pursuit equipment for over fifty years. Karrimor has many firsts to its name, including the first internal frame rucksack in 1976, the first textured and waterproof fabric in 1978, and the first self-adjustment system for rucksacks in 1983. Karrimor equipment has been used in many successful ascents of Everest, including the first ascent of the south-west face, the first ascent without oxygen, the first female ascent, and the first ascent of the Kangshung face. Over the years, many famous names worldwide have thus used Karrimor equipment and have had close associations with the company. Indeed, many of these adventurers are close friends of Mike Parsons, the son of the founder and the then managing director.

It is not only through such expeditions that Karrimor maintained close contact with its customers: the company also sponsored and, in some cases, organized a number of events, such as the Karrimor International Mountain Marathon and the Karrimor Three Peaks Yacht Race.

To aid the innovation process more directly, Karrimor drew from the intensive input of a think tank and a test team, made up of a variety of highly experienced and semi-professional users. The five or six individuals who made up the think tank were carefully selected for their all-round knowledge of mountain activities, training, and outdoor education, and were instrumental in providing input at the concept stage of the development process. The test team, of a similar size, drawn from various walks of life and involved in a variety of outdoor activities, tested both prototypes and production models thoroughly and independently, before reporting back its findings to the company.[83]

Needs may only surface during 'field tests' or post-commercial use. This is particularly the case with novel or complex innovations. We now turn to the role of usage in the innovation and re-innovation processes.

10.3.3 Usage and adaptation—the role of users in prototype testing and re-innovation

Usage, whether through the field testing of prototypes or everyday use of commercialized innovations by users, plays an important part in the processes of innovation and re-innovation. This is perhaps not at all surprising for non-consumer goods, because, as Habermeier argues, the user organization *'constitutes a complex system, embedded in an often equally complex environment'*.[84] Such usage within a variety of contexts and environments allows for embedded assumptions to be tested and weaknesses to be revealed in the original design, which can be particularly important for novel or complex innovations.[85] As a result, the process of learning through usage is often one of learning from mistakes or oversights, or even from abject failure. This can be crucial to the incremental refinement and development of robust solutions, because it not only provides the incentive to improve the innovation, but also offers essential feedback on the most suitable direction to proceed.[86] Indeed, Petroski, who has written extensively on the role of failure in successful design, contends that '*success and failure in design are intertwined*'.[87]

Innovators employ a range of approaches in undertaking pre-commercial tests by users. In many instances, the approach is very informal, often employing friends who can be trusted,[88] or ad hoc, as is common with computer software, where users are typically invited to download software from the Internet and provide feedback online. In other cases, the process is more formal and structured, as in the use of test teams by Karrimor (see **Illustration 10.4**). In some sectors, new products are typically required to proceed through formal trials, such as with pharmaceuticals and agricultural pesticides.

Even where the innovator has a clear understanding of user needs and requirements, usage still has an important role to play, due to the '*definite and almost insurmountable limits*' on what can be learnt through theory, simulation, and laboratory experimentation.[89] Furthermore, many modern technologies are composed of numerous interacting components and, even when the scientific principles underlying each component are well understood, the behaviour of the system as a whole may show unexpected properties. This is as true for innovative new systems and processes, as evidenced by the fiasco that accompanied the opening of Terminal 5 at Heathrow in March 2008, as it is for products, such as new versions of applications software: not only is Windows 2000 comprised of some 20 million lines of code and Windows XP around 40 million, for example, but this software must operate seamlessly alongside various versions of a plethora of other applications and across a wide range of hardware configurations.

Sometimes, the scientific principles underlying a technology are not well understood, forcing usage to the foreground of innovation and re-innovation. This is well exemplified by 'frontier' projects in the aviation, space, and construction industries. On this basis, Petroski argues that:

> every new structure which projects into new fields of magnitude involves problems for the solution of which neither theory nor practical experience [are] an adequate guide. It is then that we must rely largely on judgement and if, as a result, errors or failures occur, we must accept them as a price for human progress.[90]

As an innovation matures, however, user needs and scientific principles generally become much better understood, and thus the reliance on usage declines.

In addition to filling knowledge gaps and revealing design faults, usage also has the potential to uncover new and unexpected needs and applications, and ultimately lead to re-innovation through adapting the original innovation for new uses. In some instances, re-innovation might be undertaken by users during the process of adoption and implementation of an innovation.[91] A good example of this is provided by Slaughter, in her study of innovation by 'user-builders' in the US residential construction sector. Here, she found that:

'Learning-by doing' and user innovation appear to be inextricably linked processes during the implementation of a technology. Testing the limits and requirements of a technology by using it can provide opportunities for significant changes which alter its application or performance.[92]

The intangible nature of service and administrative innovations potentially offers greater scope for user re-innovation than that for technological product or process innovations. Indeed, many of the key examples of re-innovation provided by Rogers in his seminal work on diffusion centre around innovations in educational programmes and computer-based administrative tools.[93] But evidence also suggests that product innovations can be designed so as to be 'user-flexible'—that is, amenable to adaptation and improvement by innovative users.[94]

Finally, usage can also act as the 'battleground' for competing and emergent designs. In an interesting historical case study of the co-evolution of two competing dinghy designs that brought affordable sailing to post-war Britain, Blundel describes how:

> sailing clubs acted as the primary testing-ground for dinghies manufactured under the new technologies. With many new classes [of dinghy] being introduced competition on the water mirrored that taking place in the marketplace.[95]

Similar patterns have been observed in other outdoor pursuit activities, such as mountain biking and kayaking.[96]

10.3.4 Transferring user-innovations from users to producers

Although users have been shown to innovate in a wide range of industrial and consumer markets, from medical and scientific equipment, to mountain bikes and kayaks, this is no guarantee that such innovations will be commercialized by either the users themselves or upstream producer organizations. In some cases, the producer has been approached by the user innovator, or is aware of the innovation, but is not interested in commercializing it; in others, the producer is simply unaware and has not been approached. Indeed, many user innovators show little or no real interest in commercializing their innovations or transferring them to producers, simply being '*satisfied with the realization of their ideas for personal use*'. In many instances, however, the innovation is 'freely revealed and diffused' within the user community, whether that be librarians, snowboarders, or software gamers.[97] Where the producer is able to identify such cases of user innovation, both of these circumstances offer substantial opportunities for innovating organizations to source user innovations, and at a minimum, detailed and tested user need and solution data. Informal networking is an important mechanism adopted by innovators to aid this process of identification, allowing them to plug into the relevant user communities, whether through attending conferences or sporting events, or subscribing to online communities, for example.

But the identification of user innovation is not enough. Innovating organizations must also transfer the innovation from the user, and assimilate the accompanying need, use, and solution data. Studies of scientific instruments, semiconductor and electronic subassembly manufacturing equipment, and medical equipment have found that the most common pattern of transfer of user innovations to manufacturers was through multiple and continuous user–producer interaction. Such interactions may be either formal or informal, or a combination of both.[98] In some cases, such interaction is initiated and coordinated by the user.[99]

Intermediaries can also play an important role in bridging the gap between innovative users and innovators (see **10.7.2** for further discussion of the role of intermediaries). In his UK study of user innovation in medical equipment, Shaw outlines the critical intermediary role played by governmental bodies in providing a framework for the commercialization of innovations by doctors and clinicians:

> the MRC [Medical Research Council] personnel . . . assist in the choice of manufacturer, develop jointly with the manufacturer pre-production models and then give active support in the marketing of the equipment. The other intermediaries are the Department of Health and Social Security (DHSS) consultants and the Department of Industry (DoI). The DHSS has a consultant in every medical discipline who links into the networking system and who enjoys and is encouraged to have good contacts with equipment manufacturers . . . a useful link between the user and the manufacturer.[100]

We noted in **10.3.1** that innovators are increasingly employing integrated toolkits that allow their users to create and test designs for customized products or services that can then be transferred seamlessly back to the innovator to deliver. Thomke and Hippel, for example, reveal how Bush Boake Allane (BBA), a supplier of speciality flavours to companies such as Nestlé, have built a tool that enables users to develop their own flavours, which can then be produced by BAA. To encourage such user contribution by consumers, Jeppesen and Molin argue that '[user] *development and use of the product should be equally playful*'.[101]

Illustration 10.5 The Pillsbury biannual baking competition

Pillsbury—now owned by two large US food product companies, General Mills and J. M. Smucker—is a well-known flour and bakery products brand. Established in 1949, this biannual contest, held in the USA, is designed to publicize Pillsbury's products. Entrants develop recipes under one of five categories, including 'Breakfast and Brunches', 'Pizza Creations', and 'Sweet Treats', each of which must incorporate two or more listed ingredients, consisting of Pillsbury-branded products and those from sponsor organizations. Tens of thousands of entries are received and judged each year. Winners receive a prize (US$1m for the overall winner in the forty-third competition judged in April 2008) and their recipes are published. Although intended as a publicity vehicle, Pillsbury has commercialized a number of the competition recipes.[102]

Other less interactive and indirect modes of transferring ideas from user innovators to producers include, for example, the licensing of patents, conference papers, and journal articles. The Pillsbury baking competition discussed in **Illustration 10.5** provides a good example of a novel method for identifying and transferring user innovations. Inter-net-based versions of such idea competitions are also beginning to emerge.[103]

10.3.5 **Lead users and tough customers**

In the above sections, we have discussed the important role played by users in the innovation process—but central to the success of this role is the careful selection of users.

So on what basis should innovators make this selection? Drawing upon a range of research into individual problem solving and the effectiveness of relatively sophisticated marketing research techniques, such as the 'multiattribute mapping' of product perceptions and preferences, Hippel concludes that *'Users steeped in the present are . . . unlikely to generate novel product concepts that conflict with the familiar'*.[104] This he sees as being particularly problematic in fast-changing hi-tech industries. Hippel proposes that, in such sectors, 'lead users' can play an important role in identifying novel needs or product ideas, because although they too will be constrained by the familiar, they are *'familiar with conditions that lie in the future'* for the typical user.

Hippel identifies lead users as those that face similar needs to other users, only months or years ahead of the typical user in the marketplace, and who are positioned to benefit substantially from the meeting of those needs. Lead users are broadly akin to the 'innovator' and 'early adopter' categories as outlined by Rogers (see **4.7**).[105] To this categorization of the lead user should be added the notion of the 'tough customer'—that is, the technologically demanding customer who insists on high-quality, high-reliability solutions.[106]

Taking these various features in turn, we can summarize as follows.

- *Users that are innovative*—innovative users are those who face needs months, maybe years, ahead of the general marketplace. This has been termed 'leading-edge status'.[107] This aspect is particularly important in fast-moving hi-tech fields, in which the experience of ordinary users is often rendered obsolete very rapidly; for such users, their insights into new product needs and potential solutions are constrained by this obsolete experience, hence making it unlikely that they will help to generate novel product concepts.[108] Where domestic users are technically unprogressive or undemanding, this can present the innovator with considerable problems. For example, it has been argued—most notably by Porter, in his well-known text *The Competitive Advantage of Nations*—that the comparative weakness of domestic users in an industry can go some way to explaining the weakness of their domestic suppliers.[109]

- *Users that are representative*—lead users must be representative of the broader user base, otherwise any resulting designs will have little general market appeal.[110] In **4.7**, we discussed different types of adopter, and we noted that whilst 'innovators' have a keen interest in new ideas and were the first to adopt an innovation in a given target market or social system, they were not necessarily representative of the broader market or social system. As Rogers notes: *'This interest in new ideas leads them out of a local circle of peer networks and into more cosmopolite social relationships.'* In contrast, 'early adopters' are embedded in their local social system; they are 'localites', rather than 'cosmopolites', and are thus likely to be more representative of the broader population and, as a result of their central position in the communication networks of a social system, can have a major impact on the diffusion of an innovation within that social system.[111]

- *Users who benefit significantly*—lead users should be positioned to benefit significantly by obtaining a solution to a given need, because the effort expended by the user to find a solution is likely to be directly proportional to the benefit obtainable. Such users are thus also likely to have devoted the most resources and hence obtained the deepest understanding of the need.[112]

- *Users that are demanding*—selected users should also be demanding, requiring high-quality, high-reliability products and services that provide the innovator with a stringent design stimulus. Companies involving their most demanding customers tend to introduce viable innovations at a faster rate than their immediate competitors. Gardiner and Rothwell also argue that:

> The tougher and more demanding the customers are in their requirements, the better and more *robust* the designs will be along with their probability of re-innovation and propensity for successful long-term commercial exploitation.[113]

Both experimental and 'real-world' (e.g. that at 3M) applications of the lead user method, which involves the identification of lead users, the generation of novel needs and solutions by these users, and the testing of these ideas among a broader sample of users, have demonstrated that the concept can be successfully operationalized.[114] This research has also highlighted the efficacy of the concept and its application. For example, in a study of users of CAD systems for printed circuit boards, those identified as lead users were found to have unique and valuable ideas regarding new product needs and potential solutions to those needs, and that these ideas were strongly preferred by a broader and representative sample of users.[115] A study of adaptation and re-innovation among librarians of a computerized information search system, meanwhile, found a strong correlation between user innovation and lead user status.[116]

Perhaps most interesting is an evaluation of the application of the lead user approach by 3M by Lilien et al., who conclude:

> ideas generated by LU [lead user] processes had forecast sales in Year 5 that were more than eight times higher than the sales of contemporaneously funded projects . . . significantly more novelty . . . addressed more original newer customer needs, and also had significantly higher forecasted market share in Year 5 (on average, 68% vs. 33% for non-LU ideas).[117]

It would appear from such studies that the lead user method has great potential relative to traditional idea-generation techniques, which typically would collect customer need information from a random or representative sample of customers. Because innovative organizations are increasingly attempting to tap into user communities for novel ideas, it is important to recognize that such communities are often divided into sub-communities. In such cases, it follows that organizations must become adept at identifying and tapping into 'lead communities'.[118]

10.4 The role of universities in the innovation process

Towards the start of this chapter, we noted that studies from the 1960s and 1970s had found little evidence to support the then commonly held belief that 'science feeds technology'. Indeed, studies of the sources of innovation from this period had found a rather insignificant contribution from academic and scientific communities. Research from the 1990s onwards, however, has highlighted the increasingly significant role played by science and universities in a number of key and emerging sectors, such as biotechnology, telecommunications, aerospace, and computer technologies,[119] and in particular in the USA.[120]

The diverse ways in which science and universities 'feed' industry is succinctly summarized by Sampat:

> The economically important 'outputs' of university research have come in different forms, varying over time and across industries. They include, among others: scientific and technological information . . . equipment and instrumentation . . . skills or human capital . . . networks of scientific and technological capabilities . . . and prototypes for new products and processes. The relative importance of the different channels through which these outputs diffuse to . . . industry also has varied over industry and time. The channels include, inter alia labour markets (hiring students and faculty), consulting relationships . . . publications, presentations at conferences, informal communications with industrial researchers, formation of firms by faculty members and licensure of patents by universities.[121]

This range of university involvement in the innovation process maybe distilled into the following spectrum.

- *The university as a source of technical information and expertise*—innovative organizations are able to tap into the knowledge and expertise located within universities in a number of ways, from the passive, such as academic publications, to the active, through the interaction between academics and individuals within innovative organizations. These interactions range from one-off, formal, consultancy arrangements, in which the university is engaged to undertake a specific of work, perhaps around problem solving or validation, through to informal, ongoing networking between academics and industrial engineers and technologists. Such informal interactions should not be underestimated.[122] Indeed, in their study of university–industry interaction in the microelectronics sector in Italy, Balconi and Laboranti concluded that:

> > networks of academics and industrial researchers are a fundamental instrument of collaboration between the two worlds and seem quite effective in enhancing productivity in terms of both discoveries and inventions.[123]

Emerging evidence suggests that geographical proximity can promote knowledge spillovers from universities to industry.[124]

- *The university as innovator and a source of technology*—following the lead of US research universities, universities around the world are increasingly seeking to patent in-house research and generate revenue through licensing, often through increasingly sophisticated and business-orientated technology-transfer departments or university 'spin-out' companies. Innovative organizations are clearly tapping into this source of technology: statistics for the USA, for example, show a near sevenfold increase in university licensing income during the 1990s, from under US$200m in 1990 to nearly US$1,300m in 2000, although a small number of universities account for the majority of this revenue.[125] Whilst the rise of such patenting and licensing activity has financially benefited the elite research universities, however, this process has been criticized '*as representing a socially inefficient "privatization" of academic research and as a threat to the ethos of science itself*'.[126]

- *The university as co-developer*—innovators are not only increasingly relying on joint R&D projects with supplier, competitor, and user organizations, but they are also becoming increasing engaged in joint research activity with universities. This is particularly the case in sectors such as biotechnology, IT, and aerospace.[127] Whilst this route allows innovators to draw upon highly specialized and state-of-the-art knowledge and expertise, the clash in cultures between research universities and private sector companies can create major tensions in university–industry partnerships. (We outline these tensions later.)

- *The university as entrepreneur*—universities and academics are also increasingly creating technological 'start-up' companies to develop and commercialize novel products and processes, although these tend to originate largely from top research universities. Such start-ups are concentrated within the life sciences and IT fields, where they have made significant contributions to the development of these sectors. Although spin-out activity is accelerating around the world, there are major variations between different countries, flourishing where there are positive political, cultural, and legal 'signals' and availability of finance, such as in the USA and the UK.[128]

Whilst there is great innovative potential in university–industry partnerships, research universities and private sector companies generally operate under very different cultures, manifested in divergent goals, conflicting time orientations, and differences in language and underlying assumptions.[129] At the extreme, there remains '*cultural predispositions against academic involvement with commerce*', particularly within a number of European countries, and in relation to applied and commercial research as opposed to basic science and 'blue sky' research. This predisposition often results in indifference or resistance to the commercialization of university research. Indeed, a recent report by a body representing twenty elite research universities across Europe—the League of European Research Universities (LERU)—strongly challenges the emerging view among European governments that universities have a major role to play in driving innovation within economies.[130]

Equally as important is the tension that exists in relation to disclosure limitations that often accompany industry funding of university research. Restrictions on the publication of research not only creates a fundamental mismatch between the reward and recognition systems within the university sector, where the mantra remains 'publish or perish', but falls foul of well-entrenched academic norms of disclosure and the diffusion of knowledge. Whilst this may be seen as one of the unintended consequences of university–industry partnerships, it plays a major role in discouraging academic engagement with industry. These important issues remain to be fully resolved.[131]

10.5 The role of suppliers in the innovation process

Although suppliers have long been identified as one of the key sources of inputs into the innovation process, it is really only since the late 1990s that there has been a substantive body of research that has centred on the role and integration of the supplier into the innovation process. Seminal work in this area was undertaken during the 1980s, triggered by an attempt to understand the success of Japanese manufacturing companies—and automotive firms, in particular.[132] Today, partnerships and alliances with suppliers are seen to be an increasingly important component of developing and maintaining competitive advantage.

Close relationships may emerge between suppliers and producers in the innovation process where there is technological uncertainty,[133] and where the producer has limited internal expertise or knowledge is 'sticky'.[134] Petersen et al. also note the value of involving suppliers in the process of project goal setting: *'Suppliers, because of their technical knowledge and expertise, may have more realistic technology goals and information on tradeoffs involved in achieving particular goals.'*[135] But there are also risks, such as the potential for loss of proprietary knowledge, reduced control over the innovation process, and the costs and resources associated with managing collaboration with suppliers.[136]

As with user involvement in the innovation process, supplier involvement also varies widely. Once again with reference to earlier research, we are able to identify a series of categories of involvement that are similar to those of the user, as follows.

- *The supplier as a source of ideas*—the supplier is approached by the innovating organization with, for example, queries regarding the tolerances or appropriateness of certain materials or components. In such cases, the role of the supplier is often limited and reactive, only 'speaking when spoken to'. This is perhaps also the classic perception of the role of the supplier in the innovation process.[137]

- *Continuous and early interaction with supplier*—the innovating organization actively seeks to involve the supplier at stages throughout the innovation process, from idea generation, specification setting, and problem solving.[138]

- *The supplier as co-developer*—suppliers are formally integrated into the innovation process through invitation onto the development team, which might involve the co-location of engineers of the supplier and innovating organization.[139]

- *The supplier as innovator*—driven by its own objectives, the supplier determines the specification for the innovation, and develops and tests fully functional prototypes, although it may fall short of commercializing the innovation itself, often approaching an innovator to fine tune the design and bring it to the marketplace.[140]

Each of the above categories of supplier involvement require some degree of supplier–innovator integration. Despite the importance of suppliers to the innovation projects of many organizations, however, managers perceive the processes of integration as a 'black box'.[141]

Nevertheless, recent studies have indicated a number of critical factors for successful supplier integration:[142]

- early involvement of the supplier in the development process, preferably during the concept exploration and definition stages;

- establishment of integrated supplier–innovator project teams, although co-location has been found to be beneficial, but not critical;

- ensuring the free flow of technology and cost information between innovator and supplier; and

- long-term commitment to the supplier.

There are, however, some indications that issues of trust arising from close integration that need to be overcome—as Petersen et al. found:

> the majority of the [US and European] engineers interviewed . . . expressed their initial and acute discomfort at having an external supplier participate on an NPD team, where sensitive technical information is being discussed.[143]

Such issues of trust appear to be far less prevalent in studies of Japanese supplier–producer relationships.[144] Furthermore, much of the work on supplier integration has focused on the supplier–innovator interface, largely ignoring the importance of bridging the intra-organizational (internal) boundaries of the innovator. Indeed, drawing upon data on the Japanese automotive sector, Takeishi concludes that:

> outsourcing [of development activity] does not work effectively without extensive internal effort . . . external coordination [with suppliers] needs effective internal co-ordination [between various engineering and purchasing functions].[145]

Other researchers have sought to uncover variation in the coordination mechanisms adopted between supplier and innovator in the development process. For example, a study of the coordination of supplier involvement in innovation projects within the Swedish food packaging sector revealed a typology of three approaches:

- *project integration*—where there is extensive and continuous exchange of information between supplier and innovator;

- *ad hoc contact*—where interaction between supplier and innovator only occurs when problems arise and communication is required; and

- *disconnected sub-projects*—where the supplier plays a relatively independent role, performing the development in-house.

It was found that the choice or suitability of these three different levels of supplier integration were linked to the novelty and complexity of the task, and the need for convergent expectations and long-term objectives, with greater task novelty and complexity, and the need for greater convergence requiring greater supplier integration.[146]

We now turn to the phenomenon of supplier innovation: why might a commodity, material, or component supplier develop new equipment, processes, or products, downstream? The fairly limited research in this area suggests that each of the following three objectives would provide an incentive, because each seeks to increase the demand for the 'core' product of the supplier.[147]

- *Product switching*—the supplier innovates downstream to encourage manufacturers to switch from existing competing commodities, materials, or components. It does this by designing equipment to handle, or products or processes to incorporate the commodity, material, or component, that it sells. Alcoa, for example—the world's leading aluminium producer—employed downstream innovation to promote the switching to aluminium in the construction of trailers for trucks in the early days of market development for aluminium.[148] This process is more likely to occur early on in the product life cycle of a new material or component, where the supplier may be forced to innovate downstream simply because the material at this stage of the life cycle is unproven, for example, and is hence a more risky prospect for those downstream to adopt. In addition, the expertise in applying the new material or component is likely to reside initially with the supplier. This hypothesis is supported by research into innovation within the US thermoplastics and gas (e.g. nitrogen, oxygen, and argon) industries, which found that suppliers undertook the majority of new process innovations, but a minority of the subsequent minor improvement innovations.[149]

- *Product growth*—this involves the supplier devising new products that incorporate the material or component that it supplies.[150] The example later, in **Illustration 10.7**, provides an interesting alternative, although related, route for commodity suppliers to create derived demand for their own products through the funding of R&D into new applications by organizations 'downstream'—in this case, the development of platinum-based anti-cancer drugs and the advancement of fuel cells.

- *Product tying*—here, the part or component supplier will innovate downstream to provide equipment, products, or processes of such a design and on such contractual terms so as to tie the user to the supplier's own parts or components. The relationship between same-brand parts and equipment is reinforced over time by co-designing subsequent part and equipment innovations. In order to induce users to adopt, it is common for such 'part equipment' suppliers to provide their equipment at a level close to cost. They are able to do this because the ultimate objective is to generate revenue and profit from the materials or parts that they supply, rather than from their equipment. Such market behaviour is likely to force out 'pure' equipment manufacturers, which are unable to price their machinery at cost, because the equipment does not generate alternative revenue streams. Such product-typing innovation behaviour was observed by Vanderwerf in the US wire preparation industry; in his conclusions, he postulates that the ability of the supplier to tie part and equipment '*tends to be self-perpetuating once it begins*', but that where part or component suppliers are unable to tie in their products, then the 'pure' equipment supplier is more likely to be the source of innovation.[151]

Suppliers can thus play a vital role in the innovation process, whether as a source of ideas and technical expertise, as co-developer, or as 'downstream' innovator.

We now turn to the role of competitors.

10.6 **The role of competitors in the innovation process**

Competitors have long been recognized as a potential source of ideas and inputs into the innovation process.[152] But the 'rules of engagement' between competing organizations is somewhat more tricky as compared to other organizations, such as suppliers and customers: in short, trust is often more fragile, and the risks of knowledge 'leakage' or 'spillover' and the resultant potential for loss of competitive advantage are much higher. Nevertheless, in many hi-tech sectors, such as biotechnology, and information and communication technology (ICT), there is an increasing trend of partnering and organizational networking with competitor organizations.[153]

Interestingly, this pattern has been accompanied by alliances between what may be considered to be fierce rivals, such as IBM and Apple in the PC industry.[154] In large part, this is driven by the desire to share risk, resource, and competence in the face of shortening product development times and product life cycles, and ambiguity over the future of competing technologies and the emergence of standards or dominant designs.[155] Perhaps more intriguing are the less 'visible' patterns, such as the informal networking of individuals between competing organizations, but also to the murkier and less researched side of organizations and markets, such as counterfeiting and industrial espionage.

The following provides a flavour of the range of mechanisms through which competitors might be a source of ideas and inputs into the innovation process.

- *Alliances, co-development, and organizational networks*—as we noted in **Chapters 3** and **6**, there has been a clear pattern of growth of R&D partnering since the 1960s, accompanied by a notable shift since the mid-1970s from equity-based partnerships (e.g. joint ventures) to non-equity contractual forms of partnership (e.g. joint development agreements), which are seen as more suited to periods of technological instability and uncertainty. Since the 1980s, an increasingly larger proportion of R&D partnerships have occurred in hi-tech sectors, and in particular within the biotechnology and ICT sectors, now accounting for over 80 per cent of the total.[156] There is also emerging evidence that there has been a similar trend toward partnerships and networking in a variety of knowledge-intensive service sectors, such as banking, and capital-intensive service sectors, such as hotels.[157] Such formal partnering activity has been accompanied by a rapid growth of 'looser' and more dynamic organizational networking activity, especially in hi-tech sectors where expertise is widely dispersed, and the knowledge base is complex and expanding.[158] Although the international knowledge flows have traditionally been between organizations in Europe, the USA, Canada, and Japan, there are emerging patterns, for example, between hi-tech multinational companies in China and India.[159]
- *Licensing*—this generally represents a much more distant, arm's-length relationship between two organizations than the range of mechanisms outlined above, although the greater the complexity of the technology, the greater the need for interaction between source and receipt organizations to transfer the accompanying tacit knowledge and embed the technology. For Hippel:

agreements to trade or license know-how involve firms in less uncertainty than do agreements to perform R&D collectively. This is because the former deals with *existing* knowledge of known value that can be exchanged quickly and certainly. In contrast, agreements to perform R&D offer *future* know-how conditioned by important uncertainties as to its value and the likelihood that it will be delivered at all.[160]

- *Informal networking*—in **Chapter 3**, we highlighted the importance of informal networks for the transfer of knowledge between firms, especially within clusters such as Silicon Valley.[161] Research that has focused explicitly on informal knowledge flow between competitors and rivals has reinforced the importance of this mechanism.[162] Such social networks are built through the career histories of individuals whose paths may have crossed during their education, past research projects, conferences, or previous employment, and appear to be especially important for the transfer of tacit knowledge[163] and for the flow of knowledge flow between geographically distant locations.[164] Whilst such informal exchange activity may have positive benefits, it is very difficult to monitor and contain. Carter, for example, argues that '*because knowhow trading is informal and* off-the-books *such trading is difficult for the firm to evaluate and to manage,*'[165] whilst Kreiner and Schultz, in their study of the Danish biotechnology sector, found that '*the norms governing the interaction seem to reside in the network itself rather than in any of the participating organisations*'.[166]

- *Imitation and reverse engineering*—imitation is a common feature across a wide variety of sectors, such as banking, electronics, airlines, automotives, soft drinks, and television programming.[167] A common technique employed has been that of 'reverse engineering'—a process that involves a systematic and detailed analysis of the features and functioning of a product, and the design of a similar, and often improved, imitation that does not infringe the intellectual property rights of the original innovator. For many organizations, this process has proved an effective form of learning, in the medium term, leading to significant improvements in design, quality, and cost, and in the longer term, bringing about the development of in-house capabilities that have enabled such firms to shift from imitator to innovator, and from follower to leader.[168]

- *Counterfeiting*—whilst the process of counterfeiting may be similar to that of imitation discussed above, the outcome is clearly different—that of the intentional infringement of intellectual property rights of the original innovator. Counterfeiting is big business, with estimates of around US$1 trillion in sales per annum worldwide, with a large proportion undertaken in China.[169] In an interesting study of three Chinese companies engaged in counterfeiting activities, Minagawa et al. reveal some of the practices and incentives of such illicit activity.[170] In the medium to long term, counterfeiting, as with legal imitation, offers the potential for organizations to learn and ultimately innovate.

- *Industrial espionage*—although clearly an extremely difficult activity to uncover, the US Federal Bureau of Investigation (FBI) estimated that the value of stolen formulae, process information, blueprints, and customer lists was US$250 billion per annum in the late 1990s, whilst another study concluded that propriety information stolen from the Fortune 1,000 companies was valued at US$45 billion in 1999 and US$53–59 billion in 2001.[171] There is also plenty of historical evidence to suggest that organizations and nations have long sourced information and technology through illicit means.[172]

It is clear from this brief overview that there are a multitude of different mechanisms and channels for the sourcing of external ideas and knowledge from competitors and rivals. Many of these, whether formal or informal, represent two-way flows of information and knowledge between organizations. Others, whether legitimate or illicit, facilitate unidirectional flows of information and knowledge away from the original innovator organizations to imitating firms.

10.7 The role of distributors, intermediaries, and consultancies in the innovation process

In this section, we will explore the nature and extent of a range of other potential sources of ideas and inputs into the innovation process. Each is interesting in their own right: distributors, because of their unique position within the supply chain between innovator and user; intermediaries, due to their important knowledge and relationship brokering activities; and consultancies, because of their crucial role in supplementing internal technical and innovative capacity, especially within SMEs. Other sources not discussed further here include independent inventors: some suggest that inventors continue to play a role as a source of innovative ideas leading to successful innovations in larger firms,[173] whilst others argue that inventors and research units of large firms 'inhabit different worlds and speak different languages', and that 'the independent inventor's greatest difficulty lies in their lack of recognition of, and linkage with, the innovative process'.[174]

10.7.1 The role of distributors in the innovation process

Although recognized as one of a range of potential sources of inputs into the innovation process,[175] there has been relatively little research that has focused specifically on the role of the distributor. Nevertheless, the distributor, situated 'upstream' between the innovator and the customer, is in a unique position to identify customer needs and market opportunities, and to channel these back up the supply chain. Furthermore, in recent years, there has been a trend toward larger and more technologically sophisticated distributors, especially in relation to industrial products such as chemicals, engineering equipment, and office equipment. This is, in part, a response to the rising levels of customer support required as a result of the technical complexity of many such industrial products.

This combination of market and technical knowledge is a potent mix. It is perhaps not surprising, then, that, in a study of Australian industrial product manufacturers and distributors, a high incidence of product innovation among distributors was revealed. This pattern of innovative activity was particularly noticeable among distributors who had undergone 'backward vertical integration'—that is, those that had acquired or established a manufacturing capability. But the study also indicated that 40 per cent of distributor innovations were imitative, as compared to 25 per cent for manufacturers, with fewer than 10 per cent that might be described as 'innovative'.[176] Given the important position that distributors occupy in the supply chain and the changing nature of their role, more research is needed to fully understand their potential contribution to the innovation process.

10.7.2 The role of intermediaries in the innovation process

Interest in the role of intermediaries (often referred to as 'brokers') in the innovation process has grown over the last two decades among both academics[177] and policymakers.[178] In his review and synthesis of the literature, however, Howells notes that it remains 'surprising disparate', employing a multiplicity of terms (for example, 'third party', 'broker', 'bridge builder', 'boundary organization', and 'bricoleur') and emphasizing a diversity of roles, from that of simply diffusing and transferring existing innovation or knowledge, at one extreme, to initiating change within and through networks, at the other. Evidence suggests that the roles and functions of intermediaries are becoming more numerous, their relationships longer term rather than one-off, and their interactions increasingly complex as they shift from intermediaries between pairs of organizations (i.e. within 'one-to-one' linkages), such as between supplier and customer, to intermediaries within networks (i.e. within 'many-to-many' linkages). Furthermore, as Howells notes, '*organizations identified as providing intermediary roles are complex and multiple entities . . . whose primary role may often not be as an intermediary*'.[179]

The literature on intermediaries explores both the range of intermediary activities and the types of organization that act as intermediaries. Intermediary activities include, for example, knowledge sourcing, processing, and validation, the articulation and/or diagnosis of needs, the provision of information concerning potential collaborators and possible funding or support for innovation projects, project management and mediation between collaborators, and the evaluation of project outcomes.[180] Howells also highlights the systemic value of intermediaries, for example, through '*improving connectedness within a system, particularly through bridging ties, but also in the "animateur" role of creating new possibilities and dynamism in the system*'. There are two key network positions (as discussed in **3.5.4**) that can add value, in different ways, to the intermediary role: 'boundary-spanners' or 'brokers', who bridge structural holes in a network, and can bring together disparate organizations, resources, and ideas; and highly connected 'stars', who can improve the connectedness or density of relationships within a network, for example. A wide range of organizations have been found to undertake these intermediary activities, such as consultancies, industry trade associations, standards bodies, professional associations, local chambers of commerce, and various governmental agencies. Many such organizations are in the public realm or bridge the public and private spheres, and are run as 'not-for-profit' companies.

Attempts have also been made to categorize different types of intermediary in relation to their brokerage activities. Although the labels are not particularly helpful, a distinction is emerging in the literature between 'innovation brokers', who are concerned with facilitating innovation, and 'knowledge' or 'technology brokers', who are involved in the transfer and/or development of knowledge and technology. For example, Winch and Courtney define an 'innovation broker' as an organization that:

> act[s] in a liaison role between the sources of new ideas and the users of those ideas in innovation networks . . . [is] set up specifically to perform this broking role . . . [and is] focused neither on the generation nor the implementation of innovations, but on enabling other organisations to innovate.[181]

This definition attempts to distinguish the 'pure' innovation broker from other knowledge-intensive business service (KIBS) organizations, many within the private sector, which may act as innovation broker alongside or as a by-product of their principal activities of knowledge transfer and production. It is argued that such a dual function has the potential to undermine the (actual and perceived) impartiality and objectivity of the innovation broker in key roles such as idea validation and the identification of best practice.[182]

Illustration 10.6 Business Link—connecting SMEs to expertise and funding

Established in 1992, Business Link is a public-funded government initiative, designed to offer impartial advice and brokering services to SMEs in England. Similar initiatives exist in Scotland (Business Gateway), and Wales (Business Connect). Business Link operates through a number of local offices throughout England, and is managed regionally by nine regional development agencies (RDAs). Whilst each Business Link office provides information, advice, support, and brokering services across a wide range of business issues, from sale and marketing, to IT and health and safety, of particular relevance here are services around patenting, building prototypes, identifying potential R&D funding sources, and facilitating knowledge transfer through linking SMEs to expertise within universities, research institutes, and consultancies. The initiative is considered an important component of the government's campaign to promote enterprise and, according to a recent high-profile government innovation policy document, *'will become the primary access route for individuals and businesses seeking* [business] *support'*.[183]

Studies have focused on a range of sectors, such as construction, agriculture, and flight simulation,[184] and across a number of countries, including, the UK,[185] the USA,[186] the Netherlands,[187] and Canada, France, Sweden, Germany, and Australia,[188] demonstrating the widespread prevalence of intermediaries. Given the importance of their brokering activities to the innovativeness of firms and sectors, however, further research is required of these emergent, diverse, and complex organizations and their activities.[189]

10.7.3 The role of consultancies in the innovation process

As noted above, whilst consultancies—or KIBS organizations, as they are increasingly referred to—may act as intermediaries in both relationship and knowledge or technology brokering between firms and universities, for example, they are also an increasingly important source of ideas and knowledge in their own right, especially for SMEs, which have limited resources and technical capacity.[190]

Muller and Zenker, for example, from their large-scale cross-sector survey of SMEs and KIBS within a number of French and German regions, reveal that KIBS not only transmit knowledge, but also play a crucial role in relation to knowledge re-engineering. Indeed, they argue that KIBS might often be considered 'co-innovators'. Interestingly, Muller and Zenker also found that interactions between SMEs and KIBS had the potential to contribute mutually to the innovative capacity of both the consultancy and the client, because the knowledge bases of both were extended through such activity, leading to a 'virtuous circle of innovation'.[191] They conclude that *'KIBS may be of crucial importance for the support of innovating SMEs'*, and *'constitute a "relevant subject" for both innovation and regional policies'*.[192]

From their study of a cross-sector sample of small, medium, and large UK firms, Tether and Tajar found that:

> links with specialist knowledge providers tend to complement rather than substitute for a firm's own internal innovation activities and to complement firm's sourcing of information from other sources such as suppliers, customers and competitors.

They nevertheless conclude by highlighting the importance of private sector consultants and research organizations to systems of innovation.[193]

10.8 Variations in the locus of innovation

10.8.1 'Upstream' and 'downstream' innovation

Research centred on investigating the sources and origins of innovation has also highlighted variations in the locus of innovative activity between business sectors—that is, in some sectors, the innovation process is dominated by suppliers, innovating 'upstream' of those organizations that might be expected to be the 'natural' innovator within the supply chain, or by distributors and users, innovating 'downstream'. In **Figure 10.3**, we can see how the various phases of the innovation process (located along the vertical access) are dominated by organizations or individuals at different points in the supply chain depending on the locus of innovation. Thus, we see, in contrast to instances of manufacturer-dominated innovation, supplier-dominated innovation (i.e. upstream innovation) is typically characterized by a supplier dominating the idea-generation and problem-solving phases of the innovation process, whilst user-dominated innovation (i.e. downstream innovation) is characterized by users dominating the idea-generation, problem-solving, and pre-commercialization field-testing phases of the innovation process.

Research over the years across a variety of industrial sectors has shown the user to be a significant source of innovation within the oil industry and chemical industry,[194] the semiconductor, scientific instruments and electronic subassembly industries, medical equipment, and residential construction sectors,[195] as well as in a range of outdoor leisure activities;[196] studies in various materials sectors, and subsequent investigations of the

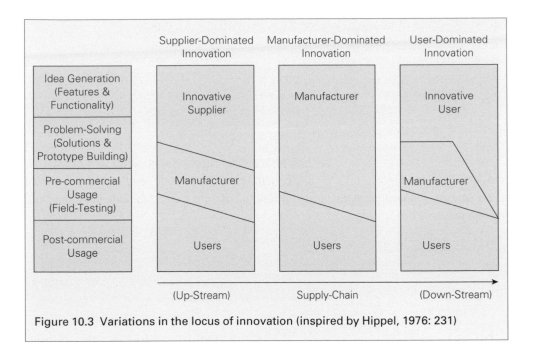

Figure 10.3 Variations in the locus of innovation (inspired by Hippel, 1976: 231)

industrial gas, thermoplastic, and wire preparation equipment sectors, have highlighted the importance of suppliers as a source of innovation.[197]

10.8.2 Explaining variations in the locus of innovation

Having identified that the locus of innovation may vary, it is important to attempt to explain why different patterns exist between different business sectors. It has been hypothesized that the observed variations in the locus of innovative activity across business sectors could be explained by differences in economic incentives.[198] Other factors may include variations in the locus of state-of-the-art expertise and the 'stickiness' of knowledge.

Locus of economic benefit of the innovation

Hippel argues that the observed variations in the source of innovation across business sectors can be explained by differences in economic incentives. Simply stated, the 'appropriability of innovation benefit' hypothesis asserts that whoever can make the greatest profit from an innovation is the most likely to originate and develop it.[199] The essence of this argument would help to explain, for example, why commodity suppliers, such as Rustenburg Platinum Mines (a primary producer of platinum) might undertake or fund 'downstream' R&D into new applications, or why Anglo Platinum (which now own Rustenburg) recently bought an equity stake in Johnson Matthey's Fuel Cell subsidiary, which utilizes platinum—see **Illustration 10.7**. The principal incentive for suppliers in these cases was the desire to create derived demand for the materials that they supply, through a combination of 'product switching' and 'product growth', as discussed in **10.5**.

Illustration 10.7 Creating 'derived' demand for platinum through R&D in its applications—explaining the long-term relationship between Johnson Matthey and Rustenburg Platinum Mines

The strong and enduring relationship between Rustenburg Platinum Mines (the world's largest primary producer of platinum) and Johnson Matthey (the world's leading distributor of platinum, with businesses in a wide range of platinum-based products) goes back to the founding of the mining company in South Africa in 1931. One interesting aspect of this relationship has been the joint funding of R&D into applications for platinum at various Johnson Matthey technology centres.[200] A good example of this is provided by the joint funding in the 1970s of research by Professor Rosenberg at Michagen State University, in the USA, into the use of platinum in anti-cancer drugs. This ultimately led to the development of cisplatin and carboplatin, two of the most widely used anti-cancer drugs in the world today.[201]

More recently, in November 2002, Anglo Platinum, which now owns Rustenburg Platinum Mines, announced that it had acquired a 17.5 per cent stake in Johnson Matthey Fuel Cells Ltd, the fuel cell components subsidiary of Johnson Matthey. The executive chairman of Anglo Platinum, Barry Davison, made the following comment on this acquisition:

> The development of fuel cells is a logical extension to our strategy of growing the market for platinum group metals. Fuel cells will drive the long-term demand for platinum and this strategic holding will provide commercial returns as well as stimulate demand.[202]

Whilst financial benefit appears to be a strong incentive for supplier organizations to innovate, this relationship appears much weaker for 'professional' users within organizations and among consumers. For example, the key benefit cited by users in developing scientific instrumentation was '*to create and report on scientific advances*'.[203] In relation to consumer goods, Lüthje, for example, concluded from his study of user innovation in climbing/mountaineering, hiking, cross-country skiing, and mountain biking in Germany that '*It is not the financial benefit, but the chance to execute their sports more effectively, which motivates the outdoor users* [to innovate]'.[204] There are examples of consumer products, however, where user-entrepreneurship, and thus financial incentives, are more prevalent (as discussed in **10.3.1**).[205]

Firm concentration along the supply chain

The degree of concentration of firms along the supply chain of a particular sector is likely to have an impact on the concentration of economic benefit and therefore the locus of innovation. Thus, whilst the total economic benefit of an innovation may be greatest among users, for example, the existence of many users would carve up the economic benefit to such an extent that no single user would have the incentive to undertake the required inventive activity. Vanderwerf, for example, in his study of innovation in the US thermoplastics and gas industries mentioned earlier, concluded that suppliers were more likely to innovate downstream when there is concentration among suppliers, when such innovation was likely to create new demand for the commodities that they supplied, and when the group of potential user organizations was diffuse.[206] It is perhaps not surprising, then, that the high concentration among suppliers in commodities such as aluminium and platinum, and the relative diffusion of manufacturers and users of products containing these materials, pushes innovation upstream.

Locus of state-of-the-art expertise

Where the locus of state-of-the-art expertise in a given specialism relevant to a given innovation resides outside of the innovator, the more likely it is that others, such as users or suppliers, will play an important, or even dominant, role in the development process of that innovation. This would appear to be at least part of the explanation for Alcoa's continued role in downstream innovation, because, over the years, it has accumulated extensive expertise in the use of aluminium in a wide range of specialist applications, from canning to aerospace, and transportation to building and construction.[207]

In relation to users, Shaw notes the dominance of the user in the development process for medical equipment, particularly with respect to radical and major improvement innovations. This he believes:

> should be expected where *state-of-the-art* clinical and diagnostic knowledge resides in the user, and the user has a high probability of deriving *output-embodied* benefit from the innovation . . . The closer these benefits are to the state-of-the-art advances, the greater the benefit to the clinician and therefore, the more involved he will want to be. Where these benefits are not present, he will try not to get involved.[208]

The importance of the locus of expertise was also revealed by Slaughter, in her study of innovation by 'user-builders' in the US residential construction sector. Here, she found that not only did users draw upon their expertise to innovate on-site in response to non-trivial problems, but manufacturers seemed to:

> rely upon the builders to solve the problems which appear in the implementation process, depending upon the learning and problem-solving abilities of the builders to develop the technology [in this case, 'stressed skin' panelling] further.[209]

Recent studies of user-innovation in outdoor leisure and sports equipment also highlights the phenomenon of user experience and expertise running ahead of innovators, especially in emerging sports such as rodeo kayaking, mountain biking, and snowboarding.[210]

The 'stickiness' of information and knowledge

Knowledge is often costly to transfer, due to the nature of the knowledge itself (e.g. tacitness), the amount of knowledge, or attributes of the knowledge provider or recipient (e.g. absorptive capacity of recipient). Such knowledge or information is sometimes termed 'sticky'. Hippel hypothesizes that where knowledge or information of user needs is more 'sticky' than knowledge or information of possible technical solutions, for example, then the locus of innovation is more likely to reside with the user, and where the reverse is true, with the innovating organization.[211] There is some empirical evidence to support this notion, such as studies of innovations in the Japanese convenience store sector, gaming software, and outdoor pursuits.[212]

10.8.3 Implications for the management of innovation

What is of particular value and interest in the research concerning the 'sources of innovation' is not so much the identification of specific sectors in which users or suppliers dominant the innovation process, but: firstly, that such a phenomenon exists; secondly, the explanation that has emerged as to why variations in the locus of innovation may arise; and thirdly, the insights that this provides to managers and policymakers, alike. Indeed, for Hippel, '*if we can understand the cause(s) of such variations, we may be able to predict and manage the innovation process much better*'.[213]

One important message from this research relates to the way in which organizations organize and orientate their innovation activities. On this point, Hippel argues:

> Firms organize and staff their innovation-related activities based on their assumptions regarding the sources of innovation. Currently, I find that most firms organize around the conventional assumption that new products are—or should be—developed by the firm that will manufacture them for commercial sale. This leads manufacturers to form R&D departments capable of fulfilling the entire job of new product development in-house and to organise the market research departments to search for needs instead of innovations. Indeed, if a manufacturer depends on in-house development of innovations for its new products, then such arrangements can serve well. But if users, suppliers, or others are the typical sources of innovation prototypes that a firm may wish to analyze and possibly develop, then these same arrangements can be dysfunctional . . . New sources of innovation demand new management tools as well as new organization.[214]

10.9 Emerging trends

Two related and complementary models of innovation have recently been articulated in an attempt to capture emerging trends in the locus and distribution of innovation activity, on the one hand, and the manner in which many successful organizations are undertaking and exploiting innovation, on the other. Reflecting on the case of open-source software, Eric von Hippel and Georg von Krogh have developed a 'private–collective' model of innovation, which falls between, and combines elements of, both the 'private investment' and 'collective action' models of innovation.[215]

Perhaps more widely known is the term 'open innovation', devised by Henry Chesbrough of Harvard Business School.[216] The term was coined to capture the new ways in which organizations are harnessing external ideas and technology in their innovative activities, alongside the adoption of more open ways in which they are seeking to leverage value from these in-house activities.

10.9.1 The 'private–collective' model of innovation

The 'private investment' model of innovation is characterized by the assumption that innovation is derived through private investments by organizations and individuals, who are then able to appropriate 'private returns' from these investments. Such investments are encouraged through the existence of intellectual property rights, such as patent and copyright laws, that mitigate against the appropriation of these returns by others. Under this model, innovators attempt to minimize leakage or spillovers of propriety knowledge to others, and refrain from 'free-revealing'. In contrast, under the 'collective action' model of innovation, innovators do not retain control over the knowledge that they produce; instead, it becomes a 'public good' that may be freely used by all. The downside is that this requirement for free-revealing has a negative impact on the incentive for innovators to invest, because it becomes difficult for them to appropriate the returns from their investments. Furthermore, it offers the possibility that 'freeriders' may be able to benefit as much as innovators from the resulting public goods. Whilst free-revealing has long been at the heart of the norms operating within the scientific community, this is, in large part, made possible by public investment and alternative means of appropriating returns by innovators, such as through reputation.[217]

For Hippel and Krogh, open-source software projects occupy the middle ground between these two traditional models of innovation: private investment, often in personal time, is accompanied by free-revealing. This apparent paradox may be explained by the non-financial benefits that are accrued by innovators in open-source software projects—personal enjoyment, technical learning, personal control over the direction of the project that is often not evident in commercial settings, and being part of, and regarded by, a community—none of which are available to freeriders and, in fact, some of which are inextricably linked to free-revealing, such as learning from feedback and feelings of belonging associated with community membership.[218] Whilst companies have attempted to exploit open-source software, this is associated with a range of costs and risks, and runs counter to the culture of such communities.[219] Hippel and Krogh argue that '*open-source software has emerged as a major cultural and economic phenomenon*'.[220] Whilst this case is widely known and cited, however, similar phenomena also exist across a wide range of outdoor pursuits and sports, such as climbing/mountaineering, hiking, cross-country skiing, and mountain biking, medical equipment, and library information systems.[221]

The private–collective model of innovation offers very interesting possibilities for society—a process that Hippel has referred to as the 'democratization of innovation'.[222]

10.9.2 'Open innovation' versus 'closed innovation'

The notions of 'open' and 'closed innovation' relate to the orientation of individual organizations, rather than the pattern of innovation within a sector, as is the case above, although the free-revealing associated with the private–collective model is seen as a basic ingredient of open innovation.[223]

Chesbrough describes the 'closed innovation' model as one in which an organization '*generates, develops, and commercializes its own ideas*' and is dominated by a '*philosophy of self-reliance*'. In contrast, he describes the 'open innovation' model as one in which the boundaries of the organization are 'porous' and in which it '*commercializes both its own ideas as well as innovations from other firms and seeks ways to bring its in-house ideas to market by deploying pathways outside its current businesses* [e.g. licensing]'. This is not a dichotomy—that is, organizations are considered to be located somewhere along a spectrum from 'essentially closed' to 'completely open'. Chesbrough argues that the closed model of innovation was tacitly viewed as the 'right way' for organizations to undertake and exploit their innovative activities, and that this 'worked well' for most of the twentieth century, until factors such as the upsurge in mobility of knowledge workers and the growth in availability of venture capital undermined its efficacy.[224]

Echoing the sentiment of Chesbrough, Gassmann argues that *'open innovation is not an imperative for every company and every innovator. Instead, there is a need for a contingency approach regarding the management of innovation'*. The more global and technologically intense an industry, and the more important interdisciplinary research is for that industry, for example, the more appropriate, it is argued, is the model of open innovation for an organization.[225] Whilst Chesbrough recognizes that the open innovation model has been operating in a number of sectors for 'some time', to a certain extent, this underplays the wealth of empirical evidence since the 1970s, much of it referred to in this chapter, that has highlighted the increasing role of external sources of ideas and innovation across a wide range of sectors. In this regard, the model might be considered to be a repackaging and relabelling of the 'interactive' or 'network' model of innovation. But it is argued that the notion of 'open innovation' embraces *'a more comprehensive and systematic'* articulation of the ways in which organizations manage increasingly external orientated innovation processes, and more broadly, how organizations *'engage in* technological *innovation'*.[226]

10.10 Concluding comments

In this chapter, we have sought to highlight the role and importance of external sources to the innovation process from both historical and contemporary perspectives. From this review, it is possible to distil that:

- the importance of external inputs into the innovation process has long been recognized;

- these external inputs are sourced from a multiplicity of external entities, from users to suppliers, and universities to competitors, and through a plurality of channels and mechanisms, from formal to informal, and passive to active;

- external sources and external relationships are becoming ever more important across a wider set of sectors, as competition grows, development and product life cycles shorten, the costs and risks of innovation rise, and knowledge, expertise, and innovation become more distributed;

- instances of 'user innovation' and 'supplier innovation' appear to be emerging across a broader set of sectors—earlier studies tended to link the phenomenon to business and industrial markets, such as design and information systems software, and medical and scientific equipment, whilst recent research has found increasing evidence of user innovation, for example, among a range of consumer products and services, and in particular among a wide range of outdoor pursuit and sports;

- to increase the chances of success, innovators should seek out 'lead users', and increasingly 'lead sub-communities', as partners in the innovation process;

- innovative organizations are being increasingly sophisticated and open in the manner in which they manage and exploit both internally and externally generated ideas and innovation—this approach has been termed the 'open innovation' model.

Whilst external inputs are clearly important to the innovation process, a number of academics have argued that external sources should be viewed as an important complementary source of information, ideas, knowledge, and technologies, rather than as a substitute for indigenous innovative activity. For example, Freeman contends that: *'the successful exploitation of imported technology is strongly related to the capacity to adapt and improve this technology through indigenous R&D'*, whilst Cohen and Levinthal argue *'that R&D not only generates new information . . .* [but] *also develops the firm's ability to identify, assimilate, and exploit existing knowledge from the environment'*—what they term the 'two faces of R&D'.[227]

In the next chapter, we will focus on the manner in which the innovative capability and behaviour of organizations is shaped by the context in which they are embedded. To this end, we will be looking at concepts such as regional and national 'systems of innovation'.

CASE STUDY 10.1 BRITISH GAS AND HOTWORK—THE DEVELOPMENT OF THE REGENERATIVE BURNER

Introduction

Traditional industrial burners and furnaces usually deliver only a small proportion of the available energy in the fuel used as heat. Efficiency can be improved by recovering heat from the combustion and using it either to preheat the furnace load (as with regenerative burners) or to preheat incoming combustion air (as with recuperative burners). Regenerative burners deliver much greater fuel efficiency than recuperative burners, but earlier designs were very large scale and could not be retrofitted. The British Gas/Hotwork regenerative burner is small (one fiftieth of the size of existing regenerative burners), very fuel-efficient, and capable of retrofit. Developed in the early to mid-1980s and brought to the market in the late 1980s, the innovation won the Queen's Award for Technological Achievement in 1992.

Privatized in the 1980s, British Gas plc is one of the largest integrated gas businesses in the world. At the time of the regenerative burner development, the Midlands Research Station (MRS) of British Gas employed 450 staff and commanded an annual budget of around £20m. Hotwork manufactures and installs a range of industrial burners. It is the largest subsidiary of Hotwork International Ltd, founded in 1962. At the time of the development, the group had an annual turnover of around £7m and employed 200 people.

The stimuli for innovation

By the end of the 1970s, MRS had been developing recuperative burners for a number of years and already had a recuperative burner licensed and on the market. Around this time, MRS was trying to push forward gas burner performance to compete better with other fuels—particularly electricity, which was becoming increasingly prominent in industrial markets. Rising fuel prices and increased competition during the 1970s had also prompted many industrial users to examine the ways in which their heating requirements could be provided cost-effectively. As a consequence, the development of energy-efficient combustion equipment had already attained considerable momentum, particularly in the area of industrial high-temperature process heating. The question was how to go forward, because MRS recognized that the energy efficiency of the recuperative burner was around 35–40 per cent, and thus there was still a great deal of energy wastage in the existing burner technology.

MRS and the development of the regenerative burner prototype

At the time, MRS was employing a placement student from Bath University, who was looking at regenerators as a project. It suddenly occurred to engineers at MRS that regenerators might be the answer. The concept of regenerators had been around for some time, but they were only in use in very large furnaces (the MRS regenerative burner was to eventually require a 'thermal bed' only one fiftieth of the size of old-style regenerative burners). During 1981, engineers at MRS started thinking about how they might employ the regenerative concept in new burner designs. The driving force behind the first three years of the project was a new engineer called Julian, who undertook a number of sophisticated calculations and built a prototype. Initially, the project team was very small: including Julian on a full-time basis and Roger (the project leader) on a part-time basis. Even at its peak of activity, the project team rose to only three or four people.

The performance of regenerative burners is governed by a number of parameters, including bed geometry, overall heat transfer co-efficients, storage characteristics of the thermal bed, the flow rates of the gases, and the cycle time. A mathematical model to describe the behaviour of a compact regenerative burner was developed with the help of expertise from academics at Leeds and York universities. The basic mathematical models were already around, but required an element of modification. The design of a prototype system was finalized using this mathematical model. The prototype was installed in a laboratory furnace at MRS in 1983 and subjected to an extensive test programme, an important part of which was to assess the long-term effects of cyclical thermal stress on the various components. This work was crucial in proving the high-temperature effectiveness of the small-scale, compact regenerative burner concept.

Commercializing the innovation—enter Hotwork Development Ltd

The next step for MRS was to find a suitable industrial partner to help bring the development to the marketplace. Although British Gas (through MRS) had established a worldwide reputation for developing state-of-the-art technology in gas-fired burners, as a company, it was not so much interested in the technology itself, but in the improvement in the derived demand for gas. Thus, British Gas sought to transfer such technology developments to licensees which were better able to manufacture the industrial units and service the marketplace. The development work was publicized in a publication called *Gas R&D News*. This was picked up by Trevor (the then chairman of Hotwork Development Ltd) in 1983, who had also been working on recuperative burners and had been thinking about future opportunities.

Further development was then undertaken in collaboration with Hotwork, which was brought in to collaborate with product design and to manufacture the end product. When the new design had been proven in the laboratory, it was considered appropriate to undertake a field trial to demonstrate its performance under industrial conditions; this took place in the glass industry. The work that occurred following this first trial was mainly concerned with the development of a range of burner sizes and with increasing the range of applications of the system. Subsequent installations with cooperative lead users in new application areas were encouraged by grants from the Energy Technology Support Unit, linked to the Department of Energy. These field trials were

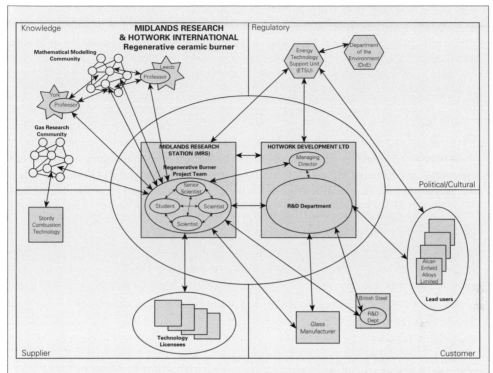

Figure 10.4 Actors and links mobilized in the development of the regenerative burner (adapted from Conway, 1994)

important, in that they confirmed the performance of the regenerator, proved the robustness in harsh environments, and highlighted the cost-effectiveness of the regenerative burner system.

Questions

1. What is the locus of innovation in this case study and what might best explain this locus?

2. What were the key features of the network mobilized in the development of the regenerative burner?

3. Identify the nature of the relationships and transfer mechanisms employed in bringing the innovation to the point of commercialization by Hotwork?

FURTHER READING

1. A key text in this area is *Democratizing Innovation* by Eric von Hippel (2005), which focuses on the role of users, lead users, communities of users, and free-revealing in the innovation process. The article by Hippel and Krogh (2003), concerning open-source software and their 'private–collective' model of innovation, is a useful complementary paper.

2. The seminal work by Henry Chesbrough (2003a; 2003b) on 'open innovation' provides a good introduction to this emerging concept. This can be usefully supplemented by reference to a

special issue in 2006 of *R&D Management*—36(3)—which incorporates, for example, papers that look at the challenges and paradox of open innovation (West and Gallagher, 2006) and the role of technology in the shift toward open innovation, exemplified through the case of Proctor and Gamble's 'Connect and Develop' strategy (Dodgson et al., 2006).

NOTES

[1] Allen et al. (1983: 201).

[2] For an excellent overview of this work, see Hippel (1988; 2005).

[3] Allen (1977: 157).

[4] Szulanski (1996).

[5] Information found online at <http://www.ministryofsound.com> and <http://www.blacks.co.uk> (accessed on 21 May 2008).

[6] Sherwin and Isenson (1967).

[7] Illinois Institute of Technology (1968); Battelle Memorial Institute (1973).

[8] Myers and Marquis (1969).

[9] Achilladelis et al. (1971); Rothwell et al. (1974); Langrish et al. (1972), respectively.

[10] DeSolla Price (1969).

[11] Allen (1977: 49).

[12] Department of Trade and Industry (1993; 2000).

[13] Department for Innovation, Universities, and Skills (2008: 13).

[14] Van Vianen et al. (1990); Cyert and Goodman (1997); Hall et al. (2003); Balconi and Laboranti (2006).

[15] Mowery and Nelson (1999); Owen-Smith et al. (2002).

[16] Lord Sainsbury of Turville (2007).

[17] NESTA (2006); for Europe see, e.g., Commission of the European Communities (2003).

[18] Rosenberg and Nelson (1994); Stokes (1997); Balconi and Laboranti (2006: 1629).

[19] Hippel (1976); Spital (1979); Shaw (1985); Voss (1985); Vanderwerf (1992), respectively

[20] Håkansson (1987); Hippel (1988; 2005); Kreiner and Schultz (1993); Conway (1995); Hagedoorn (2002).

[21] Gemunden et al. (1992: 373).

[22] Pittaway et al. (2004: 150).

[23] Chesbrough (2003a; 2003b).

[24] Carter and Williams (1957); Myers and Marquis (1969); Utterback (1971); Langrish et al. (1972).

[25] Utterback (1971); Rothwell et al. (1974).

[26] Utterback (1971).

[27] Allen et al. (1983).

[28] Baker et al. (1985).

[29] Fritsch and Lukas (2001).

[30] Rothwell and Gardiner (1983); Shaw (1993); Tidd (1993).

[31] Pittaway et al. (2004).

[32] Belderbos et al. (2004).

[33] Arundel et al. (1995).

[34] Sowrey (1987).

[35] Easingwood (1986).

[36] Grangsjo and Gummesson (2006); Ul-Haq and Howcroft (2007); Tether and Tajar (2008b).

[37] Myers and Marquis (1969); Utterback (1971); Allen (1977); Conway (1995).

[38] Conway (1995).

[39] Roberts (2001: 33).

[40] Owen-Smith et al. (2002).

[41] Biggs and Clay (1981); Biggs (1990); Klerkx and Leeuwis (2008).

[42] Voss (1985); Jordan and Segelod (2006); Assimakopoulos and Yan (2006); West and Gallagher (2006).

[43] McAdam et al. (2008).

[44] Hippel (1976; 1977a; 1977b); Riggs and Hippel (1994).

[45] Shaw (1985; 1993); Lettl et al. (2006).

[46] Liebeskind et al. (1996).

[47] Backman et al. (2007).

[48] Ogawa (1998).

[49] Franke and Shah (2003); Lüthje (2004); Hienerth (2006).

[50] Tether and Tajar (2008b).

[51] Hagedoorn (2002); Owen-Smith et al. (2002).

[52] Kelley and Brooks (1991).

[53] Hagedoorn (2002).

[54] Shaw (1993); Conway (1995); Jeppesen and Molin (2003); Franke and Shah (2003).

[55] Freeman (1991: 503).

[56] Jewkes et al. (1969); Myers and Marquis (1969); Rothwell et al. (1974); Langrish et al. (1972).

[57] Hippel (1976; 1977a; 1977b; 1977c; 1978a; 1978b; 1978c; 1982b; 1986).

[58] Neale and Corkindale (1998); Franke and Shah (2003); Lüthje (2004); Hippel (2005); Hienerth (2006).

[59] Department of Innovation, Universities, and Skills (2008).

[60] Achilladelis et al. (1971); Rothwell et al. (1974); Hippel (1978b).

[61] Conway (1995).

[62] Rothwell et al. (1974); Parkinson (1981; 1982); Gardiner and Rothwell (1985); Shaw (1985); Hippel (1988; 2005).

[63] Leonard (1995); Neale and Corkindale (1998); Lamming et al. (2002).

[64] Jeppesen and Molin (2003).

[65] Lettl et al. (2006).

[66] Hippel and Katz (2002); Thomke and Hippel (2002); Jeppesen and Molin (2003); Hippel (2005).

[67] Hippel (1988; 1998; 2005).

[68] Hippel (1976); Riggs and Hippel (1994); Shaw (1985); Hippel (1977b), respectively.

[69] Voss (1985); Urban and Hippel (1988); Morrison et al. (2000).

[70] Franke and Shah (2003); Lüthje (2004); Lüthje et al. (2005); Hienerth (2006).

[71] Hippel and Krogh (2003; 2006); Hippel (2005).

[72] Jones-Evans (1995).

[73] See Dyson (1997).

[74] See <http://www.menitest.co.uk/profile.html>.

[75] Hienerth (2006); Shah and Tripas (2007), respectively.

[76] 'Social entrepreneurship' may be usefully defined as *ventures that address social issues as their prime strategic objective and do so in an innovative and creative fashion*' (Nicholls, 2006: 223).

[77] See <http://www.nas.org.uk>.

[78] Rothwell et al. (1974); Hippel (1978b); Conway (1995).

[79] Lavidge (1984).

[80] Hippel (1978b); Riggs and Hippel (1994).

[81] Riggs and Hippel (1994); Jeppesen and Molin (2003).

[82] Rothwell and Gardiner (1983); Hippel (1986; 1988); Jeppesen and Molin (2003); Hienerth (2006).

[83] Conway (1994).

[84] Habermeier (1990: 275).

[85] Habermeier (1990); Douthwaite et al. (2001); Petroski (2006: 4–5).

[86] Petroski (2006: 63).

[87] Petroski (2006: 3).

[88] Conway (1995: 335–6).

[89] Habermeier (1990: 276).

[90] Petroski (1982: 164).

[91] Rice and Rogers (1980); Hippel (1986); Habermeier (1990); Rogers (2003).

[92] Slaughter (1993: 92–3).

[93] Rogers (2003).

[94] Hippel and Finkelstein (1979); Rothwell (1986); Jeppesen and Molin (2003).

[95] Blundel (2006: 325).

[96] Lüthje (2004); Hienerth (2006).

[97] Morrison et al. (2000); Franke and Shah (2003); Lüthje (2004: 692); Hippel (2005); Hienerth (2006).

[98] Hippel (1976; 1977a; 1977b; 1988); Shaw (1985).

[99] Lettl et al. (2006).

[100] Shaw (1985: 287).

[101] Hippel and Katz (2002); Thomke and Hippel (2002); Jeppesen and Molin (2003: 378); Hippel (2005).

[102] See <http://www.pillsbury.com/BakeOff> and Hippel (1982b).

[103] Piller and Walcher (2006).

[104] Hippel (1988: 102).

[105] Rogers (2003).

[106] Gardiner and Rothwell (1985: 7).

[107] Morrison et al. (2000: 1520).

[108] Rothwell and Gardiner (1983); Hippel (1986; 1988).

[109] Porter (1990).

[110] Gardiner and Rothwell (1985); Hippel (1986; 1988).

[111] Rogers (1995: 263).

[112] Hippel (1986; 1988).

[113] Gardiner and Rothwell (1985: 7).

[114] For example, Urban and Hippel (1988), and Lilien et al. (2002), respectively.

[115] Urban and Hippel (1988).

[116] Morrison et al. (2000).

[117] Lilien et al. (2002: 1055).

[118] Blundel (2006); Hienerth (2006).

[119] Van Vianen et al. (1990); Cyert and Goodman (1997); Hall et al. (2003); Balconi and Laboranti (2006).

[120] Mowery and Nelson (1999); Owen-Smith et al. (2002).

[121] Sampat (2006: 773).

[122] Liebeskind et al. (1996); Kaufmann and Todtling (2001); Audretsch et al. (2005); Balconi and Laboranti (2006).

[123] Balconi and Laboranti (2006: 1629).

[124] Bottazzi and Peri (2003); Audretsch et al. (2005).

[125] Association of University Technology Managers (2002).

[126] Sampat (2006: 772).

[127] Cyert and Goodman (1997); Hall et al. (2003); Meeus et al. (2004).

[128] Owen-Smith et al. (2002); Di Gregorio and Shane (2003); Shane (2004); Lawton-Smith and Ho (2006).

[129] Cyert and Goodman (1997).

[130] Boulton and Lucas (2008).

[131] Amabile et al. (2001); Owen-Smith et al. (2002); Garrett-Jones et al. (2005); Markman et al. (2005).

[132] Abernathy et al. (1983); Clark (1989); Clark and Fujimoto (1991).

[133] Bensaou and Venkatraman (1995).

[134] Petersen et al. (2003).

[135] Petersen et al. (2003: 288).

[136] Bruce et al. (1995); Gadde and Snehota (2000).

[137] Håkansson (1987); Biemans (1991); Conway (1995).

[138] Clark (1989).

[139] Rothwell (1992); Petersen et al. (2003).

[140] Hippel (1988); Vanderwerf (1990; 1992).

[141] Handfield et al. (1999).

[142] Bozdogan et al. (1998); Petersen et al. (2003: 286).

[143] Petersen et al. (2003: 286).

[144] Bensaou and Venkatraman (1995).

[145] Takeishi (2001: 419).

[146] Lakemond et al. (2006).

[147] Corey (1956); Peck (1962); Hippel (1988); Vanderwerf (1990; 1992).

[148] Hippel (1976).

[149] Vanderwerf (1992).

[150] Vanderwerf (1992: 331).

[151] Vanderwerf (1990: 95).

[152] Langrish et al. (1972); Myers and Marquis (1969); Hippel (1987); Hamel et al. (1989); Biemans (1991).

[153] Hagedoorn (2002).

[154] Hagedoorn et al. (2001).

[155] Pisano (1990); Contractor and Lorange (2002); Das and Teng (2002); Oshri and Weeber (2006).

[156] Hagedoorn (2002).

[157] Grangsjo and Gummesson (2006); Ul-Haq and Howcroft (2007).

[158] Powell et al. (1996).

[159] Hagedoorn (2002); Teagarden et al. (2008).

[160] Hippel (1988: 89).

[161] Rogers and Larson (1984); Saxenian (1985; 1991; 1994); Castilla et al. (2000); Assimakopolous et al. (2003).

[162] Allen et al. (1983); Hippel (1987); Carter (1989); Schrader (1991); Kreiner and Schultz (1993); Conway (1995).

[163] Conway (1995); Fleming et al. (2007).

[164] Bell and Zaheer (2007).

[165] Carter (1989: 155).

[166] Kreiner and Schultz (1993: 24).

[167] Lieberman and Asaba (2006); Minagawa et al. (2007).

[168] Freeman (1987); Church (1999); Hobday et al. (2004); Minagawa et al. (2007).

[169] Hung (2003).

[170] Minagawa et al. (2007).

[171] American Society for Industrial Security and PricewaterhouseCoopers, cited by Ferdinand and Simm (2007).

[172] Samli and Jacobs (2003); Ferdinand and Simm (2007).

[173] Udell (1990); Whalley (1991).

[174] Macdonald (1986).

[175] Hippel (1988); Biemans (1991).

[176] Yoon and Lilien (1988).

[177] Watkins and Horley (1986); Hargadon and Sutton (1997); Howells (2006); Winch and Courtney (2007).

[178] For example, Department of Trade and Industry (1993) and Department of Innovation, Universities, and Skills (2008) in the UK, and Commission of the European Communities (2003) in the EU.

[179] Howells (2006).

[180] Bessant and Rush (1995); Hargadon and Sutton (1997); Howells (2006); Winch and Courtney (2007).

[181] Winch and Courtney (2007).

[182] Winch and Courtney (2007); Klerkx and Leeuwis (2008).

[183] Department of Innovation, Universities, and Skills (2008: 34).

[184] Winch and Courtney (2007); Klerkx and Leeuwis (2008); Rosenkopf and Tushman (1998), respectively.

[185] Howells (2006).

[186] Hargadon and Sutton (1997); Rosenkopf and Tushman (1998).

[187] Klerkx and Leeuwis (2008).

[188] Winch and Courtney (2007).

[189] Pittaway et al. (2004); Howells (2006).

[190] Muller and Zenker (2001); Alam (2003b); Tether and Tajar (2008a).

[191] Muller and Zenker (2001).

[192] Muller and Zenker (2001: 1505, 1514).

[193] Tether and Tajar (2008a: 1092).

[194] Enos (1962); Freeman (1968), respectively.

[195] Hippel (1976; 1977b); Spital (1979); Shaw (1985; 1993); Slaughter (1993), respectively.

[196] Franke and Shah (2003); Lüthje (2004); Hienerth (2006).

[197] Corey (1956); Peck (1962); Hippel (1988); Vanderwerf (1990; 1992), respectively.

[198] Hippel (1982a; 1988); Vanderwerf (1990; 1992).

[199] Hippel (1982a; 1988).

[200] Bruce (1996).

[201] Interview on 9 March 1992 by Steve Conway with the principal scientist at the Johnson Matthey BioMedical Technology Centre.

[202] Anglo Platinum press release dated 8 November 2002, available online at <http://www.angloplatinum.com>.

[203] Riggs and Hippel (1994: 465).

[204] Lüthje (2004: 693).

[205] Hienerth (2006); Shah and Tripas (2007).

[206] Vanderwerf (1992: 332).

[207] See Alcoa's 2007 annual report, available online at <http://www.alcoa.com> for examples.

[208] Shaw (1985: 290).

[209] Slaughter (1993: 92).

[210] Franke and Shah (2003); Lüthje (2004); Lüthje et al. (2005); Hienerth (2006).

[211] Hippel (1994).

[212] Ogawa (1998); Jeppesen and Molin (2003); Hienerth (2006), respectively.

[213] Hippel (1988: 5).

[214] Hippel (1988: 8–9).

[215] Hippel and Krogh (2003; 2006).

[216] Chesbrough (2003a; 2003b).

[217] Hippel and Krogh (2003).

[218] Hippel and Krogh (2003: 216).

[219] West and Gallagher (2006: 329).

[220] Hippel and Krogh (2003: 210).

[221] Franke and Shah (2003); Lüthje (2004); Hippel and Finkelstein (1979); Morrison et al. (2000), respectively.

[222] Hippel (2005).

[223] Hippel and Krogh (2006).

[224] Chesbrough (2003b).

[225] Gassmann (2006: 223–4).

[226] Christensen et al. (2005: 1534).

[227] Freeman (1991: 501); Cohen and Levinthal (1989: 569), respectively.

11

The Transformative Capacity of Innovation and Innovation Systems

Chapter overview

Learning objectives

This chapter will enable the reader to:

- outline the key features of a 'system of innovation', and appreciate the strengths and weaknesses of the approach;

- distinguish between national, regional, sectoral, and technological systems of innovation, and assess their relative utility in relation to policymaking;

- outline the different policy approaches for promoting innovation and innovative capacity;

- distinguish between 'conventional', 'participatory', and 'constructive' technology assessment, and assess the relative utility of each for the social shaping of innovation and technology; and

- identify the various ways in which the 'transformative' potential of innovation and technology might be realized.

11.1 **Introduction**

In this concluding chapter, we explore the transformative capacity of innovation and of innovation systems. We start by seeking to reveal the mechanisms by which innovation outputs are shaped during the innovation process through interventions and influences from outside of the organization; here, we will build on earlier discussions regarding socio-technical paradigms and landscapes (see **Chapter 5**), and the external sources of innovation (see **Chapter 10**). We are also interested in the manner in which the innovative capacity of an organization is shaped by external factors. In this regard, we have already discussed the role and nature of inter-organizational networks and regional clusters (see **Chapter 3**), and the notion of relational assets (see **Chapter 6**). We will extend this discussion to look at the influence of broader contextual factors.

Whilst we are interested in developing an appreciation of the manner in which innovative organizations interact with their environment, we are also seeking in this chapter to open up the discussion by shifting the emphasis away from the organization towards a 'systemic' perspective of innovation. In part, this is driven by an increasing recognition within the literature of the 'distributed' nature of innovation—that is, that innovation is often the result of a process that is distributed across multiple organizations rather than within one.[1] Furthermore, this distributed process is shaped and nourished by a multitude of broader contextual factors, such as cultural attitudes to entrepreneurship and innovation, the prevailing education system, rules and regulations, and a wide range of organizations within the private, public, and not-for-profit sectors.

The 'systemic' perspective of innovation and technology is informed by the belief that '*the* direction *as well as rate of innovation*', '*the* form *of technology*', and '*the* outcomes *of technological change for different groups*' are shaped by '*social, institutional, economic and cultural factors*'.[2] The possibility for such shaping occurs because:

> Every stage in the generation and implementation of new technologies involves a set of choices between different technical options. Alongside narrowly 'technical' considerations, a range of 'social' factors affect which options are selected—thus influencing the content of technologies, and their social implications.[3]

There are a range of concepts and approaches within the innovation literature that adopt a 'systemic' perspective. We have touched upon a number of these in earlier chapters, such as, 'actor network theory' (see **3.4.5**), 'socio-technical regimes' and 'landscapes' (see **5.2.2** and **5.2.3**), and 'techno-economic paradigms' (see **5.5.1**). In this chapter, we will draw largely upon the 'systems of innovation' concept. This choice is made for two reasons: firstly, it is an approach that has been widely adopted by both academics and policymakers alike; secondly, because there have been attempts to apply the approach explicitly to different levels of analysis—most notably to national, regional, sectoral, and technological systems. We will explore each of these. Our discussion attempts to address a number of important issues, such as the identification of the key organizations, contextual factors, and activities of successful innovation systems, the extent to which such systems may be consciously constructed and mobilized, and whether or not there is 'one best way' in which they may be managed and organized.

We also wish to touch upon the ways in which innovation outputs can address social or environmental problems, for example; here, we will draw upon the notion of 'transformative' innovation or technology. 'Transformative' innovation or technology is an emergent concept and, as such, is expressed in a variety of ways.[4] Broadly speaking, it relates to the impact, rather than the nature, of an innovation or technology, such as the economic, social or environmental changes that it brings about. The impact of innovation and technology is in itself a vast subject area and largely beyond the scope of this text. Thus, the intention here is only to signal the transformative capacity of innovation and to highlight possible ways in which this potential might be realized.

11.2 Systems of innovation

Since the late 1980s, there has been a growing interest among academics and policy-makers in the 'systems of innovation' (SI) approach. This has been accompanied by a rise in the use of 'systemic instruments' by policymakers to nurture, build, and organize systems of innovation.[5] Although initially adopted in relation to analysing and understanding systems of innovation at the national level,[6] it has been increasingly employed in research and policymaking focused at the regional[7] and sectoral[8] levels. Research concerning regional clusters is essentially a combination of the regional and sectoral level of analysis (see **3.6.3**).[9] Sharif notes that:

> there is much disagreement among academics and policymakers about whether it is efficacious to identify the 'national' unit as the delimiting criterion of an innovation system as opposed to choosing the regional, sectoral, technological innovation system as the chief object of study and measurement.[10]

For Edquist:

> All these approaches may be fruitful, but for different purposes . . . Generally, the variants . . . complement rather than exclude each other and it is useful to consider sectoral and regional SIs in relation to—and often parts of—national ones.[11]

This view is supported by Teixeira, who argues that:

> the *systemic approach* to innovation in general—regardless of the analytically selected boundary of the system—has been established and proved as a useful framework to study technical change and its determinants.[12]

Following a discussion of the generic approach, we will consider the innovation systems concept at the national, regional, sectoral, and technological levels.

Broadly speaking, a 'system of innovation' embraces '*all the important economic, social, political, organizational, institutional, and other factors that influence the development, diffusion, and use of innovations*'.[13] As such, it includes a wide variety of organizations, such as innovative firms, intermediaries, universities, and government departments, the behaviour and interactions of which are shaped by '*sets of common habits, norms, routines, established practices, rules, or laws*'[14]—or what are sometimes termed 'the rules of the game'.[15] But Edquist points out that:

the array of determinants of innovations and the relations among them can be expected to vary over time and space, i.e. between innovation systems, as well as among different categories of innovation.[16]

Table 11.1 provides examples of organizations, contextual factors, and activities that are of importance to a 'typical' system of innovation.

Table 11.1 Examples of organizations, contextual factors, and activities embraced by the systems of innovation concept

Organizations	Contextual factors	Activities (Edquist and Hommen, 2008a, 10)
Concerning education: e.g. schools, colleges, and universities. Concerning research: e.g. universities, consultancies, and research laboratories. Concerning the competitive arena: e.g. users, suppliers, and competitors. Concerning laws and regulation: e.g. government and opposition, courts, regulatory authorities, and professional bodies. Concerning public opinion: e.g. media organizations and pressure groups.	Culture and norms: e.g. those concerning attitudes to progress, novelty, success, failure, risk, the purpose of the education system, and entrepreneurship. Laws and regulations: e.g. those pertaining to protection of intellectual property, competition, and foreign investment. Established practices: e.g. those concerning the degree of government intervention, cooperation vs competition between organizations, and the extent and nature of the interactions between science and technology/universities and organizations.	Creation of new knowledge. Competence building via training and education. Formation of new markets and the articulation of quality and functional requirements. Creating new firms and changing existing firms around emerging sectors. Facilitating networking and learning between a diverse range of organizations. Creating new laws and regulations and/or altering existing laws and regulations to promote innovation. Provision of support services, such as incubators for new firms, consultancies, and intermediaries.

Fagerberg argues that the nature and structure of an innovation system *'will facilitate certain patterns of interaction and outcomes (and constrain others)'*, which has the potential for 'locking' the system *'into a specific path of development'*. On the one hand, this might be viewed as an advantage, because it nudges the various actors within the system in a common direction, providing increased opportunities for synergy; on the other hand, however, this may blind the actors to fruitful alternatives.[17]

Although the 'system' is seen to have an important impact upon the innovative performance of firms, Nelson and Rosenberg argue that *'There is no presumption that the system was, in some sense, consciously designed, or even that the set of institutions involved work together smoothly and coherently'*.[18] As the concept is increasingly taken up by policymakers, however, we are increasingly seeing policy interventions aimed at directing, shaping, and influencing the configuration and activities of various systems of innovation, at the supranational, national, sectoral, and regional levels. Indeed, in the summing-up from a comparative analysis of national systems of innovation, Nelson argues that:

> At present nations seem to be conscious as never before of their 'innovation systems' and how they differ from those of their peers . . . it is leading to attempts on the part of nations to adopt aspects of other systems that they see as lending them strength.[19]

Fagerberg, meanwhile, highlights the important role of 'system managers', such as policymakers, in ensuring the 'openness' of the innovation system to lower the potential of it *'being "locked out" from promising paths of development that emerge from outside of their system'*.[20] Fagerberg and Srholec, however, warn that *'An innovation system is something that is built incrementally over many years'* and can be hampered by *'unfavourable aspects relating to geography, nature, and history'*.[21]

Given the rapid and widespread adoption of the systems of innovation perspective, it is clear that it is perceived to have utility in understanding the broader influences of innovative capacity. Edquist highlights a number of generally accepted strengths of the systems approach, including:[22]

- it is *holistic*—in that it attempts to embrace *'all of the important determinants of innovation'*;

- it is *interdisciplinary*—in that it draws upon a wide range of disciplines, such as economics, sociology, and economic history, for example;

- it emphasizes *interdependence*—through its recognition that organizations do not 'innovate in isolation' (this position is clearly supported by the extensive evidence discussed in **Chapter 3**, concerning the role of inter-organizational networks, and **Chapter 10**, concerning the external sources of innovation);

- it emphasizes *non-linearity*—through its recognition of innovation as an interactive (see **3.2.2**) and distributed process; and

- it emphasizes *institutions*—through its recognition of the importance of factors, such as norms, routines, rules, and laws, within the system that forms the context in which organizations innovate.

There are, however, a number of weaknesses also, largely around what Edquist has termed 'conceptual diffuseness':[23]

- *variations in terminology*—this differs between various proponents of the systems of innovation approach.[24] Most notable is that around the meaning of 'institution',[25] which, for some, concerns different types of organization or 'player' in the system, whilst for others, it relates to 'the rules of the game';[26]

- *a vagueness of element inclusion*—whilst studies of innovation systems provide a clear pointer to the various important components or elements of an innovation system, the approach is not accompanied by a definitive list of what should be included in empirical work, although, as noted above, this might be expected to vary 'over time and space' and between different types of innovation;[27]

- *a vagueness of system activities and functions*—there is also vagueness as to the activities and functions that a system of innovation might be expected to perform. Recently, a number of scholars have attempted to outline the key functions and activities, but these have not as yet lead to a consensus;[28] and

- it is *under-theorized*—although there are differences among scholars as to whether this is constitutes a problem.[29]

Conceptual diffuseness or vagueness is not simply an academic issue: policymakers within the Organisation for Economic Co-operation and Development (OECD), for example, have argued that *'There are still concerns in the policy making community that the NIS approach has too little operational value and is difficult to implement'*.[30]

11.2.1 National systems of innovation

Since the mid-1990s, there has been significant interest in the national systems of innovation (NSI) concept among academics and policymakers, such as those within the European Union (EU) and OECD.[31] Two notable books have played an important role in stimulating the adoption of the concept: Bengt-Åke Lundvall's *National Systems of Innovation: Towards a Theory of Innovation and Interactive Learning,* and Richard Nelson's edited collection *National Innovation Systems: A Comparative Analysis.*[32] The two are complementary. As the titles suggest, the former text focuses on the theoretical development of the concept, whilst the latter provides a comparative study of fifteen different national innovation systems. The two texts also adopt distinctly different approaches in relation to their scope: Lundvall is seen to provide a *'broad conception of NSI—embedded in a wider socioeconomic system'*, whilst Nelson is viewed as advancing *'a more narrow approach, focusing on national R&D systems and organizations supporting R&D'*.[33]

The studies collated by Nelson cover the national innovation systems of three groupings of nations: the USA, Japan, Germany, Britain, France, and Italy (as examples of 'large high-income countries'); Denmark, Sweden, Canada, and Australia (as examples of 'smaller high-income countries'); and finally, South Korea, Taiwan, Brazil, Argentina and Israel (as examples of 'lower income countries'). In comparing these different innovation systems, Nelson and Rosenberg found that *'although there are many areas of similarity*

Table 11.2 Examples of three different national systems of innovation (data drawn from Nelson, 1993b)

USA	UK	Japan
High % of R&D expenditure on military/mission research	High % of R&D expenditure on military/ mission research	Government funding of R&D relatively very low
Strong funding of basic science in academia	Short-termist nature of British capital markets	Key role by MITI in directing and coordinating technology
Anti-trust policy, leading to an emphasis on organic growth rather acquisition and the protection of start-ups	Growth often through acquisition rather than R&D	Good supply of highly trained research personnel, with a relatively large age in industry
Emergence of giant firms with high R&D spending	Persistent under-resourcing and undervaluation of education and training by industry and government	Rapid increase in patenting activity (relative to Europe)
Entrepreneurial culture and the acceptance of risk and failure	Loss of 'technological' culture—rise of managerialism and the decline of status of engineers	Few capital constraints from capital market: reciprocal shareholdings between banks and firms
Availability of venture capital	Weak link between science base and technology firms	Good employee–employer relations: mutual loyalty
Strong link between 'Ivy League' universities and hi-tech clusters, e.g. MIT, Stanford, etc.	Rise of entrepreneurial culture, but failure still problematic	Carefully organized employee rotation and training schemes
Technical career still ladder available		Higher proportion of managers from R&D or production

between the systems of countries in comparable economic settings, there still are striking differences as well', and that *'these differences reside, to a significant degree, in differences in national histories and cultures . . . [and] have profoundly shaped national institutions, laws, and policies'.*[34]

Subsequent work on national innovation systems has built on both these theoretical and empirical foundations.[35] A recent text by Charles Edquist and Leif Hommen, for example, extends both the empirical base, with studies of ten 'small country' innovation systems (e.g. Singapore, Ireland, Hong Kong, Norway, Finland, and the Netherlands), and seeks to *'refine elaborate, and operationalize the SI approach'.*[36] As Sharif observes, however: *'Although the concept . . . has been in use for the past 20 years, even today it is subject to a remarkable variety of interpretations.'*[37] In large part, this variation may be attributable to the under-theorization of the concept, as well as ambiguity surrounding components of the system—features noted above in relation to weaknesses of the generic systems of innovation approach. From his interviews with various academics involved in the development of the concept, Sharif notes that there is disagreement between those that believe that *'the concept should be more deeply theorized and explained in greater detail in order to make it more precisely applicable'* and those that view *'the approach's usefulness is a product of it being "loose" and "flexible"'*.[38]

Niosi notes, however, that *'Although no single definition has yet imposed itself, there is a semantic core that appears in most of the definitions used'*.[39] The 'variety' and 'semantic core' in the various definitions of the 'national innovation system' concept is revealed by the following selection from among key proponents of the approach:

> [the] set of institutions whose interactions determine the innovative performance . . . of national firms.[40]

> The elements and relationships which interact in the production, diffusion and use of new, and economically useful knowledge . . . and are either located within or rooted inside the borders of a nation state.[41]

> The national system of innovation is constituted by the institutions and economic structures affecting the rate and direction of technological change in the society. Obviously, the national system of innovation is larger than the R&D system . . . [but] is, of course, less comprehensive than the economy/society as a whole.[42]

> A national system of innovation is the system of interacting private and public firms (either large or small), universities, and government agencies aiming at the production of science and technology within national borders. Interaction among these units may be technical, commercial, legal, social, and financial, in as much as the goal of the interaction is the development, protection, financing or regulation of new science and technology.[43]

Although there are alternative delimiters, as we noted above, for bounding an innovation system, the nation state is often considered to be one of the most salient, not least because public policies in relation to innovation are still predominantly *'designed and implemented at the national level'*.[44]

For Nelson and Rosenberg:

> the policies and programs of national governments, the laws of a nation, and the existence of a common language and a shared culture define an inside and outside that can broadly affect how technical advance proceeds.[45]

This view is supported by a recent analysis of the relationship between national innovation systems and economic development in 115 countries between 1992 and 2004, from which Fagerberg and Srholec conclude that:

> The empirical analysis suggests that a well-developed innovation system is essential for countries that wish to succeed in catch-up. There is a strong, significant and robust statistical relationship between (level and change of) GDP per capita on the one hand, and (level and change of) the innovation system on the other. Historical and descriptive evidence also suggest that countries that have succeeded in catch-up have given a high priority to this dimension of development.[46]

But what might a 'well-developed' innovation system look like? And might there be a 'one best way' in which to organize a nation system of innovation?

One useful starting point is provided by Niosi, who focuses on the factors that impact the efficiency and effectiveness of key activities of a national innovation system. These include the adequacy of legislation that protects intellectual property, the presence of key organizations, such as research universities, and entrepreneurial and new hi-tech firms, the coordination among different organizations, to promote, for example, the match between the supply and demand of university graduates, and cooperation between universities and industry, and the extent of knowledge flow and 'spillovers' between organizations and sectors.[47] Niosi contends that *'Most of the inefficiencies and ineffectiveness of national innovation systems may be related to path-dependence and lock-in situations'*.[48] Niosi goes on to argue that such indicators of efficiency and effectiveness might be employed to benchmark one national innovation system against another.

But whilst research may have highlighted a number of broad characteristics of '(in)efficient' and '(in)effective' national systems of innovation, it also suggests that there is no single best way in which these characteristics may be manifested. In reflecting upon their collection of case studies, for example, Nelson and Rosenberg note that they *'have been impressed by the diversity of "national systems" that seem to be compatible with relatively strong . . . economic performance'*; they attribute this diversity to the *'variety of alternative arrangements for accomplishing basically the same thing'*.[49]

Despite the utility and resilience of the nation state as the predominant unit of analysis for innovation systems, the last five years or so have seen a growing interest in regional and sectoral systems of innovation. This is perhaps not too surprising given that Nelson and Rosenberg note, for example, that, on the one hand, *'there are important interindustry differences in the nature of technical change, the sources, and how the involved actors are connected to each other'*,[50] and, on the other, that *'There is good reason to believe that in recent years, just as the idea of national innovation systems has become widely accepted, technological communities have become transnational as never before'*.[51]

We now turn attention to these alternative levels of analysis.

11.2.2 Regional systems of innovation

There has been increasing interest in regional systems of innovation since the late 1990s.[52] In large part, this interest has been fuelled by the emerging recognition of the importance of 'industrial districts'[53] and 'regional clusters'.[54] These, as we noted in **Chapter 3**, are geographical concentrations of interconnected companies and associated institutions, including end producers, universities, research laboratories, service providers, and a highly specialized pool of labour, often focused around a specialized area of economic activity. We also noted that they are frequently characterized by extensive inter-firm social networks and job mobility. For Asheim and Gertler, *'the regional innovation system can be thought of as the institutional infrastructure supporting innovation within the production structure of a region'*.[55]

In support of a regional system of innovation approach, Ashiem and Gertler argue that:

innovative activity is not uniformly or randomly distributed across the geographical landscape. Indeed, the more knowledge-intensive the economic activity, the more geographically clustered it tends to be.[56]

Exemplars of such geographical or regional clusters include Silicon Valley in the USA (for information technology), Motor Sport Valley in the UK (for racing car production), the Baden-Wurttemburg region in southwestern Germany (for advanced production technology and electronics), the Hsinchu Science-Based Industry Park located southwest of Taipei, Taiwan (for information technology), and the Emilia Romagna region in north central Italy (for knitwear, ceramic tiles, motorcycles, and food-processing equipment).[57] A wide range of other examples from across the various states of the USA are provided by Porter (see **Figure 3.13**).[58]

Whilst the above examples provide a clear indication of the importance of regional systems of innovation, a number of scholars have attempted to explain why this might be so. Cooke et al. argue that regions (whether defined 'culturally', such as the Basque region, or 'administratively' such as the German *Länder*), evolve along distinct trajectories, leading to distinct regionalized 'rules of the game', 'collective identities', and 'collective social orders'. This is likely to lead to regional variations in innovative capacity.[59] Similarly for Oughton et al., the rationale for focusing on regional rather than national systems of innovation:

> lies in the fact that the factors that the national innovation systems theory identifies as important, such as the institutional framework, the nature of inter-firm relationships, learning capability, R&D intensity and innovation activity all differ significantly across regions.

They support their argument by an analysis of variance in innovative activity between 178 regions and twelve nation states within the EU, which reveals that '*for all indicators of innovation activity the variation across regions is significantly greater than the variation across nation states*'.[60]

Alternative explanations that account for the importance of regional or localized systems of innovation relate to organizational 'embeddedness' (see **3.5.3**), the 'stickiness' of knowledge (see **10.8.2**), and knowledge 'spillovers'. Highly embedded organizations, for example, can accrue benefits in relation to organizational survival, learning, risk sharing, and the speed of bringing new products and services to market. Organizational activity is said to be embedded in social networks to the extent that interactions within and between organizations are shaped by friendship or kinship ties, rather than the '*economic logic of the market*'.[61]

With regard to knowledge acquisition and transfer, Asheim and Isaken argue that '*The best way for firms to acquire . . . "sticky" knowledge is to be located . . . in areas where learning processes that develop new and economically useful knowledge takes place*',[62] whilst Feldman, from her review of the empirical work, concludes that:

> knowledge spillovers from science-based activities are localized and contribute to higher rates of innovation, increased entrepreneurial activity and increased productivity within geographically bounded areas.[63]

These processes are self-reinforcing, and promote the local accumulation of knowledge and expertise.

Interestingly, it has been argued that *'globalization and regionalization go hand in hand'*.[64] Indeed, for Porter, *'the enduring competitive advantages in a global economy lie increasingly in local things—knowledge, relationships, motivations—that distant rivals cannot match'*.[65]

11.2.3 Sectoral and technological systems of innovation

Whilst it is common, as illustrated by the examples provided in the previous section, for regional systems of innovation to be dominated by one or a small number of sectors, sectoral systems of innovation are typically not limited to a single region. Sectoral systems, in turn, may be dominated by one or, more typically, several technologies.[66]

As with regional systems of innovation, interest in sectoral systems of innovation has also grown over the last decade.[67] A key proponent of this approach is Franco Malerba, who has researched and written extensively in the area.[68] He defines a sectoral system of innovation as:

> a set of new and established products for specific uses and the set of agents carrying out market and non-market interactions for the creation, production, and sale of those products . . . [it] has a knowledge base, technologies, inputs . . . The agents composing the sectoral system are organizations and individuals . . . characterized by specific learning processes, competencies, beliefs, objectives, organizational structures and behaviors. They interact through processes of communication, exchange, cooperation, competition, and command, and their interactions are shaped by institutions (rules and regulations).[69]

In support of the adoption of a sectoral system of innovation approach, Malerba argues that *'innovation takes place in quite different sectoral environments, in terms of sources, actors and institutions. These differences are striking'*.[70] To illustrate these differences, his recent edited text, *Sectoral Systems of Innovation*, provides a detailed study of six distinct sectors within Europe—that is, those of pharmaceuticals, chemicals, mobile telecommunications, software, machine tools, and services. These studies reveal fundamental differences between the sectors in relation to the size of incumbent firms (e.g. the chemical sector is characterized by large firms, whilst the software industry is characterized by both global and local firms), regulation (e.g. regulation in the pharmaceutical sector can play a major role in directing technology and, in some cases, retarding innovation, whilst in the software industry, standardization has played an important role in diffusion), and nature of innovation (e.g. innovation in the pharmaceutical sector is based on scientific knowledge, whilst it is based on application knowledge within the machine-tool industry). This research complements earlier work, which has highlighted sectoral differences in relation to technological regimes,[71] patterns of innovation,[72] and the sources and locus of innovation.[73] Importantly, these sectoral differences imply that different policies are required to support and stimulate different sectors.

An overlapping approach to sectoral systems of innovation is that of 'technological systems'. Key proponents are Bo Carlsson, Rikard Stankiewicz, and Staffan Jacobsson.[74] This approach focuses on the *'network of agents interacting in a specific economic/industrial area under a particular institutional infrastructure . . . and involved in the generation, diffusion, and utilization of technology'.*[75] Technological systems typically span multiple sectors.

11.3 Innovation, innovation systems, and government intervention

From the above discussion of the various systems of innovation, it is clear that government has the potential to intervene in a variety of ways in order to influence the innovative capacity of organizations, regions, sectors, and the nation as a whole. Governments also have the possibility of directly shaping the nature and trajectory of a technology, or of enabling a broader set of stakeholders to become involved in the innovation process.

We explore each of these two themes briefly below.

11.3.1 Innovation policies

Lundvall and Borrás provide a useful distinction between 'science' policy, in which the focus is on the *'production of scientific knowledge'*, 'technology' policy, which centres on the *'advancement and commercialization of sectoral technical knowledge'*, and 'innovation' policy, which seeks to improve the *'overall innovative performance of the economy'*.[76] Mani further distinguishes between innovation policies based on financial measures, such as tax incentives for research and development (R&D), the funding of R&D projects, the provision of venture capital or the enhancement of financial market mechanisms, from those based on non-financial measures, such as those aimed at protecting intellectual property, developing industry standards, promoting education and skills, and facilitating the diffusion of technology and innovation.[77]

From a systems of innovation perspective, Smits and Kuhlmann argue that:

> actors involved in innovation processes not only need instruments that focus on individual organisations (e.g. financial and managerial instruments) or on the relation between two organisations (e.g. diffusion and mobility orientated instruments), but also instruments that focus on the system level.[78]

They contend that 'systemic' innovation policy instruments must fulfil five generic functions:

- manage the interface between different subsystems, such as those relating to science and technology, and exploration and exploitation (see **Illustration 11.1**);
- build and organize innovation systems and prevent 'lock-in';
- provide the conditions for learning and experimentation;
- provide the infrastructure for developing strategic intelligence below; and
- stimulate and facilitate the development of a vision and the search for novel applications.[79]

There is plenty of evidence to suggest that governments are increasingly employing systemic instruments in relation to innovation policy to address such systemic issues.[80] This pattern is directly related to the widespread recognition among policymakers across a wide range of countries that efficient and effective national systems of innovation are important for economic performance.

Illustration 11.1 The Technology Strategy Board of the UK government

The Technology Strategy Board (TSB) was founded in 2003. It currently operates under the UK government's Department for Innovation, Universities, and Skills (DIUS). Its stated vision is:

> For the UK to be seen as a global leader in innovation and a magnet for technology-intensive companies, where new technology is applied rapidly and effectively to wealth creation.

Under this vision, it has a number of major goals, including: to help leading UK sectors and organizations maintain their global position in the face of increasing competition; to stimulate those UK sectors and organizations with the potential to be among the best in the world; and to ensure that the emerging technologies of today become the UK growth sectors of tomorrow. These are currently delivered through a number of key programmes, including 'innovation platforms', which operates through *'bringing together organisations focused on a particular social challenge . . .* [and enabling] *the integration of a range of technologies along with better co-ordination of policy, regulations, standards and* [public] *procurement'*, and 'knowledge transfer networks' (KTNs), designed to *'bring together a variety of organizations . . . to enable the exchange of knowledge and the stimulation of innovation in the network'*—these exist in cyber security, industrial mathematics, bioscience, aerospace, and nanotechnology, for example.[81]

But the evidence also suggests that such innovation policies vary greatly from one nation to another. This is not surprising, given the distinct characteristics of different national systems of innovation.[82] The picture is further complicated by the different policy requirements of different regional and sectoral systems of innovation, and the differing needs between the national, regional, sectoral, and technological levels. Oughton et al., for example, highlight what they refer to as the 'regional innovation paradox', which arises because of:

> the apparent contradiction between the comparatively greater need to spend on innovation in lagging regions and their relatively lower capacity to absorb public funds earmarked for the promotion of innovation.

Furthermore, they argue that:

> Whereas industrial policy . . . is targeted at lagging regions and aims to promote convergence, technology policy appears to be reinforcing regional inequalities as more public resources . . . are absorbed by richer regions.[83]

11.3.2 Constructive technology assessment

Schot distinguishes between 'conventional', 'participatory', and 'constructive' technology assessment. In doing so, he notes that, during the 1980s and 1990s, conventional technology assessment was *'widely adopted for mapping the probable consequences of various technological options'*. Although this assessment was typically conducted by experts, Schot argues that it *'has proven ineffective at predicting social responses or unexpected consequences'*. Participatory technology assessment, which attempts to bring 'public values and opinions' into the assessment process, has been one approach adopted to solve these problems. But since both of these approaches are aimed at evaluating technology and innovation post hoc, they are simply aimed at shaping public policy in relation to the application, diffusion, and adoption of a technology, rather than shaping the technology itself.[84]

The more recent emergence of constructive technology assessment (CTA) is a bolder attempt both to solve the inherent problems of prevailing technology assessment approaches, and to create a space for new forms of participation in the development of new technologies and innovation. The approach was developed in the Netherlands driving the mid-1990s. Key proponents include Arie Rip and Johan Schot.[85] Schot argues that CTA *'is based on different assumptions about the nature of technology development, social values and subsequent outcomes'*, and involves technology being *'assessed from many points of view throughout the entire process of design and redesign'*, so that *'the interests of all parties can be incorporated in the design'*.[86] The CTA process typically involves technology developers, societal actors, such as users, citizens, and workers, and regulators, and offers great potential for the formalization of societal participation in the shaping of technology and innovation.

But whilst the approach has been increasingly adopted among policy organizations in the EU and OECD, for example, Genus and Coles warn that:

> the orientation of CTA towards co-construction of technology underplays differences in agenda- or rule-setting capacity, influence and resources, which tilt influence over decision-making about new technology toward the already powerful in society. Here, making interventions for the (democratic) better requires structural adjustments in representation, resources or influence enjoyed by the respective parties, or greater clarity about the role of deliberation in technological choice.[87]

11.4 Transformative innovation and innovation processes

The term 'transformative' technology or innovation is generally reserved for innovation that brings about deep and pervasive change. Bright et al., for example, define transformative innovation as:

> activities that increase the mutual benefit to both business and society, with the potential to create a deep shift in the values, assumptions, and behaviours of people, organizations, industry, and the global society.[88]

Steward argues that *'Transformative innovation is about radical change of a more generic kind. It is about the implementation of paradigm-breaking, system-wide novelty'*;[89] whilst for Philips, *'Transformative technologies involve disjointed, step adjustments in our productive and institutional capacity, displacing, destabilizing or overturning precursor systems'*.[90] Philips also makes two further useful observations of transformative technology: that they *'are of sufficiently high profile to attract the attention, interest and risk perception of social movements, citizens, politicians and regulators'*, and *'often emerge wrapped in an optimistic fervour . . . [but] in spite of the rhetoric . . . [seldom] overwhelmed society quickly'*.[91]

From such definitions, transformative innovation may be considered to equate to the pervasive technologies discussed in relation to economic long waves (see **5.5.1**), such as steam power, rail, iron, electricity, automobiles, chemicals, computers, electronics, pharmaceuticals, and the Internet.

The transformative theme may be employed more broadly, however, in relation to:

- giving 'voice' to disenfranchized, marginalized, or disadvantaged individuals or groups within the innovation process, such as the young or old, the poor or disabled, in order that innovation outputs are appropriately shaped. This might be undertaken informally by innovative organizations or more formally through CTA processes;

- providing a partnership role in the innovation process for disenfranchized, marginalized, or disadvantaged individuals and organizations, such that innovation activity is undertaken 'with' rather than 'for' such actors, in order that they develop the necessary skills and capacity to innovate independently (see **Illustration 11.2**);[92]

- directing innovative activity towards neglected problems, particularly where commercial benefits may not be clear, such as in relation to addressing environmental or sustainability issues, or encouraging the development of drugs for illnesses that predominate in developing countries, as exemplified by the Drugs for Neglected Diseases Initiative (DNDi) (see **Illustration 11.3**);[93] and

- distributing the benefits of innovation more widely, either through the encouragement and extension of 'free-revealing' and 'open source', or through new thinking concerning intellectual property rights, such as those promoted by Médecins Sans Frontières.[94]

Each of these novel possibilities and processes has the potential for sharing the benefits of innovation more widely within and between societies.

Illustration 11.2 UK scientists working alongside scientists in Ghana and Tanzania

The Leverhulme Trust has established a £3.3m programme to fund collaborative projects between university scientists in the UK, and those in Ghana and Tanzania. A key objective of the programme is to build research capacity at the grass-roots levels, through aiding the development of research skills of individual scientists in Africa.[95]

Illustration 11.3 Drugs for Neglected Diseases Initiative (DNDi)

The Drugs for Neglected Diseases Initiative (DNDi) was founded in July 2003, through the activity of seven organizations from around the world: the Oswaldo Cruz Foundation, Brazil; the Indian Council for Medical Research; the Kenya Medical Research Institute; the Ministry of Health of Malaysia; the Pasteur Institute, France; Médecins sans Frontières (MSF); and the United Nations Development Programme (UNDP)/World Bank/World Health Organization (WHO) Special Programme for Research and Training in Tropical Diseases (TDR), which acts as an observer to the initiative.

The vision of DNDi '*is to improve the quality of life and the health of people suffering from neglected diseases by developing new drugs or new formulations of existing drugs for patients suffering from these diseases*'. The initiative currently runs approximately twenty projects and:

fosters collaboration both amongst developing countries and between developing and developed countries . . . [and] capitalizes on existing, fragmented R&D capacity, especially in the developing world, and complements it with additional expertise as needed.[96]

11.5 **Concluding comments**

In this concluding chapter, we have sought to draw together a number of the themes developed in earlier chapters, such as those concerning inter-organizational networks, regional clusters, socio-technical regimes and landscapes, relational assets, and the external sources of innovation. The key vehicle for this process has been the 'systems of innovation' perspective. This choice was made for two reasons: firstly, the SI approach has been widely adopted by both academics and policymakers alike; secondly, the approach was been applied explicitly to various levels of innovation system—most notably, to national, regional, sectoral, and technological-level systems.

Through the adoption of a systemic view of innovation, it has been possible to demonstrate the multitude of mechanisms through which the innovative capacity and innovative outputs of organizations are shaped. It has, for example, highlighted the diversity of actors, such as suppliers, users, universities, regulators, pressure groups, and competitors, the breadth of contextual factors, such as culture, norms, laws, regulations, and practices, and the range of potential governmental interventions that have the potential to shape the innovation process.

FURTHER READING

1. For an overview of the 'social shaping of technology' perspective, see Williams and Edge (1996).

2. For an interesting paper on the emergence and development of the national systems of innovation (NSI) concept, see Sharif (2006). This article is based on interviews with key developers of the concept, such as Christopher Freeman, Bengt-Åke Lundvall, Richard Nelson, and Charles Edquist. Also see Liu and White (2001), for a framework for comparing national innovation systems.

3. For a discussion of regional systems of innovation, see Cooke et al. (1997) and Cooke (2001), and sectoral systems of innovation, see Malerba (2002), and Storz (2008), who illustrates the approach through a focus on the Japanese game software sector.

4. For an overview of science, technology, and innovation policy, see Lundvall and Borrás (2007).

5. For a discussion and overview of CTA, see Schot (2001), and Genus and Coles (2005).

NOTES

[1] Coombs et al. (2003); Ramlogan et al. (2006).

[2] Williams and Edge (1996: 868).

[3] Williams and Edge (1996: 866).

[4] Bright et al. (2006); Philips (2007); Sovacool (2008); Steward (2008).

[5] Smits and Kuhlmann (2004).

[6] Lundvall (1992); Nelson (1993b).

[7] Cooke et al. (1997); Cooke (2001); Ashiem and Isaksen (2002); Turpin et al. (2002).

8 Malerba (2002; 2004a; 2005); Tether and Metcalfe (2004); Storz (2008).

9 Saxenian (1985); Henry et al. (1996); Porter (1990; 1998).

10 Sharif (2006: 756).

11 Edquist (2005: 199).

12 Teixeira (2008: 30)

13 Edquist (1997a: 14).

14 Edquist and Johnson (1997: 46).

15 Edquist (2005: 182).

16 Edquist (2005: 202).

17 Fagerberg (2005: 13).

18 Nelson and Rosenberg (1993: 4).

19 Nelson (1993a: 520).

20 Fagerberg (2005: 13).

21 Fagerberg and Srholec (2008: 1427).

22 Edquist (1997b); Edquist (2005).

23 Edquist (2005: 186).

24 Edquist (2005); Sharif (2006).

25 Nelson and Rosenberg (1993).

26 Lundvall (1992).

27 Edquist (2005).

28 Liu and White (2001); Bergek and Jacobsson (2003); Edquist and Hommen (2008a).

29 Edquist (2005); Sharif (2006).

30 OECD (2002: 11).

31 Balzat and Hanusch (2004); Edquist (2005); Sharif (2006); Lundvall (2007); Teixeira (2008).

32 Lundvall (1992); Nelson (1993b).

33 Edquist and Hommen (2008a: 5).

34 Nelson and Rosenberg (1993: 18).

35 For a review, see Balzat and Hanusch (2004); Sharif (2006); Teixeira (2008).

36 Edquist and Hommen (2008b).

37 Sharif (2006: 756).

38 Sharif (2006: 757).

39 Niosi (2002: 291).

40 Nelson and Rosenberg (1993: 4).

41 Lundvall (1992).

42 Edquist and Lundvall (1993: 267).

43 Niosi et al (1993).

44 Edquist (2005: 199).

45 Nelson and Rosenberg (1993: 16).

46 Fagerberg and Srholec (2008: 1427).

47 Niosi (2002: 296).

48 Niosi (2002: 293).

[49] Nelson and Rosenberg (1993: 20).

[50] Nelson and Rosenberg (1993: 13).

[51] Nelson and Rosenberg (1993: 17).

[52] Cooke et al. (1997); Ashiem and Isaksen (2002); Oughton et al. (2002); Turpin et al. (2002); Fleming et al. (2007).

[53] Piore and Sabel (1984).

[54] Rogers and Larson (1984); Saxenian (1985; 1991; 1994); Castilla et al. (2000); Casper (2007).

[55] Asheim and Gertler (2005: 299).

[56] Asheim and Gertler (2005: 291).

[57] Saxenian (1985); Henry et al. (1996); Sabel et al. (1987); Mathews (1997); Brusco (1982).

[58] Porter (1998).

[59] Cooke et al. (1997).

[60] Oughton et al. (2002: 99).

[61] Granovetter (1985).

[62] Asheim and Isaksen (2002: 86).

[63] Feldman (1999: 20).

[64] Oughton et al. (2002: 99).

[65] Porter (1998: 78).

[66] Malerba (2004b: 18).

[67] Breschi and Malerba (1997); Malerba (2002; 2004a; 2005); Tether and Metcalfe (2004); Storz (2008).

[68] Breschi and Malerba (1997); Malerba (2002; 2004a; 2005).

[69] Malerba (2002: 250).

[70] Malerba (2004b: 9).

[71] Malerba and Orsenigo (1996).

[72] Pavitt (1984)

[73] Hippel (1988).

[74] Carlsson (1995; 1997); Carlsson and Stankiewicz (1995); Carlsson and Jacobsson (1997).

[75] Carlsson and Stankiewicz (1995: 49).

[76] Lundvall and Borrás (2005).

[77] Mani (2002: 30).

[78] Smits and Kuhlmann (2004: 11).

[79] Smits and Kuhlmann (2004: 11–12).

[80] Smits and Kuhlmann (2004); Balzat and Hanusch (2004); Sharif (2006); Lundvall (2007); Teixeira (2008).

[81] Lord Sainsbury of Turville (2007: 47).

[82] Nelson (1993a); Mani (2002); Edquist and Hommen (2008a).

[83] Oughton et al. (2002: 98).

[84] Schot (2001: 39).

[85] Rip et al. (1995); Schot and Rip (1996); Schot (2001).

[86] Schot (2001: 40).

87 Genus and Coles (2005: 440).

88 Bright et al. (2006: 28).

89 Steward (2008: 15).

90 Philips (2007: 13).

91 Philips (2007: 13–14).

92 Chataway et al. (2005).

93 See <http://www.dndi.org>.

94 See <http://www.msf.org>.

95 Corbyn (2008).

96 See <http://www.dndi.org>.

■ REFERENCES

A

Aa, W. and Elfring, T. (2002) 'Realizing innovation in services', *Scandinavian Journal of Management*, 18: 155–71.

Abernathy, W. (1978) *The Productivity Dilemma: Roadblock to Innovation in the Automobile Industry*, London: John Hopkins University Press.

Abernathy, W. and Clark, K. (1985) 'Innovation: mapping the winds of creative destruction', *Research Policy*, 14(1): 3–22.

Abernathy, W. and Utterback, M. (1978) 'Patterns of industrial innovation', *Technology Review*, 80(7): 41–7.

Abernathy, W. and Utterback, M. (1988) 'Patterns of industrial innovation' in M. Tushman and W. Moore (eds) *Readings in the Management of Innovation*, 2nd edn, New York: Harper Business, 25–36.

Abernathy, W., Clark, K., and Kantrow, A. (1981) 'The new industrial competition', *Harvard Business Review*, 59(5): 68–81.

Abernathy, W., Clark, K., and Kantrow, A. (1983) *Industrial Renaissance: Producing a Competitive Future for America*, New York: Basic Books.

Achilladis, B., Robertson, A., and Jervis, P. (1971) *Project SAPPHO: A Study of Success and Failure in Innovation*, two vols, Brighton: Science Policy Research Unit, University of Sussex.

Adler, P. and Kwon, S. (2002) 'Social capital: prospects for a new concept', *Academy of Management Review*, 27(1): 17–40.

Adner, R. and Levinthal, D. (2001) 'Demand heterogeneity and technology evolution: implications for product and process innovation', *Management Science*, 47(5): 611–28.

Afuah, A. (1998) *Innovation Management: Strategies, Implementation, and Profits*, Oxford: Oxford University Press.

Afuah, A. and Bahram, N. (1995) 'The hypercube of innovation', *Research Policy*, 24: 51–76.

Alam, I. (2003a) 'Innovation strategy, process, and performance, in the commercial banking industry', *Journal of Marketing Management*, 19(9–10): 973–99.

Alam, I. (2003b) 'Commercial innovations from consulting engineering firms: an empiricial exploration of a novel source of new product ideas', *Journal of Product Innovation Management*, 20: 300–13.

Alba, R. (1982) 'Taking stock of network analysis: a decade's results' in S. Bacharach (ed.) *Research in the Sociology of Organizations: A Research Annual, Vol. 1*, Stamford, CT: JAI Press, 39–74.

Albertini, S. and Butler, J. (1995) 'R&D networking in a pharmaceutical company: some implications for human resource management', *R&D Management*, 25(4): 377–93.

Alder, P. and Kwon, S. (2002) 'Social capital: prospect for a new concept', *Academy of Management Review*, 27: 17–40.

Aldrich, H. (1979) *Organizations and Environments*, New Jersey: Prentice-Hall.

Aldrich, H. and Herker, D. (1977) 'Boundary spanning roles and organization structure', *Academy of Management Review*, 2(3): 217–30.

Aldrich, H. and Whetten, D. (1981) 'Organisation-sets, action-sets, and networks: making the most of simplicity' in P. Nystrom and W. Starbuck (eds) *Handbook of Organizational Design, Vol. 1*, New York: Oxford University Press, 385–408.

Aldrich, H. and Zimmer, C. (1986) 'Entrepreneurship through social networks' in D. Sexton and R. Smilor (eds) *The Art and Science of Entrepreneurship*, Cambridge, MA: Ballinger, 3–23.

Allen, J., James, A., and Gamlen, P. (2007) 'Formal versus informal knowledge networks in R&D: a case study using social network analysis', *R&D Management*, 37(3): 179–96.

Allen, T. (1970) 'Communication networks in R&D laboratories', *R&D Management*, 1(1): 14–21.

Allen, T. (1977) *Managing the Flow of Technology: Technology Transfer and the Dissemination of Technological Information within the R&D Organization*, Cambridge, MA: MIT Press.

Allen, T. (2007) 'Architecture and communication among product design engineers', *California Management Review*, 49(2): 23–41.

Allen, T. and Henn, G. (2007) *The Organization and Architecture of Innovation: Managing the Flow of Technology*, Burlington, MA: Butterworth-Heinemann.

Allen, T., Hyman, D., and Pinckney, D. (1983) 'Transferring technology to the small manufacturing firm: a study of technology transfer in three countries', *Research Policy*, 12(2): 199–211.

Alvesson, M. (1993) 'Organisation as rhetoric: knowledge-intensive firms and the struggle with ambiguity', *Journal of Management Studies*, 30(6): 997–1015.

Amabile, T., Conti, R., Coon, H., Lazenby, J., and Herron, M. (1996) 'Assessing the work environment for creativity', *Academy of Management Journal*, 39(5): 1154–84.

Amabile, T., Patterson, C., Mueller, J., Wojcik, T., Odomorik, P., Marsh, M., and Kramer, S. (2001) 'Academic–practitioner collaboration in management research: a case of cross profession collaboration', *Academy of Management Journal*, 44(2): 418–31.

Ancona, D. and Caldwell, D. (1992) 'Bridging the boundary: external activity and performance in organisational teams', *Administrative Science Quarterly*, 37(4): 634–65.

Ancona, D. and Caldwell, D. (1998) 'Rethinking team composition from the outside in' in M. Neale and E. Mannix (eds) *Research on Managing Groups and Teams, Vol. 1*, Stamford, CT: JAI Press, 21–37.

Anderson, P. and Tushman, M. (1991) 'Managing through cycles of technological change', *Research Technology Management*, May/June: 26–31.

Andrews, F. (1979) *Scientific Productivity*, Cambridge: Cambridge University Press.

Antonacopoulou, E. (1993) *The Contribution of the Competency Framework to an Organisation Which is Changing*, Research Paper No. 102, Coventry: Warwick Business School.

Argyris, C. (1990) *Overcoming Organizational Defenses*, Boston, MA: Allyn and Bacon.

Arthur, M. (2008) 'Examining contemporary careers: a call for interdisciplinary inquiry', *Human Relations*, 61(2): 163–86.

Arundel, A., Van de Paal, G., and Soete, L. (1995) *Innovation Strategies of Europe's Largest Firms: Results of the PACE Survey for Information Sources, Public Research, Protection of Innovations, and Government Programmes*, EIMS Report No. 23, Brussels: European Commission.

Ashby, W. (1952) *Design for a Brain*, New York: John Wiley.

Ashby, W. (1960) *An Introduction to Cybernetics*, London: Chapman and Hall.

Asheim, B. and Gertler, M. (2005) 'The geography of innovation: regional innovation systems' in J. Faberberg, D. Mowery, and R. Nelson (eds) *The Oxford Handbook of Innovation*, Oxford: Oxford University Press, 291–317.

Asheim, B. and Isaksen, A. (2002) 'Regional innovation systems: the integration of local "sticky" and global "ubiquitous" knowledge', *Journal of Technology Transfer*, 27: 77–86.

Ashkanasy, N., Hartel, C., and Zerbe, W. (eds) (2000) *Emotions in the Workplace: Research, Theory, and Practice*, Westport, CT: Quorum Books.

Assimakopoulos, D. and Yan, J. (2006) 'Sources of knowledge acquisition for chinese software engineers', *R&D Management*, 36(1): 97–106.

Assimakopoulos, D., Everton, S., and Tsutsui, K. (2003) 'The semiconductor community in the Silicon Valley: a network analysis of the SEMI genealogy chart (1947–1986)', *International Journal of Technology Management*, 25(1–2): 181–99.

Association of University Technology Managers (2002) *The AUTM Licensing Survey: FY 2001*, Norwalk, CT: AUTM.

Audretsch, D., Lehmann, E., and Warning, S. (2005) 'University spillovers and new firm location', *Research Policy*, 34: 1113–22.

Auster, E. (1990) 'The interorganizational environment: network theory, tools, and applications' in F. Williams and D. Gibson (eds) *Technology Transfer: A Communication Perspective*, London: Sage, 63–89.

Axelrod, W., Mitchell, R., Thomas, D., Bennett, D., and Bruderer, E. (1995) 'Coalition formation in standards-setting alliances', *Management Science*, 41(9): 1493–508.

B

Backman, M., Börjesson, S., and Setterberg, S. (2007) 'Working with concepts in the fuzzy front end: exploring the context for innovation for different types of concepts at Volvo cars', *R&D Management*, 37(1): 17–28.

Badaracco, J. (1991) *The Knowledge Link: How Firms Compete Through Strategic Alliances*, Boston, MA: Harvard Business School Press.

Baden Fuller, C. and Stopford, J. (1992) *Rejuvenating the Mature Business*, London: Routledge.

Bailyn, L. (1985) 'Autonomy in the industrial R&D laboratory', *Human Resource Management*, 24: 129–46.

Baker, N., Green, S., and Bean, A. (1985) 'How management can influence the generation of new ideas', *Research Management*, 28: 35–42.

Baker, T. and Nelson, R. (2005) 'Creating something from nothing: resource construction through entrepreneurial bricolage', *Administrative Science Quarterly*, 50(3): 329–66.

Balconi, M. and Laboranti, A. (2006) 'University–Industry interactions in applied research: the case of microelectronics', *Research Policy*, 35: 1616–30.

Balzat, M. and Hanusch, H. (2004) 'Recent trends in the research on national innovation systems', *Journal of Evolutionary Economics*, 14: 197–210.

Baptista, R. (1999) 'The diffusion of process innovations: a selective review', *International Journal of the Economics of Business*, 6(1): 107–29.

Barber, B. (1952) *Science and the Social Order*, New York: Collier Books.

Barnes, J. (1979) 'Network analysis: orienting notion, rigorous technique, or substantive field of study' in P. Holland and S. Leinhardt (eds) *Perspectives on Social Research*, New York: Academic Press, 403–23.

Barnes, S. and Dolby, R. (1971) 'The scientific ethos: a deviant viewpoint', *European Journal of Sociology*, 11(1): 3–35.

Barney, J. (1991) 'Firm resources and sustained competitive advantage', *Journal of Management*, 17: 99–120.

Barras, R. (1986) 'Towards a theory of innovation in services', *Research Policy*, 15: 161–73.

Barras, R. (1990) 'Interactive innovation in financial and business services: the vanguard of the service revolution', *Research Policy*, 19: 215–37.

Barrell, A. and Image, S. (1993) *Executive Networking: Making the Most of Your Business Contacts*, Hemel Hempstead: Director Books.

Bartlett, C. and Ghoshal, S. (1998) *Managing Across Borders: The Transnational Solution*, 2nd edn, Boston. MA: Harvard Business School Press.

Bate, P. (1994) 'Tales from the rails: the APT fiasco' in P. Bate (ed.) *Strategies for Cultural Control*, Oxford: Butterworth-Heinemann, 102–25.

Battelle Memorial Institute (1973) *Interactions of Science and Technology in the Innovation Process*, NSF-C667, Columbus, OH: National Science Foundation.

Baum, J. (1996) 'Organizational ecology' in S. Clegg, C. Hardy, and W. Nord (eds) *Handbook of Organization Studies*, Thousand Oaks, CA: Sage, 77–114.

Baum, J., Calabrese, T., and Silverman, B. (2000) 'Don't go it alone: alliance network composition and startups' performance in Canadian biotechnology', *Strategic Management Journal*, 21: 267–94.

Becker, F. and Sims, W. (2000) *Offices that Work: Balancing Cost, Flexibility, and Communication*, Ithaca, NY: Cornell University International Workplace Studies Program (IWSP).

Becker, R. and Speltz, L. (1983) 'Putting the S-curve concept to work', *Research Management*, 26(5): 31–3.

Becker, R. and Speltz, L. (1986) 'Working the S-curve: making more explicit forecasts', *Research Management*, 29(4): 21–3.

Behrens, T. and Gray, D. (2001) 'Unintended consequences of cooperative research: impact of industry sponsorship on climate for academic freedom and other graduate student outcome', *Research Policy*, 30: 179–99.

Belbin, R. (1981) *Management Teams: Why They Succeed or Fail*, London: Heinemann.

Belderbos, R., Carree, M., and Lokshin, B. (2004) 'Cooperative R&D and firm performance', *Research Policy*, 33: 1477–92.

Bell, G. and Zaheer, A. (2007) 'Geography, networks, and knowledge flow', *Organization Science*, 18(6): 955–72.

Benner, M. and Tushman, M. (2003) 'Exploitation, exploration, and process management: the productivity dilemma revisited', *Academy of Management Review*, 2: 238–56.

Bensaou, M. and Venkatraman, N. (1995) 'Configurations of interorganizational relationships: a comparison between US and Japanese automakers', *Management Science*, 41(9): 1471–92.

Bergek, A. and Jacobsson, S. (2003) 'The emergence of a growth industry: a comparative analysis of the German, Dutch and Swedish wind turbine industries' in S. Metcalfe and U. Cantner (eds) *Transformation and Development: Schumpeterian Perspectives*, Heidelberg: Physica/Spinger, 197–228.

Bessant, J. and Rush, H. (1995) 'Building bridges for innovation: the role of consultants in technology transfer', *Research Policy*, 24(1): 94–114.

Betz, F. (1987) *Managing Technology: Competing Through New Ventures, Innovation, and Corporate Research*, Englewood Cliffs, NJ: Prentice-Hall.

Biemans, W. (1991) 'User and third-party involvement in developing medical equipment innovations', *Technovation*, 11(3): 163–82.

Biggs, S. (1990) 'A multiple source of innovation model of agricultural research and technology promotion', *World Development*, 18(11): 1481–99.

Biggs, S. and Clay, E. (1981) 'Sources of innovation in agricultural technology', *World Development*, 9(4): 321–36.

Bijker, W. (1987) 'The social construction of Bakelite: toward a theory of invention' in W. Bijker, T. Hughes, and T. Pinch (eds) *The Social Construction of Technological Systems*, Cambridge, MA: MIT Press, 159–87.

Birley, S. (1985) 'The role of networks in the entrepreneurial process', *Journal of Business Venturing*, 1: 107–17.

Blackler, F. (1995) 'Knowledge, knowledge work and organisations: an overview and interpretation', *Organization Studies*, 16(6): 1021–47.

Blau, P. (1982) 'Structural sociology and network analysis: an overview', in P. Marsden and N. Lin (eds) *Social Structure and Network Analysis*, Beverly Hills, CA: Sage, 273–97.

Blau, P. and Scott, W. (1962) *Formal Organizations: A Comparative Approach*, San Francisco, CA: Chandler.

Blauner, R. (1964) *Alienation and Freedom: The Factory Worker and His Industry*, Chicago, IL: University of Chicago Press.

Blundel, R. (2006) '"Little Ships": the co-evolution of technological capabilities and industrial dynamics in competing innovation networks', *Industry and Innovation*, 13(3): 313–34.

Boden, M. and Miles, I. (eds) (2000) *Services and the Knowledge-Based Economy*, London: Continuum.

Boissevain, J. (1974) *Friends of Friends: Networks, Manipulators and Coalitions*, Oxford: Basil Blackwell.

Bonnett, D. (1986) 'Nature of the R&D/marketing co-operation in the design of technologically advanced new industrial products', *R&D Management*, 16(2): 117–26.

Boorman, S. and White, H. (1976) 'Social structure from multiple networks II: role structures', *American Journal of Sociology*, 81(6): 1384–446.

Borredon, L. and Ingham, M. (2005) 'Mentoring and organisational learning in research and development', *R&D Management*, 35(5): 493–500.

Bottazzi, L. and Peri, G. (2003) 'Innovation and spillovers in regions: evidence from European patent data', *European Economic Review*, 47: 687–710.

Boud, D. and Garrick, J. (eds) (1999) *Understanding Learning at Work*, London: Routledge.

Boulding, W. and Christen, M. (2001) 'First-mover disadvantage', *Harvard Business Review*, 79(9): 20–1.

Boulton, G. and Lucas, C. (2008) *What Are Universities For?*, Leuven, Belgium: League of European Research Universities (LERU).

Bounfour, A. (2003) *The Management of Intangibles: The Organisation's Most Valuable Assets*, London: Routledge.

Bouty, I. (2000) 'Interpersonal and interaction influences on informal resource exchanges between R&D researchers across organizational boundaries', *Academy of Management Journal*, 43(1): 50–65.

Bower, J. and Christensen, C. (1995) 'Disruptive technologies: catching the wave', *Harvard Business Review*, 73(1): 43–53.

Boyatzis, R. (1982) *The Competent Manager: A Model for Effective Performance*, Chichester: John Wiley.

Bozgodan, K., Deyst, J., Hoult, D., and Lucas, M. (1998) 'Architectural innovation in product development through early supplier integration', *R&D Management*, 28(3): 163–73.

Brass, D. (1984) 'Being in the right place: a structural analysis of individual influence in an organization', *Administrative Science Quarterly*, 29: 518–39.

Breschi, S. and Malerba, F. (1997) 'Sectoral innovation systems: technological regimes, Schumpeterian dynamics, and spatial boundaries' in C. Edquist (ed.) *Systems of Innovation: Technologies, Institutions and Organizations*, London: Routledge, 130–56.

Bright, D., Fry, R., and Cooperrider, D. (2006) 'Transformative innovations for the mutual benefit of business, society, and environment', *BAWB Interactive Working Paper Series*, 1(1): 17–33.

Bromiley, P. (1991) 'Testing a causal model of corporate risk taking and performance', *Academy of Management Journal*, 34: 37–59.

Brown, J. and Duguid, P. (2002) 'Organizing knowledge' in S. Little, Quintas, P., and Ray, T. (eds) *Managing Knowledge: A Essential Reader*, London: Sage, 19–40.

Brown, R. (2000) 'Contemplating the emotional component of learning: the emotions and feelings involved when undertaking an MBA', *Management Learning*, 31(3): 275–93.

Bruce, T. (1996) 'Rustenburg and Johnson Matthey: an enduring relationship', *Platinum Metals*, 40(1): 2–7.

Bruce, M., Leverick, F., Littler, D., and Wilson, D. (1995) 'Success factors for collaborative product development: a study of suppliers of information and communication technology', *R&D Management*, 25(1): 33–44.

Brusco, S. (1982) 'The Emilian model: productive decentralization and social integration', *Cambridge Journal of Economics*, 6: 167–84.

Bryans, P. and Smith, R. (2000) 'Beyond training: reconceptualising learning at work', *Journal of Workplace Learning*, 12(6): 228–35.

Buchanan, D. and Huczynski, A. (2006) *Organizational Behaviour: An Introductory Text*, 6th edn, Harlow: FT Prentice Hall.

Burgelman, R. (2001) 'Managing the internal corporate venturing process: some recommendations for practice' in R. Burgelman, M. Maidique, and S. Wheelwright (eds) *Strategic Management of Technology and Innovation*, 3rd edn, New York: McGraw-Hill, 692–702.

Burgelman, R. (2002) 'Strategy as vector and the inertia of coevolutionary lock-in', *Administrative Science Quarterly*, 47: 325–57.

Burgelman, R. and Sayles, L. (1986) *Inside Corporate Innovation: Strategy, Structure, and Managerial Skills*, New York: Free Press.

Burns, T. (1969) 'Models, images and myths' in W. Gruber and D. Marquis (eds) *Factors in the Transfer of Technology*, Cambridge, MA: MIT Press, 11–23.

Burns, T. and Flam, H. (1987) *The Shaping of Social Organization: Social Rule System Theory with Applications*, London: Sage.

Burns, T. and Stalker, G. (1961) *The Management of Innovation*, London: Tavistock Publications.

Burns, T. and Stalker, G. (1966) *The Management of Innovation*, 2nd edn, London: Tavistock Publications.

Burns, T. and Stalker, G. (1994) *The Management of Innovation*, 3rd edn, London: Tavistock Publications.

Burt, R. (1992) *Structural Holes: The Social Structure of Competition*, Cambridge, MA: Harvard University Press.

Burt, R. (2005) *Brokerage and Closure*, Oxford: Oxford University Press.

C

Calantone, R. and Cooper, R. (1981) 'New product scenarios: prospects for success', *Journal of Marketing*, 45: 48–60.

Callon, M. (1986) 'The sociology of an actor-network: the case of the electric vehicle' in M. Callon, J. Law, and A. Rip (eds) *Mapping the Dynamics of Science and Technology: Sociology of Science in the Real World*, London: Macmillan, 19–34.

Callon, M. (1992) 'The dynamics of techno-economic networks' in R. Coombs, P. Saviotti, and V. Walsh (eds) *Technological Change and Company Strategies: Economic and Sociological Perspective*, London: Academic Press, 72–102.

Callon, M. and Latour, B. (1981) 'Unscrewing the big Leviathan: how actors macrostructure reality and how sociologists help them to do so' in K. Knorr-Cetina and A. Cicourel (eds) *Advances in Social Theory and Methodology: Toward an Integration of Micro- and Macro-Sociologies*, Boston, MA: Routledge and Kegan Paul, 277–303.

Camagni, R. (ed.) (1991) *Innovation Networks: Spatial Perspective*, London: Belhaven.

Cancer Research UK (2004) 'Shaping new treatments', *Together*, Winter: 3–4.

Canto, J. and González, I. (1999) 'A resource-based analysis of the factors determining a firm's R&D activities', *Research Policy*, 28: 891–905.

Carlsson, B. (ed.) (1995) *Technological Systems and Economic Performance: The Case of Factory Automation*, Dordrecht, Netherlands: Kluwer Academic Publishers.

Carlsson, B. (ed.) (1997) *Technological Systems and Industrial Dynamics*, Dordrecht, Netherlands: Kluwer Academic Publishers.

Carlsson, B. and Jacobsson, S. (1995) 'Diversity creation and technological systems: a technology policy perspective' in C. Edquist (ed.) *Systems of Innovation: Technologies, Institutions and Organizations*, London: Routledge, 266–93.

Carlsson, B. and Stankiewicz, R. (1995) 'On the nature, function, and composition of technological systems' in B. Carlsson (ed.) *Technological Systems and Economic Performance: The Case of Factory Automation*, Dordrecht,

Netherlands: Kluwer Academic Publishers, 21–56.

Carter, A. (1989) 'Knowhow trading as economic exchange', *Research Policy*, 18(2): 155–63.

Carter, C. and Williams, B. (1957) *Industry and Technical Progress: Factors Affecting the Speed and Application of Science*, Oxford: Oxford University Press.

Casper, S. (2007) 'How do technology clusters emerge and become sustainable? Social network formation and inter-firm mobility within the San Diego biotechnology cluster', *Research Policy*, 36: 438–55.

Casson, M. (1982) *The Entrepreneurs: An Economic Theory*, Oxford: Martin Robertson.

Castellacci, F. (2006) 'Innovation, diffusion and catching up in the fifth long wave', *Futures*, 38: 841–63.

Castells, M. (2000) *The Rise of the Network Society: The Information Age—Economy, Society and Culture, Vol. 1*, 2nd edn, Oxford: Blackwell.

Castilla, E., Hwang, H., Granovetter, E., and Granovetter, M. (2000) 'Social networks in Silicon Valley' in C. Lee, W. Miller, M. Hancock, and H. Rowen (eds) *The Silicon Valley Edge*, Palo Alto, CA: Stanford University Press, 218–47.

Chakrabarti, A. (1974) 'A process model of internal corporate venturing in the major diversified firm', *Administrative Science Quarterly*, 28: 223–44.

Chan, J., Beckman, S., and Lawrence, P. (2007) 'Workplace design: a new managerial imperative', *California Management Review*, 49(2): 6–22.

Chandler, A. (1962) *Strategy and Structure*, Cambridge, MA: MIT Press.

Chataway, J., Smith, J., and Wield, D. (2005) *Partnerships and Building Capabilities For Science, Technology, Innovation and Development in Africa*, Innovation, Knowledge and Development Research Centre Working Paper No. 2, Milton Keynes: Open University Press.

Chesbrough, H. (2003a) *Open Innovation*, Boston, MA: Harvard University Press.

Chesbrough, H. (2003b) 'The era of open innovation', *MIT Sloan Management Review*, 44(3): 35–41.

Child, J. (1972) 'Organizational structure, environment, and performance: the role of strategic choice', *Sociology*, 6(1), 1–22.

Child, J. (1997) 'Strategic choice in the analysis of action, structure, organizations and environment: retrospective and prospect', *Organization Studies*, 18(1): 43–76.

Chow, C., Deng, F., and Ho, L. (2000) 'The openness of knowledge sharing within organisations: a comparative study of the United States and the People's Republic of China', *Journal of Management Accounting Research*, 12: 65–95.

Chow, C., Harrison, G., McKinnon, J., and Wu, A. (1999) 'Cultural influences on informal information sharing in Chinese and Anglo-American organizations: an exploratory study', *Accounting, Organizations and Society*, 24(5): 561–82.

Christensen, C. (1997) *The Innovator's Dilemma: When New Technologies Cause Great Firms to Fail*, Boston, MA: Harvard Business School Press.

Christensen, C. (2001a) 'Exploring the limits of the technology S-curve: Part 1—Component technologies' in R. Burgelman, M. Maidique, and S. Wheelwright (eds) *Strategic Management of Technology and Innovation*, 3rd edn, New York: McGraw-Hill, 124–42.

Christensen, C. (2001b) 'Exploring the limits of the technology S-curve: Part 2—Architectural technologies' in R. Burgelman, M. Maidique, and S. Wheelwright (eds) *Strategic Management of Technology and Innovation*, 3rd edn, New York: McGraw-Hill, 142–9.

Christensen, C. and Rosenbloom, R. (1995) 'Explaining the attacker's advantage: technological paradigms, organizational dynamics, and the value network', *Research Policy*, 24: 233–57.

Christensen, C., Anthony, S., and Roth, E. (2004) *Seeing What's Next: Using the Theories of Innovation to Predict Industry Change*, Boston, MA: Harvard Business School Press.

Christensen, J., Olesen, M., and Kjær, J. (2005) 'The industrial dynamics of open innovation —evidence from the transformation of consumer electronics', *Research Policy*, 34: 1533–49.

Chryssochoidos, G. and Wong, V. (1998) 'Rolling out new products across country markets: an empirical study of causes of delays', *Journal of Product Innovation Management*, 15(1): 16–41.

Church, R. (1999) 'New perspectives on the history of products, firms, marketing, and consumers in Britain and United States since the mid-nineteenth century', *Economic History Review*, LII: 405–35.

Clark, J., Freeman, C., and Soete, L. (1981) 'Long waves, inventions, and innovations', *Futures*, 13: 308–22.

Clark, K. (1985) 'The interaction of design hierarchies and market concepts in technological evolution', *Research Policy*, 14: 235–51.

Clark, K. (1989) 'Project scope and project performance: the effect of parts strategy and supplier involvement on product development', *Management Science*, 35(10): 1247–63.

Clark, K. and Fujimoto, T. (1991) *Product Development Performance: Strategy, Organization, and Management in the World Auto Industry*, Boston, MA: Harvard Business School Press.

Clark, P. (2000) *Organisations in Action: Competition Between Contexts*, London: Routledge.

Clark, P. and Starkey, K. (1988) *Organization Transitions and Innovation Design*, London: Pinter Publishers.

Clark, P. and Staunton, N. (1989) *Innovation in Technology and Organisation*, London: Routledge.

Clarke, N. (2004) 'HRD and the challenges of assessing learning in the workplace', *International Journal of Training and Development*, 8(2): 140–56.

Claxton, G. (1997) *Hare Brain, Tortoise Mind: Why Intelligence Increases When You Think Less*, London: Fourth Estate.

Clegg, S., Cunha, J., and Cunha, M. (2002) 'Management paradoxes: a relational view', *Human Relations*, 55(5): 483–503.

Cobbenhagen, J., Philips, G., and Friso, H. den (1991) 'Managing innovations. coping with complexity and dilemmas', Paper presented at the First Advances in the Social and Economic Analysis of Technology (ASEAT) Conference, April, Manchester.

Coff, R., Coff, D., and Eastvold, R. (2006) 'The knowledge-leveraging paradox: how to achieve scale without making knowledge imitable', *Academy of Management Review*, 31(2): 452–65.

Coffield, F. (2000) *The Necessity of Informal Learning*, Bristol: Polity Press.

Cohen, S. and Fields, G. (1999) 'Social capital and capital gains in Silicon Valley', *California Management Review*, 41(2): 108–30.

Cohen, W. and Levinthal, D. (1989) 'Innovation and learning: the two faces of R&D', *The Economic Journal*, 99: 569–96.

Cohen, W. and Levinthal, D. (1990) 'Absorptive capacity: a new perspective on learning and innovation', *Administrative Science Quarterly*, 35(1): 128–52.

Cohendet, P. and Simon, L. (2007) 'Playing across the playground: paradoxes of knowledge creation in the videogame firm', *Journal of Organizational Behaviour*, 28: 587–605.

Collins, J. and Moore, D. (1970) *The Organization Makers*, New York: Appleton-Century-Crofts.

Commission of the European Communities (2003) *Innovation Policy: Updating the Union's Approach in the Context of the Lisbon Strategy*, Report No. COM(2003) 112, Brussels: Commission of the European Communities.

Constable, J. and McCormick, R. (1987) *The Making of British Managers*, London: CBI/BIM.

Constant, E. (1980) *The Origins of the Turbojet Revolution*, Baltimore, MD: The John Hopkins University Press.

Constant, E. (2002) 'Why evolution is a theory about stability: constraint, causation, and ecology in technological change', *Research Policy*, 31: 1241–56.

Contractor, F. and Lorange, P. (2002) *Cooperative Strategies and Alliances*, Amsterdam: Elsevier.

Contu, A. and Willmott, H. (2003) 'Re-embedding situatedness: the importance of power relations in learning theory', *Organization Science*, 14(3): 283–96.

Conway, S. (1994) 'Informal boundary-spanning links and networks in successful technological innovation', Unpublished doctoral thesis, Birmingham: Aston Business School.

Conway, S. (1995) 'Informal boundary-spanning networks in successful technological innovation', *Technology Analysis and Strategic Management*, 7(3): 327–42.

Conway, S. (1997) 'Strategic personal links in successful innovation: link-pins, bridges, and liaisons', *Creativity and Innovation Management*, 6(4): 226–33.

Conway, S. and Steward, F. (1998) 'Mapping innovation networks', *International Journal of Innovation Management*, 2(2): 223–54.

Conway, S., Jones, O., and Steward, F. (2001) 'Realising the potential of the social network perspective in innovation studies' in O. Jones, S. Conway, and F. Steward (eds) *Social Interaction and Organisational Change: Aston Perspectives on Innovation Networks*, London: Imperial Press, 349–66.

Cook, S. and Brown, J. (1999) 'Bridging epistemologies: the generative dance between organizational knowledge and organizational Learning', *Organization Science*, 10(4): 381–400.

Cooke, P. (2001) 'Regional innovation systems, clusters, and the knowledge economy', *Industrial and Corporate Change*, 10(4): 945–74.

Cooke, P., Uranga, M., and Etxebarria, G. (1997) 'Regional innovation systems: institutional and organisational dimensions', *Research Policy*, 26: 475–91.

Cookson, G. (1997) 'Family firms and business networks: textile engineering in Yorkshire, 1780–1830', *Business History*, 39(1): 1–20.

Coombs, R. (1996) 'Core competencies and the strategic management of R&D', *R&D Management*, 26(4): 345–55.

Coombs, R., Harvey, M., and Tether, B. (2003) 'Analysing distributed processes of provision and innovation', *Industrial and Corporate Change*, 12(6): 1125–55.

Cooper, R. (1993) *Winning at New Products: Accelerating the Process from Idea to Launch*, Reading, MA: Addison-Wesley.

Cooper, R. (1994) 'Third generation new product processes', *Journal of Product Innovation Management*, 11(1): 3–14.

Cooper, R. and Brentani, U. (1984) 'Criteria for screening new industrial products', *Industrial Marketing Management*, 13: 149–56.

Cooper, R. and Kleinschmidt, E. (1986) 'An investigation into the new product process:

steps, deficiencies, and impact', *Journal of Product Innovation Management*, 3: 71–85.

Cooper, R. and Kleinschmidt, E. (1990) *New Products: The Key Factors in Success*, Chicago, IL: American Marketing Association.

Cooper, R. and Kleinschmidt, E. (1994) 'Determinants of timeliness in product development', *Journal of Product Innovation Management*, 11(5): 381–96.

Cooper, R. and Kleinschmidt, E. (1995) 'Benchmarking the firm's critical success factors in new product development', *Journal of Product Innovation Management*, 12(5): 374–91.

Corbyn, Z. (2008) 'The scramble to help Africa', *Times Higher Education*, 13 November, available online at <http://www.timeshighereducation. co.uk> (accessed 24 November 2008).

Corey, E. (1956) *The Development of Markets for New Materials: A Study of Building New End-Product Markets for Aluminium, Fibrous Glass and the Plastics*, Cambridge, MA: Harvard University Press.

Crane, D. (1972) *Invisible Colleges: Diffusion of Knowledge in Scientific Communities*, Chicago, IL: University of Chicago Press.

Crawford, C. (1992) 'The hidden costs of accelerated product development', *Journal of Product Innovation Management*, 9: 188–99.

Criscuolo, P. (2005) 'On the road again: researcher mobility inside the R&D network', *Research Policy*, 34: 1350–65.

Cross, R., Borgatti, S., and Parker, A. (2002a) 'Making invisible work visible', *California Management Review*, 44(2): 25–46.

Cross, R, Nohria, N., and Parker, A. (2002b) 'Six myths about informal networks—and how to overcome them', *Sloan Management Review*, 43(3): 67–75.

Cunningham, M. and Homse, E. (1984) *The Role of Personal Contacts in Supplier–Customer Relationships*, Occasional Paper No. 8410. Manchester: UMIST.

Curtis, R., Leon, D., and Miller, R. (2002) 'Supporting knowledge work with physical design', *Knowledge Management Review*, 5(5): 26–29.

Cusumano, M., Mylonadis, Y., and Rosenbloom, R. (1997) 'Strategic maneuvering and mass-market dynamics: the triumph of VHS over Beta' in M. Tushman and P. Anderson (eds) *Managing Strategic Innovation and Change: A Collection of Readings*, New York: Oxford University Press, 75–98.

Cyert, R. and Goodman, P. (1997) 'Creating effective university–industry alliances: an organizational learning perspective', *Organizational Dynamics*, 25(4): 45–57.

Cyert, R. and March, J. (1963) *A Behavioural Theory of the Firm*, Englewood Cliffs, NJ: Prentice-Hall.

D

Daft, R. (2008) *Management*, 8th edn, Mason, OH: Thomson South-Western.

Dahl, M. and Pedersen, C. (2004) 'Knowledge flows through informal contacts in industrial clusters: myth or reality?', *Research Policy*, 33: 1673–86.

Damanpour, F. (1991) 'Organizational innovation: a meta-analysis of effects of determinants and moderators', *Academy of Management Journal*, 34: 355–90.

Danneels, E. (2007) 'The process of technological competence leveraging', *Strategic Management Journal*, 28: 511–33.

Das, T. and Teng, B. (2002) 'A resource-based theory of strategic alliances', *Journal of Management*, 26(1): 31–60.

Davenport, T. (2005) *Thinking for a Living: How to Get Better Performance and Results from Knowledge Workers*, Boston, MA: Harvard Business School Press.

Davenport, T. and Prusak, L. (1998) *Working Knowledge: How Organizations Manage What They Know*, Boston, MA: Harvard Business School Press.

David, P. (1985) 'Clio and the economics of QWERTY', *American Economic Review*, 75(2): 332–6.

Dean, J. (1950) 'Pricing policies for new products', *Harvard Business Review*, 28(6): 45–53.

Debackere, K., Clarysse, B., Wijnberg, N., and Rappa, M. (1994) 'Science and industry: a theory of networks and paradigms', *Technology Analysis and Strategic Management*, 6(1): 21–37.

DeBresson, C. and Amesse, F. (1991) 'Networks of innovators: a review and introduction to the issue', *Research Policy*, 20(5): 363–79.

DeBresson, C. and Townsend, J. (1981) 'Multivariate models for innovation: looking at the Abernathy-Utterback model with other data', *Omega*, 9(4): 429–36.

DeFillippi, R. and Arthur, M. (1994) 'The boundaryless career: a competency-based perspective', *Journal of Organizational Behavior*, 15: 307–24.

DeFillippi, R., Grabher, G., and Jones, C. (2007) 'Paradoxes of creativity: managerial and organizational challenges in the cultural economy', *Journal of Organizational Behaviour*, 28: 511–21.

Department of Innovation, Universities, and Skills (2008) *Innovation Nation*, Report No. CM7345, London: HMSO.

Department of Trade and Industry (1993) *Realising Our Potential: A Strategy for Science, Engineering and Technology*, Report No. CM2250. London: HMSO.

Department of Trade and Industry (2000) *Excellence and Opportunity—A Science and Innovation Policy for the 21st Century*, Report No. CM4814. London: HMSO.

DeSolla Price, D. (1963) *Little Science, Big Science*, New York: Columbia University Press.

DeSolla Price, D. (1969) 'The structures of publication in science and technology' in W. Gruber and D. Marquis (eds) *Factors in the Transfer of Technology*, Cambridge, MA: MIT Press, 91–104.

DeSolla Price, D. and Beaver, D. (1966) 'Collaboration in an invisible college', *American Psychologist*, 21: 1011–18.

Dess, G. and Shaw, J. (2001) 'Voluntary turnover, social capital and organizational performance', *Academy of Management Review*, 26(3): 446–56.

Devezas, T., LePoire, D., Matias, J., and Silva, A. (2008) 'Energy scenarios: toward a new energy paradigm', *Futures*, 40: 1–16.

Devezas, T., Linstone, H., and Santos, H. (2005) 'The growth dynamics of the Internet and the long wave theory', *Technological Forecasting and Social Change*, 72: 913–35.

Dewick, P., Green, K., and Miozzo, M. (2004) 'Technological change, industry structure and the environment', *Futures*, 36: 267–93.

De Wit, B. and Meyer, R. (2004) *Strategy: Process, Content, Context*, 3rd edn, London: Thomson Learning.

Dhal, M. and Pedersen, C. (2004) 'Knowledge flows through informal contacts in industrial clusters: myth or reality?', *Research Policy*, 33: 1673–86.

Dhanaraj, C. and Parkhe, A. (2006) 'Orchestrating innovation networks', *Academy of Management Review*, 31(3): 659–69.

Dietrich, M. (1994) *Transaction Cost Economics and Beyond*, London: Routledge.

Dierickx, L. and Cool, K. (1989) 'Asset stock accumulation and sustainability of competitive advantage', *Management Science*, 35(12): 1504–11.

Di Gregorio, D. and Shane, S. (2003) 'Why do some universities generate more start-ups than others?', *Research Policy*, 32: 209–27.

Dill, W. (1958) 'Environment as an influence on managerial autonomy', *Administrative Science Quarterly*, 2: 409–43.

Divall, C. (2006) 'Technological networks and industrial research in Britain: the London, Midland & Scottish Railway, 1926–47', *Business History*, 48(1): 43–68.

Dodgson, M. (1993) 'Learning, trust, and technological collaboration', *Human Relations*, 46(1): 77–95.

Dodgson, M., Gann, D., and Salter, A. (2006) 'The role of technology in the shift towards open innovation: the case of Proctor & Gamble', *R&D Management*, 36(3): 333–46.

Dong, J. (2002) 'The rise and fall of the HP way', *Palo Alto Weekly Online*, 10 April, available online at <http://www.paloaltoonline.com/weekly/>.

Dosi, G. (1982) 'Technological paradigms and technological trajectories: a suggested interpretation of the determinants and directions of technical change', *Research Policy*, 11: 147–62.

Dougherty, D. (1996) 'Organizing for innovation' in S. Clegg, C. Hardy, and W. Nord (eds) *Handbook of Organisation Studies*, Thousand Oaks, CA: Sage, 424–39.

Dougherty D. and Bowman, E. (1995) 'The effects of organisational downsizing on product innovation', *California Management Review*, 37(4): 28–42.

Dougherty, D. and Hardy, C. (1996) 'Sustained product innovation in large, mature organizations: overcoming innovation-to-organization problems', *Academy of Management Journal*, 39(5): 1120–53.

Douthwaite, B., Keatinge, J., and Park, J. (2001) 'Why promising technologies fail: the neglected role of user innovation during adoption', *Research Policy*, 30: 819–36.

Drucker, P. (1969) *The Age of Discontinuity*, New York: Harper and Row.

Drucker, P. (1985) *Innovation and Entrepreneurship*, New York: Harper and Row.

Drucker, P. (1993) *Post-Capitalist Society*, New York: Harper Collins.

Duijn, J. van (1981) 'Fluctuations in innovations overtime', *Futures*, 13: 264–75.

Duijn, J. van (1983) *The Long Wave in Economic Life*, London: George Allen and Unwin.

Duncan, R. (1972) 'Characteristics of organisational environments and perceived environmental uncertainty', *Administrative Science Quarterly*, 17: 313–27.

Duncan, R. (1976) 'The ambidextrous organization: designing dual structures for innovation' in R. Kilmann, L. Pondy, and D. Slevin (eds) *The Management of Organization Design, Vol. 1*, New York: Elsevier North-Holland, 167–88.

Duncan, S. (1974) 'The isolation of scientific discovery: indifference and resistance to a new idea', *Science Studies*, 4: 109–34.

Duysters, G. and Vanhaverbeke, W. (1996) 'Strategic interactions in DRAM and RISC technology: a network approach', *Scandinavian Journal of Management*, 12(4): 437–61.

Dvir, D., Lipovetsky, A., Shenhar, A., and Tishler, A. (1998) 'In search of project classification: a non-universal approach to project success', *Research Policy*, 27: 915–35.

Dyer, J. and Chu, W. (2003) 'The role of trustworthiness in reducing transaction costs and improving performance: empirical evidence from the United States, Japan, and Korea', *Organization Science*, 14: 57–68.

Dyer, J. and Singh, H. (1998) 'The relational view: cooperative strategies and sources of interorganizational competitive advantage', *Academy of Management Review*, 23(4): 660–79.

Dyson, J. (1997) *Against the Odds: An Autobiography*, London: Orion Business Books.

E

Eagle, N. (2004) 'Can serendipity be planned?', *Sloan Management Review*, 4: 10–13.

Easingwood, C. (1986) 'New product development for service companies', *Journal of Product Innovation Management*, 3(4): 264–75.

Edquist, C. (1997a) 'Systems of innovation approaches: their emergence and characteristics' in C. Edquist (ed.) *Systems of Innovation: Technologies, Institutions and Organizations*, London: Routledge, 1–35.

Edquist, C. (ed.) (1997b) *Systems of Innovation: Technologies, Institutions and Organizations*, London: Routledge.

Edquist, C. (2005) 'Systems of innovation: perspectives and challenges' in J. Faberberg, D. Mowery, and R. Nelson (eds) *The Oxford Handbook of Innovation*, Oxford: Oxford University Press, 181–208.

Edquist, C. and Hommen, L. (2008a) 'Comparing national systems of innovation in Asia and Europe: theory and comparative framework' in C. Edquist and L. Hommen (eds) *Small Country Innovation Systems: Globalisation, Change and Policy in Asia and Europe*, Cheltenham: Edward Elgar, 1–28.

Edquist, C. and Hommen, L. (eds) (2008b) *Small Country Innovation Systems: Globalisation, Change and Policy in Asia and Europe*, Cheltenham: Edward Elgar.

Edquist, C. and Johnson, B. (1997) 'Institutions and organizations in systems of innovation' in C. Edquist (ed.) *Systems of Innovation: Technologies, Institutions and Organizations*, London: Routledge, 41–63.

Edquist, C. and Lundvall, B-Á. (1993) 'Comparing the Danish and Swedish systems of innovation' in R. Nelson (ed.) *National Innovation Systems: A Comparative Analysis*, Oxford: Oxford University Press, 265–98.

Ehrnberg, E. and Sjöberg, N. (1995) 'Technological discontinuities, competition and firm performance', *Technology Analysis and Strategic Management*, 7: 93–107.

Eisenhardt, K. and Westcott, B. (1988) 'Paradoxical demands and the creation of excellence' in R. Quinn and K. Cameron (eds) *Paradox and Transformation: Toward a Theory of Change in Organisation and Management*, New York: Ballinger, 169–94.

Elsbach, K. and Bechky, B. (2007) 'It's more than a desk: working smarter through leveraged office design', *California Management Review*, 49(2): 80–101.

Ende, J. van den and Kemp, R. (1999) 'Technological transformations in history: how the computer regime grew out of existing computing regimes', *Research Policy*, 28: 833–51.

Enos, J. (1962) 'Invention and innovation in the petroleum refining industry' in R. Nelson (ed.) *The Rate and Direction of Inventive Activity: Economic and Social Factors*, Princeton, NJ: Princeton University Press, 299–321.

Evans, P. (1990) 'International management development and the balance between generalism and professionalism', *Personnel Management*, December: 46–50.

F

Fagerberg, J. (2005) 'Innovation: a guide to the literature' in J. Faberberg, D. Mowery, and R. Nelson (eds) *The Oxford Handbook of Innovation*, Oxford: Oxford University Press, 1–26.

Faberberg, J. and Srholec, M. (2008) 'National innovation systems, capabilities and economic development', *Research Policy*, 37: 1417–35.

Farson, R. and Keyes, R. (2002) *Whoever Makes the Most Mistakes Win: The Paradox of Innovation*, New York: Free Press.

Feldman, M. (1999) 'The new economics of innovation, spillovers and agglomeration: a review of empirical studies', *Economics of Innovation and New Technology*, 8: 5–25.

Ferdinand, J. and Simm, D. (2007) 'Re-theorizing external learning: insights from economic and industrial espionage', *Management Learning*, 38(3): 297–317.

Festinger, L. (1957) *A Theory of Cognitive Dissonance*, New York: Row Peterson.

Feyerabend, P. (1974) 'Consolations for a specialist' in I. Lakatos and A. Musgrave (eds) *Criticism and the Growth of Knowledge*, Cambridge: Cambridge University Press, 197–230.

Feyerabend, P. (1975) *Against Method*, London: Verso.

Fildes, J. (2008) 'Supercomputer sets petaflop pace', 9 June, available online at <http://news.bbc.co.uk> (accessed 30 September 2008).

Fine, G. and Deegan, J. (1996) 'Three principle of serendip: the role of chance in ethnographic research', *International Journal of Qualitative Studies in Education*, 9: 434–47.

Fineman, S. (ed.) (1993) *Emotion in Organizations*, London: Sage.

Fineman, S. (1997) 'Emotions and management learning', *Management Learning*, 28(1): 13–25.

Fineman, S. (ed.) (2000) *Emotion in Organizations, Vol. 2*, London: Sage.

Fleck, J. (2000) 'Artefact-activity: the coevolution of artefacts, knowledge and organisation in technological innovation' in J. Ziman (ed.) *Technological Innovation as an Evolutionary Process*, Cambridge: Cambridge University Press, 248–66.

Fleming, L., King, C., and Juda, A. (2007) 'Small worlds and regional innovation', *Organization Science*, 18(6): 938–54.

Fombrun, C. (1982) 'Strategies for network research in organisations', *Academy of Management Review*, 7(2): 280–91.

Foss, N. (1999) 'Networks, capabilities, and competitive advantage', *Scandinavian Journal of Management*, 15(1): 1–15.

Foster, R. (1986) *Innovation: The Attacker's Advantage*, New York: Summit Books.

Foster, A. and Ford, N. (2003) 'Serendipity and information seeking: an empirical study', *Journal of Documentation*, 59(3): 321–40.

Franke, N. and Shah, S. (2003) 'How communities support innovative activities: an exploration of assistance and sharing among end-users', *Research Policy*, 32(1): 157–78.

Freeman, C. (1968) 'Chemical process plant: innovation and the world market', *National Institute Economic Review*, 45: 29–57.

Freeman, C. (1974) *The Economics of Industrial Innovation*, 1st edn, Harmondsworth: Penguin.

Freeman, C. (1986) 'Successful industrial innovation' in R. Roy and Wield, D. (eds) *Product Design and Technological Innovation*, Milton Keynes: Open University Press, 29–33.

Freeman, C. (1987) *Technology Policy and Economic Performance: Lessons From Japan*, London: Printer Publishers.

Freeman, C. (1991) 'Networks of innovators: a synthesis of research issues', *Research Policy*, 20(5): 499–514.

Freeman, C. and Soete, L. (1997) *The Economics of Industrial Innovation*, 3rd edn, London: Continuum.

Freeman, C., Clark, J., and Soete, L. (1982) *Unemployment and Technical Innovation: A Study of Long Waves and Economic Development*, London: Frances Pinter.

Freeman, L., Freeman, S., and Michaelson, A. (1988) 'On human social intelligence', *Journal of Social and Biological Structures*, 11: 415–25.

Freeman, S., Cray, D., and Sandwell, M. (2007) 'Networks and Australian professional services in newly emerging markets of Asia', *International Journal of Service Industry Management*, 18(2): 152–66.

Fritsch, M. and Lukas, R. (2001) 'Who cooperates on R&D?', *Research Policy*, 30: 297–312.

Frost, P. and Egri, C. (2002) 'The political process of innovation' in S. Clegg (ed.) *Central Currents in Organization Studies II: Contemporary Trends, Vol. 5*, London: Sage; reprinted from Frost, P. and Egri, C. (1991) 'The political process of innovation', in L. Cummings and B. Staw (eds), *Research in Organisational Behaviour, Vol. 13*, Greenwich, CT: JAI Press, 229–95.

Frost, P. and Whitley, R. (1971) 'Communication patterns in a research laboratory', *R&D Management*, 1(2): 71–79.

Fuld, L. (1991) 'A recipe for business intelligence success', *The Journal of Business Strategy*, 12(1): 12–17.

G

Gadde, L. and Snehota, I. (2000) 'Making the most of supplier relationships', *Industrial Marketing Management*, 29: 305–16.

Galambos, L. and Sewell, J. (1995) *Networks of Innovation: Vaccine Development at Merck, Sharp and Dohme, and Mulford, 1895–1995*, Cambridge: Cambridge University Press.

Gallouj, C. and Gallouj, F. (2000) 'Neo-Schumpeterian perspectives on innovation in services' in M. Boden and I. Miles (eds) *Services and the Knowledge-Based Economy*, London: Continuum, 21–37.

Gallouj, F. (1998) 'Innovating in reverse: services and the reverse product cycle', *European Journal of Innovation Management*, 1(3): 123–38.

Gallouj, F. (2000) 'Beyond technological innovation: trajectories and varieties of services innovation' in M. Boden and I. Miles (eds) *Services and the Knowledge-Based Economy*, London: Continuum, 129–45.

Gambardella, A. (1992) 'Competitive advantages from in-house scientific research: the US pharmaceutical industry in the 1980s', *Research Policy*, 21: 391–407.

Gardiner, P. and Rothwell, R. (1985) 'Tough customers: good designs', *Design Studies*, 6(1): 7–17.

Gargiulo, M. and Benassi, M. (2000) 'Trapped in your own net? Network cohesion, structural holes, and the adaptation of social capital', *Organization Science*, 11(2): 183–96.

Garrett-Jones, S., Turpin, T., Burns, P., and Diment, K. (2005) 'Common purpose and divided loyalties: the risks and rewards of cross-sector collaboration for academic and government', *R&D Management*, 35(5): 535–44.

Gassmann, O. (2006) 'Editorial: opening up the innovation process—towards an agenda', *R&D Management*, 36(3): 223–28.

Gay, B. and Dousset, B. (2005) 'Innovation and network structural dynamics: study of the alliance network of a major sector of the biotechnology industry', *Research Policy*, 34: 1457–75.

Geels, F. (2002) 'Technological transitions as evolutionary reconfiguration processes: a multi-level perspective and a case study', *Research Policy*, 31: 1257–74.

Geels, F. and Schot, J. (2007) 'Typology of sociotechnical transition pathways', *Research Policy*, 36: 399–417.

Genus, A. and Coles, A-M. (2005) 'On constructive technology assessment and limitations on public participation in technology assessment', *Technology Analysis and Strategic Management*, 17(4): 433–43.

Genus, A. and Coles, A-M. (2008) 'Rethinking the multi-level perspective of technological transitions', *Research Policy*, 37: 1436–45.

Gemunden, H., Heydebreck, P., and Herden, R. (1992) 'Technological interweavement: a means of achieving innovation success', *R&D Management*, 22: 359–76.

George, G. (2005) 'Slack resources and the performance of privately held firms', *Academy of Management Journal*, 48(4): 661–676.

Ghoshal, S. and Bartlett, C. (1988a) 'Creation, adoption, and diffusion of innovations by subsidiaries', *Journal of International Business Studies*, 19: 365–88.

Ghoshal, S. and Barlett, C. (1988b) 'Innovation processes in multinational corporations' in M. Tushman and W. Moore (eds) *Readings in the Management of Innovation*, 2nd edn, New York: Harper Business, 499–518.

Giddens, A. (1984) *The Constitution of Society: Outline of the Theory of Structuration*, Cambridge: Polity.

Glasmeier, A. (1991) 'Technological discontinuities and flexible production networks: the case of Switzerland and the world watch industry', *Research Policy*, 20(5): 469–85.

Goerzen, A. (2007) 'Alliance networks and firm performance: the impact of repeated partnerships', *Strategic Management Journal*, 28: 487–509.

Goldstein, J. (1988) *Long Cycles: Prosperity and Ware in the Modern Age*, New Haven, MA: Yale University Press.

Goleman, D. (1995) *Emotional Intelligence*, New York: Bantam Books.

Gouldner, A. (1964) *Patterns of Industrial Bureaucracy*, New York: Free Press.

Grandori, A. (1997) 'An organizational assessment of interfirm coordination modes', *Oranization Studies*, 18: 897–925.

Grangsjo, Y. and Gummesson, E. (2006) 'Hotel networks and social capital in destination marketing', *International Journal of Service Industry Management*, 17(1): 58–75.

Granovetter, M. (1973) 'The strength of weak ties', *American Journal of Sociology*, 78(6): 1360–80.

Granovetter, M. (1983) 'The strength of weak ties: a network theory revisited', *Sociological Theory*, 1: 201–33.

Granovetter, M. (1985) 'Economic action and social structure: the problem of embeddedness', *American Journal of Sociology*, 91: 481–510.

Grant, R. (1991) 'The resource-based theory of competitive advantage: implications for strategy formulation', *California Management Review*, 33(3): 114–35.

Grant, R. (2002) *Contemporary Strategy Analysis: Concepts, Techniques, Applications*, 4th edn, Oxford: Blackwell.

Gray, J. and Starke, F. (1984) *Organizational Behaviour: Concepts and Applications*, 3rd edn, Columbus, OH: Charles Merrill.

Greve, A. and Salaff, J. (2003) 'Social networks and entrepreneurship', *Entrepreneurship Theory and Practice*, 28(4): 1–22.

Grieco, M. and Lilja, K. (1996) ' "Contradictory couplings" culture and the synchronisation of opponents', *Organisation Studies*, 17(1): 131–37.

Griffin, A. (1992) 'Evaluating QFD's use in US firms as a process for developing products', *Journal of Product Innovation Management*, 9(3): 171–87.

Griffin, A. (1993) 'Metrics for measuring product development cycle time', *Journal of Product Innovation Management*, 10(2): 112–25.

Griffin, A. (1997) 'PDMA research on new product development practices: updating trends and benchmarking best practices', *Journal of Product Innovation Management*, 14(6): 429–58.

Griffin, A. and Page, A. (1993) 'An interim report on measuring product development success and failure', *Journal of Product Innovation Management*, 10(4): 291–308.

Griliches, Z. (1957) 'Hybrid corn: an exploration in the economics of technological change', *Econometrica*, 25: 501–22.

Gronroos, C. (1990) *Service Management and Marketing*, Lexington, MA: Lexington Books.

Gulati, R. (1998) 'Alliances and networks', *Strategic Management Journal*, 19(4): 293–317.

Gulati, R. (1999) 'Network location and learning: the influence of network resources and firm capabilities on alliance formation', *Strategic Management Journal*, 20(5): 397–420.

Gupta, A. and Wilemon, D. (1988) 'The credibility-cooperation connection at the R&D–marketing interface', *Journal of Product Innovation Management*, 5(1): 20–31.

Gupta, A., Smith, K., and Shalley, C. (2006) 'The interplay between exploration and exploitation', *Academy of Management Journal*, 49(4): 693–706.

Gwynne, P. (1997) 'Skunkworks, 1990s style', *Research Technology Management*, 40(4): 18–23.

H

Haas, E. and Drabek, T. (1973) *Complex Organisations: A Sociological Perspective*, London: Macmillan.

Habermeier, K. (1990) 'Product use and product improvement', *Research Policy*, 19(3): 271–83.

Hagedoorn, J. (1993) 'Understanding the rationale of strategic technology partnering: interorganizational modes of cooperation and sectoral differences', *Strategic Management Journal*, 14: 371–85.

Hagedoorn, J. (2002) 'Inter-firm R&D partnerships: an overview of major trends and patterns since 1960', *Research Policy*, 31: 477–92.

Hagedoorn, J. and Schakenraad, J. (1992) 'Leading companies and networks of strategic alliances in information technologies', *Research Policy*, 21(2): 163–90.

Hagedoorn, J., Carayannis, E., and Alexander, J. (2001) 'Strange bedfellows in the personal computer industry: technology alliances between IBM and Apple', *Research Policy*, 30: 837–49.

Håkansson, H. (1987) *Industrial Technological Development. A Network Approach*, London: Croom Helm.

Håkansson, H. (1989) *Corporate Technological Behaviour, Cooperation and Networks*, London: Routledge.

Håkansson, H. (1990) 'Technological collaboration in industrial networks', *European Management Journal*, 8(3): 371–79.

Håkansson, H. and Johanson, J. (1990) 'Formal and informal co-operation strategies in international industrial networks' in D. Ford (ed.) *Understanding Business Markets*, London: Academic Press, 459–67.

Håkansson, H. and Snehota, I. (1989) 'No business is an island: the network concept of business strategy', *Scandinavian Journal of Management*, 5(3): 187–200.

Hall, B., Link, A., and Scott, J. (2003) 'Universities as research partners', *The Review of Economics and Statistics*, 85(2): 485–91.

Hall, R. (1992) 'The strategic analysis of intangible resources', *Strategic Management Journal*, 13(2): 135–44.

Hall, R. (1999) *Organizations: Structures, Processes, and Outcomes*, Englewood Cliffs, NJ: Prentice Hall.

Hamel, G. and Prahalad, C. (1994) *Competing for the Future*, Boston, MA: Harvard Business School Press.

Hamel, G., Doz, Y., and Prahalad, C. (1989) 'Collaborate with your competitors—and win', *Harvard Business Review*, 67(1): 133–39.

Hampden-Turner, C. (1990) *Charting the Corporate Mind: From Dilemma to Strategy*, London: Basil Blackwell.

Handfield, R., Ragatz, G., Monczka, R., and Petersen, K. (1999) 'Involving suppliers in new product development', *California Management Review*, 42(1): 59–82.

Handley, K., Sturdy, A., Fincham, R., and Clark, T. (2006) 'Within and beyond communities of practice: making sense of learning through participation, identity and practice', *Journal of Management Studies*, 43(3): 641–53.

Handy, C. (1994) *The Age of Paradox*, Boston, MA: Harvard University Business Press.

Hannan, M. and Freeman, J. (1984) 'Structural inertia and organizational change', *American Sociological Review*, 49: 259–85.

Hansen, M. (1999) 'The search-transfer problem: the role of weak ties in sharing knowledge across organization subunits', *Administrative Science Quarterly*, 44: 82–111.

Hansen, M. (2002) 'Knowledge networks: explaining effective knowledge sharing in multiunit companies', *Organization Science*, 13(3): 232–48.

Hansen, M., Mors, M., and Lovas, B. (2005) 'Knowledge sharing in organizations: multiple networks, multiple phases', *Academy of Management Journal*, 48(5): 776–93.

Hansen, M., Podilny, J., and Pfeffer, J. (2001) 'So many ties, so little time: a task contingency perspective on corporate social capital in organizations', *Social Capital of Organizations*, 18: 21–57.

Hargadon, A. and Sutton, R. (1997) 'Technology brokering and innovation in a product development firm', *Administrative Science Quarterly*, 42: 718–49.

Hart, S., Hultink, E., Tzokas, N., and Commandeur, H. (2003) 'Industrial companies' evaluation criteria in new product development gates', *Journal of Product Innovation Management*, 20: 22–36.

Hasenfeld, Y. (1972) 'People-processing organisations: an exchange approach', *American Sociological Review*, 37(3): 256–63.

Hauser, J. and Clausing, D. (1988) 'The house of quality', *Harvard Business Review*, 66(3): 63–73.

Hayes, N. and Walsham, G. (2000) 'Safe enclaves, political enclaves and knowledge working' in C. Prichard, R. Hull, M. Chumer, and H. Willmott (eds) *Managing Knowledge: Critical Investigations of Work and Learning*, Basingstoke: Macmillan Business, 69–87.

Hayton, J. (2005) 'Competing in the new economy: the effect of intellectual capital on corporate entrepreneurship in high-technology new ventures', *R&D Management*. 35(2): 137–55.

Heimer, C. (1992) 'Doing your job and helping your friends: universalistic norms about obligations to particular others in networks' in N. Nohria and R. Eccles (eds) *Networks and Organizations: Structures, Form, and Action*, Boston, MA: Harvard Business School Press, 143–64.

Hellgren, B. and Stjernberg, T. (1995) 'Design and implementation in major investments: a project network approach', *Scandinavian Journal of Management*, 11(4): 377–94.

Henderson, R. (1994) 'The evolution of integrative capability: innovation in cardiovascular drug discovery', *Industrial and Corporate Change*, 3(3): 607–30.

Henderson, R. and Clark, K. (1990) 'Architectural innovation: the reconfiguration of existing product technologies and the failure of established firms', *Administrative Science Quarterly*, 35: 9–30.

Henry, N. and Pinch, S. (2000) 'Spatialising knowledge: placing knowledge community of Motor Sport Valley', *Geoforum*, 31: 191–208.

Henry, N., Pinch, S., and Russell, S. (1996) 'In pole position? Untraded interdependencies, new industrial spaces and the british motorsport industry', *Area*, 28(1): 25–36.

Hertzfeld, A. (2004) *Revolution in the Valley: The Insanely Great Story of How the Mac was Made*, Santa Clara, CA: O'Reilly.

Hienerth, C. (2006) 'The commercialization of user innovations: the development of the Rodeo kayak industry', *R&D Management*, 36(3): 273–94.

Hill, S. (1974) 'Questioning the influence of a social system of science: a study of Australian scientists', *Science Studies*, 4: 135–63.

Hippel, E. von (1976) 'The dominant role of users in the scientific instrument innovation process', *Research Policy*, 5(3): 212–39.

Hippel, E. von (1977a) 'Has a customer already developed your next product?', *Sloan Management Review*, 18(2): 63–74.

Hippel, E. von (1977b) 'The dominant role of the user in semi-conductor and electronic subassembly process innovation', *IEEE Transactions on Engineering Management*, 24(2): 60–71.

Hippel, E. von (1977c) 'Transferring process equipment innovations from user-innovators to equipment manufacturing firms', *R&D Management*, 8(1): 13–22.

Hippel, E. von (1978a) 'A customer-active paradigm for industrial product idea generation', *Research Policy*, 7(3): 240–66.

Hippel, E. von (1978b) 'Successful industrial products from customer ideas', *Journal of Marketing*, 42(1): 39–49.

Hippel, E. von (1978c) 'Users as innovators', *Technology Review*, 8(1): 30–9.

Hippel, E. von (1982a) 'Appropriability of innovation benefit as a predictor of the functional locus of innovation', *Research Policy*, 11(2): 95–115.

Hippel, E. von (1982b) 'Get new products from customers', *Harvard Business Review*, 60(2): 117–22.

Hippel, E. von (1986) 'Lead users: a source of novel product concepts', *Management Science*, 32(7): 791–805.

Hippel, E. von (1987) 'Cooperation between rivals: informal know-how trading', *Research Policy*, 6(6): 291–302.

Hippel, E. von (1988) *The Sources of Innovation*, Oxford: Oxford University Press.

Hippel, E. von (1994) '"Sticky" information and the locus of problem solving: implications for innovation', *Management Science*, 40: 429–39.

Hippel, E. von (1998) 'Economics of product development by users: the impact of 'sticky' local information', *Management Science*, 44(5): 629–44.

Hippel, E. von (2005) *Democratizing Innovation*, Cambridge, MA: MIT Press.

Hippel, E. von and Finkelstein, S. (1979) 'Related innovation by users: an analysis of innovation in automated clinical chemistry analyzers', *Science and Public Policy*, 6(1): 24–37.

Hippel, E. von and Katz, R. (2002) 'Shifting innovation to users via toolkits', *Management Science*, 48: 821–33.

Hippel, E. von and Krogh, G. von (2003) 'Open source software and the "private-collective" model: issues for organization science', *Organization Science*, 14(2): 209–23.

Hippel, E. von and Krogh, G. von (2006) 'Free revealing and the private-collective model for innovation incentives', *R&D Management*, 36(3): 295–306.

Hislop, D. (2005) *Knowledge Management in Organizations: A Critical Introduction*, Oxford: Oxford University Press.

Hobday, M. (1994) 'The limits of Silicon Valley: a critique of network theory', *Technology Analysis and Strategic Management*, 6(2): 231–44.

Hobday, M., Rush, H., and Bessant, J. (2004) 'Approaching the innovation frontier in Korea: the transition phase to leadership', *Research Policy*, 33(10): 1433–57.

Hochschild, L. (1983) *The Managed Heart: Commercialization of Human Feeling*, Berkeley, CA: University of California Press.

Holden, N. (2002) *Cross-Cultural Management: A Knowledge Management Perspective*, Harlow: FT Prentice Hall.

Hoskisson, R., Hitt, M., Johnson, R., and Grossman, W. (2002) 'Conflicting voices: the effects of institutional ownership heterogeneity and internal governance on corporate innovation strategies', *Academy of Management Journal*, 45(4): 697–716.

Howells, J. (2006) 'Intermediation and the role of intermediaries in innovation', *Research Policy*, 35: 715–28.

Howells, J. and Higgins, C. (1990) 'Champions of technological innovation', *Administrative Science Quarterly*, 35: 317–41.

Huff, A. (1988) 'Politics and argument as a means of coping with ambiguity and change' in L. Pondy, R. Boland, and H. Thomas, (eds) *Managing Ambiguity and Change*, Chichester: John Wiley, 79–92.

Hughes, T. (1987) 'The evolution of large technological systems' in W. Bijker, T. Hughes, and T. Pinch (eds) *The Social Construction of Technological Systems*, Cambridge, MA: MIT Press, 51–82.

Hultink, E. and Robben, H. (1995) 'Measuring new product success: the difference that time perspective has', *Journal of Product Innovation Management*, 12: 392–402.

Hung, C. (2003) 'The business of product counterfeiting in China and the post-WTO membership environment', *Asia Pacific Business Review*, 10(1): 58–77.

Huy, Q. (1999) 'Emotional capability, emotional intelligence, and radical change', *Academy of Management Review*, 24(2): 325–45.

I

Iansiti, M. and Clark, K. (1994) 'Integration and dynamic capability: evidence from product development in automobiles and mainframe computers', *Industrial and Corporate Change*, 3(3): 557–605.

Iansiti, M. and MacCormack, A. (1997) 'Developing products on the Internet', *Harvard Business Review*, 75(5): 108–17.

Illinois Institute of Technology (1968) *Technology in Retrospect and Critical Events in Science*, Report to the National Science Foundation, NSF-C235, Chicago, IL: Illinois Institute of Technology.

Inkpen, A. and Pien, W. (2006) 'An examination of collaboration and knowledge transfer: China–Singapore Suzhou Industrial Park', *Journal of Management Studies*, 43(4): 779–811.

Inkpen, A. and Tsang, E. (2005) 'Social capital, networks, and knowledge transfer', *Academy of Management Review*, 30(1): 146–65.

International Accounting Standards Committee (1998) *IAS 38: Intangible Assets*, London: IASC.

J

Jaffee, D. (2001) *Organization Theory: Tension and Change*, New York: McGraw-Hill.

Jarillo, J. (1988) 'On strategic networks', *Strategic Management Journal*, 9(1): 31–41.

Jeffreys, A. (2003) 'A century of human genetics: exploring variation and mutation in the human genome', E-Bulletin, Leicester: University of Leicester (accessed online at <http://at www.le.ac.uk> on 16 September 2008).

Jelinek, M. and Schoonhoven, C. (1990) *The Innovation Marathon*, New York: Blackwell.

Jeppesen, L. and Molin, M. (2003) 'Consumers as co-developers: learning and innovation outside of the firm', *Technology Analysis and Strategic Management*, 15: 363–83.

Jewkes, J., Sawers, D., and Stillerman, R. (1969) *The Sources of Invention*, 2nd edn, London: MacMillan.

Johannessen, J-A., Olsen, B., and Lumpkin, G. (2001) 'Innovation as newness: what is new, how new, and new to whom?', *European Journal of Innovation Management*, 4(1): 20–31.

Johannisson, B. (2000) 'Networking and entrepreneurial growth' in D. Sexton and H.

Landström (eds) *The Blackwell Handbook of Entrepreneurship*, Oxford: Blackwell, 368–86.

Johannisson, B. and Peterson, R. (1984) 'The personal networks of entrepreneurs', Paper presented at the Third Canadian Conference of the International Council for Small Business, May, Toronto.

Johne, A. and Storey, C. (1998) 'New service development: a review of the literature and annotated bibliography', *European Journal of Marketing*, 32(3–4): 184–251.

Jones, O. and Conway, S. (2004) 'The international reach of entrepreneurial social networks: the case of James Dyson in the UK' in H. Etemad (ed.) *International Entrepreneurship in Small and Medium-sized Enterprises: Orientation, Environment and Strategy, Vol. 3*, McGill International Entrepreneurship Series, Cheltenham: Edward Elgar, 87–106.

Jones, O. and Stevens, G. (1999) 'Evaluating failure in the innovation process: the micropolitics of new product development', *R&D Management*, 29(2): 167–78.

Jones, O., Conway, S., and Steward, F. (1998) 'Introduction: social interaction and innovation networks', *International Journal of Innovation Management*, 2(2): 123–36.

Jones, O., Edwards, T., and Beckinsale, M. (2000) 'Technology management in a mature firm: structuration theory and the innovation process', *Technology Analysis and Strategic Management*, 12(2): 161–77.

Jones, T. (2002) *Innovating at the Edge: How Organizations Evolve and Embed Innovation Capability*, Oxford: Butterworth-Heinemann.

Jones-Evans, D. (1995) 'A typology of technology-based entrepreneurs: a model based on previous occupational background', *International Journal of Entrepreneurial Behaviour and Research*, 1(1): 26–47.

Jordan, G. and Segelod, E. (2006) 'Software innovativeness: outcomes on project performance, knowledge enhancement, and external linkages', *R&D Management*, 36(2): 127–42.

Joshi, A. (2006) 'The influence of organizational demography on the external networking behavior of teams', *Academy of Management Review*, 31(3): 583–95.

Joshi, A. and Jackson, S. (2003) 'Understanding work team diversity: challenges and opportunities' in M. West, D. Tjosvold, and K. Smith (eds) *The International Handbook of Organizational Teamwork and Cooperative Working*, Chichester: John Wiley, 277–96.

K

Kanter, R. (1972) *Commitment and Community*, Cambridge, MA: Cambridge University Press.

Kanter, R. (1985) *The Change Masters: Corporate Entrepreneurs at Work*, London: Taylor and Francis.

Kanter, R. (1989a) 'Swimming in newstreams: mastering innovation dilemmas', *California Management Review*, 31: 45–69.

Kanter, R. (1989b) *When Giants Learn to Dance*, New York: Simon and Schuster.

Kaplan, S. and Tripas, M. (2008) 'Thinking about technology: applying a cognitive lens to technical change', *Research Policy*, 37: 790–805.

Katila, R. and Shane, S. (2005) 'When does lack of resources make new firms innovative?' *Academy of Management Journal*, 48(5): 814–29.

Katz, M. and Shapiro, C. (1986) 'Technology adoption in the presence of network externalities', *Journal of Political Economy*, 94: 822–41.

Katz, R. and Allen, T. (1982) 'Investigating the not invented here (NIH) syndrome: a look at the performance, tenure, and communication patterns of 50 R&D projects', *R&D Management*, 12(1): 7–19.

Kaufmann, A. and Todtling, F. (2001) 'Science–industry interaction in the process of innovation: the importance of boundary-crossing between systems', *Research Policy*, 30: 791–804.

Kelley, M. and Brooks, H. (1991) 'External learning opportunities and the diffusion of process innovations to small firms: the case of programmable automation', *Technological Forecasting and Social Change*, 39: 103–25.

Kelly, P., Kranzberg, M., Rossini, F., Baker, N., Tarpley, F., and Mitzner, M. (1986) 'Introducing innovation' in R. Roy and D. Wield (eds) *Product Design and Technological Innovation*, Milton Keynes: Open University Press, 18–28.

Kemp, R., Schot, J., and Hoogma, R. (1998) 'Regime shifts to sustainability through processes of niche formation; the approach of strategic niche management', *Technology Analysis and Strategic Management*, 10(2): 175–98.

Kessler, E. and Chakrabarti, K. (1999) 'Speeding up the pace of new product development', *Journal of Product Innovation Management*, 16: 231–47.

Kilduff, M. and Tsai, W. (2003) *Social Networks and Organizations*, London: Sage.

Kim, T., Oh, H., and Swaminathan, A. (2006) 'Framing interorganizational network change: a network inertia perspective', *Academy of Management Review*, 31(3): 704–20.

Kirzner, I. (1973) *Competition and Entrepreneurship*, Chicago, IL: University of Chicago Press.

Kleinknecht, A. (1981) 'Observations on the Schumpeterian swarming of innovations', *Futures*, 13: 293–307.

Klerkx, L. and Leeuwis, C. (2008) 'Balancing multiple interests: embedding innovation intermediation in the agricultural knowledge infrastructure', *Technovation*, 28: 364–78.

Knights, D. and Murray, F. (1994) *Managers Divided: Planning, Implementation and Control*, 8th edn, Chichester: John Wiley.

Knott, A., Bryce, D., and Posen, H. (2003) 'On the strategic accumulation of intangible assets', *Organization Science*, 14(2): 192–207.

Kodama, R. (1992) 'Technology fusion and the new R&D', *Harvard Business Review*, 70(4): 70–8.

Koka, B. and Madhavan, R. (2006) 'The evolution of interfirm networks: environmental effects on patterns of network change', *Academy of Management Review*, 31(3): 721–37.

Kondratiev, N. (1925) 'The long waves in economic life', *Review of Economic Statistics*, 17: 105–15.

Kotler, P. (2000) *Marketing Management*, Millennium edn, London: Prentice-Hall International.

Kotler, P. (2008) *Marketing Management*, 13th edn, London: Pearson Education.

Kotter, J. (1982) *The General Managers*, New York: Free Press.

Krackhardt, D. (1990) 'Assessing the political landscape: structure, cognition, and power in organizations', *Administrative Science Quarterly*, 35: 342–69.

Krackhardt, D. (1992) 'The strength of strong ties: the importance of philos in organizations' in N. Nohria and R. Eccles (eds) *Networks and Organizations: Structures, Form, and Action*, Boston, MA: Harvard Business School Press, 216–39.

Krackhardt, D. and Hanson, J. (1993) 'Informal networks: the company behind the chart', *Harvard Business Review*, 71(4): 104–11.

Kraut, R., Fish, R., Root, R., and Chalfonte, B. (1990) 'Informal communication in organizations: form, function, and technology' in S. Oskamp and S. Spacapan (eds) *People's Reactions to Technology in Factories, Offices, and Aerospace*, Thousand Oaks, CA: Sage, 145–99.

Kreiner, K. and Schultz, M. (1993) 'Informal collaboration in R&D: the formation of networks across organizations', *Organization Studies*, 14(2): 189–209.

Krishnan, R. and Martin, X. (2006) 'When does trust matter to alliance performance?' *Academy of Management Journal*, 49(5): 894–917.

Kroeber, A. (1957) *Style and Civilizations*, New York: Cornell University Press; cited by D. Crane (1972) *Invisible Colleges: Diffusion of Knowledge in Scientific Communities*, Chicago: University of Chicago Press.

Kuhl, J. (1992) 'A theory of self-regulation: action versus state orientation, self-discrimination, and some applications', *Applied Psychology: An International Review*, 41: 97–129.

Kuhn, T. (1962) *The Structure of Scientific Revolutions*, Chicago: University of Chicago Press.

Kuhn, T. (1996) *The Structure of Scientific Revolutions*, 3rd edn, Chicago: University of Chicago Press.

Kuper, G. and Sterken, E. (2003) 'Endurance in speed skating: the development of world records', *European Journal of Operational Research*, 148: 293–301.

Kuznets, S. (1930) *Secular Movements in Production and Prices*, New York: Houghton Mifflin.

Kuznets, S. (1940) 'Schumpeter's business cycles', *American Economic Review*, 30(2): 257–71.

L

Lado, A., Boyd, N., Wright, P., and Kroll, M. (2006) 'Paradox and theorizing within the resource-based view', *Academy of Management Review*, 31(1): 115–131.

Lakatos, I. (1978) *The Methodology of Scientific Research Programmes*, Cambridge: Cambridge University Press.

Lakemond, N., Berggren, C., and Weele, A. van (2006) 'Coordinating supplier involvement in product development projects: a differentiated coordination typology', *R&D Management*, 36(1): 55–66.

Lambe, C. and Spekman, R. (1997) 'Alliances, external technology acquisition, and discontinuous technological change', *Journal of Product Innovation Management*, 14: 102–16.

Lamming, R., Hajee, D., Horrill, M., Kay, G., and Staniforth, J. (2002) 'Lessons from co-development of a single vessel processor: methodologies for managing innovation in customer–supplier networks', *International Journal of Technology Management*, 23: 21–39.

Lampel, J., Shamsie, J., and Lant, T. (2006) 'Toward a deeper understanding of cultural industries' in J. Lampel, J. Shamsie, and T. Lant (eds) *The Business of Culture*, Mahwah, NJ: Lawrence Erlbaum Associates Publishers, 3–14; cited by F. Perretti and G. Negro (2007) 'Mixing genres and matching people: a study in innovation and team composition in Hollywood', *Journal of Organizational Behaviour*, 28: 563–86.

Langrish, J., Gibbons, M., Evans, W., and Jevons, F. (1972) *Wealth From Knowledge: A Study of Innovation in Industry*, London: Macmillan.

Laumann, E. and Pappi, F. (1976) *Networks of Collective Action: A Perspective on Community Influence Systems*, New York: Academic Press.

Lave, J. and Wenger, E. (1991) *Situated Learning: Legitimate Peripheral Participation*, Cambridge: Cambridge University Press.

Lavidge, R. (1984) 'Nine tested ways to mislead product planners', *Journal of Product Innovation Management*, 1(1): 101–5.

Lavie, D. (2006) 'The competitive advantage of interconnected firms: an extension of the resource-based view', *Academy of Management Review*, 31(3): 638–58.

Law, J. and Callon. M. (1988) 'Engineering and sociology in a military aircraft project: a network analysis of technological change', *Social Problems*, 35: 284–97.

Law, J. and Callon, M. (1992) 'The life and death of an aircraft: a network analysis of technical change' in W. Bijker and J. Law (eds) *Shaping Technology/Building Society: Studies in Sociotechnical Change*, Cambridge, MA: MIT Press, 21–52.

Lawrence, P. and Lorsch, J. (1967) 'Differentiation and integration in complex organisations', *Administrative Science Quarterly*, 12: 1–47.

Lawton-Smith, H. and Ho, K. (2006) 'Measuring the performance of Oxford University, Oxford Brookes University and the government laboratories' spin-off companies', *Research Policy*, 35: 1554–68.

Lawton-Smith, H., Dickson, K., and Smith, S. (1991) 'There are two sides to every story: innovation and collaboration within networks of large and small firms', *Research Policy*, 20(5): 457–68.

Lee, G. (2007) 'The significance of network resources in the race to enter emerging product markets: the convergence of telephony communications and computer networking, 1989–2001', *Strategic Management Journal*, 28(1): 17–37.

Leinhard, J. (2006) *How Invention Begins: Echos of Old Voices in the Rise of New Machines*, Oxford: Oxford University Press.

Leitner, K. (2005) 'Managing and reporting intangible assets in research technology organisations', *R&D Management*, 35(2): 125–36.

Leonard, D. (1995) *Wellsprings of Knowledge: Building and Sustaining the Sources of Innovation*, Boston, MA: Harvard Business School Press.

Leonard-Barton, D. (1984) 'Interpersonal communication patterns among Swedish and Boston-area entrepreneurs', *Research Policy*, 13(2): 101–14.

Leonard-Barton, D. (1992) 'Core capabilities and core rigidities: a paradox in managing new product development', *Strategic Management Journal*, 13: 111–25.

Lessig, L. (2001) *The Future of Ideas: The Fate of the Commons in a Connected World*, New York: Random House.

Lettl, C., Herstatt, C., and Gemuenden, G. (2006) 'Users' contributions to radical innovation: evidence from four cases in the field of medical equipment technology', *R&D Management*, 36(3): 251–72.

Levinthal, D. (1997) 'Adaptation on rugged landscapes', *Management Science*, 43: 934–50.

Levinthal, D. (1998) 'The slow pace of rapid technological change: gradualism and punctuation in technological change', *Industrial and Corporate Change*, 7(2): 217–47.

Levitt, B. and March, J. (1988) 'Organisational learning', *Annual Review of Sociology*, 14: 319–40.

Levitt, T. (1965) 'Exploit the product life cycle', *Harvard Business Review*, 43(6): 81–94.

Levy, S. (1994) *Insanely Great: The Life and Times of Mackintosh, the Computer that Changed Everything*, New York: Viking.

Lewis, M. (2000) 'Exploring paradox: toward a more comprehensive guide', *Academy of Management Review*, 25(4): 760–76.

Lieberman, M. (1987) 'Excess capacity as a barrier to entry: an empirical appraisal', *Journal of Industrial Economics*, 35: 607–27.

Lieberman, M. and Asaba, S. (2006) 'Why do firms imitate each other?', *Academy of Management Journal*, 31(2): 366–85.

Lieberman, M. and Montgomery, D. (1988) 'First-mover advantages', *Strategic Management Journal*, 9: 41–58.

Lieberman, M. and Montgomery, D. (1998) 'First mover (dis)advantages: retrospective and link with the resource-based view', *Strategic Management Journal*, 19: 1111–25.

Liebeskind, J., Porter, O., Zucker, L., and Brewer, M. (1996) 'Social networks learning and flexibility: sourcing scientific knowledge in new biotechnology firms', *Organization Science*, 7: 428–43.

Lievens, A., Ruyter, K., and Lemmink, J. (1999) 'Learning during new banking service development', *Journal of Service Research*, 2(2): 145–63.

Lilien, G., Morrison, P., Searls, K., Sonnack, M., and Hippel, E. von (2002) 'Performance assessment of the lead user idea-generation process for new product development', *Management Science*, 48(8): 1042–59.

Lin, N., Garvey, W., and Nelson, C. (1970) 'A study of the communication structure in science' in C. Nelson and D. Pollock (eds) *Communication Among Scientists and Engineers*, Lexington, MA: Heath Lexington Books, 23–60.

Lincoln, J. (1982) 'Intra- (and inter-) organizational networks' in S. Bacharach (ed.) *Research in the Sociology of Organizations: A Research Annual, Vol. One*, Stamford, CT: JAI Press, 1–38.

Lindsay, V., Chadee, D., Mattsson, J., Johnston, R., and Millett, B. (2003) 'Relationships, the role of individuals and knowledge flow in the internationalisation of service firms', *International Journal of Service Industry Management*, 14(1): 7–35.

Linstone, H. (2002) 'Corporate planning, forecasting, and the long wave', *Futures*, 34: 317–36.

Little, S. and Ray, T. (2005) *Managing Knowledge: A Essential Reader*, 2nd edn, London: Sage.

Liu, X. and White, S. (2001) 'Comparing innovation systems: a framework and application to China's transitional context', *Research Policy*, 30: 1091–1114.

Lord, R., Klimoski, R., and Kanfer, R. (eds) (2002) *Emotions in the Workplace: Understanding the Structure and Role of Emotions in Organizational Behaviour*, San Francisco, CA: Jossey-Bass.

Lord Sainsbury of Turville (2007) *The Race to the Top: A Review of Government's Science and Innovation Policies*, London: HMSO.

Lorsch, J. (1965) *Product Innovation and Organization*, New York: MacMillan Publishing.

Lovallo, D. and Kahneman, D. (2003) 'Delusions of success: how optimism undermines executive's decisions', *Harvard Business Review*, 81(7): 56–63.

Lundvall, B-Å. (ed.) (1992) *National Systems of Innovation: Towards a Theory of Innovation and Interactive Learning*, London: Pinter.

Lundvall, B-Å. (2005) 'National innovation systems: analytical concept and development tool', *Industry and Innovation*, 14(1): 95–119.

Lundvall, B-Å. and Borrás, S. (2007) 'Science, technology, and innovation policy' in J. Faberberg, D. Mowery, and R. Nelson (eds) *The Oxford Handbook of Innovation*, Oxford: Oxford University Press, 599–631.

Lüthje, C. (2004) 'Characteristics of innovating users in a consumer goods field: an empirical study of sport-related product consumer', *Technovation*, 24: 683–95.

Lüthje, C., Herstatt, C., and Hippel. E. von (2005) 'User-innovators and "local" information: the case of mountain biking', *Research Policy*. 34: 951–65.

M

Macdonald, S. (1986) 'The distinctive research of the individual inventor', *Research Policy*, 15: 199–210.

Macdonald, S. and Piekkari, R. (2005) 'Out of control: personal networks in European collaboration', *R&D Management*, 35(4): 441–53.

Macdonald, S. and Williams, C. (1993a) 'Beyond the boundary: an information perspective on the role of the gatekeeper in the organisation', *Journal of Product Innovation Management*, 10: 417–27.

Macdonald, S. and Williams, C. (1993b) 'The survival of the gatekeeper', *Research Policy*, 23: 123–32.

MacKenzie, D. (1992) 'Economic and sociological explanations of technical change' in R. Coombs, P. Saviotti, and V. Walsh (eds) *Technological Change and Company Strategies*, London: Academic Press, 25–48.

Mahajan, V., Muller, E., and Bass, F. (1990) 'New product diffusion models in marketing: a review and directions for research', *Journal of Marketing*, 54(1): 1–26.

Maidique, M. and Patch, P. (1988) 'Corporate strategy and technology policy' in M. Tushman and W. Moore (eds) *Readings in the Management of Innovation*, 2nd edn, New York: HarperBusiness, 236–48.

Maidique, M. and Zirger, B. (1985) 'The new product learning cycle', *Research Policy*, 14: 299–313.

Maier, S. and Seligman, M. (1976) 'Learned helplessness: theory and evidence', *Journal of Experimental Psychology: General*, 105: 3–46.

Malerba, F. (1993) 'The national system of innovation: Italy' in R. Nelson (ed.) *National Innovation Systems: A Comparative Analysis*, Oxford: Oxford University Press, 230–59.

Malerba, F. (2002) 'Sectoral systems of innovation and production', *Research Policy*, 31: 247–64.

Malerba, F. (ed.) (2004a) *Sectoral System of Innovation: Concepts, Issues and Analyses of Six Major Sectors in Europe*, Cambridge: Cambridge University Press.

Malerba, F. (2004b) 'Sectoral systems of innovation: basic concepts' in F. Malerba (ed.) *Sectoral System of Innovation: Concepts, Issues and Analyses of Six Major Sectors in Europe*, Cambridge: Cambridge University Press, 9–41.

Malerba, F. (2005) 'Sectoral systems: how and why innovation differs across sectors' in J. Faberberg, D. Mowery, and R. Nelson (eds) *The Oxford Handbook of Innovation*, Oxford: Oxford University Press, 380–406.

Malerba, F. and Orsenigo, L. (1993) 'Technological regimes and firm behaviour', *Industrial and Corporate Change*, 2: 45–74.

Malerba, F. and Orsenigo, L. (1996) 'Schumpeterian patterns of innovation are technology-specific', *Research Policy*, 25: 451–78.

Mani, S. (2002) *Government, Innovation and Technology Policy: An International Comparative Analysis*, Cheltenham: Edward Elgar.

Mansfield, E. (1961) 'Technical change and the rate of innovation', *Econometrica*, 29: 741–66.

Mansfield, E. (1985) 'How rapidly does new industrial technology leak out?', *Journal of Industrial Economics*, 34(2): 217–23.

March, J. (1991) 'Exploration and exploitation in organizational learning', *Organization Science*, 2: 71–87.

March, J. (1996) 'Continuity and change in theories of organisational action', *Administrative Science Quarterly*, 41: 278–87.

Markman, G., Gianiodis, P., Phan, P., and Balkin, D. (2005) 'Innovation speed: transferring university technology to the market', *Research Policy*, 34: 1058–75.

Marks, A. and Lockyer, C. (2004) 'Producing knowledge: the use of the project team as a vehicle for knowledge and skill acquisition for software employees', *Economic and Industrial Democracy*, 25(2): 219–45.

Markus, M. (1990) 'Toward a "critical mass" theory of interactive media' in J. Fulk and C. Steinfield (eds) *Organizations and Communication Technology*, Newbury Park, CA: Sage, 194–218.

Marquis, D. (1988) 'The anatomy of successful innovation' in M. Tushman and W. Moore (eds) *Readings in the Management of Innovation*, 2nd edn, New York: HarperBusiness, 79–87; reprinted from D. Marquis (1969) 'The anatomy of successful innovation', *Innovation Magazine*, 1(Nov): 28–37.

Marquis, D. and Allen, T. (1966) 'Communication patterns in applied technology', *American Psychologist*, 21: 1052–60.

Marshall, A. (1961) *Principles of Economics*, 9th edn, London: MacMillan.

Marsick, V. and Watkins, K. (1997) *Informal and Incidental Learning in the Workplace*, London: Routledge.

Martin, C. and Horne, D. (1992) 'Restructuring toward a service orientation: the strategic challenges', *International Journal of Service Industry Management*, 3(1): 28–38.

Martin, C. and Horne, D. (1993) 'Services innovation: successful versus unsuccessful firms', *International Journal of Service Industry Management*, 4(1): 49–65.

Martin, C. and Horne, D. (1995) 'Level of success inputs for service innovations in the same firm', *International Journal of Service Industry Management*, 6(4): 40–56.

Martins, L., Gilson, L., and Maynard, M. (2004) 'Virtual teams: what do we know and where do we go from here?', *Journal of Management*, 30(6): 805–35.

Mathews, J. (1997) 'A Silicon Valley of the East: creating Taiwan's semiconductor industry', *California Management Review*, 39(4): 26–54.

McAdam, R., O'Hare, T., and Moffett, S. (2008) 'Collaborative knowledge sharing in composite new product development: an aerospace study', *Technovation*, 28: 245–56.

McClelland, D. (1961) *The Achieving Society*, Princeton, NJ: Van Nostrand.

McDermott, C. and O'Connor, G. (2002) 'Managing radical innovation: an overview of emergent strategy issues', *The Journal of Product Innovation Management*, 19: 424–38.

McFadyen, M. and Cannella, A. (2004) 'Social capital and knowledge creation: diminishing returns of the number and strength of exchange relationships', *Academy of Management Journal*, 47(5): 735–46.

McGrath, R. and MacMillan, I. (1992) 'More like each other than anyone else? A cross-cultural study of entrepreneurial perceptions', *Journal of Business Venturing*, 7(5): 419–29.

McInerney, C. and LeFevre, D. (2000) 'Knowledge managers: history and challenges' in C. Prichard, R. Hull, M. Chumer, and H. Willmott (eds) *Managing Knowledge: Critical Investigations of Work and Learning*, Basingstoke: Macmillan Business, 1–19.

McIntyre, H. (1988) 'Market adaptation as a process in the product life cycle of radical innovations and high technology products', *Journal of Product Innovation Management*, 5: 140–9.

McKinlay, A. (2000) 'The bearable lightness of control: organisational reflexivity and the politics of knowledge management' in C. Prichard, R. Hull, M. Chumer, and H. Willmott (eds) *Managing Knowledge: Critical Investigations of Work and Learning*, Basingstoke: Macmillan Business, 107–21.

Meeus, M., Oerlemans, L., and Hage, J. (2004) 'Industry-public knowledge infrastructure interaction: intra- and inter-organizational explanations of interactive learning', *Industry and Innovation*, 11(4): 327–52.

Megginson, D. and Clutterbuck, D. (1995) *Mentoring in Action*, London: Kogan Page.

Mensch, G. (1979) *Stalemate in Technology*, New York: Ballinger.

Meyer, M. and Utterback, J. (1993) 'The product family and the dynamics of core capability', *Sloan Management Review*, 34(3): 29–47.

Michailova, S. and Husted, K. (2003) 'Knowledge sharing hostility in Russian firms', *California Management Review*, 45(3): 59–77.

Michailova, S. and Hutchings, K. (2006) 'National cultural influences on knowledge sharing: a comparison of China and Russia', *Journal of Management Studies*, 43(3): 383–405.

Miles, R. and Snow, C. (1978) *Organizational Strategy, Structure, and Process*, New York: McGraw-Hill.

Milgram, S. (1967) 'The small-world problem', *Psychology Today*, 1(1): 60–7.

Miller, D. and Friesen, P. (1984) *Organisations: A Quantum View*, Englewood Cliffs, NJ: Prentice-Hall.

Milson, M., Raj, S., and Wilemon, D. (1992) 'A survey of major approaches to accelerating product development', *Journal of Product Innovation Management*, 9: 53–69.

Min, S., Kalwani, M., and Robinson, W. (2006) 'Market pioneer and early follower survival risks: a contingency analysis of really new versus incrementally new product markets', *Journal of Marketing*, 70(1): 15–33.

Minagawa, T., Trott, P., and Hoecht, A. (2007) 'Counterfeit, imitation, reverse engineering and learning: reflections from Chinese manufacturing firms', *R&D Management*, 37(5): 455–67.

Mintzberg, H. (1973) *The Nature of Managerial Work*, New York: Harper and Row.

Mintzberg, H. (1987) 'The strategy concept I: Five "P"s for strategy', *California Management Review*, 30(1): 11–24.

Mintzberg. H. (1988) 'The structuring of organizations' in J. Quinn, H. Mintzberg, R. James (eds) *The Strategy Process: Concepts, Contexts, and Cases*, Englewood Cliffs, NJ: Prentice-Hall, 276–304.

Mintzberg, H., Quinn, J., and Ghoshal, S. (2002) *The Strategy Process*, 4th edn, London: Prentice-Hall.

Mitchell, J. (1969) 'The concept and use of social networks' in J. Mitchell (ed.) *Social Networks in Urban Situations*, Manchester: Manchester University Press, 1–50.

Mitroff, I. (1974) 'Norms and counter-norms in a select group of the Apollo Moon scientists: a case study of the ambivalence of scientists', *American Sociological Review*, 39(4): 579–95.

Miyazaki, K. (1994) 'Search, learning and accumulation of technological competences: the case of optoelectronics', *Industrial and Corporate Change*, 3(3): 631–54.

Moenaert, R., Meyer, A., Souder, W., and Deschoolmeester, D. (1994a) 'R&D–marketing

communications during the fuzzy front end', *IEEE Transactions on Engineering Management*, 42(3): 243–58.

Moenaert, R., Souder, W., Meyer, A., and Deschoolmeester, D. (1994b) 'R&D–marketing integration mechanisms, communication flows, and innovation success', *Journal of Product Innovation Management*, 11(1): 31–45.

Monge, P. and Eisenberg, E. (1987) 'Emergent communication networks' in F. Jablin, L. Putman, K. Roberts, and L. Porter (eds) *Handbook of Organisational Communication*, London: Sage, 304–42.

Moore, G. (1965) 'Cramming more components onto integrated circuits', *Electronics*, 38(8): 114–17.

Moreno, J. (1934) *Who Shall Survive?: A New Approach to the Problem of Human Interrelations*, Nervous and Mental Disease Monograph Series 58, Washington, DC: Nervous and Mental Disease Publishing Co.

Moreno, J. (1953) *Who Shall Survive?: Foundations of Sociometry, Group Psychotheraphy and Sociodrama*, New York: Beacon House.

Morgan, G. (1981) 'The schismatic metaphor and its implications for organizational analysis', *Organization Studies*, 2: 23–44.

Morgan, G. (1986) *Images of Organization*, Beverly Hills, CA: Sage.

Morrison, P., Roberts, J., and Hippel, E. von (2000) 'Determinants of user innovation and innovation sharing in a local market', *Management Science*, 46(12): 1513–27.

Mouzelis, N. (1967) *Organization and Bureaucracy*, Chicago, IL: Aldine.

Mowery, D. and Nelson, R. (1999) *Sources of Industrial Leadership. Studies of Seven Industries*, New York: Cambridge University Press.

Mueller, R. (1986) *Corporate Networking: Building Channels for Information and Influence*, New York: Free Press.

Mulec, K. and Roth, J. (2005) 'Action, reflection, and learning: coaching in order to enhance the performance of drug development project management teams', *R&D Management*, 35(5): 483–91.

Muller, E. and Zenker, A. (2001) 'Business services as actors of knowledge transformation: the role of KIBS in regional and national innovation systems', *Research Policy*, 30: 1501–16.

Mullins, L. (2007) *Management and Organisational Behaviour*, 8th edn, Harlow: FT Prentice Hall.

Munro, F. and Slaven, T. (2001) 'Networks and markets in Clyde shipping: the Donaldsons and the Hogarths, 1870–1939', *Business History*, 43(2): 19–50.

Mutch, A. (2003) 'Communities of practice and habitus: a critique', *Organization Studies*, 24(3): 383–401.

Myers, S. and Marquis, D. (1969) *Successful Commercial Innovations*, Washington, DC: National Science Foundation.

N

Nadler, D. and Tushman, M. (1988) 'Strategic linking: designing formal coordination mechanisms' in M. Tushman and W. Moore (eds) *Readings in the Management of Innovation*, 2nd edn, New York: HarperBusiness, 469–86.

Nahapiet, J. and Ghoshal, S. (1998) 'Social capital, intellectual capital, and the organizational advantage', *Academy of Management Review*, 23(2): 242–66.

Neale, M. and Corkindale, D. (1998) 'Codeveloping products: involving customers earlier and more deeply', *Long Range Planning*, 31(3): 418–25.

Nelson, R. (1989) 'The strength of strong ties: social networks and inter-group conflict in organizations', *Academy of Management Journal*, 32: 377–401.

Nelson, R. (1993a) 'A retrospective' in R. Nelson (ed.) *National Innovation Systems: A Comparative Analysis*, Oxford: Oxford University Press, 505–23.

Nelson, R. (ed.) (1993b) *National Innovation Systems: A Comparative Analysis*, Oxford: Oxford University Press.

Nelson, R. and Rosenberg, N. (1993) 'Technical innovation and national systems' in R. Nelson (ed.) *National Innovation Systems: A Comparative Analysis*, Oxford: Oxford University Press, 3–21.

Nelson, R. and Winter, S. (1977) 'In search of useful theory of innovation', *Research Policy*, 6: 36–76.

Nelson, R. and Winter, S. (1982) *An Evolutionary Theory of Economic Change*, Cambridge, MA: Harvard University Press.

NESTA (2006) *The Innovation Gap: Why Policy Needs to Reflect the Reality of Innovation in the UK*, London: National Endowment for Science, Technology, and the Arts (NESTA).

Neumann, A. (1988) 'Making mistakes: error and learning in the college presidency', Paper presented at the American Educational Research Association Annual Conference, April, New Orleans, LA; cited by V. Marsick and K. Watkins (1990) *Informal and Incidental Learning in the Workplace*, London: Routledge.

Newell, S. and Swan, J. (2000) 'Trust and inter-organisational networking', *Human Relations*, 53(10): 1287–328.

Newell, S., Robertson, M., Scarbrough, H., and Swan, J. (2002) *Managing Knowledge Work*, London: Palgrave Macmillan.

Nicholls, A. (2006) 'Social tntrepreneurship' in S. Carter and D. Jones-Evans (eds) *Enterprise and Small Business: Principles, Policy and Practice*, 2nd edn, London: FT Prentice Hall, 220–42.

Nicolaou, N. and Birley, S. (2003) 'Academic networks in a trichotomous categorisation of university spinouts', *Journal of Business Venturing*, 18: 333–59.

Nieto, M. (1997) 'The basis for the S-curve model' in D. Bennett and F. Steward (eds) *Technological Innovation and Global Challenges: Proceedings of the First IAMOT European Conference on Management of Technology*, Birmingham: Aston Business School, 213–20.

Niosi, J. (2002) 'National systems of innovations are "X-efficient" (and X-effective): why some are slow learners', *Research Policy*, 31: 291–302.

Niosi, J., Saviotti, P., Bellon, B., and Crow, M. (1993) 'National systems of innovations: in search of a workable concept', *Technology in Society*, 15: 207–27.

Nobelius, D. (2000) *An Ambidextrous Organization in Practice: Strategic Actions in Ericsson's Management of 'Bluetooth'*, IMIT Working Paper 2000/113, Gothenburg, Sweden: Chalmers University of Technology.

Nobelius, D. and Trygg, L. (2002) 'Stop chasing the front end process: management of the early phases in product development projects', *International Journal of Project Management*, 20: 331–40.

Nohria, N. and Gulati, R. (1996) 'Is slack good or bad for innovation?', *Academy of Management Journal*, 39(5): 1245–64.

Nonaka, I. and Takeuchi, H. (1995) *The Knowledge Creating Company: How Japanese Companies Create the Dynamics of Innovation*, Oxford: Oxford University Press.

Nord, W. and Tucker, S. (1987) *Implementing Routine and Radical Innovations*, Lexington, MA: Lexington Books.

O

Oakley, M. (1984) *Managing Product Design*, London: Weidenfeld and Nicolson.

Ogawa, S. (1998) 'Does sticky information affect the locus of innovation? evidence from the Japanese convenience store industry', *Research Policy*, 26(7–8): 777–90.

Okamura, K. and Vonortas, M. (2006) 'European alliance and knowledge networks', *Technology Analysis and Strategic Management*, 18(5): 535–60.

Olin, T. and Wickenberg, J. (2001) 'Rule breaking in new product development: crime or necessity', *Creativity and Innovation Management*, 10(1): 15–25.

Organisation for Economic Co-operation and Development (2002) *Innovative Clusters: Drivers of National Innovation Systems*, Paris: OECD.

Orlikowski, W. (2002) 'Knowing in practice: enacting a collective capability in distributed organizing', *Organization Science*, 13(3): 249–73.

Orlikowski, W. and Gash, D. (1994) 'Technological frames: making sense of information technology in organizations', *ACM Transactions on Information Systems*, 2(2): 174–207.

Orr, J. (1990) 'Sharing knowledge, celebrating identity: community memory in a service culture' in D. Middleton and D. Edwards (eds) *Collective Remembering*, London: Sage, 169–89.

Oshri, I. and Weeber, C. (2006) 'Cooperation and competition standards-setting activities in the digitization era: the case of wireless information devices', *Technology Analysis and Strategic Management*, 18(2): 265–83.

Oughton, C., Landabaso, M., and Morgan, K. (2002) 'The regional innovation paradox: innovation policy and industrial policy', *Journal of Technology Transfer*, 27: 97–110.

Owen-Smith, J., Riccaboni, M., Pammolli, F., and Powell, W. (2002) 'A Comparison of US and European university–industry relations in the life sciences', *Management Science*, 48(1): 24–43.

P

Page, A. (1993) 'Assessing new product development practices and performance: establishing crucial norms', *Journal of Product Innovation Management*, 8(1): 18–31.

Page, A. and Rosenbaum, H. (1987) 'Redesigning product lines with conjoint analysis: how sunbeam does it', *Journal of Product Innovation Management*, 4: 120–37.

Panne, G. van der, Beers, C. van, and Kleinknecht, A. (2003) 'Success and failure of innovation: a literature review', *International Journal of Innovation Management*, 7(3): 309–38.

Parkinson, R. (1972) 'The Dvorak simplified keyboard: forty years of frustration', *Computers and Automation*, November: 18–25.

Parkinson, S. (1981) 'Successful new product development: an international comparative study', *R&D Management*, 11(2): 79–85.

Parkinson, S. (1982) 'The role of the user in successful new product development', *R&D Management*, 12(3): 123–31.

Partridge, C. (1985) *Electonic Times*, 2 May.

Pascale, R. (1990) *Managing on the Edge: How Successful Companies Use Conflict to Stay Ahead*, London: Viking.

Paulus, P. (2000) 'Groups, teams and creativity: the creative potential of idea-generating groups', *Applied Psychology: An International Review*, 49: 237–62.

Pavitt, K. (1984) 'Sectoral patterns of technical change: towards a taxonomy and a theory', *Research Policy*, 13: 343–73.

Peck, M. (1962) 'Inventions in the postwar American aluminum industry' in R. Nelson (ed.) *The Rate and Direction of Inventive Activity: Economic and Social Factors*, Princeton, NJ: Princeton University Press, 299–322.

Perez, C. (1983) 'Structural change and assimilation of new technologies in the economic and social systems', *Futures*, 15(5): 357–75.

Perez, C. (1985) 'Microelectronics, long waves and world structural change: new perspectives for developing countries', *World Development*, 13(3): 441–63.

Perretti, F. and Negro, G. (2007) 'Mixing genres and matching people: a study in innovation and team composition in Hollywood', *Journal of Organizational Behaviour*, 28: 563–86.

Perry-Smith, J. (2006) 'Social yet creative: the role of social relationships in facilitating individual creativity', *Academy of Management Journal*, 49(1): 85–101.

Peters, T. (1988) 'The mythology of innovation, or a skunkwork's tale, Part II' in M. Tushman and W. Moore (eds) *Readings in the Management of Innovation*, 2nd edn, New York: HarperBusiness, 138–47; reprinted from T. Peters (1983) 'The mythology of innovation, or a skunkwork's tale, Part II', *The Stanford Magazine*, Summer: 13–21.

Peters, T. (1997) *The Circle of Innovation*, New York: Alfred Knopf.

Peters, T. and Waterman, R. (1982) *In Search of Excellence: Lessons from America's Best-Run Companies*, New York: Harper and Row.

Petersen, K., Handfield, R., and Ragatz, G. (2003) 'A model of supplier integration into new product development', *Journal of Product Innovation Management*, 20: 284–99.

Petroski, H. (1982) *To Engineer is Human: The Role of Failure in Successful Design*, New York: St. Martins.

Petroski, H. (2006) *Success Through Failure: The Paradox of Design*, Princeton, NJ: Princeton University Press.

Pettigrew, A. (1973) *The Politics of Organizational Decision-Making*, London: Tavistock.

Pfeffer, J. (1981) *Power in Organizations*, Boston, MA: Pitman.

Philips, P. (2007) *Governing Transformative Technological Innovation: Who's in Charge?*, Cheltenham: Edward Elgar.

Pierce, J. and Delbecq, A. (1977) 'Organization structure, individual attitudes, and innovation', *Academy of Management Review*, 2(1): 27–37.

Piller, F. and Walcher, D. (2006) 'Toolkits for idea competitions: a novel method to integrate users in new product development', *R&D Management*, 36(3): 307–18.

Pinch, T. and Bijker, W. (1987) 'The social construction of facts and artifacts: or how the sociology of technology might benefit each other' in W. Bijker, T. Hughes, and T. Pinch (eds) *The Social Construction of Technological Systems*, Cambridge, MA: MIT Press, 17–50.

Pinto, J. (2000) 'Understanding the role of politics in successful project management', *International Journal of Project Management*, 18(2): 85–91.

Piore, M. and Sabel, C. (1984) *The Second Industrial Divide*, New York: Basic Books.

Pisano, G. (1990) 'The R&D boundaries of the firm: an empirical analysis', *Administrative Science Quarterly*, 35: 153–77.

Pittaway, L., Robertson, M., Munir, K., Denyer, D., and Neely, A. (2004) 'Networking and innovation: a systematic review of the evidence', *International Journal of Management Reviews*, 5–6(3–4): 137–68.

Poel, I. van de (2003) 'The transformation of technological regimes', *Research Policy*, 32: 49–58.

Pogany, G. (1986) 'Working the S-curve: cautions about using S-curves', *Research Management*, 29(4): 24–25.

Polanyi, M. (1966) *The Tacit Dimension*, Garden City, NY: Doubleday.

Polanyi, M. (1969) *The Logic of Tacit Inference*, London: Routledge and Kegan Paul.

Poole, M. and Van de Ven, A. (1989a) 'Toward a general theory of innovation processes' in A. Van de Ven, H. Angle and M. Poole (eds) *Research on the Management of Innovation: the Minnesota Studies*, New York: Ballinger, 637–62.

Poole, M. and Van de Ven, A. (1989b) 'Using paradox to build management and organization theories', *Academy of Management Review*, 14(4): 562–78.

Popper, K. (1974) 'Normal science and its dangers' in I. Lakatos and A. Musgrave (eds) *Criticism and the Growth of Knowledge*, Cambridge: Cambridge University Press, 51–8.

Porter, M. (1980) *Competitive Strategy: Techniques for Analyzing Industries and Competitors*, New York: Free Press.

Porter, M. (1985) *Competitive Advantage: Creating and Sustaining Superior Performance*, New York: Free Press.

Porter, M. (1990) *The Competitive Advantage of Nations*, New York: Free Press.

Porter, M. (1998) 'Clusters and the new economics of competition', *Harvard Business Review*, 76(6): 77–90.

Powell, W. (1990) 'Neither market nor hierarchy: network forms of organisation', *Research in Organisational Behaviour*, 12: 295–336.

Powell, W., Koput, K., and Smith-Doerr, L. (1996) 'Interorganizational collaboration and the locus of innovation: networks of learning in biotechnology', *Administrative Science Quarterly*, 41: 116–45.

Prahalad, C. and Hamel, G. (1990) 'The core competence of the corporation', *Harvard Business Review*, 68(3): 79–91.

Prior, A. and Kirby, M. (1993) 'The Society of Friends and the family firm', *Business History*, 35(4): 66–85.

Pugh, D., Hickson, D., and Hinings, R. (1969) 'The context of organisational structures', *Administrative Science Quarterly*, 14(1): 91–114.

Q

Quinn, J. (1985) 'Managing innovation: controlled chaos', *Harvard Business Review*, 63(3): 73–84.

Quinn, J. (1988a) 'Innovation and corporate strategy: managed chaos' in M. Tushman and W. Moore (eds) *Readings in the Management of Innovation*, 2nd edn, New York: HarperBusiness, 123–37; reprinted from J. Quinn (1986) 'Innovation and corporate strategy: managed chaos' in M. Horwitch (ed.) *Technology in the Modern Corporation: A Strategic Perspective*, New York: Pergamon Press, 167–83.

Quinn, J. (1988b) 'Strategic change: "logical incrementalism"' in J. Quinn, H. Mintzberg, and R. James (eds) *The Strategy Process: Concepts, Contexts, and Cases*, Englewood Cliffs, NJ: Prentice-Hall, 94–104; excerpted from J. Quinn (1978) 'Strategic change: "logical incrementalism"', *Sloan Management Review*, 20: 7–21.

Quinn, J. (1992) 'The intelligent enterprise: a new paradigm', *Academy of Management Executive*, 6(4): 48–63.

Quinn, J. and Voyer, J. (1998) 'Logical incrementalism: managing strategy formation' in H. Mintzberg, J. Quinn, and S. Ghoshal (eds) *The Strategy Process: Concepts, Contexts, and Cases*, rev'd European edn, London: Prentice-Hall, 103–10; based on J. Quinn (1978) 'Strategic change: "logical incrementalism"', *Sloan Management Review*, 20: 7–21.

Quinn, R. (1988c) *Beyond Rational Management: Mastering the Paradoxes and Competing Demands of High Performance*, London: Jossey Bass.

Quinn, R. and Cameron, K. (eds) (1988) *Paradox and Transformation: Toward a Theory of Change in Organisation and Management*, New York: Ballinger.

R

Raelin, J. (2000) *Work-based Learning: The New Frontier of Management Development*, Englewood Cliffs, NJ: Prentice-Hall.

Ramlogan, R., Mina, A., Tampubolon, G., and Metcalfe, J. (2006) *Networks of Knowledge: The Distributed Nature of Medical Innovation*, Centre for Research on Innovation and Competition (CRIC) Discussion Paper No. 74, Manchester: University of Manchester.

R&D Magazine (2006) 'Record-breaking chip speed set', *R&D Magazine*, 48(7): 10.

Rappa, M. and Debackere, K. (1992) 'Technological communities and the diffusion of knowledge', *R&D Management*, 22: 209–20.

Rathmell, J. (1974) *Marketing in the Service Sector*, Cambridge, MA: Winthrop Publishers.

Read, D. (1999) *The Power of News: The History of Reuters*, 2nd edn, Oxford: Oxford University Press.

Reddy, N., Cort, S., and Lambert, D. (1989) 'Industrywide technical product standards', *R&D Management*, 19(1): 13–25.

Reid, S. and Brentani, U. (2004) 'The fuzzy front end of new product development for discontinuous innovations: a theoretical model', *Journal of Product Innovation Management*, 21(3): 170–84.

Rhodes, R. and Marsh, D. (1992) 'Policy networks in british politics: a critique of existing approaches' in D. Marsh and R. Rhodes (eds) *Policy Networks in British Government*, Oxford: Clarendon Press, 1–26.

Rice, R. and Rogers, E. (1980) 'Reinvention in the innovation process', *Knowledge: Creation, Diffusion, Utilization*, 1: 499–514.

Riggs, W. and Hippel, E. von (1994) 'Incentives to innovate and the sources of innovation: the case of scientific instruments', *Research Policy*, 23: 459–69.

Rink, D. and Swan, J. (1979) 'Product life cycle research: a literature review', *Journal of Business Research*, 3: 219–42.

Rip, A. and Kemp, R. (1998) 'Technological change' in S. Raynor and L. Malone (eds) *Human Choice and Climate Change: Vol. 2—Resources and Technology*, Washington, DC: Batelle Press, 327–400.

Rip, A., Misa, T., and Schot, J. (1995) *Managing Technology in Society*, London: Pinter.

Rivkin, J. (2001) 'Reproducing knowledge: replication without imitation at moderate complexity', *Organization Science*, 12: 274–93.

Roberts, E. (1986) 'Generating effective corporate innovation' in R. Roy and D. Wield (eds) *Product Design and Technological Innovation*, Milton Keynes: Open University Press, 124–7.

Roberts, E. (2001) 'Benchmarking global strategic management of technology', *Research Technology Management*, March–April: 25–36.

Roberts, J. (2006) 'Limits to communities of practice', *Journal of Management Studies*, 43(3): 623–63.

Robinson, W. (1988) 'Sources of market pioneer advantages: the case of industrial goods industries', *Journal of Marketing Research*, 25(1): 87–94.

Robinson, W. and Fornell, C. (1985) 'The sources of market pioneer advantages in consumer goods industries', *Journal of Marketing Research*, 22(3): 305–17.

Roethlisberger, F. and Dickson, W. (1939) *Management and the Worker*, Cambridge, MA: Harvard University Press.

Rogers, E. (1962) *Diffusion of Innovations*, New York: Free Press.

Rogers, E. (1982) 'Information exchange and technological change' in D. Sahal (ed.) *The Transfer and Utilisation of Technical Knowledge*, Lexington, MA: Lexington Books, 105–23.

Rogers, E. (1987) 'Progress, problems and prospects for network research: investigating relationships in the age of electronic communication', Paper presented at the VII Sunbelt Social Networks Conference, 12–15 February, Clearwater Beach, FL.

Rogers, E. (1995) *Diffusion of Innovations*, 4th edn, New York: Free Press.

Rogers, E. (2003) *Diffusion of Innovations*, 5th edn, New York: Simon and Schuster International.

Rogers, E. and Bhowmik, D. (1971) 'Homophily–heterophily: relational concepts for communication research', *Public Opinion Quarterly*, 34: 523–38.

Rogers, E. and Kincaid, D. (1981) *Communication Networks*, New York: Free Press.

Rogers, E. and Larson, J. (1984) *Silicon Valley Fever*, New York: Basic Books.

Roijakkers, N. and Hagedoorn, J. (2006) 'Inter-firm R&D partnering in pharmaceutical biotechnology since 1975: trends, patterns, and networks', *Research Policy*, 35: 431–46.

Rose, B. (2000) *Firms, Networks and Business Values: The British and American Cotton Industries Since 1750*, Cambridge: Cambridge University Press.

Rosenbaum, D. (ed.) (1998) *Market Dominance*, Westport, CT: Praeger.

Rosenberg, N. (1969) 'The direction of technological change: inducement mechanisms and focusing devices', *Economic Development and Cultural Change*, 18(1): 1–24.

Rosenberg, N. (1979) 'Technological interdependence in the American economy', *Technology and Culture*, 1(1): 25–50.

Rosenberg, N. (1982) *Inside the Black Box: Technology and Economics*, Cambridge: Cambridge University Press.

Rosenberg, N. (1990) 'Why do firms do basic research (with their own money)?', *Research Policy*, 19: 165–74.

Rosenberg, N. and Nelson, R. (1994) 'American universities and technical advance in industry', *Research Policy*, 23: 323–48.

Rosenkopf, L. and Tushman, M. (1998) 'The co-evolution of community networks and technology: lessons from the flight simulation industry', *Industrial and Corporate Change*, 7(2): 311–46.

Rothaermel, F. (2000) 'Technological discontinuities and the nature of competition', *Strategic Analysis and Strategic Management*, 12(2): 149–60.

Rothenburg, A. (1979) *The Emerging Goddess*, Chicago, IL: University of Chicago Press.

Rothwell, R. (1983) *Information and Successful Innovation*, Report No. 5782, London: British Library.

Rothwell, R. (1986) 'Innovation and re-innovation: a role for the user', *Journal of Marketing Management*, 2(2): 109–23.

Rothwell, R. (1992) 'Successful industrial innovation: critical factors for the 1990s', *R&D Management*, 22(3): 221–38.

Rothwell, R. (2002) 'Towards the fifth-generation innovation process' in J. Henry and Mayle, D. (eds) *Managing Innovation and Chance*, 2nd edn, London: Sage, 115–35.

Rothwell, R. and Gardiner, P. (1983) 'Invention, innovation, re-innovation and the role of the user: a case study of British hovercraft development', *Technovation*, 3: 167–86.

Rothwell, R. and Zegveld, W. (1985) *Reindustrialization and Technology*, Harlow: Longman.

Rothwell, R., Freeman, C., Horsley, A., Jervis, P., Robertson, A., and Townsend, J. (1974) 'SAPPHO updated: Project SAPPHO Phase II', *Research Policy*, 3(3): 258–91.

Roussel, P. (1984) 'Technological maturity proves a valid and important concept', *Research Management*, 1(1): 29–34.

Roy, R. (1986a) 'Introduction: design evolution, technological evolution and economic growth' in R. Roy and D. Wield (eds) *Product Design and Technological Innovation*, Milton Keynes: Open University Press, 252–56.

Roy, R. (1986b) 'Introduction: meanings of design and innovation' in R. Roy and D. Wield (eds) *Product Design and Technological Innovation*, Milton Keynes: Open University Press, 2–7.

Roy, R. and Cross, N. (1983) *Bicycles: Invention and Innovation*, T263: Design Processes and Products (Units 5–7), Milton Keynes: Open University Press.

Ryan, B. and Gross, N. (1943) 'The diffusion of hybrid corn seed in two Iowa communities', *Rural Sociology*, 8(1): 15–24.

S

Sabel, C., Herrigel, R., Kazis, R., and Deeg, R. (1987) 'How to keep mature industries innovative', *Technology Review*, 90(3): 26–35.

Sahal, D. (1981) *Patterns of Technological Innovation*, London: Addison-Wesley.

Sakakibara, M. and Dodgson, M. (2003) 'Strategic research partnerships: empirical evidence from Asia', *Technology Analysis and Strategic Management*, 15: 223–41.

Sako, M. (1991) 'The role of trust in Japanese buyer–supplier relationships', *Ricerche Economiche*, 45(2–3): 375–99.

Samli, C. and Jacobs, L. (2003) 'Counteracting global industrial espionage: a damage control strategy', *Business and Society Review*, 108(1): 95–113.

Sampat, B. (2006) 'Patenting and US academic research in the 20th century: the world before and after Bayh-Dole', *Research Policy*, 35: 772–89.

Sanderson, S. and Uzumeri, M. (1995) 'Managing product families', *Research Policy*, 24(5): 761–82.

Saren, M. (1984) 'A classification and review of models of the intra-firm innovation process', *R&D Management*, 14(1): 11–24.

Saren, M. (1994) 'Reframing the process of new product development: from stage models to blocks', *Journal of Marketing Management*, 10(7): 633–44.

Sawhney, M., Verona, G., and Prandelli, E. (2005) 'Collaborating to create: the Internet as a platform for customer engagement in product innovation', *Journal of Interactive Marketing*, 19(4): 4–17.

Saxenian, A. (1985) 'Genesis of Silicon Valley' in P. Hall and A. Markusen (eds) *Silicon Landscapes*, Boston, MA: Allen and Unwin, 20–34.

Saxenian, A. (1991) 'The origins and dynamics of production networks in the Silicon Valley', *Research Policy*, 20(5): 423–37.

Saxenian, A. (1994) *Regional Advantage: Culture and Competition in Silicon Valley and Route 128*, Cambridge, MA: Harvard University Press.

Scarbrough, H., Swan, J., and Preston, J. (1999) *Knowledge Management: A Literature Review*, London: Institute of Personnel and Development.

Schilling, M. (2003) 'Technological leapfrogging: lessons from the US video game console industry', *California Management Review*, 45(3): 6–32.

Schmookler, J. (1966) *Invention and Economic Growth*, Boston, MA: Harvard University Press.

Schoenecker, T. and Cooper, A. (1998) 'The role of firm resources and organizational attributes in determining entry timing: a cross-industry study', *Strategic Management Journal*, 19: 1127–43.

Schon, D. (1963) 'Champions for radical new inventions', *Harvard Business Review*, 41(2): 77–86.

Schon, D. (1966) 'The fear of innovation', *International Science and Technology*. November: 70–8.

Schon, D. (1967) *Technology and Change*, Oxford: Pergamon.

Schoonhoven, C. and Jelinek, M. (1990) 'Dynamic tensions in innovative firms: managing rapid technological change through organisational structure' in M. von Glinow and A. Mohrman (eds) *Managing Complexity in High Technology Organisations*, New York: Oxford University Press, 90–118.

Schot, J. (2001) 'Towards new forms of participatory technology development', *Technology Analysis and Strategic Management*, 13(1): 39–52.

Schot, J. and Rip, A. (1996) 'The past and future of constructive technology assessment', *Technological Forecasting and Social Change*, 54: 251–68.

Schrader, S. (1991) 'Informal technology transfer between firms: cooperation through information trading', *Research Policy*, 20(2): 153–70.

Schrage, M. (1993) 'The culture(s) of prototyping', *Design Management Journal*, 4(1): 55–65.

Schrage, M. (1999) 'What's that bad odor at innovation skunkworks?', *Fortune*, 140(12): 338.

Schumacher, E. (1974) *Small is Beautiful*, Reading: Abacas.

Schumpeter, J. (1939) *Business Cycles: A Theoretical, Historical and Statistical Analysis of the Capitalist Process*, two vols, New York: McGraw-Hill.

Schumpeter, J. (1942) *Capitalism, Socialism, and Democracy*, New York: Harper and Row.

Schwartz, D. and Jacobson, E. (1977) 'Organizational communication network analysis: the liaison communication role', *Organizational Behaviour and Human Performance*, 18: 158–74.

Scott, J. (2000) *Social Network Analysis: A Handbook*, 2nd edn, London: Sage.

Shah, S. and Tripas, M. (2007) 'The accidental entrepreneur: the emergent and collective process of user entrepreneurship', *Strategic Entrepreneurship Journal*, 1: 123–40.

Shamsie, J., Phelps, C., and Kuperman, J. (2004) 'Better late than never: a study of late entrants in household electrical appliances', *Strategic Management Journal*, 25(1): 69–84.

Shane, S. (1994) 'Are champions different from non-champions?', *Journal of Business Venturing*, 9: 397–421.

Shane, S. (2004) *Academic Entrepreneurship: University Spinoffs and Wealth Creation*, Cheltenham: Edward Elgar.

Shankar, V., Carpenter, G., and Krishnamurthi, L. (1998) 'Late mover advantage: how innovative late entrants outsell pioneers', *Journal of Marketing Research*, 35(1): 54–70.

Shapiro, C. and Varian, H. (1999) 'The art of standards wars', *California Management Review*, 41(1): 8–32.

Sharif, N. (2006) 'Emergence and development of the national innovation systems concept', *Research Policy*, 35: 745–66.

Shaw, B. (1985) 'The role of the interaction between the user and the manufacturer in medical equipment innovation', *R&D Management*, 15(4): 283–92.

Shaw, B. (1993) 'Formal and informal networks in the UK medical equipment industry', *Technovation*, 13: 349–65.

Sheremata, W. (2004) 'Competing through innovation in network markets: strategies for challengers', *Academy of Management Review*, 29(3): 359–77.

Sherwin, E. and Isenson, R. (1967) 'Project Hindsight', *Science*, 156: 1571–77.

Silverberg, G. (2002) 'The discrete charm of the bourgeoisie: quantum and continuous perspectives on innovation and growth', *Research Policy*, 31: 1275–89.

Simmel, G. (1955) *Conflict and the Web of Group-Affiliations*, trans. K. Wolff and R. Bendix, New York: Free Press.

Single, A. and Spurgeon, W. (1996) 'Creating and commercializing innovation inside a skunkworks', *Research Technology Management*, 39(1): 38–41.

Skyrme, D. (2000) *Knowledge Networking: Creating Collaborative Enterprise*, Oxford: Butterworth-Heinemann.

Slack, N., Chambers, S., and Johnston, R. (2006) *Operations Management*, 5th edn, Harlow: FT Prentice Hall.

Slappendel, C. (1996) 'Perspectives on innovation in organizations', *Organization Studies*, 17(1): 107–29.

Slaughter, S. (1993) 'Innovation and learning during implementation: a comparison of user and manufacturer innovations', *Research Policy*, 22: 81–95.

Slowik, E. (1998) 'Music, science, and analogies: teaching the philosophy of science with non-scientific examples', Paper presented at the Twentieth World Congress of Philosophy, 10–16 August, Boston, MA.

Smith, A., Stirling, A., and Berkhout, F. (2005) 'The governance of sustainable socio-technical transitions', *Research Policy*, 34: 1491–510.

Smith, P. and Reinersten, D. (1991) *Developing Product in Half the Time*, New York: Van Nostrand Reinhold.

Smits, R. and Kuhlmann, S. (2004) 'The rise of systemic instruments in innovation policy', *Foresight and Innovation Policy*, 1(1–2): 4–32.

Sorenson, O., Rivkin, J., and Fleming, L. (2006) 'Complexity, networks, and knowledge flow', *Research Policy*, 35: 994–1017.

Sovacool, B. (2008) 'Innovating the innovators: the case for transformational energy research and development', *International Journal of Energy Technology and Policy*, 6(4): 368–80.

Sowrey, T. (1987) *The Generation of Ideas for New Products*, London: Kogan Page.

Spender, J. (1989) *Industry Recipes*, Oxford: Blackwell.

Spender, J. and Kessler, E. (1995) 'Managing the uncertainties of innovation: extending Thompson (1967)', *Human Relations*, 48(1): 35–56.

Spital, F. (1979) 'An analysis of the role of users in the total R&D portfolios of scientific instrument firms', *Research Policy*, 8(3): 284–96.

Stacey, R. (1996) *Strategic Management and Organisational Dynamics*, 2nd edn, London: Pitman Publishing.

Stagg, C., Saunders, J., and Wong, V. (2002) 'Go/no-go criteria during grocery brand development', *Journal of Product and Brand Management*, 11(7): 459–82.

Standley, J. (2003) 'Space trips up for grabs', *BBC News*, 19 June, available online at <http://news.bbc.co.uk/>.

Steiner, C. (1995) 'A philosophy for innovation: the role of unconventional individuals in innovation success', *Journal of Product Innovation Management*, 12: 431–40.

Stephenson, K. and Krebs, V. (1993) 'A more accurate way to measure diversity', *Personnel Journal*, 72: 66–74.

Stephenson, K. and Lewin, D. (1996) 'Managing workforce diversity: macro and micro level HR implications of network analysis', *International Journal of Manpower*, 17(4–5): 168–96.

Steward, F. (2008) *Breaking the Boundaries: Transformative Innovation for the Global Good*, London: NESTA.

Steward, F. and Conway, S. (1997) *Networks of Innovative Managers: The Paths of Education and Experience*, Final project report to the ESRC Innovation Programme, Phase I.

Steward, F. and Conway, S. (1998) 'Building networks for innovation diffusion in Europe: learning from the SPRINT programme', *Enterprise and Innovation Management Studies*, 1(3): 281–301.

Steward, F., Conway, S., Teubert, W., and Townson, M. (1996) *Environmental Innovation and Corporate Communication: A UK/German Comparative Study*, Final project report to the EC Environmental Research Programme: Research Area III Economic and Social Aspects of the Environment, Contract EV5V-CT94–0381.

Stewart, R. (1976) *Contrasts in Management: A Study of Different Types of Managers Jobs—Their Demands and Choices*, New York: McGraw-Hill

Stokes, D. (1997) *Pasteur's Quadrant: Basic Science and Technological Innovation*, Washington, DC: Brooking Press.

Storey, J. (1994) 'Management development' in K. Sisson (ed.) *Personnel Management*, Oxford: Blackwell, 365–96.

Storey, J. and Barnett, E. (2000) 'Knowledge management initiatives: learning from failure', *Journal of Knowledge Management*, 4(2): 145–56.

Storz, C. (2008) 'Dynamics in innovation systems: evidence from Japan's game software industry', *Research Policy*, 37: 1480–91.

Strickberger, M. (1996) *Evolution*, Boston, MA: Jones and Bartlett; cited by D. Levinthal (1998) 'The slow pace of rapid technological change: gradualism and punctuation in technological change', *Industrial and Corporate Change*, 7(2): 217–47.

Styhre, A. (2006) 'Organization creativity and the empiricist image of novelty', *Creativity and Innovation Management*, 15(2): 143–49.

Subramaniam, M. and Youndt, M. (2005) 'The influence of intellectual capital on the types of innovative capabilities', *Academy of Management Journal*, 48(3): 450–63.

Sundbo, J. (1994) 'Modulization of service production and a thesis of convergence between service and manufacturing organizations',

Scandinavian Journal of Management, 10(3): 245–66.

Sundbo, J. (1997) 'Management of innovation in services', *The Service Industries Journal*, 17(3): 432–55.

Sundbo, J. (2000) 'Organization and innovation strategy in services' in M. Boden and I. Miles (eds) *Services and the Knowledge Based Economy*, London: Continuum, 109–28.

Sundbo, J. (2001) *The Strategic Management of Innovation: A Sociological and Economic Theory*, Cheltenham: Edward Elgar.

Swan, J. and Scarbrough, H. (2005) 'The politics of networked innovation', *Human Relations*, 58(7): 913–43.

Swann, G. (2001) 'Sales practice and market evolution: the case of virtual reality', *International Journal of Industrial Organization*, 19(7): 1119–39.

Swann, G. and Watts, T. (2000) 'Visualisation needs vision: the pre-paradigmatic character of virtual reality', *Virtual Society? Programme*, Swindon: ESRC.

Szulanski, G. (1996) 'Exploring internal stickiness: impediments to the transfer of best practice within the firm', *Strategic Management Journal*, 17: 27–43.

T

Takeishi, A. (2001) 'Bridging inter- and intrafirm boundaries: management of supplier involvement in automobile product development', *Strategic Management Journal*, 22(5): 403–33.

Tarde, G. (1903) *The Laws of Imitation*, trans. E. Parsons, New York: Holt.

Tate, W. (1995) *Developing Managerial Competence: A Critical Guide to Methods and Materials*, Aldershot: Gower.

Tchijov, I. and Norov, E. (1989) 'Forecasting methods for CIM technologies', *Engineering Costs and Production Economics*, 15: 323–89.

Teagarden, M., Meyer, J., and Jones, D. (2008) 'Knowledge sharing among high-tech MNCs in China and India', *Organizational Dynamics*, 37(2): 190–202.

Teece, D. (1980) 'The diffusion of an administrative innovation', *Management Science*, 26: 464–70.

Teece, D. (2007) 'Explicating dynamic capabilities: the nature and microfoundations of (sustainable) enterprise performance', *Strategic Management Journal*, 28: 1319–50.

Teece, D. and Pisano, G. (1994) 'The dynamic capabilities of firms: an introduction', *Industrial and Corporate Change*, 3(3): 537–56.

Teece, D., Pisano, G., and Shuen, A. (1997) 'Dynamic capabilities and strategic management', *Strategic Management Journal*, 18(7): 509–33.

Teixeira, A. (2008) *National Systems of Innovation: A Bibliometric Appraisal*, Faculdade De Economia (FEP) Working Papers No. 271, Porto, Portugal: Universidade Do Porto.

Tether, B. and Metcalfe, S. (2004) 'Services and systems of innovation' in F. Malerba (ed.) *Sectoral System of Innovation: Concepts, Issues and Analyses of Six Major Sectors in Europe*, Cambridge: Cambridge University Press, 9–41.

Tether, B. and Tajar, A. (2008a) 'Beyond industry-university links: sourcing knowledge for innovation from consultants, private sector research organisations and the public science-base', *Research Policy*, 37: 1079–95.

Tether, B. and Tajar, A. (2008b) 'The organisational-cooperation mode of innovation and its prominence amongst European service firms', *Research Policy*, 37: 720–39.

Thomas, H. (1988) 'Policy dialogue in strategic planning: talking our way through ambiguity and change' in L. Pondy, R. Boland, and H. Thomas (eds) *Managing Ambiguity and Change*, Chichester: John Wiley, 51–78.

Thomas, R. (1994) *What Machines Can't Do: Politics and Technology in the Industrial Enterprise*, Berkeley, CA: University of California Press.

Thomke, S. (2003) 'R&D comes to services: Bank of America's pathbreaking experiments', *Harvard Business Review*, 81(4): 70–9.

Thomke, S. and Hippel, E. von (2002), 'Customers as innovators: a new way to create value', *Harvard Business* Review, April: 5–11.

Thompson, A., Strickland, A., and Gamble, J. (2007) *Crafting and Executing Strategy: The Quest for Competitive Advantage*, 16th edn, New York: McGraw-Hill.

Thompson, J. (1967) *Organizations in Action*, New York: McGraw-Hill.

Tichy, N. (1981) 'Networks in organizations' in P. Nystrom and W. Starbuck (eds) *Handbook of Organizational Design, Vol. 2*, New York: Oxford University Press, 225–47.

Tichy, N., Tushman, M., and Fombrun, C. (1979) 'Social network analysis for organisations', *Academy of Management Review*, 4(4): 507–19.

Tidd, J. (1993) 'Technological innovation, organisational linkages and strategic degrees of freedom', *Technology Analysis and Strategic Management*, 5: 273–84.

Tidd, J. and Bodley, K. (2002) 'The effects of project novelty on the new product development process', *R&D Management*, 32(2): 127–38.

Tidd, J., Bessant, J., and Pavitt, K. (2001) *Managing Innovation: Integrating Technological, Market and Organizational Change*, 2nd edn, Chichester: Wiley.

Tjepkema, S., Stewart, J., Sambrook, S., Mulder, M., Hoerst, H. ter, and Scheerens, J. (eds) (2002) *HRD and Learning Organisations in Europe*, London: Routledge.

Toker, U. and Gray, D. (2007) 'Innovation spaces: workspace planning and innovation in US university research centers', *Research Policy*, 37: 309–29.

Towner, S. (1994) 'Four ways to accelerate new product development', *Long Range Planning*, April: 57–65.

Tripsas, M. (1997) 'Unravelling the process of creative destruction: complementary assets and incumbent survival in the typesetter industry', *Strategic Management Journal*, 18: 119–42.

Tripsas, M. (2008) 'Customer preference discontinuities: a trigger for radical technical change', *Managerial and Decision Economics*, 29: 79–97.

Trott, P. (2005) *Innovation Management and New Product Development*, 3rd edn, London: FT Pitman Publishing.

Tsai, W. (2001) 'Knowledge transfer in inter-organizational networks: effects of network position and absorptive capacity on business unit innovation and performance', *Academy of Management Journal*, 44: 996–1005.

Tsai, W. (2002) 'Social structure of "coopetition" within a multiunit organization: coordination, competition, and intraorganizational knowledge sharing', *Organization Science*, 13(2): 179–90.

Tsai, W. and Ghoshal, S. (1998) 'Social capital and value creation: the role of intrafirm networks', *Academy of Management Journal*, 42(4): 464–76.

Tsoukas, H. (2003) 'Do we really understand tacit knowledge?' in M. Easterby-Smith and M. Lyles (eds) *The Blackwell Handbook of Organizational Learning and Knowledge Management*, Oxford: Blackwell, 410–27.

Turpin, T., Xielin, L., Garrett-Jones, S., and Burns P. (eds) (2002) *Innovation, Technology Policy and Regional Development: Evidence from China and Australia*, Cheltenham: Edward Elgar.

Tushman, M. (1977) 'Special boundary roles in the innovation process', *Administrative Science Quarterly*, 22: 587–605.

Tushman, M. and Anderson, P. (1986) 'Technological discontinuities and organizational environments', *Administrative Science Quarterly*, 31: 439–65.

Tushman, M. and Katz, R. (1980) 'External communication and project performance: an investigation into the role of gatekeepers', *Management Science*, 26(11): 1071–85.

Tushman, M. and O'Reilly, C. (1996) 'Ambidextrous organizations: managing evolutionary and revolutionary change', *California Management Review*, 34(4): 8–30.

Tushman, M. and Romanelli, E. (1985) 'Organizational evolution: a metamorphosis model of convergence and reorientation' in L. Cummings and B. Straw (eds) *Research in Organizational Behavior*, Greenwich, CT: JAI Press, 171–222.

Tushman, M. and Rosenkopf, L. (1992) 'Organizational determinants of technological change: towards a sociology of technological evolution', *Research in Organizational Behavior*, 14: 311–47.

Tushman, M., Anderson, P., and O'Reilly, C. (1997) 'Technology cycles, innovation streams,

REFERENCES

465

and ambidextrous organizations: organization renewal through innovation streams
and strategic change' in M. Tushman and
P. Anderson (eds) *Managing Strategic Innovation
and Change: A Collection of Readings*, New York:
Oxford University Press, 3–23.

U

Udell, G. (1990) 'It's still caveat, inventor',
Journal of Product Innovation Management, 7(3):
230–43.

Uehara, E. (1990) 'Dual exchange theory,
social networks, and informal social support',
American Journal of Sociology, 96(3): 521–57.

Ul-Haq, R. and Howcroft, B. (2007) 'An examination of strategic alliances and the origins of
international banking in Europe', *International
Journal of Service Industry Management*, 18(2):
120–29.

Urban, G. and Hauser, J. (1993) *Design and
Marketing of New Products*, Englewood Cliffs,
NJ: Prentice-Hall.

Urban, G. and Hippel, E. von (1988) 'Lead
user analyses for the development of new industrial products', *Management Science*, 34(5):
569–82.

Utterback, J. (1971) 'The process of innovation: a study of the origination and development of ideas for scientific instruments', *IEEE
Transactions on Engineering Management*, 18(4):
124–31.

Utterback, J. (1994) *Mastering the Dynamics
of Innovation*, Boston, MA: Harvard Business
School Press.

Utterback, J. and Abernathy, W. (1975) 'A
dynamic model of process and product
innovation', *Omega*, 3(6): 639–56.

Uzzi, B. (1996) 'The sources and consequences
of embeddedness for the economic performance of organizations: the network effect',
American Sociological Review, 61(4): 674–98.

V

Vanderwerf, P. (1990) 'Product tying and innovation in US wire preparation equipment',
Research Policy, 19(1): 83–96.

Vanderwerf, P. (1992) 'Explaining downstream
innovation by commodity suppliers with expected innovation benefit', *Research Policy*,
21(4): 315–33.

Van de Ven, A. (1983) 'Review of Peters and
Waterman', *Administrative Science Quarterly*,
28: 621–24.

Van de Ven, A. (1988) 'Central problems in the
management of innovation' in M. Tushman
and W. Moore (eds) *Readings in the Management
of Innovation*, 2nd edn, New York: HarperBusiness, 103–22; reprinted from A. Van de Ven
(1986) 'Central problems in the management
of innovation', *Management Science*, 32(5):
590–607.

Van de Ven, A. and Garud, R. (1993)
'Innovation and industry emergence: the case
of cochlear implants', *Research on Technological
Innovation, Management, and Policy*, 5: 1–46.

Van de Ven, A. and Poole, M. (1988) 'Paradoxical requirements for a theory of change'
in R. Quinn and K. Cameron (eds) *Paradox
and Transformation: Toward a Theory of Change
in Organisation and Management*, New York:
Ballinger, 19–63.

Van de Ven, A. and Poole, S. (1995) 'Explaining development and change in organisations',
Academy of Management Review, 20(3): 510–40.

Van de Ven, A., Angle, H., and Poole, M.
(1989) *Research on the Management of Innovation*, New York: Harper and Row.

Van de Ven, A., Polliy, D., Garud, R., and
Venkataraman, S. (1999) *The Innovation Journey*,
New York: Oxford University Press.

Van Vianen, B., Moed, H., and Van Raan, A.
(1990) 'An exploration of the science base of
recent technology', *Research Policy*, 19: 61–81.

Van Wyk, R., Haour, G., and Japp, S. (1991)
'Permanent magnets: a technological analysis',
R&D Management, 34: 301–8.

Von Krogh, G. (1998) 'Care in knowledge
creation', *California Management Review*, 40(3):
133–53.

Von Krogh, G., Ichijo, K., and Nonaka, I.
(2000) *Enabling Knowledge Creation*, New York:
Oxford University Press.

Versluis, C. (2005) 'Innovations on thin ice',
Technovation. 25: 1183–92.

Voss, C. (1985) 'The role of users in the development of applications software', *Journal of Product Innovation Management*, 2(2): 113–21.

W

Walker, G. and Weber, D. (1984) 'A transaction cost approach to make or buy decisions', *Administrative Science Quarterly*, 29: 373–91.

Warner, W. and Lunt, P. (1941) *The Social Life of a Modern Community*, New Haven, CT: Yale University Press.

Warr, P. and Downing, J. (2000) 'Learning strategies, learning anxiety, and knowledge acquisition', *British Journal of Psychology*, 91: 311–33.

Watkins, D. and Horley, G. (1986) 'Transferring technology from large to small firms: the role of intermediaries' in T. Webb, T. Quince, and D. Watkins (eds) *Small Business Research*, Aldershot: Gower, 215–51.

Watts, T. (2001) 'Conveying tacit knowledge using virtual reality technology', Unpublished teaching case study, Birmingham: Aston Business School.

Watts, T., Swann, G., and Pandit, N. (1998) 'Virtual reality and innovation potential', *Business Strategy Review*, 9(3): 45–54.

Weber, M. (1947) *The Theory of Social and Economic Organization*, trans. A. Henderson and T. Parsons. New York, Free Press.

Weekakkody, N. (2007) 'The present and the future of digital TV in Australia', *Proceedings of the Computer Science and IT Education Conference*, Santa Rosa, CA: Informing Science Institute, 703–715.

Weick, K. (1969) *The Social Psychology of Organising*, London: Addison-Wesley.

Weick, K. and Frances, W. (1996) 'Organisational learning: affirming an oxymoron' in S. Clegg, C. Hardy, and W. Nord (eds) *Handbook of Organisation Studies*, Thousand Oaks, CA: Sage, 440–58.

Wellman, B. (1983) 'Network analysis: some basic principles' in R. Collins (ed.) *Sociological Theory*, San Francisco, CA: Jossey-Bass, 155–200.

Wenger, E. (1998) *Communities of Practice: Learning, Meaning, and Identity*, Cambridge: Cambridge University Press.

Wenger, E. (2000) 'Communities of practice and social learning systems', *Organization*, 7(2): 225–46.

Wenger, E. and Snyder, W. (2000) 'Communities of practice: the organizational frontier', *Harvard Business Review*, Jan–Feb: 139–45.

Wenger, E., McDermott, R., and Snyder, W. (2001) *Cultivating Communities of Practice: A Guide to Managing Knowledge*, Cambridge, MA: Harvard Business School Press.

West, J. and Gallagher, S. (2006) 'Challenges of open innovation: the paradox of firm investment in open-source software', *R&D Management*, 36(3): 319–31.

West, M. (1987) 'Role innovation in the world of work', *British Journal of Social Psychology*, 26: 305–15.

West, M. (1990) 'The social psychology of innovation in groups' in M. West and J. Farr (eds) *Innovation and Creativity at Work: Psychological and Organizational Strategies*, Chichester: John Wiley, 309–33.

West, M. (2002) 'Sparkling fountains or stagnant ponds: an integrative model of creativity and innovation implementation in work groups', *Applied Psychology: An International Review*, 51(3): 355–424.

Whalley, P. (1991) 'The social practice of independent inventing', *Science, Technology and Human Values*, 16(2): 208–32.

Wheelwright, S. and Clark, K. (1992) *Revolutionizing Product Development: Quantum Leaps in Speed, Efficiency, and Quality*, New York: Free Press.

White, H., Boorman, S., and Breiger, R. (1976) 'Social structure from multiple networks I: blockmodels of roles and positions', *American Journal of Sociology*, 81(4): 730–80.

Whitley, R. (1977) 'Concepts of organization and power in the study of organizations', *Personnel Review*, 6(1): 54–59.

Whitley, R., Thomas, A., and Marceau, J. (1981) *Masters of Business? Business Schools and Business Graduates in Britain and France*, London: Tavistock.

Whittington, R. (1988) 'Environmental structure and theories of strategic choice', *Journal of Management Studies*, 25: 521–36.

Williams, R. and Edge, D. (1996) 'The social shaping of technology', *Research Policy*, 25: 865–99.

Williamson, O. (1975) *Markets and Hierarchies: Analysis and Antitrust Implications*, New York: Free Press.

Williamson, O. (1985) *The Economic Institutions of Capitalism*, New York: Free Press.

Winch, G. and Courtney, R. (2007) 'The organization of innovation brokers: an international review', *Technology Analysis and Strategic Management*, 19(6): 747–63.

Wittink, D., Vriens, M., and Burhenne, W. (1994) 'Commercial uses of conjoint analysis in Europe: results and critical reflections', *International Journal of Research in Marketing*, 11(1): 41–52.

Wolek, F. and Griffith, B. (1974) 'Policy and informal communication in applied science and technology', *Science Studies*, 4: 411–20.

Wollard A. (2001) 'An industry under threat: the music industry and digital technology', Unpublished MBA dissertation, Birmingham: Aston Business School.

Wong, V. (2002) 'Antecedents of international new product rollout timeliness', *International Marketing Review*, 19(2): 120–32.

XYZ

Yoon, E. and Lilien, G. (1988) 'Characteristics of the industrial distributor's innovation activities: an exploratory study', *Journal of Product Innovation Management*, 5(3): 227–40.

Youndt, M., Subramaniam, M., and Snell, S. (2004) 'Intellectual capital profiles: an examination of investments and returns', *Journal of Management Studies*, 41: 335–62.

Zaltman, G., Duncan, R., and Holbeck, J. (1973) *Innovations and Organizations*, New York: Wiley-Interscience.

Zander, U. and Kogut, B. (1995) 'Knowledge and the speed of the transfer and imitation of organizational capabilities: an empirical test', *Organization Science*, 6: 76–92.

Zipf, G. (1949) *Human Behaviour and the Principle of Least Effort: An Introduction to Human Ecology*, Cambridge, MA: Addison-Wesley; cited by E. Rogers and D. Kincaid (1981) *Communication Networks*, New York: Free Press.

■ INDEX